Silicon Carbide Devices and Technology

Silicon Carbide Devices and Technology

Edited by **Bill Fraley**

NY Research Press
New York

Published by NY Research Press,
23 West, 55th Street, Suite 816,
New York, NY 10019, USA
www.nyresearchpress.com

Silicon Carbide Devices and Technology
Edited by Bill Fraley

International Standard Book Number: 978-1-63238-414-0 (Hardback)

Printed in China.

Contents

Preface

This book elaborately discusses the silicon carbide devices and their technology. Lately, a few Silicon Carbide (SiC) power devices like metal-oxide-semiconductor field-effect-transistors (MOSFETs), Schottky-barrier diodes (SBDs), junction FETs (JFETs), and their combined modules have been introduced in the market. However, to securely supply them and decrease their cost, further enhancements for material characterizations as well as for device processing are still required. This book comprehensively elucidates current technologies on processing, modeling, characterization, manufacturing, and other important aspects of SiC devices. The aim of this book is to serve as a helpful source of information for advancements in SiC devices.

All of the data presented henceforth, was collaborated in the wake of recent advancements in the field. The aim of this book is to present the diversified developments from across the globe in a comprehensible manner. The opinions expressed in each chapter belong solely to the contributing authors. Their interpretations of the topics are the integral part of this book, which I have carefully compiled for a better understanding of the readers.

At the end, I would like to thank all those who dedicated their time and efforts for the successful completion of this book. I also wish to convey my gratitude towards my friends and family who supported me at every step.

Editor

Section 1

Characterization for Device Application

Chapter 1

Characterization of Defects Evolution in Bulk SiC by Synchrotron X-Ray Imaging

T. S. Argunova, M. Yu. Gutkin, J. H. Je,
V. G. Kohn and E. N. Mokhov

Additional information is available at the end of the chapter

1. Introduction

Optimization of Silicon Carbide (SiC) bulk growth by physical vapor transport (PVT) has allowed commercial realization of SiC wafers with low defect density. Dislocated micropipes (MPs), known as the most killing defects in SiC devices, have been reduced to the density levels as low as 0.7–0.5 cm^{-2}. However applications of SiC as a material for high-power electronics are still hampered. High-performance high-power electric systems require SiC devices capable of handling the current capacity of at least 100 A, which means that the MP density of SiC wafer should be less than 0.5 cm^{-2} [16]. Fabrication of such high quality wafers, in particular, in low cost is indispensable for manufacturing of high-current devices.

Various methods such as the enhancement of polytype stability [26, 31], the reduction of screw dislocations [22], the restriction of inclusions [23], *etc.*, help control MP density. Polytype transformation, dislocations, and MPs, and their interaction or conversion are complicated phenomena which are of particular interest to SiC crystal growers and device engineers. Dislocations are largely induced by foreign polytype inclusions (FPIs) [27], but disappear when the inclusions transform back to the initial polytype [26]. The improvement of the polytype stability immediately and dramatically decreases the MP densities [29]. When accumulated at the boundaries of FPIs, MPs may coalesce each other, forming pores [10, 11, 13].

Nowadays, dislocation processes in SiC single crystals are studied by means of Synchrotron Radiation (SR) X-ray topography [24, 25, 33] combined with defect selective etching [30] or computer simulation of Bragg-diffraction images [3, 18]. A very good methodology to study hollow defects in SiC — to map them and to evaluate their sizes — is phase contrast X-ray imaging [19], which is available due to high spatial coherency of third generation SR sources.

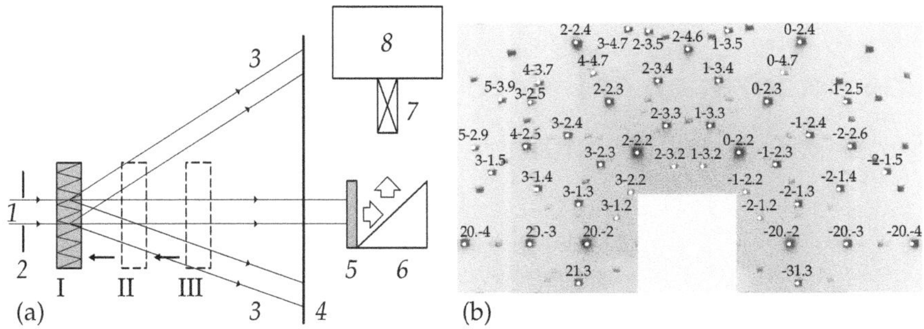

Figure 1. (a) Scheme for obtaining diffraction and phase-contrast images on-line: (*1*) SR beam from storage ring, (*2*) beam defining slits, (*3*) diffracted beams, (*4*) film, (*5*) mirror, (*6*) scintillator, (*7*) objective, (*8*) detector. I, II and III correspond to the succession of 6*H*-SiC axial-cut slices investigated in turn. (b) Laue pattern of the (0001) 6*H*-SiC wafer (direction [1$\bar{2}$10] horizontal) fixed perpendicular to SR beam.

The purpose of the present paper is to characterize the evolution of pores and MPs during SiC growth using SR white beam phase contrast imaging and Bragg diffraction topography. We document defect reduction mechanisms and suggest theoretical and computer models to explain them.

2. Experimental

The samples used in this study were prepared from a 4*H* and 6*H*-SiC boules grown in Ar by the sublimation sandwich technique [31]. The Ar pressure in the growth chamber was 80 mbar. The growth temperature was 2180°C, and the growth rate was 0.5 mm $\times$ h^{-1}. The crystals were N-doped and had a *n*-type conductivity with a donor concentration 2 $\times$ 10^{18} cm^{-3}. They contained B to a concentration (1–2) $\times$ 10^{17} cm^{-3}. Besides, doping of SiC by Al to a concentration of approximately (2–7) $\times$ 10^{17} cm^{-3} occurred because of the Al presence in the polycrystalline SiC source. After polished and treated to eliminate damaged layers on both sides, the sample thicknesses were controlled as about 500 μm.

The foreign polytype inclusions located close to the surface were revealed with the aid of a photoluminescence (PL) microscope in visible as well as in ultraviolet light under the magnifications of 50$\times$ – 200$\times$. The polytype was identified by the color of PL [28]. The opening of pores and MPs on molten KOH etched sample surfaces was detected using optical and scanning electron microscopies (SEM), performed with Zeiss universal microscope and JEOL JSM-6330F FESEM, respectively.

Imaging experiments were done at the 7B2 X-ray microscopy beamline of Pohang Light Source (PLS), Korea [2]. The 7B2 bending magnet port of the PLS storage ring provided an effective source size of the order of 60 μm (160 μm) in the vertical (horizontal) direction at the distance 34 m from the sample. Unmonochromatized ('white') radiation with the spectrum ranged from 6 to 40 keV was propagated through a polished beryllium window of 2 mm thick and then through a specimen with no optical elements in between.

X-ray phase contrast in the white beam is formed because of a small angular size and a high spatial coherence of third generation SR source. The white beam spectrum has a curve shape with a maximum at a certain energy in the case at hand. Decreases in the radiation intensity at high and low energies are caused by the SR generation's nature and the absorption in the sample, respectively. Therefore, even 'white' SR is partially coherent, which is quite sufficient for recording phase-contrast images of small objects such as MPs in SiC [19].

White rather than monochromatic beam gives higher intensity, larger exposed area, the use of various research methods during a single experiment, but has the attendant disadvantage that obtained information is mostly qualitative. The task of obtaining reliable information can be solved by developing numerical simulation of phase-contrast images. Such an approach was recently proposed [1, 19] based on Kirchhoff integral calculation for monochromatic SR harmonics followed by summation over an actual spectrum, taking into account the absorption in sample. In particular, the Fit Microtube Image (FIMTIM) program was developed for automatic determination of the parameters of the MP cross section in two dimensions from the good-fit condition for calculated and experimental profiles of relative intensity. From this it was shown that MP cross section can vary not only in its size, but also in its orientation, during its growth [1].

In our studies, MPs are assumed to be screw dislocations with hollow cores [17], as made certain by combining phase contrast imaging and topography. Evolution of pores and MPs during SiC growth presents a substantial challenge, because a SiC boule is not transparent as a whole even for hard synchrotron X-rays. The series of slices were cut perpendicular (on-axis) and parallel (axial-cut) to the growth direction of 6*H* and 4*H*-SiC boules. Between on-axis 6*H*-SiC slices, the shapes and distributions of defects were variable. It was essential to have some special means of finding the same area of interest between the slices. Taking into account the microscopic sizes of defects, this area was controlled as rather small as 1.5 mm $\times$ 1.5 mm. X-ray phase-contrast micro-imaging was utilized for the area mapping, and the Laue pattern served for orientation.

Fig. 1(a) shows the experimental set-up combining phase contrast imaging and topography techniques. In a diffraction mode, topographs were recorded on a photographic film. Phase-contrast images were taken with a charge coupled device camera (14-bit gray scale and 1600 $\times$ 1200 pixels range). Before recorded X-ray image was converted into visible lights by a 150 μm thick $CdWO_4$ scintillator and magnified by a lens system with a magnification from 1$\times$ to 50$\times$.

Six on-axis 6*H*-SiC wafers were numbered as 'wafer I' (adjacent to the seed), 'wafer II' (the next to grow), and so the last wafer near the top of the boule as 'wafer VI'. The area mapping was preceded by the following orientation procedure. Each wafer was mounted with a miscut line, that is, an intersection between the surface and the basal plane (0001), vertical and the surface perpendicular to the beam. Each wafer was aligned to have the same Laue pattern of the orientation. For instance, the wafers I, II, III in the succession of one to another are sketched on Fig. 1(a). Fig. 1(b) shows an indexed Laue pattern for the (0001) orientation obtained from the wafer I at a distance of 9 cm. Similar patterns were then recorded from the wafer II, and so on.

3. The evolution of defects during SiC growth

Defect evolution from wafer to wafer was investigated on the area of interest from the matched overlap of the wafer images. For example, Fig. 2(a) shows the matched overlap of the wafer I (gray) cut off near the seed and the next wafer II (black). The region 1 is the area of interest, while the region 2 is a reference point to refine the wafer match. Here a group of MPs undeviatingly propagated through all the wafers, as demonstrated in Fig. 2(b). Pores and MPs in the region of interest 1 in the wafers I and II are displayed in Figs. 2(c) and (d),

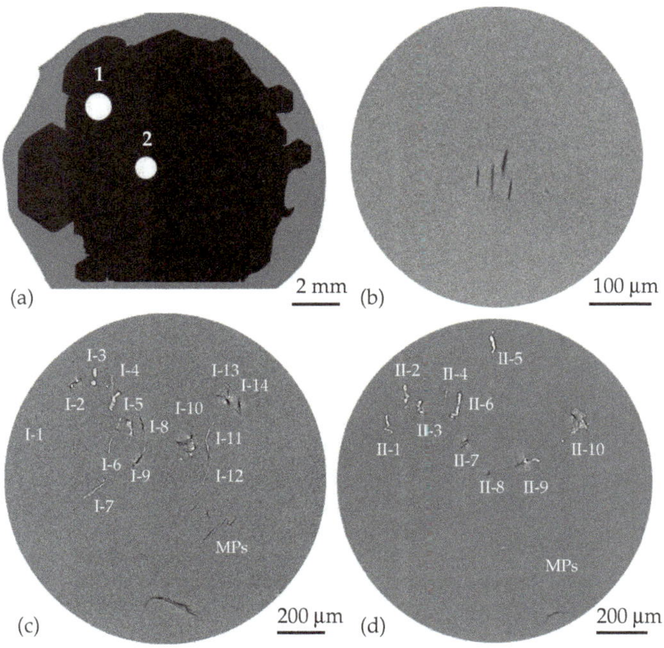

Figure 2. (a) The outline of the wafers I (gray) and II (black) imaged after the Laue pattern adjustment is completed. The area 1 is the area of interest. The wafer images are lapped over in such a way that the positions of propagating MPs encircled by the area 2 overlie in SR phase contrast images of (b). Comparison of SR phase contrast images of pores in the wafers I (c) and II (d).

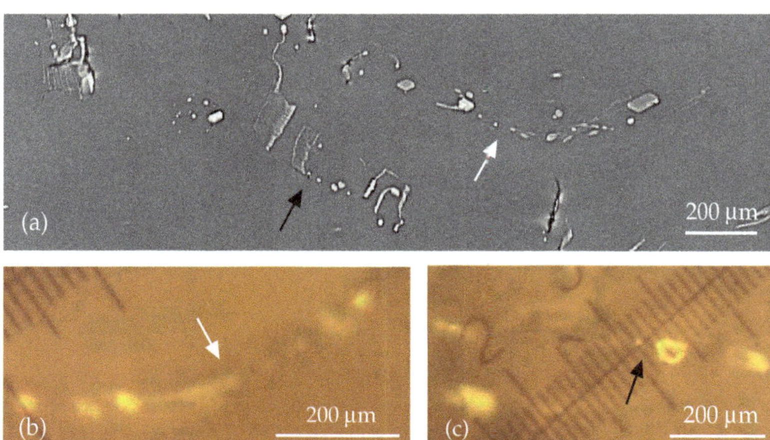

Figure 3. (a) SR phase-contrast image of the pores in the wafer I. (b) PL micrographs of 4*H*-SiC inclusions in 6*H*-SiC matrix; the white in (b) and the black in (c) arrows point to the boundaries of the inclusions where the pores are located in (a).

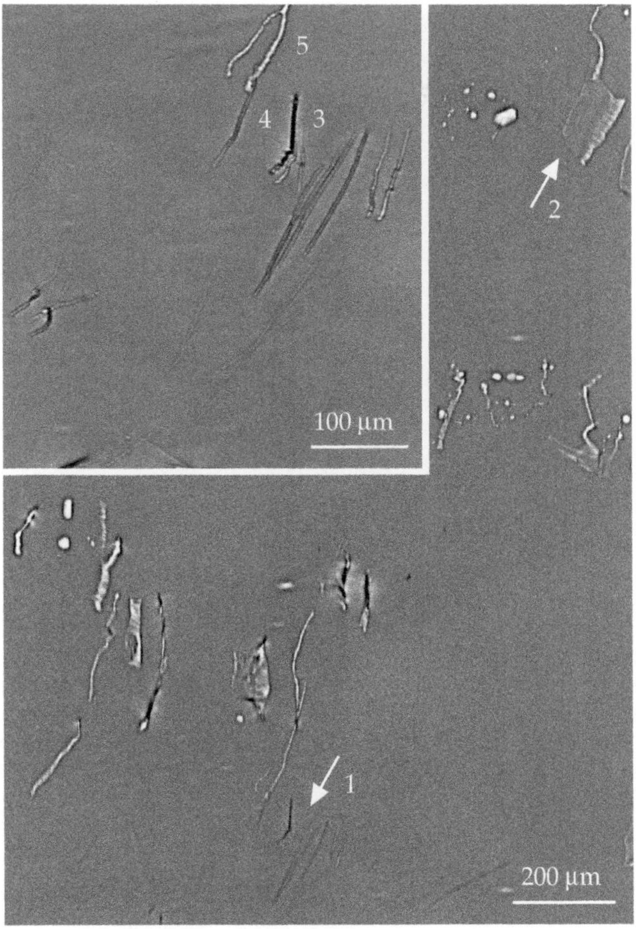

Figure 4. SR phase contrast image of the pores and MPs in the wafer I. The arrows point to the group of MPs (*1*) and the pore (*2*). The group of MPs is magnified in the inset: (*3, 4*) — MPs, (*5*) — pore.

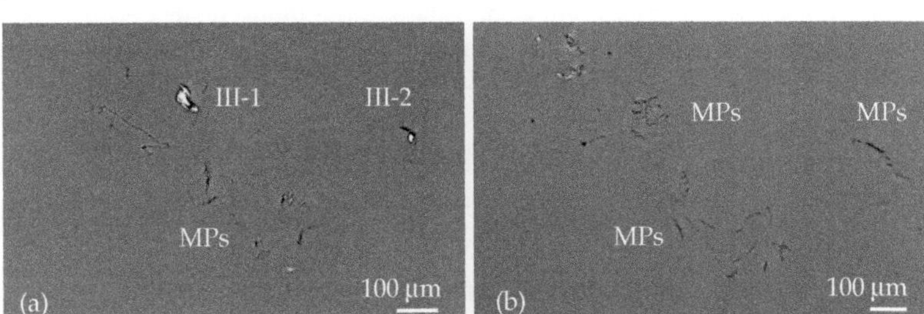

Figure 5. Pores and MPs in the wafers III, IV. The pores are counted III-1 and III-2 (a). MPs in the wafer IV (b).

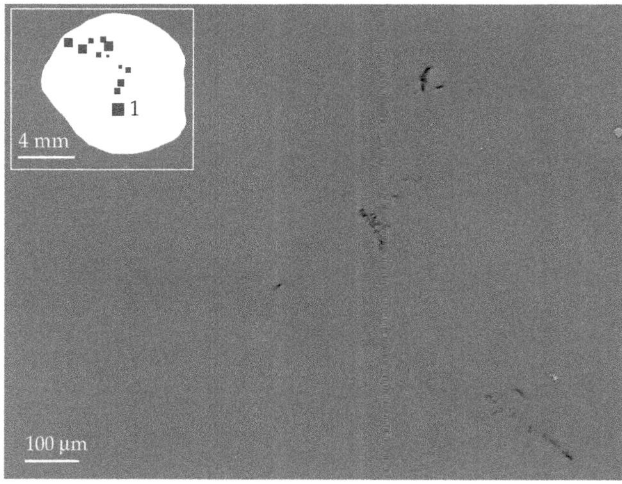

Figure 6. SR phase-contrast image of MPs in the latest-to-grow wafer VI. Schematic of the wafer with black squares highlighting the locations of MPs is shown in the inset. MPs are located on the square #1.

respectively. One can see that the wafer I has higher density of pores. In addition, Figs. 3(a), (b) and (c) taken in SR X-rays and PL, respectively, show pores and FPIs in the wafer I located just outside the scope of the image in Fig. 2(c). The high resolution of Fig. 3(a) demonstrates that the groups of pores (marked by black and white arrows) consist of short tube-shaped or slit-like segments. The morphology of such pores was investigated and attributed to the elastic interaction between MPs and boundaries of FPIs, resulting in coalescence of MPs into larger pores elongated along the boundaries [11, 13]. FPIs were indeed observed on the same location, as revealed by the yellow PL images [28] of *n*-type 4*H*-SiC containing N and B in Fig. 3(b) and (c), recorded at 77 K. The comparison with the pore images in the optical micrographs taken prior to the excitation of PL (data not shown), indicates that the pores trace the FPI boundaries.

A group of pores in Fig. 3(a) is partly displayed in the phase contrast image of Fig. 4. One pore from the group is marked by the arrow 2. Notice that the arrow 1 points to the group of MPs located at the place of interest in Fig. 2(c). The magnified image of the group is shown in the inset to Fig. 4. MPs appear as line segments, of which lengths are dependent on the wafer thickness, the miscut angle, and the sample tilting relative to the beam. The MPs run in different directions instead of lying parallel to the growth direction, similar to the observation described in an earlier paper [32]. In fact, the majority of MPs deviated from the growth direction and inclined toward one another or other defects in all six wafers studied.

Interestingly, one can see that the sign of contrast changes per every MP as well as along its axis. For instance, a typical MP image with black edges and a white inside shows a reversal to light edges and a black inside, as is the case in the upper right corner of the inset of Fig. 4. The MP #3 from the group of three MPs and the MP #4 nearby demonstrate the features of black contrast. The effect was well explained by the simulation of MP images measured in a white SR beam [20]. At high angles between the MP axis and the beam, the section size of MP along the beam is small. At low angles, the section size increases, and the wave field intensity inside the MP starts oscillating. The black contrast appears as a result of average, because in the white radiation fringes are not resolved. In such a way, the image simulation confirms that MPs can move parallel to the growth front towards one another and other defects.

A detailed investigation of MPs shows that the MP #3 and, especially, the MP #4 sharply inclined toward the slit-shaped pores nearby (for instance, to the pore #5 in the inset to Fig. 4). The attraction of the MPs to the pores is confirmed by their contrast behavior and verified by observing the density change of MPs during the SiC growth. The MP density in the group 1 reduced in the wafer II compared to I. The attraction of the MPs to the pore was suggested and explained by their absorption to pores [11, 13]. We believe that, at the initial stage of growth, the pores are generated by the attraction of MPs by FPIs, followed by MP coalescence [10, 11].

The number of pores reduced during the growth, as shown in Fig. 2(c) and (d). Fourteen pores, labeled in the sequence from I-1 to I-14, are identified from the wafer I as shown in Fig. 2(c). They reduce to ten pores on the same location of the wafer II, as shown in Fig. 2(d). We note that pores migrate and noticeably change their configuration. With further growth, the number of pores much more reduces from ten pores in the wafer II to two pores [III-1 and III-2 in Fig. 5(a)] in the wafer III. It is worth noting that new MPs appear in the wafer III. As the growth proceeds, pores are overgrown: no pores detected in the wafer I are seen in the place of interest of wafer IV [Fig. 5(b)]. However, one sees more MPs which obviously group together. They are bent and irregular in shape, size and propagation direction. In the last wafer VI, MPs density remains very inhomogeneous: it is high in some local places, as highlighted in the inset to Fig. 6 by black squares, but low on average. Phase-contrast image of the figure shows the distribution of MPs in a representative location #1. One can see short projection lengths of the MPs running in various directions inside the sample fixed perpendicular to the beam. The imaged area in Fig. 6 is 1.2 mm $\times$ 1.2 mm. Taking into account that the area displayed in Fig. 5(b) is slightly smaller (1.2 mm $\times$ 0.8 mm), and the number of MPs is noticeably bigger, one can conclude that the density of MPs decreases with growth.

The observations of MPs in this 6*H*-SiC boule and many other similar 6*H* and 4*H*-SiC boules provided strong evidences for the reactions of MPs [7–9]. Reaction between MPs is always a positive process in view of their elimination during the crystal growth.

4. Reduction of micropipes density via self-reactions

4.1. Radii reduction and gradual healing

4.1.1. Micropipe ramification at the front of a growing crystal

Ramification of MPs is accompanied by the partitioning of their Burgers vectors. According to the Frank rule [6], the equilibrium MP radius is proportional to the squared magnitude of its Burgers vector. As a result, ramification of MPs results in a decrease of their radii. The sketch of Fig. 7(a) represents a dislocated MP with the radius r_0 and the Burgers vector b_0 splitted into two smaller ones with the radii and the Burgers vectors r_1, r_2 and b_1, b_2, respectively. A typical phase-contrast image of a ramified dislocation with a hollow core is shown in Fig. 7(b).

The ramification was modeled by Gutkin *et al.* to determine when a MP is favorable to split into a pair of MPs with smaller radii and Burgers vectors [7]. The angle between ramifying segments of the MP 'tree' was assumed to be small enough to consider these segments as a pair of parallel MPs. Also, it was assumed that the ramification required the repulsion of the ramifying segments. It is worth noting that the reverse process of the MP merging may follow the MP split. If the MPs formed by the split stay in contact, they are energetically favored to coalesce and produce a single MP. In this case, the split and following coalescence results only in a decrease in the MP radius. However the merging of the MP segments generated by the split does not occur if these segments, once formed, repulse each other. It was shown that the MPs may either repulse at any distance or repulse at distances d greater than the critical distance d_c and attract each other at $d < d_c$. The ramification of the MP segments (not followed by their coalescence) is possible if they repulse at any distance or the critical distance d_c is small enough. The parameter region where the ramifying MP segments repulse at any distance is shown in Fig. 7(c). The system state diagram is depicted in the coordinate space $(b_1/b_2, r_1/r_2)$. The upper and lower curves separate regions I and III, where the MPs attract each other at short distances while they repel each other at long distances, and the corresponding critical distance $h_c = d_c - r_1 - r_2$ between the MP free surfaces exists, from region II, where the MPs repulse at any distance. As is seen, an attraction area may exist for two same-sign MPs if b_1/b_2 and r_1/r_2 differ by more than 15%–25%. Ramification which does not require overcoming an energetic barrier is possible if the MPs are of the same sign, the radius of the original MP exceeds the equilibrium one, the total MP surface area reduces due to the ramification, and the radii of the ramifying MPs are approximately in the same ratio as the magnitudes of their dislocation Burgers vectors.

4.1.2. Merging of micropipes

Many morphologies of coalesced MPs were observed experimentally in SiC crystals by means of x-ray phase-contrast imaging. To explain these phenomena, a computer simulation of MP evolution was performed [9]. It has been shown that the reaction of MP coalescence gives rise to the generation of new MPs with smaller diameters and Burgers vectors, which again

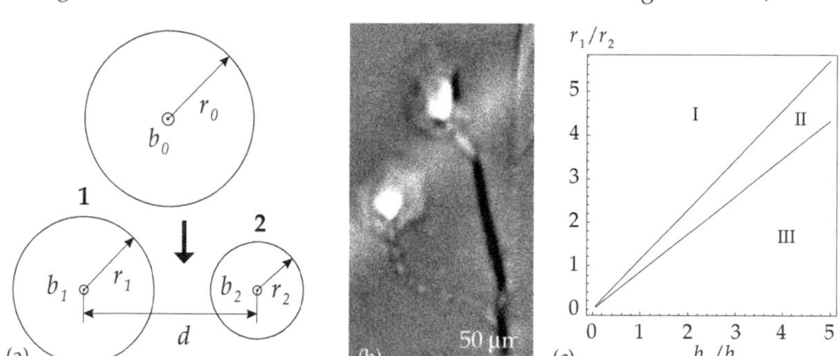

Figure 7. (a) Initial MP with the radius r_0 and the Burgers vector b_0 splits into two parallel dislocated MPs 1 and 2 with the radii and the Burgers vectors r_1, r_2 and b_1, b_2, respectively located at a distance d from each other. (b) SR phase-contrast image of a ramified configuration. Halos correspond to etch pits on the wafer surface. (c) The system state diagram in the coordinate space $(b_1/b_2, r_1/r_2)$. The curves separate parameter regions I and III, where both attraction area and unstable equilibrium position exist for two MPs, from parameter region II, where the MPs repulse at any distance.

leads to diminishing their average density. In particular, when the interacting pair of MPs has incomparable diameters or is located far (at the distance of more than about 5 average MP diameters) from other MPs, the coalescing MPs come directly to each other along the shortest way between them. In this case their behavior is governed by elastic interaction between them only. In some other cases, the ends of two MPs with opposite-sign Burgers vectors start to move around one another. One of the reasons for this is the action of neighboring MPs. The winding of MPs around each other results in coalescence of their subsurface segments. The examples of twisted configurations and the modeling can be found in a review [14].

4.1.3. Correlated reduction in micropipe cross sections

Reactions of ramification, merging and twisting of MPs occur when MPs come in contact, touching each other by their surfaces. However, phase-contrast imaging shows that MPs which are not in a direct contact may have variable cross-sections. In our paper [12], we demonstrated the experimental evidence of two neighboring MPs reducing their diameters (approximately by half) one after another, at different distances from the surface of a grown crystal. This can be treated as an indirect proof of the contact-free reaction. Such an idea appeared when we studied the axial-cut slice of 4*H*-SiC boule with the orientation ($\bar{1}1.0$). MPs located almost parallel to the growth axis were nearly parallel to the sample surface. The sample was fixed on the holder with its surface perpendicular to the beam and rotated to achieve a horizontal position of MP axes. So the images were measured by using the more coherent vertical projection of the source. In the topography mode, the sample was tuned to obtain a symmetrical Laue pattern. An example of an indexed Laue pattern for the ($\bar{1}1.0$) sample orientation is displayed in Fig. 8(a). One can see many diffraction vectors **g** available for the characterization of dislocations.

For few individual MPs, the distances between the edges modified along their axes showing that their cross-sections changed through the crystal. A good example is the MP pair shown in Fig. 8(b) and (c). An optical micrograph of Fig. 8(c) displays vertical lines of black contrast which are non-uniformly distributed over the sample area and uneven horizontal lines of light contrast. The former are attributed to MPs and the latter to layered inclusions of foreign polytypes (6*H* and 15*R*). We note that MPs agglomerate along the boundaries of the inclusions. The group of closely spaced MPs labeled as MP1 and MP2 is located distant from the other MPs. This particular group was examined in detail in synchrotron light. To indicate the positions of the MPs in figures 8 and 9, we used the features of non-variable contrast, namely, the morphological defect [visible in Fig. 8(c) to the left from the MP1] and the MP bundle located at a distance of 950 μm and 1100 μm, respectively.

Fig. 9(a) and (b) demonstrate X-ray topographs of the MPs recorded in the $\overline{21}.\bar{2}$ and $01.\overline{10}$ reflections, respectively. The white arrow indicates the MP1 located at a distance of 950 μm from the mark (that is the morphological defect) shown in Fig. 8(c). The diagram in Fig. 9(c) represents the diffraction geometry. Synchrotron beam propagates along the surface normal [$\bar{1}1.0$], while the basal plane normal [00.1] lies perpendicular to the beam. We suspect that the dislocation Burgers vector **b** of the MP1 is parallel to [00.1]. The diffraction vectors $\mathbf{g}_1 = \overline{21}.\bar{2}$, and $\mathbf{g}_2 = 01.\overline{10}$ and the Burgers vector $\mathbf{b} \| [00.1]$ are projected onto the film. The angles between the projections, $\angle \mathbf{g}_1, \mathbf{b}$ and $\angle \mathbf{g}_2, \mathbf{b}$, are also displayed. We note that, for a pure screw dislocation, its image is invisible when $\mathbf{g} \cdot \mathbf{b} = 0$. In Fig. 9 one notices that the contrast increases with the decrease of the angle. The MP1 is invisible in the $\overline{21}.\bar{2}$ topograph when $\angle \mathbf{g}_1, \mathbf{b} = 78.8°$, as in Fig. 9(a), but discernible in the $01.\overline{10}$ reflection when $\angle \mathbf{g}_2, \mathbf{b} = 20.6°$, as in Fig. 9(b). The contrast behavior indicates that the dislocation of the MP1 is of screw type.

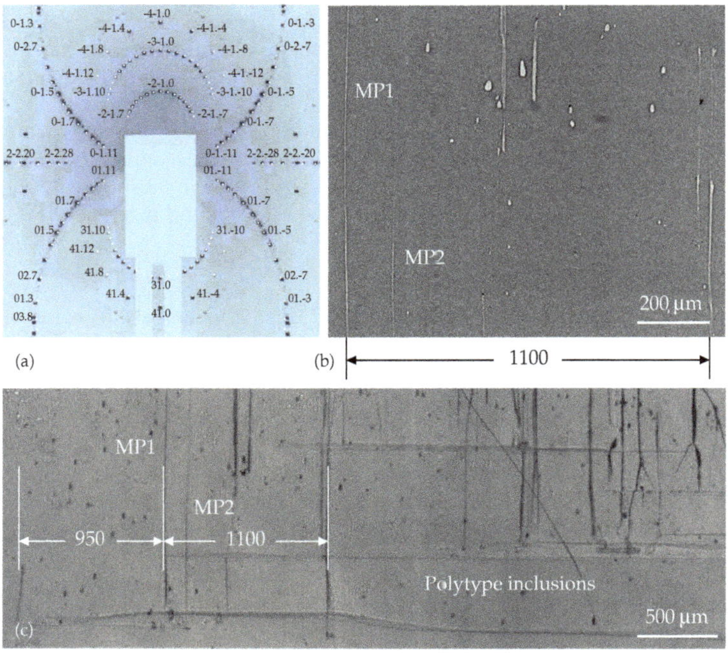

Figure 8. (a) Indexed Laue pattern for $(\bar{1}1.0)$ 4*H*-SiC wafer (direction [00 1] horizontal) fixed perpendicular to SR beam. The sample-to-film distance is 130 mm. (b) SR phase-contrast image of the studied MPs, MP1 and MP2. (c) Optical micrograph (transmission) of MPs and foreign polytype inclusions. The distances from MP1 to the morphological defect (950) and the MPs bundle (1100) are shown in microns.

Phase contrast images of the MPs 1 and 2 were measured in a series at the distances from 5 to 45 cm from the sample to detector, increasing the distance every 5 cm. The image registered at the distance of 10 cm is shown in Fig. 10(a). To determine the characteristic sizes of the MP cross sections in different points along the MP axis, we applied the method of computer simulation [1, 19] of the measured intensity profiles. For every MP cross section under investigation, the computer program calculated many profiles for various possible section sizes on the base of Kirchhoff propagation to find the profile, which gives the best fit to the experimental profile. The parameters could be varied by user as well as via automatic procedures. One procedure calculated all points over a square net, while the other one looked for the best fit at every step and arrived at the best matching point. The coincidence allowed us to determine both the longitudinal (along the beam) and transverse (across the beam) sizes of the section [Fig. 10(b)].

The transverse diameters of MP1 and MP2 are presented in Fig. 10(a) versus the distance along the pipe axes increasing in the growth direction. It is seen that the transverse size of MP1 reduces from 7.4 to 2.1 μm with growth. At the same time, the transverse size of the MP2 reduces from 4.1 to 1.6 μm. In contrast, the longitudinal diameters remain almost the same as of the order of 0.8 μm for the MP1 and 0.5 μm for the MP2 (data not shown). In the correlated decrease of the MP1 and MP2 cross-section sizes in Fig. 10(a), several features are apparent. A remarkable decrease in the MP1 cross-section size occurs in the distance

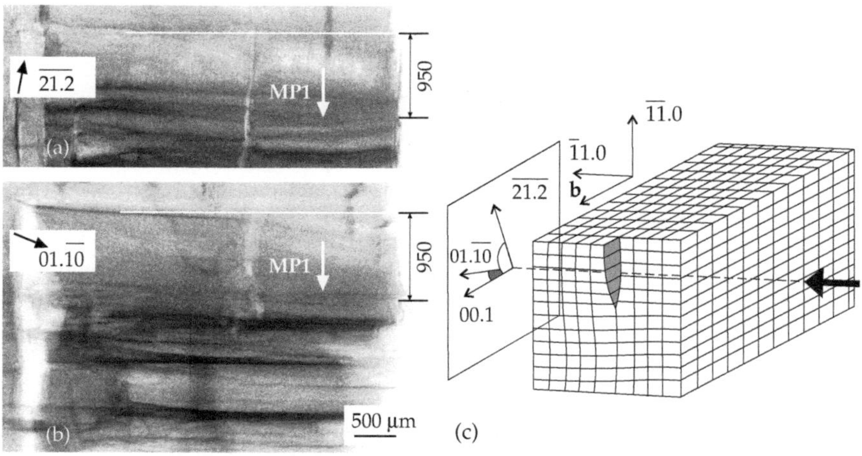

Figure 9. SR white beam topographs of MP1 and adjacent MPs obtained in the reflections $\mathbf{g}_1 = \overline{21}.\overline{2}$ (a) and $\mathbf{g}_2 = 01.\overline{10}$ (b). The position of MP1 is determined relative to the morphological defect shown in Fig. 8. Schematic diagram of the diffraction geometry (c). SR beam (a black arrow) propagates parallel to $[\overline{1}1.0]$. The dislocation Burgers vector $\mathbf{b}$ of MP1 is supposed parallel to [00.1], and the angles are $\angle \mathbf{g}_1, \mathbf{b} = 78.8°$ and $\angle \mathbf{g}_2, \mathbf{b} = 20.6°$.

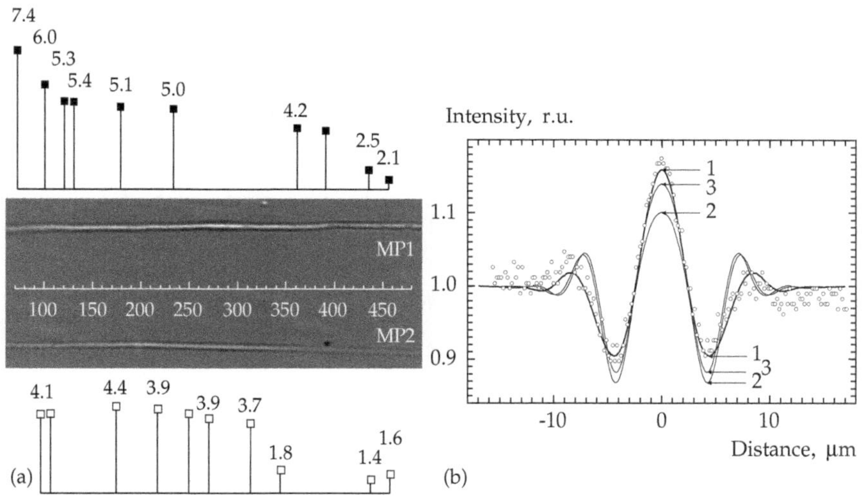

Figure 10. (a) Phase contrast image of micropipes MP1 and MP2 in the interval of 70–500 μm along the pipe axes. The sample-to-detector distance is 10 cm. Variation in transverse cross-section sizes along the lines of MP1 (black squares) and MP2 (open squares) are shown in μm. (b) The experimental (open circles) and simulated (curves 1–3) intensity profiles (the intensity in relative units via the distance across the MP image). The best agreement is achieved through a sequential adjustment in the pipe diameters perpendicular and parallel to the beam. The sample-scintillator distance is 45 cm. The transverse/longitudinal diameters are equal to 5.6/1.0 μm (curve 1, fit = 1.92×10^{-4}). For comparison, curve 2 shows the best fit (8.12×10^{-4}) for the circular cross section 1.76/1.76 μm, while curve 3 is given for the intermediate case 3.0/1.4 μm (fit = 5.71×10^{-4}).

interval from 74 to 132 μm while the transverse diameter of MP2 drastically decreases later in the distance interval from 314 to 345 μm. In addition, a rapid decrease of the transverse diameter of MP1 happens in the distance interval from 393 to 458 μm when the transverse diameter of MP2 remains almost invariable.

The radii reduction of both MPs was explained by a contact-free reaction between them [12, 14]. We suppose that the MP1 and MP2 contain superscrew dislocations with opposite Burgers vectors $\mathbf{b}_1$ and $\mathbf{b}_2$. In case of a contact-free interaction micropipe MP1 emits a full-core dislocation half-loop, which expands by gliding, reaches the surface of micropipe MP2, and reacts with its dislocation. The corresponding dislocation reactions are described by equations: $\mathbf{b}_1 - \mathbf{b}_0 = \mathbf{b}_3$ and $\mathbf{b}_2 + \mathbf{b}_0 = \mathbf{b}_4$, where $\mathbf{b}_3$ and $\mathbf{b}_4$ are new Burgers vectors of micropipes MP1 and MP2, respectively. Strong reduction in the radii can lead to their gradual healing.

4.2. Annihilation of micropipes

MP merging may lead to the annihilation of initial MPs at the growing surface. We considered the motion of subsurface segments of MPs under the action of elastic forces due to their interaction and some effective friction [9]. The latter accounted for some extra surface energy related to the steps appearing on the pipe cylindrical surfaces during lateral displacements of the pipe segments. As a result, various reactions between the subsurface pipe segments were observed. In particular, it has been shown that the reaction of MP coalescence can lead to the annihilation of initial MPs. Some typical defect configurations in a 3D space are displayed in Fig. 11. They may be subdivided into planar and twisted pipe configurations. The planar configurations arise when the interacting pair of MPs is located far from other MPs; and the coalescing MPs come to each other along the shortest way [Fig 11(a)]. The twisted configurations like double spirals form if the interacting MPs are located within dense groups of other MPs. In this case, the coalescing MPs twist [Fig 11(b)]. When the magnitudes of Burgers vectors are the same, the initial defect configuration of a dipole is transformed into a new configuration of a semiloop [Fig. 11(c)]. As a result, we expect that the density of MPs would diminish during the crystal growth. The smaller the surface energy is the faster this process would be.

5. Structural models of pore overgrowth

5.1. Generation of pores in early stage of SiC growth

From our observations of FPIs, pores, MPs and their changes during the bulk SiC crystal growth described above, we suggest the formation mechanisms of defects as follows. The nucleation of FPIs in the initial stage is suggested as a basis of massive generation of full-core dislocations [26], pores and MPs [10, 11, 13, 31]. The bottom and lateral faces of growing FPIs are possibly the formation sites of interface dislocation structures to accommodate the misfit or misorientation of the matrix and FPI's crystalline lattices. The FPI boundaries are also able to serve as easy paths and sinks for vacancies which are nucleated at the growth front and migrate towards the crystal bulk. High densities of interface dislocations and vacancies thus formed possibly lead to coagulation of vacancies, forming slit-like pores along the FPI boundaries.

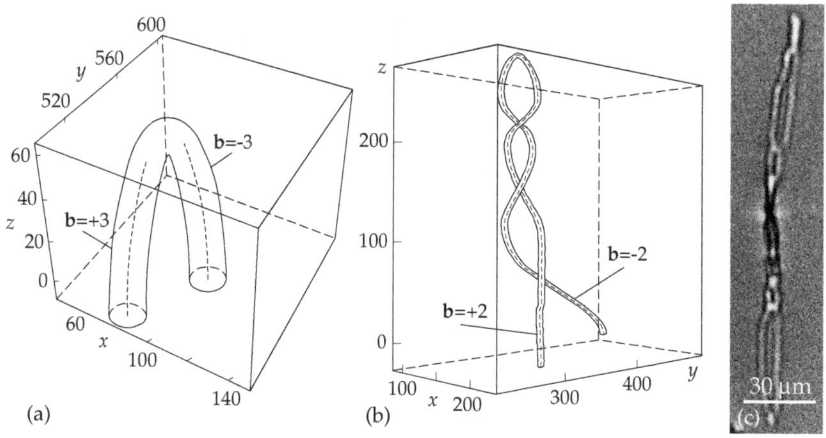

Figure 11. Coalescence of MPs with equal magnitudes of Burgers vectors results in the annihilation of the subsurface MP segments (a, b). The coalescing MPs come to each other along the shortest way (a) or twist (b). The Burgers vectors are shown in units of c, which is the lattice parameter in the growth direction. The coordinates x and y are given in units of $Gc^2/(8\pi^2\gamma)$, where G is the shear modulus and γ is the specific surface energy. The length of MPs (along the z axes) is in arbitrary units that depend on the growth rate. SR phase-contrast image of semi-loops resulting from macropipes twisting (c).

Another possible mechanism of pore formation is the attraction and agglomeration of MPs at FPI boundaries [10, 11, 13], for instance, resulting in the majority of pores observed at the boundaries of FPIs (Fig. 3). By effectively accommodating both the dilatation and orientation misfits between FPI and matrix, these pores in early stages can attract additional random full-core dislocations and MPs from neighboring regions (Fig. 4), as earlier described in detail [13]. In such a way, the interface pores can extend along FPI boundaries and accumulate dislocation charge that is the resulting Burgers vector of all the dislocations absorbed by the pore. Of course, one can not exclude the presence of the pores that have been formed by the other mechanisms and do not contain any dislocation charge.

5.2. Generation of MPs in the intermediate stage

When the FPIs stop to grow and become overgrown by the matrix, there is no reason for the pore formation as misfit defects. The disappearance of pores started to occur at this point, as seen in Figs. 2 and 5. We suggest three possible mechanisms that can explain the disappearance and its contribution to the increase in MP density.

a First, pores can dissolve through emission of vacancies, which migrate to full-core threading dislocations on the FPI/matrix interface [15, 26] and are absorbed by them, as illustrated in Fig. 12. These dislocations can reach the growth front and proceed to grow with it [Fig. 12(a)]. At the dislocation core, these vacancies are coagulated with the vacancies that migrate from the growth front along the dislocation cores, thus forming embryos of new MPs [Fig. 12(b)].

b Second, the pores can be overgrown by the lateral growth of the crystal and, if they contain dislocations with large Burgers vectors, can transform to MPs by the mechanism described earlier [21].

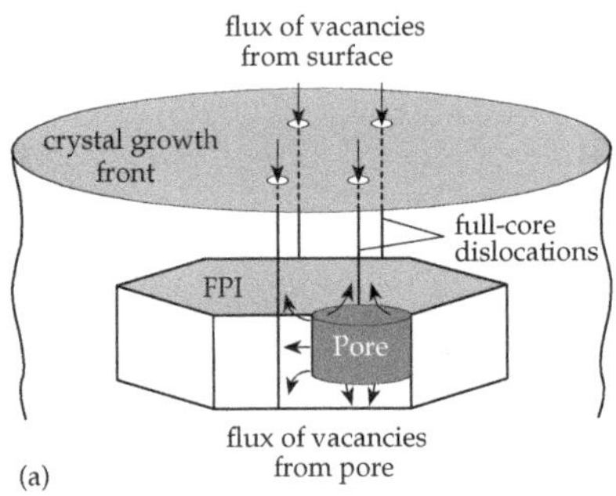

Figure 12. Mechanism of MP nucleation through coagulation of vacancies at the cores of threading dislocations. (a) Pore at the boundary of FPI starts to dissolve by emitting vacancies which migrate to neighboring full-core threading dislocations. (b) Vacancies from pore and growth surface migrate along the dislocation cores, meet and coagulate, thus transforming full-core dislocations to MPs.

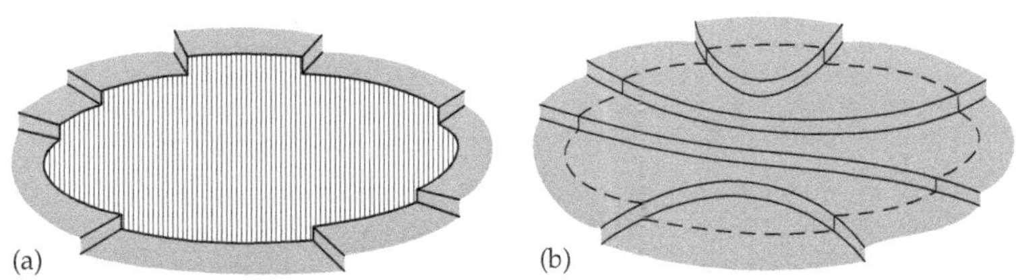

Figure 13. The pore of a convex shape has equal number of up-steps and down-steps and does not contain screw dislocation. (a) Open pore, (b) overgrown pore.

c Third, if pores have complex shapes (like those represented in Figs. 2–4, they can produce dislocated MPs during their lateral overgrowth even without full-core dislocations as described below.

The surface regions around pores always contain surface steps. If the numbers of the steps up and down (let us call them up-steps and down-steps, respectively) are equal, there would be no screw dislocation inside the pore. Otherwise there should be a screw dislocation with a Burgers vector of the magnitude equal to the difference between the sum heights of up-steps and down-steps. Let us consider the first case as shown schematically in Fig. 13 and Fig. 14 for simple convex and complex pore shapes, respectively. If the pore shape is convex (say, circular or elliptical), its overgrowth can hardly lead to the formation of screw dislocations (Fig. 13). However, if the pore shape is complex with some concave fragments [Fig. 14(a)], one can expect that the pore starts to overgrow through a bridge between two opposite concave fragments in a narrow part of the pore [Figs. 14(b) or (c)]. This bridge can separate the initial pore into smaller ones. If the two new pores have different numbers of up-steps and down-steps on the growth surface, the difference can be compensated by large (or small) steps on the bridge [it is called S-bridge here, Fig. 14(b)] with no screw dislocations. Otherwise the bridge can have a smooth surface without steps and be distorted [it is called D-bridge here, Fig. 14(c)], forming a semi-loop of superdislocation. The edge segment of the semi-loop lies under the D-bridge, while the two screw segments are within their new pores.

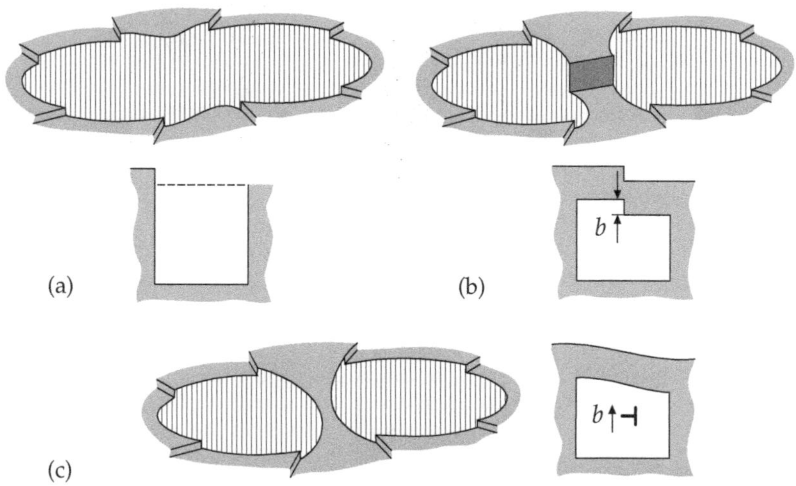

Figure 14. (a) The pore of a complex shape has equal number of up-steps and down-steps and first does not contain screw dislocation. (b) Pore overgrowth starts with a stepped bridge (S-bridge); if the large step compensates small steps around the two new pores, then no screw dislocations appear. (c) Pore overgrowth starts with a distorted bridge (D-bridge); there is no compensation of small steps around the two new pores, in which case a dislocation semi-loop arises: the edge segment of the semi-loop lies under the D-bridge, while two screw segments are within their new pores. These new pores become embryos of dislocated MPs.

These new pores can become embryos of dislocated MPs. It is worth noting that a similar mechanism for screw dislocation nucleation at foreign phase inclusions was first proposed by Chernov [4]. Dudley *et al.* experimentally observed and explained the nucleation of a pair of MPs at an inclusion in 4*H*-SiC [5]. In such a way, a pore elongated along the growth front and overgrown with the formation of many D-bridges can 'produce' many MPs of alternate dislocation signs of equal or different radii depending on the dislocation Burgers vectors. Similar separate dense arrays of MPs are demonstrated in Figs. 5 and 6.

As a result, the density of pores (and full-core dislocations) decreases at this stage of crystal growth, while the density of MPs increases, as seen in Fig. 5.

5.3. Evolution of defects in later stage

The situation at the next stage of the crystal growth depends on whether new FPIs are generated or not. If generated, then the stages described above are repeated. However, though formed as explained in the intermediate stage, MPs can be attracted to FPIs and absorbed by their boundaries, producing extended pores there by agglomeration [10, 11]. If new FPIs are not nucleated, the processes of self-organization occur in the MPs ensemble: MPs elastically interact and react with each other as well as with full-core dislocations, as described in the section 4. As a result, some MPs annihilate or diminish their Burgers vectors and are finally healed. Otherwise they form separate dense groups of MPs which proceed to grow with the crystal.

6. Conclusions

The evolution of defects was investigated by using on-axis and axial-cut slices prepared from 6*H* and 4*H*-SiC PVT grown crystals. SR phase contrast imaging enabled us to visualize not only MPs and the pores formed at the boundaries of FPIs, but also their changes during SiC growth. Detailed mechanisms for the evolution from FPIs to pores and finally to MPs were suggested. In the early growth stage, FPIs not only induce massive generation of full-core dislocations and MPs but also attract them, forming slit-type pores at the boundaries of FPIs. In the intermediate stage, when FPIs stop to grow and become overgrown by the matrix, the pore density significantly reduces, which is attributed to their transformation into new MPs. In the later stage, the MP density decreases, providing evidence for their partial annihilation and healing.

The reactions of MPs in view of their elimination during the crystal growth were briefly reviewed. The reduction of MP cross-section, which can eventually results in its overgrowth, occurs at the crystal growth when MP splits, as well as merges or interacts with another MP in a non-contact mode. The split happens if the splitting dislocation overcomes the MP attraction zone and the flat crystal surface attraction zone. Merging can occur due to collective mesoscopic effects in a random ensemble of MPs. The twisted dipoles result under the action of neighboring MPs. When the magnitudes of Burgers vectors are the same, the dipole is transformed into a new configuration of a semiloop. Such reactions of ramification and coalescence of MPs, as well as annihilation for dipoles of MPs, were observed by phase-contrast imaging. Computer simulation of phase-contrast images demonstrated the correlated reduction in the radii of two remote MPs, which provided a support of contact-free reaction between them.

This study suggests that the key point for the elimination of defects from such crystals is the suppression of FPI nucleation. The reactions of MPs are necessary for diminishing their density; and such reactions should be faster as the surface energy becomes smaller.

Acknowledgements

This work was supported by the Creative Research Initiatives (Functional X-ray Imaging) of MEST/KOSEF of Korea. The work of VGK was supported by RFBR grant No. 1002-00047-a.

Author details

T. S. Argunova[1,2], M. Yu. Gutkin[3,4,5],
J. H. Je[2,*], V. G. Kohn[6] and E. N. Mokhov[1]

[1]Ioffe Physical-Technical Institute, RAS, St. Petersburg, Russia
[2]X-ray Imaging Center, Department of Materials Science and Engineering, Pohang University of Science and Technology, Pohang, Republic of Korea
[3]Institute of Problems in Mechanical Engineering, RAS, St. Petersburg, Russia
[4]Department of Physics of Materials Strength and Plasticity, St. Petersburg State Polytechnical University, St. Petersburg, Russia
[5]Department of Theory of Elasticity, St. Petersburg State University, St. Petersburg, Russia
[6]Russian Research Center 'Kurchatov Institute', Moscow, Russia

References

[1] Argunova, T.; Kohn, V.; Jung, J-W.; Je J-H. (2009). Elliptical micropipes in SiC revealed by computer simulating phase contrast images. *Phys. Status Solidi* A, Vol. 206, 1833–1837.

[2] Baik, S.; Kim, H. S.; Jeong, M. H.; Lee, C. S.; Je, J. H.; Hwu, Y.; Margaritondo, G. (2004). International consortium on phase contrast imaging and radiology beamline at the Pohang Light Source. *Rev. Sci. Instrum.*, Vol. 75, 4355–4358.

[3] Chen, Yi; Dudley, M.; Sanchez, E.; Macmillan, M. (2008). Simulation of grazing-incidence synchrotron white beam X-ray topographic images of micropipes in 4*H*-SiC and determination of their dislocation senses. *J. Electron. Mater.*, Vol. 37, 713–720.

[4] Chernov, A. A. (1989). Formation of crystals in solutions. *Contemp. Phys.*, Vol. 30, 251-276.

[5] Dudley, M.; Huang, X. R.; Huang, W.; Powell, A.; Wang, S.; Neudeck, P.; Skowronski, M. (1999). The mechanism of micropipe nucleation at inclusions in silicon carbide. *Appl. Phys. Lett.*, Vol. 75, 784–786.

[6] Frank, F. C. (1951). Capillary equilibria of dislocated crystals. *Acta Crystallogr.*, Vol. 4, 497–501.

[7] Gutkin, M. Yu.; Sheinerman, A. G.; Argunova, T. S.; Je, J. H.; Kang, H. S.; Hwu, Y.; Tsai W-L. (2002). Ramification of micropipes in SiC crystals. *J. Appl. Phys.*, Vol. 92, 889–894.

[8] Gutkin, M. Yu.; Sheinerman, A. G.; Argunova, T. S.; Mokhov, E. N.; Je, J. H.; Hwu Y.; Tsai W-L.; Margaritondo G. (2003). Micropipe evolution in silicon carbide. *Appl. Phys. Lett.*, Vol. 83, 2157–2159.

[9] Gutkin, M. Yu.; Sheinerman, A. G.; Argunova, T. S.; Mokhov, E. N.; Je, J. H.; Hwu Y.; Tsai W-L.; Margaritondo G. (2003a). Synchrotron radiographic study and computer simulation of reactions between micropipes in silicon carbide. *J. Appl. Phys.*, Vol. 94, 7076–7082.

[10] Gutkin, M. Yu.; Sheinerman, A. G.; Argunova, T. S.; Yi, J.-M.; Kim, M.-U.; Je, J.-H.; Nagalyuk, S. S.; Mokhov, E. N.; Margaritondo, G.; Hwu, Y. (2006). Interaction of micropipes with foreign polytype inclusions in SiC. *J. Appl. Phys.*, Vol. 100, 093518.

[11] Gutkin, M. Yu.; Sheinerman, A. G.; Argunova, T. S.; Yi, J.-M.; Je, J.-H.; Nagalyuk, S. S.; Mokhov, E. N.; Margaritondo, G.; Hwu, Y. (2007). Role of micropipes in the formation of pores at foreign polytype boundaries in SiC crystals. *Phys. Rev.* B, Vol. 76, 064117.

[12] Gutkin, M. Yu.; Sheinerman, A. G.; Smirnov, M. A.; Kohn, V. G.; Argunova, T. S.; Je, J. H.; Jung J. W. (2008). Correlated reduction in micropipe cross sections in SiC growth. *Appl. Phys. Lett.*, Vol. 93, 151905.

[13] Gutkin, M. Yu.; Sheinerman, A. G.; Smirnov, M. A.; Argunova, T. S.; Je, J.-H.; Nagalyuk, S. S.; Mokhov, E. N. (2009). Micropipe absorption mechanism of pore growth at foreign polytype boundaries in SiC crystals. *J. Appl. Phys.*, Vol. 106, 123515.

[14] Gutkin, M. Yu.; Argunova, T. S.; Kohn, V. G.; Sheinerman, A. G.; Je, J. H. (2011). Micropipe Reactions in Bulk SiC Growth. *Silicon Carbide,* ISBN 978-953-307-348-4, IN-TECH. URL: *http://www.intechopen.com/articles/show/title/micropipe-reactions-in-bulk-sic-growth*

[15] Ha, S.; Nuhfer, N. T.; Rohrer, G. S.; de Graef, M.; Skowronski, M. (2000). Identification of prismatic slip bands in 4*H*-SiC boules grown by physical vapor transport. *J. Electron. Maters.,* Vol. 29, L5–L8.

[16] Hatakeyama, T.; Ichinoseki, K.; Fukuda, K.; Higuchi, N.; Arai, K. (2008). Evaluation of the quality of commercial silicon carbide wafers by an optical non-destructive inspection technique. *J. Cryst. Growth,* Vol. 310, 988-992.

[17] Huang, X. R.; Dudley, M.; Vetter, W. M.; Huang, W.; Wang, S. (1999). Direct evidence of micropipe-related pure superscrew dislocations in SiC. *Appl. Phys. Lett.,* Vol. 74, No. 3; 353–355.

[18] Kamata I.; Nagano, M.; Tsuchida, H.; Chen, Yi.; Dudley, M. (2009). Investigation of character and spatial distribution of threading edge dislocations in 4*H*-SiC epilayers by high-resolution topography. *J. Cryst. Growth,* Vol. 311, 1416–1422.

[19] Kohn, V. G.; Argunova, T. S.; Je, J. H. (2007). Study of micropipe structure in SiC by x-ray phase contrast imaging. *Appl. Phys. Let.,* Vol. 91, 171901.

[20] Kohn, V. G.; Argunova, T. S.; Je, J. H. (2012). Reason of a contrast change in SR-images of micropipes in SiC. *J. Surf. Investig-X-ra.,* Vol. 10. 60-65.

[21] Kuhr, T.; Sanchez, E.; Skowronski, M.; Vetter, W.; Dudley, M. (2001). Hexagonal voids and the formation of micropipes during SiC sublimation growth. *J. Appl. Phys.,* Vol. 89, 4625–4630.

[22] Lin, Sh.; Chen, Zh.; Yang, Y.; Liu, S.; Ba, Y.; Li L.; Yang, Ch. (2012). Formation and evolution of micropipes in SiC crystals. *CrystEngComm,* Vol. 14, 1588–1594. URL: http://pubs.rsc.org/en/content/articlepdf/2012/ce/c1ce05806a

[23] Liu, J.-L.; Gao, J.-Q.; Cheng, J.-K.; Yang, J.-F.; Qiao, G.-J. (2007). Methods for the reduction of the micropipe density in SiC single crystals. *J. Mater. Sci.,* Vol. 42, 6148–6152.

[24] Nakamura, D.; Yamaguchi, S.; Gunjishima, I.; Hirose, Y.; Kimoto, T. (2007). Topographic study of dislocation structure in hexagonal SiC single crystals with low dislocation density. *J. Cryst. Growth,* Vol. 304, 57–63.

[25] Nakamura, D.; Yamaguchi, S.; Hirose, Y.; Tani, T.; Takatori, K.; Kajiwara, K.; Kimoto, T. (2008). Direct determination of Burgers vector sense and magnitude of elementary dislocations by synchrotron white x-ray topography. *J. Appl. Phys.,* Vol. 103, 013510.

[26] Ohtani, N.; Katsuno, M.; Tsuge, H.; Fujimoto, T.; Nakabayashi, M.; Yashiro, H.; Sawamura, M.; Aigo, T.; Hoshino, T. (2006). Propagation behavior of threading dislocations during physical vapor transport growth of silicon carbide (SiC) single crystals. *J. Cryst. Growth,* Vol. 286, 55–60.

[27] Ohtani, N.; Katsuno, M.; Tsuge, H.; Fujimoto, T.; Nakabayashi, M.; Yashiro, H.; Sawamura, M.; Aigo, T.; Hoshino, T. (2006a). Dislocation processes during SiC bulk crystal growth. *Microelectron. Eng.*, Vol. 83; 142–145.

[28] Saparin G. V., Obyden, S. K.; Ivannikov, P. V.; Shishkin, E. B.; Mokhov, E. N.; Roenkov, A. D.; Hofmann, D. H. (1997). Three-dimensional studies of SiC polytype transformations. *Scanning*, Vol. 19, 269–274.

[29] Schmitt, E.; Straubinger, T.; Rasp, M.; Vogel, M.; Wohlfart, A. (2008). Polytype stability and defects in differently doped bulk SiC. *J. Cryst. Growth*, Vol. 310, 966–970.

[30] Tsuchida, H.; Kamata, I.; Nagano, M. (2007). Investigation of defect formation in 4H-SiC epitaxial growth by X-ray topography and defect selective etching. *J. Cryst. Growth*, Vol. 306, 254–261.

[31] Vodakov, Yu. A.; Roenkov, A. D.; Ramm, M. G.; Mokhov, E. N., Makarov Yu. N. (1997). Use of Ta-container for sublimation growth and doping of SiC bulk crystals and epitaxial layers. *Phys. Stat. Sol.* B, Vol. 202, 177–200.

[32] Wang, Y.; Ali, G.; Mikhov, M.; Vaidyanathan, V.; Skromme, B.; Raghothamachar, B.; Dudley, M. (2005). Correlation between morphological defects, electron beam-induced current imaging, and the electrical properties of 4*H*-SiC Schottky diodes. *J. Appl. Phys.*, Vol. 97, 013540.

[33] Wierzchowski, W.; Wieteska, K.; Balcer, T.; Malinowska, A.; Graeff, W.; Hofman, W. (2007). Observation of individual dislocations in 6H and 4H SiC by means of back-reflection methods of X-ray diffraction topography. *Cryst. Res. Technol.*, Vol. 42, 1359–1363.

Chapter 2

Nondestructive and Contactless Characterization Method for Spatial Mapping of the Thickness and Electrical Properties in Homo-Epitaxially Grown SiC Epilayers Using Infrared Reflectance Spectroscopy

Sadafumi Yoshida, Yasuto Hijikata and Hiroyuki Yaguchi

Additional information is available at the end of the chapter

1. Introduction

Silicon carbide (SiC) is one of the most promising semiconducting materials for the fabrication of high power electronic devices with extremely low loss, owing to its excellent physical properties, such as high breakdown electric field, high saturation electron drift velocity, and high thermal conductivity. Nowadays, some kinds of devices, such as SBDs, JFETs and MOSFETs have been on the market. For the fabrication of SiC devices with high yield rates, *i.e.*, for reducing the scattering of device specification, the production of high-quality, large-diameter epi-wafers with uniform thickness and electrical properties is indispensable. In order to characterize the electrical and thickness uniformity of the epi-wafers during the device process, *i.e.*, to know how the thickness, doping concentration and mobility are distributed over the SiC epi-layers, it is necessary to develop the characterization method that can perform the determinations of thickness and electrical properties simultaneously in a nondestructive and noncontact way.

To characterize the distribution of the electrical properties over SiC wafers and homo-epi-wafers, conductivity mapping is often performed [1]. However, the distribution of carrier concentration and mobility cannot provide from the conductivity mapping, because the conductivity depends both on the distribution of dopant concentration and the crystallinity and/or distribution of crystal defects. In order to characterize the distribution of carrier concentration and mobility over epi-layers, electrical measurement techniques such as Hall effect measurements and capacitance-voltage (*C–V*) measurements have been widely used. These techniques, however, are disadvantageous as a device fabrication process monitoring tool because they require the formation of electrodes on a sample. By using a mercury probe as an electrode, *C–V* measurements can be performed without the formation of electrodes on a sample. However, the problems caused by the contamination with mercury contact have been pointed out, recently.

Optical measurement techniques such as Raman scattering spectroscopy [2-5], infrared (IR) spectroscopic ellipsometry [6], optical absorption measurements [7] have been used to estimate the carrier concentration in SiC wafers as a nondestructive and contactless method. IR reflectance measurements have been used to estimate the electrical properties of GaAs [8] and SiC [9]. Macmillan *et al.* [10] reported that the thickness of homo-epitaxially grown SiC wafers can be estimated from the interference oscillations in IR reflectance spectra observed both below and upper frequency ranges of the reststrahlen band (800–1000cm^{-1} for SiC). Nowadays, the reflectance measurements in near IR spectral range (1000–4000cm^{-1}) is widely used to estimate the thickness of homo-epitaxially grown SiC layers in the SiC device process. As the thickness of epilayers used for power devices are in the range from several to several tens μm, the observation of reflectance spectra in near IR spectral range is suitable to analyze the oscillation of reflectance caused by the interference effects of light in the epilayers.

We have developed the method of obtaining the thickness and electrical properties of semiconductor wafers and epi-wafers, simultaneously, by using IR reflectance spectroscopy [11-15]. In this paper, we will summarize the development of the method, and will discuss the validity of the electrical properties derived from the IR reflectance by comparing with those estimated from Hall effect and *C–V* measurements. Finally, we will show the results of applying this method to characterize the electrical activation of impurity and crystalline damages in the ion-implanted, and post-implantation-annealed SiC epilayers.

2. Characterization method of the electrical properties in SiC wafers using IR reflectance spectroscopy

2.1. Method of obtaining carrier concentration and mobility from IR reflectance spectroscopy [11,12]

The values of dielectric constants of semiconductors in IR spectral region can be calculated as a function of wavelength or frequency using the dispersion equation. For the analysis of IR reflectance spectra, a number of dielectric function models have been proposed [16-20]. The classical dielectric function (CDF) model [16], which assumes the damping constant of the LO phonon is the same as that of the TO phonon, has been widely used. In the case of wide bandgap semiconductors with an overdamped plasmon system like SiC, the reflectance spectrum is, however, strongly dependent on LO-phonon damping because the plasmon is overdamped and the LO phonon frequency is much higher than the plasma frequency except for heavily doped cases. For these reasons, we have chosen to use the modified classical dielectric function (MDF) model taking into account the contribution of the TO phonon damping constant and the LO phonon damping constant independently [17]. Considering the contributions from phonons and plasmons, the dielectric constant is given as

$$\varepsilon(\omega)=\varepsilon_{\infty}\left(\frac{\omega_L{}^2-\omega^2-i\Gamma_L\,\omega}{\omega_T{}^2-\omega^2-i\Gamma_T\omega}-\frac{\omega_p{}^2}{\omega^2+i\gamma_p\omega}\right) \tag{1}$$

where ε_{∞} is the high frequency dielectric constant, ω_T and ω_L are the TO- and LO-phonon frequencies, respectively, Γ_T and Γ_l are the TO- and LO-phonon damping constants, respectively, γ_p is the free-carrier damping constant, and ω_p is the plasma frequency of the free carriers, which is given by

$$\omega_p=\sqrt{\frac{Ne^2}{m^*\varepsilon_{\infty}}} \tag{2}$$

where N, e, and m^* are the free carrier concentration, electron charge, and effective mass, respectively. The free-carrier damping constant γ_p is the inverse of the scattering time τ and therefore the free-carrier mobility can be derived using the following relation,

$$\mu=\frac{e}{m^*\gamma_p} \tag{3}$$

Assuming that the wafers are uniformed in the depth direction, we used the normal-incidence reflectance of a semi-infinite medium R, which is expressed as

$$R(\omega)=\frac{(n-1)^2+k^2}{(n+1)^2+k^2} \tag{4}$$

where n and k are the optical constants, derived from $\varepsilon/\varepsilon_0=(n-ik)^2$.

The carrier concentration and mobility can be determined by fitting the experimental infrared reflectance spectrum with calculated ones. To fit the spectra, we used the least-squares method based on eqs. (1) and (4), where we adopted ω_p, γ_p, and Γ_L as adjustable parameters.

2.2. Measurements of IR reflectance spectra of SiC wafers and estimation of electrical properties [12]

Single crystal wafers of commercially produced *n*-type (nitrogen doped) 6H-SiC were used. IR reflectance spectra were measured using two Fourier-transform infrared (FTIR) spectrometers, JASCO FT/IR–VM7 for the far-infrared region (30–600 cm^{-1}) and JASCO FT/IR 670-PLUS for the middle-infrared region (400–2000 cm^{-1}), respectively. For far-infrared reflectance measurements, two light sources (a mercury arc-lamp and nichrome light source), three beam splitters (4, 12, 25 μm thick Mylar films) and a p-DTGS (pyroelectric deuterated triglycine sulfate) detector were used. For middle-infrared reflectance measurements, a high intensity ceramic light source, a KBr beam splitter, and a TGS detector were employed. Each IR reflectance spectrum was measured with 1 cm^{-1} spectral resolution. The light diameters were 5 mm for far-infrared measurements, and 3 mm for middle-infrared measurements. The measurements were performed for (0001) Si-faces of 6H-SiC wafers at nearly normal incidence. An Al mirror was used as a reflectance reference.

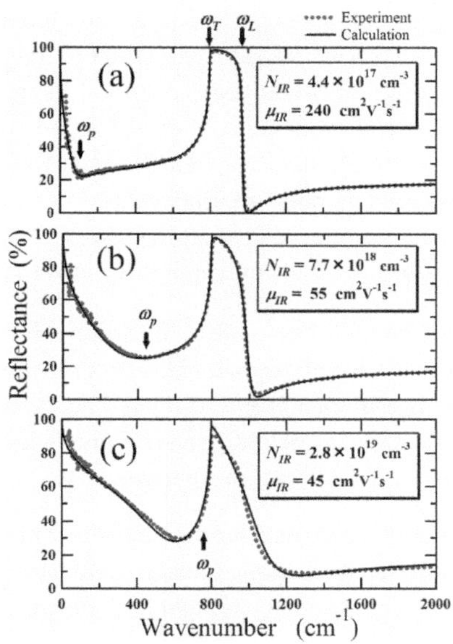

Figure 1. Infrared reflectance spectra measured for 6H-SiC wafers with various carrier concentrations at room temperature (dotted line). The solid lines show the fitted spectra calculated using MDF model. The values of carrier concentration and mobility obtained from fitting to the measured IR spectra are described in the figure [12].

The dotted lines in Figure 1 show the typical infrared reflectance spectra of several 6H-SiC wafers of different carrier concentrations at room temperature. The plasma edges and reststrahlen bands appear in the far-IR and middle-IR regions, respectively. We derived the values of carrier concentration and mobility by the curve fitting of calculated curves to the observed ones. For the curve fitting, we chose ω_p, Γ_L, γ_p as adjustable parameters. For the values of the other parameters, we employed those obtained from Raman scattering measurements, ε_∞=6.52ε_0, ω_T=797cm^{-1}, ω_L=969.4cm^{-1}, Γ_T=2cm^{-1}, and m^*=0.35m_0 for 6H-SiC[2]. Since the light is normally incident on the (0001) face of the samples, these parameters are all for the modes vibrating perpendicular to the c-axis.

From the curve fitting analysis, we obtained a good fit for each experimental spectrum, which was obtained by measuring nine samples with carrier concentrations in the range of 4×10^{17}~$3\times10^{19}cm^{-3}$. The solid lines in Figure 1 show examples of the fitted curves obtained by fitting to the typical IR reflectance spectra shown as the dotted line in each figure. The free-carrier concentration and drift mobility were derived from the best-fit parameters of ω_p and γ_p using eqs.(2) and (3), mentioned above. The values of free-carrier concentration and mobility obtained are also given in each figure.

As shown in Figure 1 (c), there is a slight discrepancy at approximately 900 cm^{-1} between the spectrum observed and that calculated using the MDF model (eq. (1)). This discrepancy increases with increasing carrier concentration in the high 10^{19} cm^{-3} range. For heavily doped SiC crystals, the CDF and MDF models would be inappropriate because the MDF model is derived considering the effects of phonons and plasmons independently. In the case of heavily doped SiC crystals, the plasma frequency is closed to the phonon frequency and the LO phonon and plasmon are strongly coupled. Therefore, though the MDF mode can approximately estimate the electrical properties of heavily doped SiC wafer, it is necessary to use another dielectric function mode that takes into account the effect of LO phonon-plasmon coupled modes [19,20] to obtain more accurate values.

2.3. Comparison with the values derived from Hall effect measurements [12]

For the confirmation of the validity of the values of carrier concentration and mobility derived from IR reflectance spectra, we performed Hall effect measurements for the same samples used for IR reflectance measurements and compared between the values obtained from the optical and electrical methods. The 6H-SiC wafers with a wide variety of carrier concentrations ranging from 3.4×10^{17} to 2.4×10^{19} cm^{-3} were used.

We cut the SiC wafers to a size of 5×5 mm^2 for the Hall effect measurements using van der Pauw method. After chemical cleaning, ohmic contacts were fabricated at the corners of each sample by the evaporation of nickel and subsequent heat treatment at 1000°C for 10 min. IR reflectance measurement and Hall measurement were carried out at room temperature.

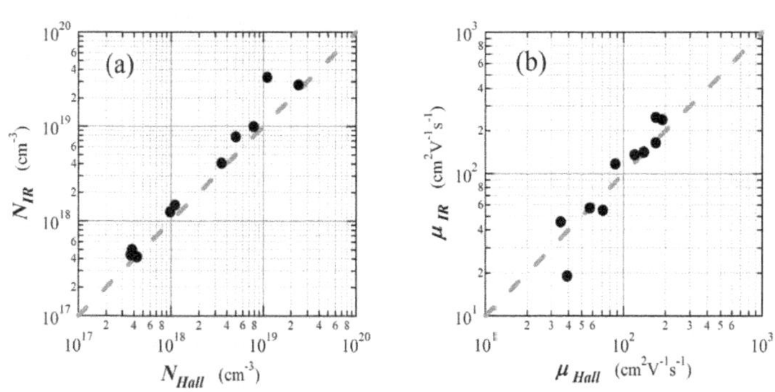

Figure 2. Comparison of (a) carrier concentration and (b) mobility values, N_{IR}, μ_{IR} obtained from IR reflectance spectroscopy measurements with those from Hall effect measurements, N_{Hall}, μ_{Hall}. The broken lines represent the case of complete agreement with each other [12].

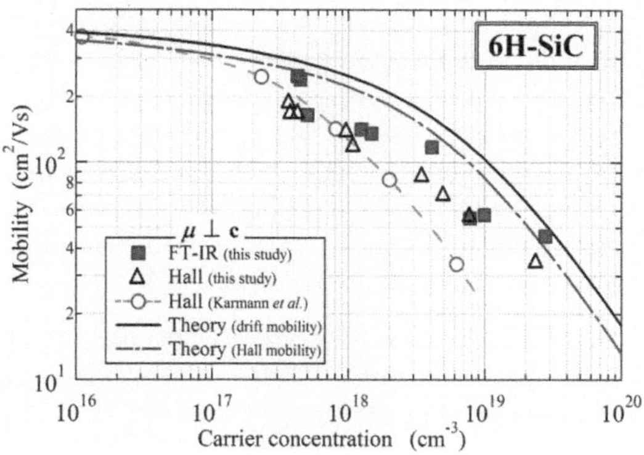

Figure 3. The variations in drift mobility evaluated from IR reflectance spectroscopy, and Hall mobility measurements plotted against carrier concentration. The reported values of Hall mobility [23] and those calculated theoretically (N_A/N_D=0) following reference [24] are also shown for comparison [12].

In Figure 2 (a) and (b), the carrier concentrations and mobilities estimated from the IR reflectance spectra are plotted against those obtained from the Hall effect measurements. As the reported Hall scattering factor r_H is approximately unity at room temperature for 6H-SiC [21,22], we assumed r_H is equal to unity for the calculation. Good agreement was obtained between the electrical characteristics obtained from IR reflectance measurements and those from Hall effect measurements.

The LO phonon damping constant Γ_L, which is one of the adjustable parameters, varies linearly with carrier concentration. This tendency is in good agreement with the results obtained by Raman scattering spectroscopy [17] in which, as was explained, the interactions between ionized impurity and LO phonon, and free carrier and LO phonon increase with increasing doping concentration.

In Figure 3, the drift mobility and Hall mobility of the 6H-SiC wafers are plotted against the determined free carrier concentration, and against those reported by Karmann *et al* [23]. The Hall mobilities obtained in the present study are a little higher than those determined by Karmann *et al.* in a high-carrier-concentration region. We calculated the drift and Hall mobility at room temperature as functions of dopant concentration following references [24-26], assuming that the compensation ratio N_A/N_D=0, and considered five carrier scattering mechanisms (acoustic phonon deformation potential scattering, polar optical phonon scattering, intervalley phonon deformation potential scattering, neutral impurity scattering, and ionized impurity scattering) as in reference [29]. The values of the mobility obtained in this work are lower than those obtained from the theoretical calculations. This result suggests that the compensation ratio is not 0 but approximately 0.2 in this study and a little higher in the case of Karmann *et al.*

Through comparison, we have ascertained that the electrical characteristics of SiC wafers can be estimated by IR reflectance spectroscopy with high credibility.

2.4. Spatial mapping of the electrical properties over SiC wafers [11,12]

To demonstrate the capability of the method proposed, we performed the spatial mapping of the distribution of the carrier concentration and mobility of a commercially produced 2 inch 6H-SiC wafer. For the spatial mapping, we employed a micro FTIR (JASCO Irtron IRT-30 infrared microscope), which was equipped with a mercury cadmium telluride (MCT) detector. The diameter of the beam was 0.1 mm and the interval between measured points was 5 mm (a total of 120 measurement points). We performed the measurements in the spectral range of 560–2000 cm^{-1} with a spectral resolution of $4cm^{-1}$.

Figure 4 shows an example of the spatial distribution of the free-carrier concentration and mobility of a commercially produced 2-inch 6H-SiC wafer obtained using this technique. This measurement technique needs no prior surface treatment, because the native oxide layer thickness and surface roughness are not more than 3 nm and their influence on the reflectance spectra is negligible in IR region. The uniformity of free-carrier concentration and mobility throughout this wafer except for 5 mm from the edge were estimated to be approximately ±9% and ±15%, respectively. The free-carrier concentration mapping shows that the free-carrier concentration in the central region is greater than that in the edge region. On the other hand, the mobility mapping shows the negative correlation of the mobility distribution with that of carrier concentration. When conductivity mapping is used as the method for the mapping of electrical properties of the wafer, it leads to the misleading conclusion that the electrical uniformity over the wafers is approximately ±5% and the wafer is almost uniform, because the conductivity is determined as the product of carrier concentration and mobility. Therefore, the proposed IR reflectance spectroscopic method is more appropriate for the characterization of the distribution of the electrical properties of SiC wafers.

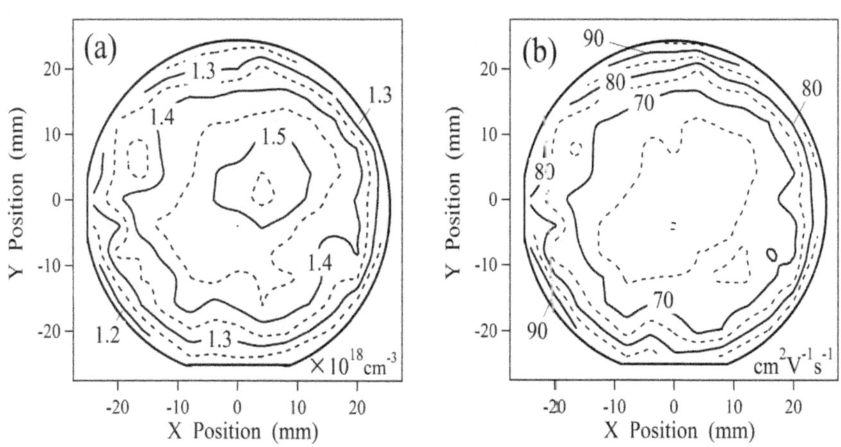

Figure 4. Spatial mapping of (a) carrier concentration and (b) mobility in a commercially produced 2 inch 6H-SiC wafer [12].

3. Characterization method of the electrical properties and thickness of epilayers using IR reflectance spectroscopy

3.1. Method of obtaining the carrier concentration, mobility, and thickness of epilayers, simultaneously [13]

In this section, we propose the method for the simultaneous determination of the electrical properties, *i.e.*, free carrier concentration and mobility, and the thickness of epilayers as well as bulk layer by IR reflectance measurements. First, we will explain the procedure of obtaining carrier concentration, mobility and thickness of the epilayers on SiC wafers. Then, we will compare the electrical properties derived from the IR reflectance analyses with those from Hall effect measurements for *n*-type epilayers grown on *p*-type substrates, and with those from *C–V* measurements in the case of *n*-type epilayers on *n*-type substrates. Finally we will discuss the validity of the obtained values from the proposed method.

The carrier concentration and mobility of epilayers and substrates, as well as the thickness of the epilayers can be determined simultaneously by fitting the calculated reflectance spectra to measured ones. The reflectance R from an air/ epilayer/substrate structure at normal incidence is given by

$$R = \left| \frac{r_1 + r_2 e^{-2i\delta}}{1 + r_1 r_2 e^{-2i\delta}} \right|^2, \quad where \quad \delta = \frac{2\pi n d}{\lambda} \tag{5}$$

where r_1 and r_2 are the Fresnel reflection coefficients at the air/epilayer and the epilayer/substrate interface, respectively, and δ is the phase shift of light in the epilayer, n and d are the refractive index and the thickness of epilayer, respectively, and λ is wavelength. The optical constants of SiC in IR spectral range are derived from the dielectric constants as a function of the frequency of the incident light, given by eq.(1) both for substrate and epilayer. As in the case of SiC bare wafers written in Section 2, we fitted the calculated spectrum to the measured one by adjusting the values of ω_p, γ_p, Γ_T and Γ_L of the epilayer and those of the substrate, and the epilayer thickness d. From these values, we can obtain the carrier concentration N and mobility μ of the epilayer and substrate using eqs. (3) and (4). In the calculation, we adopted the values ε_∞=6.56, ω_T=798cm^{-1}, ω_L=970cm^{-1}, and m^*_{MG}=0.58m_0, m^*_{MK} =0.31m_0, obtained from the Raman scattering measurements of 4H-SiC [2] and optical detection of cyclotron resonance (ODCR) [27]. Considering that the free carriers distribute themselves in proportion to the square root of each effective mass, the averaged effective mass, m^*= $(m^*_{MK} m^*_{MG})^{1/2}$, and $m^*=(1/m^{*1/2}_{MK}+1/m^{*1/2}_{MG})/(1/m^{*3/2}_{MK}+1/m^{*3/2}_{MG})$ were used for the calculation of the carrier concentration and mobility, respectively.

3.2. Measurements of IR reflectance spectra and derivation of electrical properties and thickness of SiC epi-wafers [13]

Samples used in this study were nitrogen doped *n*-type 4H-SiC epilayers grown on *n*- and *p*-type 4H-SiC substrates supplied from National Institute of Advanced Industrial Science and Technology (AIST). The epilayers were grown on 4H-SiC (0001) Si face 8 off substrates by chemical vapor deposition (CVD). The details of the epilayer growth have been described elsewhere [28]. In the case of the *n*-type epilayers on *p*-type substrates, the carrier concentration of the epilayers was in the range between 3×10^{17} and 2×10^{18} cm^{-3}, and that of the substrates was typically 4×10^{16} cm^{-3}. On the other hand, in the case of the *n*-type epilayers on *n*-type substrates, the net doping concentration ($N_D - N_A$) of the epilayers was in the range between 1×10^{17} and 8×10^{17} cm^{-3}, and that of the substrates was typically 5×10^{18} cm^{-3}. The thickness of the epilayers were 6~7 μm, measured by scanning electron microscope (SEM) observation of the cleaved facet of the samples. The IR reflectance spectra in the frequency range of 80–2000 cm^{-1} were measured at room temperature using same spectrometers mentioned in Section 2.2. Hall effect measurements were performed at room temperature using van der Pauw method for *n*-type epilayers on *p*-type substrates. Ohmic contacts were fabricated on the epilayer surfaces by the electron beam evaporation of Ni and subsequent annealing at 900°C for 30min in N_2 atmosphere. *C–V* measurements were performed at room temperature using a mercury probe as a Schottky contact.

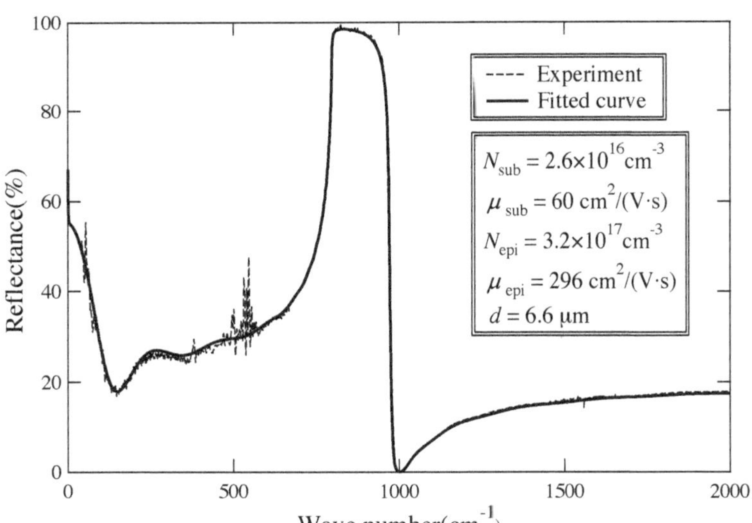

Figure 5. IR reflectance spectrum of an *n*-type epilayer on a *p*-type substrate at room temperature denoted by dashed line. The solid line shows the fitted curve calculated using the MDF model. The values estimated from this fitting are listed in the figure [13].

At first, we estimated the carrier concentration and mobility of *n*-type epilayers on *p*-type substrates from the reflectance measurements, and compared with these values obtained from Hall effect measurements. Figure 5 shows typical IR reflectance spectra measured for the *n*-type epilayers on *p*-type substrates. The solid line denotes the calculated values fitted to the experimental one shown as the dashed line. The values $N_{epi} = 3.2\times10^{17}$ cm^{-3}, $\mu_{epi} = 296$ cm^2/(V s), and $d = 6.60$ μm were obtained by a curve fitting analysis. These values are listed in the figure. In the case of *p*-type substrate whose carrier concentration is low in general, the reflectance spectrum is almost independent of the electrical properties of the substrate, thus, it is difficult to estimate the carrier concentration and mobility of the substrate. Therefore, we used the values for the *p*-type substrate without epilayer obtained from the Hall effect measurements. The thickness of the epilayers obtained from the IR reflectance measurements and those measured from the SEM observation coincide with each other within ±1%.

The carrier concentrations and mobilities obtained from the IR reflectance measurements are plotted with respect to those obtained from Hall effect measurements in Figures 6 (a) and (b), respectively. Since the Hall scattering factor r_H is reported to be approximately unity at room temperature for 4H-SiC [27,29] as in the case of 6H-SiC, we directly compared the drift mobilities estimated from the IR reflectance measurements and those of the Hall mobilities obtained from the Hall effect measurements. The error bars shown in the figures represent the accuracy of the fitting analysis, and the accuracy is about ±4% for both the carrier concentration and the mobility, whereas the accuracy of the values derived from Hall effect measurements is about ±10%. As can be seen from these figures, the electrical properties obtained from the reflectance spectra are in good agreement with those obtained from the Hall effect measurements. These results suggest that the proposed method is valid for obtaining the values of carrier concentration and mobility of the epilayers. However, careful observation confirms that the values of carrier concentration and mobility derived from the infrared reflectance measurements are slightly lower than those obtained from the Hall effect measurements. The difference can be explained by the consideration that the part of free carriers trapped in defects or bounded by dopants cannot follow in the THz frequency range used for the reflectance measurements, unlike in Hall effect measurements, where a direct current is supplied.

Next, we estimated the values of carrier concentration and mobility for *n*-type epilayers on *n*-type substrates from the IR reflectance spectra measured. Figure 7 shows a typical IR reflectance spectrum observed and its fitted curve. The values $N_{epi} = 1.3\times10^{17}$ cm^{-3}, $\mu_{epi} = 403$ cm^2/(V s) and $d = 4.45$ μm; and $N_{sub} = 7.1\times10^{18}$ cm^{-3}, $\mu_{sub} = 53$ cm^2/(V s) were obtained by curve fitting analysis as the parameters of the epilayer and substrate, respectively. The accuracy of the carrier concentration and mobility of epilayerss derived from the fitting of the IR reflectance spectrum is about ±10%. In Figure 8, the free carrier concentration estimated from the IR reflectance spectra is plotted with respect to the net doping concentrations $N_D - N_A$ derived from *C–V* measurements. We calculated the free carrier concentrations *n* from the net doping concentrations using

$$n(T) + N_A = \frac{N(h)}{1 + \{gn(T)/N_C\}\exp[\Delta E(h)/k_B T]} + \frac{N(k)}{1 + \{gn(T)/N_C\}\exp[\Delta E(k)/k_B T]} \tag{6}$$

$$N_C = 2M_C\left(\frac{m^*_{d.s.}\, k_B T}{2\pi\hbar^2}\right)^{3/2} \tag{7}$$

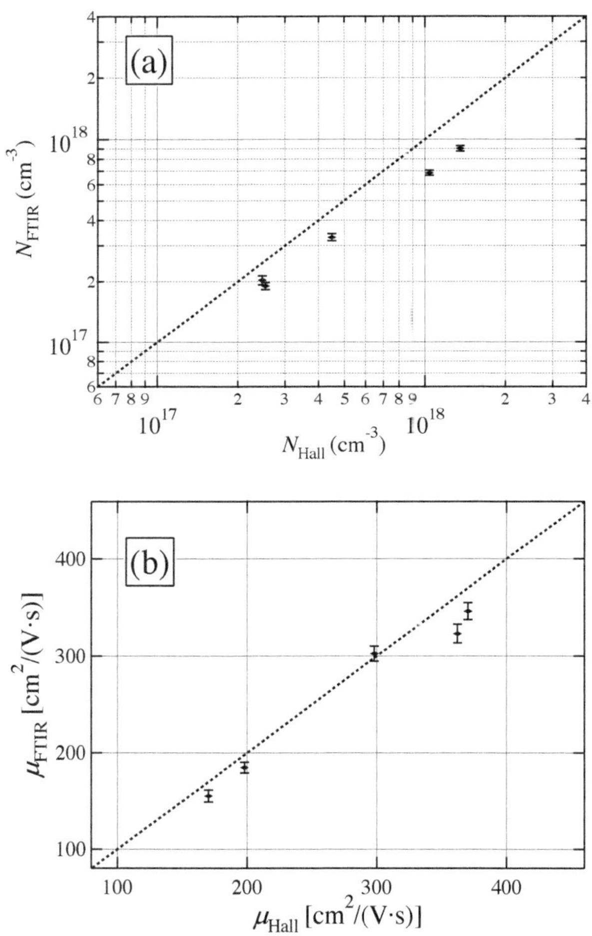

Figure 6. Values of (a) carrier concentration and (b) mobility estimated from IR reflectance measurements and Hall effect measurements. The dotted line in the figures corresponds to the case of complete agreement with each other [13].

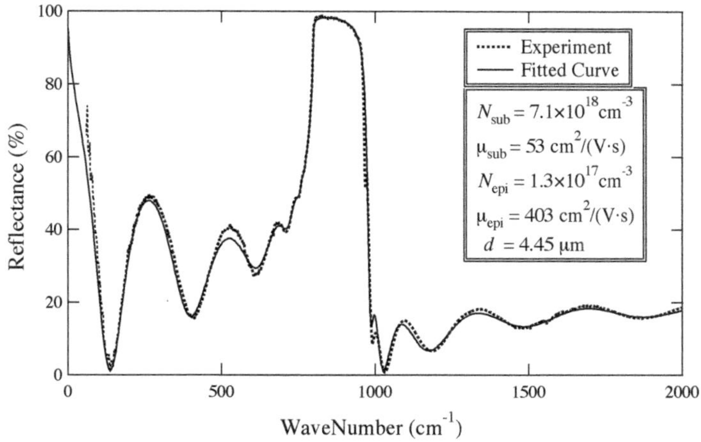

Figure 7. IR reflectance spectrum of an *n*-type epilayer on an *n*-type substrate denoted by dashed line. The solid line shows the fitted curve calculated using the MDF model. The values estimated from this fitting analysis are listed in the figure [13].

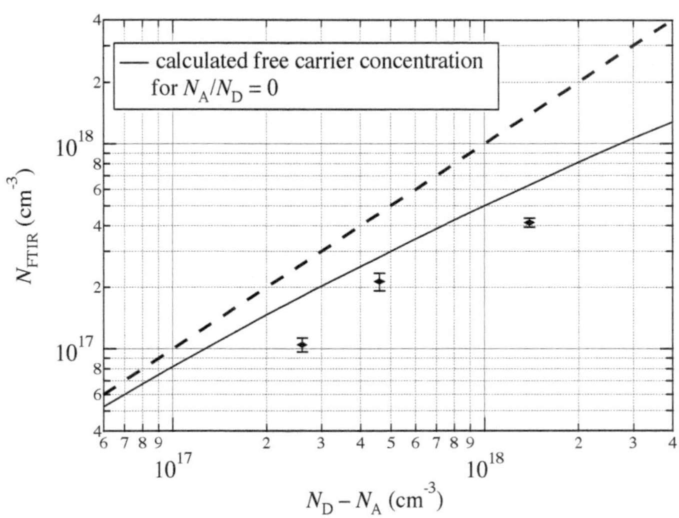

Figure 8. Values of the carrier concentration estimated from the IR reflectance for *n*-type epilayers on *n*-type substrates as a function of dopant concentration obtained from *C–V* measurements for each sample. The solid line represents the theoretical carrier concentration for T=300K assuming zero doping concentration (N_A=0) using eq. (6). The dashed line represents $N_{FTIR}=N_D-N_A$ [13].

where k_B is the Boltzmann constant, T is the temperature, N_A is the concentration of acceptors, and $N(h)$ and $N(k)$ are the concentrations of the nitrogens occupied at hexagonal and cubic lattice sites, respectively. Since the number of hexagonal sites is equal to those of cubic sites for 4H-SiC, the donor concentration N_D is given by $N(h) + N(k)$. The values of $\Delta E(h)$ and$\Delta E(k)$ are the ionization energies of the nitrogen donor at hexagonal and cubic lattice sites, respectively, and $g = 2$ is the spin degeneracy factor. Equation (7) gives the effective density of states, where $M_c = 3$ is the number of equivalent conduction band minima, and $m^*_{\mathrm{d.s.}}$ is the density-of-states effective mass. The values of $\Delta E(h)$ and $\Delta E(k)$ were set as 50 meV and 100 meV, respectively, referring to the reported experimental data [30,31]. The values of $m^*_{\mathrm{M\Gamma}} = 0.58m_0$, $m^*_{\mathrm{MK}} = 0.31m_0$, $m^*_{\mathrm{ML}} = 0.33m_0$ derived from ODCR measurements [27] were adopted. The solid line in Figure 8 shows the free carrier concentrations calculated as a function of the net doping concentration, where we assumed that $N_A/ N_D = 0$ or $N_D / (N_A + N_D) = 1$, because the epilayers we measured are hardly compensated [32]. The values obtained from the IR reflectance spectra are slightly lower than the calculated values, as in the case for the samples of n-type epilaers on p-type substrates. The value of drift mobility calculated in consideration of five carrier scattering mechanisms [24] is 340 $cm^2/(Vs)$, in the case of $N_A/ N_D = 0$ and $N_D - N_A = 2.7\times10^{17}cm^{-3}$, which is almost the same as the value obtained from the IR reflectance spectrum, *i.e.*, 403 $cm^2/(Vs)$. This result also indicates that the compensation of the sample is low.

3.3. Extension of the carrier concentration range down to $10^{16}cm^{-3}$ order using Terahertz frequency range [14]

We have shown that the carrier concentration and mobility of substrate and epilayers as well as the thickness of epilayer are obtained simultaneously from IR reflectance spectra in the frequency range of 80–2000cm^{-1}, and confirmed that the values of the carrier concentration, mobility and epilayer thickness estimated from IR reflectance spectroscopy are valid. However, it was difficult to estimate the electrical properties of homo-epilayers with carrier concentrations less than 1×10^{17} cm^{-3} without IR reflectance spectra less than 80 cm^{-1}. Figure 9 is the variation of plasma frequency with carrier concentration calculated from eq.(2) for 4H-SiC. The figure indicates the plasma frequencies are smaller than 100cm^{-1} for the carrier concentration less than 10^{17} cm^{-3}. Figure 10 shows the variations of the reflectance spectrum of epilayers with the decrease of carrier concentrations from 3×10^{18} to $3\times10^{16}cm^{-3}$. The magnified features of the calculated reflectance spectra for $1\times10^{17}cm^{-3}$, 5×10^{16}, and $1–5\times10^{15}cm^{-3}$ in Terahertz frequency range are shown in Figure 11. These figures suggest that it is necessary to measure a spectrum down to around 20cm^{-1} for extending the carrier concentration down to the order of 10^{16} cm^{-1}.

From these considerations, we extended the spectral range of the reflectance measurements down to 20 cm^{-1} (0.6 THz) by using terahertz reflectance spectroscopy to be able to apply the method for epilayers with the carrier concentrations in the range of 10^{16} cm^{-3}. Also we have compared the free carrier concentrations estimated from reflectance measurements with the net doping concentrations obtained from C–V measurements to discuss the validity of this characterization method.

Samples used in this study were nitrogen doped *n*-type 4H-SiC epilayers grown on *n*-type 4H-SiC substrates by chemical vapor deposition (CVD) [28]. The net doping concentration ($N_D - N_A$) of the epilayers was in the range between 5×10^{16} and 1×10^{18} cm^{-3}, and that of the substrates was typically 5×10^{18} cm^{-3}. The thickness of the epilayers was 6–7 μm, measured by SEM observation. *C–V* measurements were performed using gold electrodes evaporated on the samples as Schottky contacts. The reflectance spectra were measured at room temperature for the spectral region of 20–100 cm^{-1}, 80–600 cm^{-1} and 540–2000 cm^{-1} using terahertz time-domain spectroscopy (THz-TDS) (*Aispec*: pulse IRS 1000/2000), FTIR spectrometers (*JASCO*: IR-VM7) and (*JASCO*: Irtron IRT-30), respectively.

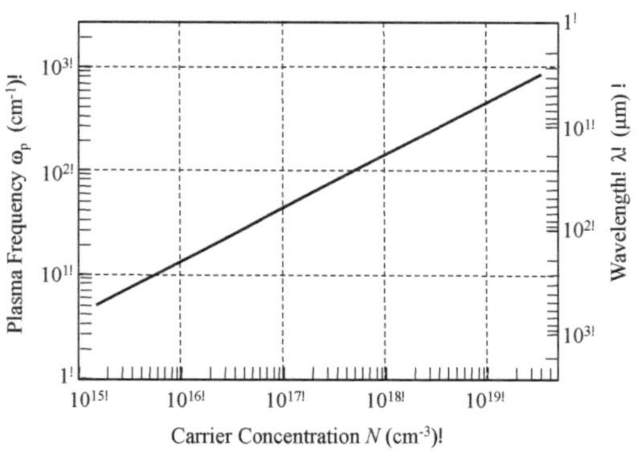

Figure 9. Variation of plasma frequency ω_p with carrier concentration *N*.

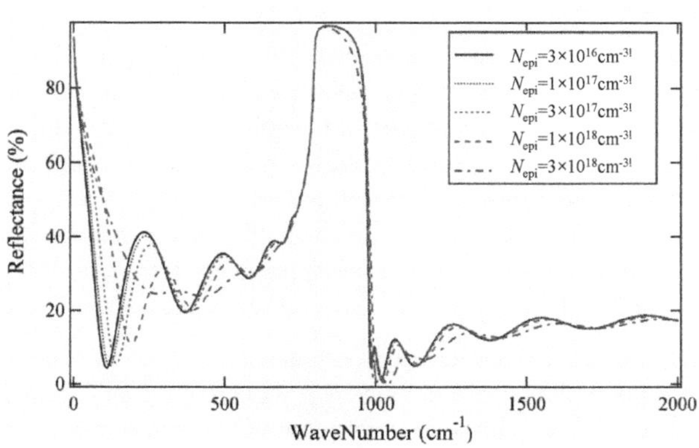

Figure 10. Calculated reflectance spectra of the SiC epilayers (3μm thick) with various carrier concentrations on a SiC wafer (N_{sub}=5.5×10^{18}, μ_{sub}=50cm^2/Vs).

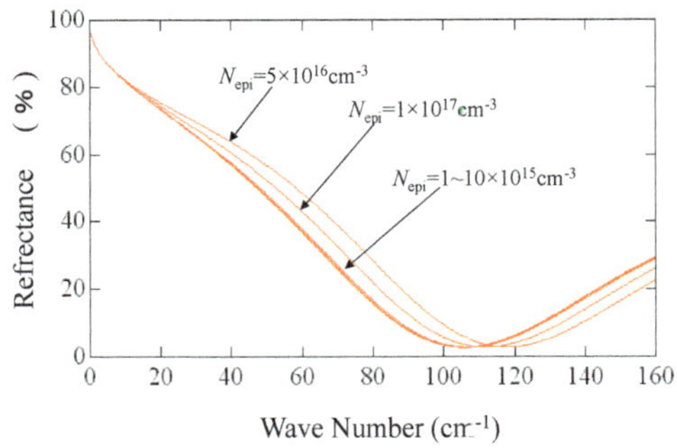

Figure 11. Calculated reflectance spectra of the SiC epilayers (5μm thick) with low carrier concentrations on a SiC wafer (N_{sub}=4×10^{18}cm^{-3}, μ_{sub}=50cm^2/Vs).

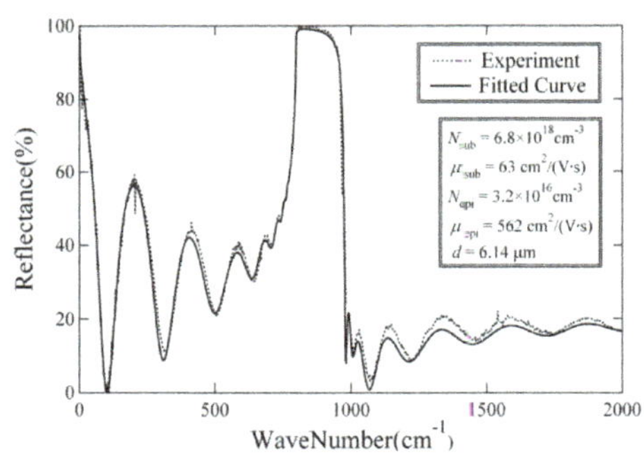

Figure 12. Measured and calculated reflectance spectra of an *n*-type ep layer on an *n*-type substrate at room temperature. The values estimated from this fitting analysis are listed in the figure [14].

We have estimated the values of carrier concentration and mobility for the samples of *n*-type epilayers on *n*-type substrates from the IR reflectance spectra measured. Figure 12 shows the measured and calculated reflectance spectra of the epilayer with a net doping concentration around 5×10^{16} cm^{-3} at room temperature. As shown in the figure, reflectance spectrum measured by THz reflectance spectroscopy are well connected with that measured by IR reflectance spectroscopy at around 100 cm^{-1}, and we obtained a good fit between the measured and the calculated spectrum. From the values of fitting parameters, the values N_{epi} = 3.2×10^{16} cm^{-3}, μ_{epi} = 562 cm^2/(V s) and d = 6.14 μm, and N_{sub} = 6.8×10^{18} cm^{-3}, μ_{sub} = 63 cm^2/(V s) were obtained.

In Figure 13, the free carrier concentrations estimated from the IR reflectance spectra are plotted against the net doping concentrations $N_D - N_A$ derived from *C–V* measurements. We calculated the free carrier concentrations *n* from the net doping concentrations using eq.(6). The solid line in the figure shows the free carrier concentrations calculated as a function of net doping concentration. We assumed that $N_A / N_D = 0$ or $N_D / (N_A + N_D) = 1$, because the epilayers we measured are hardly carrier-compensated [32]. The values obtained from the reflectance spectra are in fairly good agreement with solid line, suggesting that the values of the carrier concentrations estimated from IR reflectance spectra have a sufficient validity. However, a careful look confirms that the values of carrier concentration derived from the reflectance measurements are slightly lower than those estimated from the electrical measurements as in the case of carrier concentrations higher than $10^{17}cm^{-3}$. The same tendency was observed in the comparisons with the Hall effect measurements for the samples of *n*-type epilayers on *p*-type substrates as shown in Figure 8. This tendency is considered to be partly because of the adoption of inappropriate effective mass values for the calculation of reflectance spectra. It is also considered as a cause that the part of free carriers trapped in the defects or bounded by dopants cannot follow in the THz frequency range used for the reflectance measurements, as mentioned above.

4. Characterization of electrical properties and residual crystalline damage in ion-implanted and post-implantation-annealed 4H-SiC epilayers using IR reflectance spectroscopy

4.1. Method of obtaining the electrical properties and crystalline damage in ion-implanted SiC epilayers [15]

Ion implantation is an indispensable process for selective area doping into crystalline silicon carbide (SiC), because the doping of impurities by thermal diffusion is hard to apply for SiC device process due to very small diffusion constant of impurities in SiC. After the ion implantation, annealing at high temperatures is necessary for activating the dopants electrically as well as recovering the crystallinity of SiC damaged by ion implantation. Hall effect measurements, secondary ion mass spectroscopy (SIMS) and transmission electron microscopy (TEM) have been widely used to characterize the electrical properties, depth profile of the impurities and crystalline damage of implanted layers, respectively. These techniques are, however, inappropriate to use as device process monitoring tools because Hall effect measurement requires the formation of electric contacts, and SIMS and TEM observations result in the destruction of the samples. Recently, the short period and high temperature annealing is used in SiC device process [35]. To make clear the effect of short period high-temperature annealing, we investigated the annealing period dependence at the annealing temperature of 1700°C.

Recently, it has been reported that the crystalline damage induced by ion implantation affects the infrared (IR) reflectance spectra around the reststrahlen region (~800–1000 cm^{-1}) [33,34], and the difference of carrier concentration between epitaxial layer and substrate induces the interference oscillation in the near IR region (1000–4500 cm^{-1}). In this study, we performed the IR reflectance measurements in the spectral range between 600 and 8000 cm^{-1} for high-dose phosphorus ion implanted and post-implantation-annealed 4H-SiC wafers to characterize both the electrical properties and crystalline damage of the implanted layers without destruction and contactless.

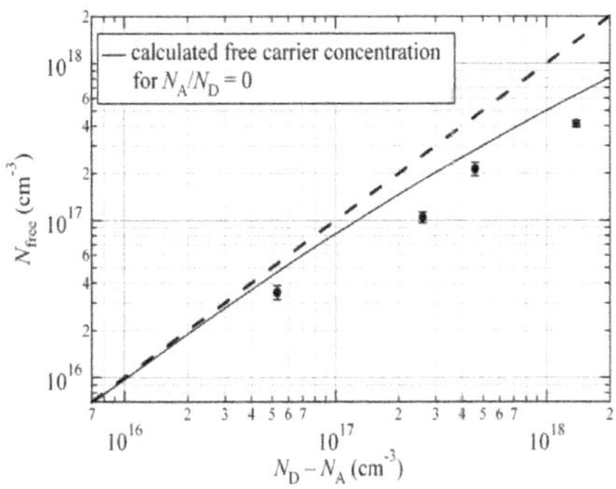

Figure 13. The carrier concentration estimated from the reflectance spectra for an *n*-type epilayer on an *n*-type substrate as a function of doping concentration obtained from *C–V* measurements for each sample. The solid line represents the theoretical carrier concentration for T=300K assuming zero doping concentration (N_A=0) using eq. (6). The dotted line represents $N_{FTIR}=N_D-N_A$ as a guide to the eye [14].

4.2. High-dose phosphorus ion implantation, post-implantation annealing and IR reflectance measurements [15]

The samples used in this study were 4H-SiC (0001) substrates with *p*-type ~5 μm thick commercially produced epitaxial layers. The multi-energy implantations of phosphorus ions at 500°C were carried out through the 10 nm thick oxide film in six steps (40–250 keV) in order to form a box-shaped profile with a thickness of 0.3 μm. The total implanted dose was 7×10^{15} cm^{-2}. After removing the oxide film by HF, the post implantation annealing was conducted in Ar atmosphere. To investigate the annealing temperature dependence of crystalline recovery and electrical properties in the implanted layers, the samples were annealed for 30 min at different temperatures of 1200°C, 1300°C, and 1400°C. In addition, to apply the IR reflectance analysis to the short-period high-temperature annealing process, we also carried out the post implantation annealing at 1700°C for various periods between 0.5 and 10 min. IR reflectance measurements were carried out at room temperature on nearly normal incidence using a micro FT-IR spectrometer (light beam diameter was 0.1 mm). The spectral resolution and range were 4 cm^{-1} and 600–8000 cm^{-1}, respectively.

4.3. Analysis of carrier concentration, mobility and crystalline damage from IR reflectance spectra [15]

Figure 14 shows the annealing temperature dependence of IR reflectance spectrum. For as-implanted samples, the reflectivity maximum and the shape in the reststrahlen band decreases and becomes blunt, respectively, as compared to those of unimplanted samples. After the high temperature annealing, the reflectivity maximum in the reststrahlen band recovers to that of unimplanted samples. This is resulted from the crystalline recovery in implanted layer. In the spectral range above ~2000 cm^{-1}, the evident interference oscillation is observed. It indicates that the implanted dopants are activated and the refractive index of an implanted layer is changed by the change of carrier concentration. We can see the tendencies that the reflectance around 1000 cm^{-1} becomes larger with increasing the annealing temperature. We analyzed the observed spectra to evaluate the damage of the ion implantation layers assuming that the implanted layers are composed of two phases, recrystallized SiC phase and defective SiC phase. We have derived the effective dielectric constants ε_{eff} of implanted layers using an effective medium approximation (EMA) [33],

$$0=(1-f)\frac{\varepsilon_c-\varepsilon_{eff}}{\varepsilon_c+2\varepsilon_{eff}}+f\frac{\varepsilon_d-\varepsilon_{eff}}{\varepsilon_d+2\varepsilon_{eff}} \tag{8}$$

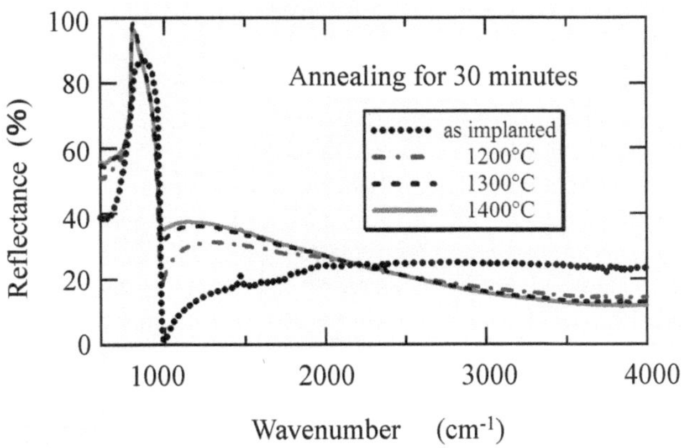

Figure 14. The IR reflectance spectra obtained from the 4H-SiC wafers high-dose implanted and post implantation annealed for 30 minutes [15].

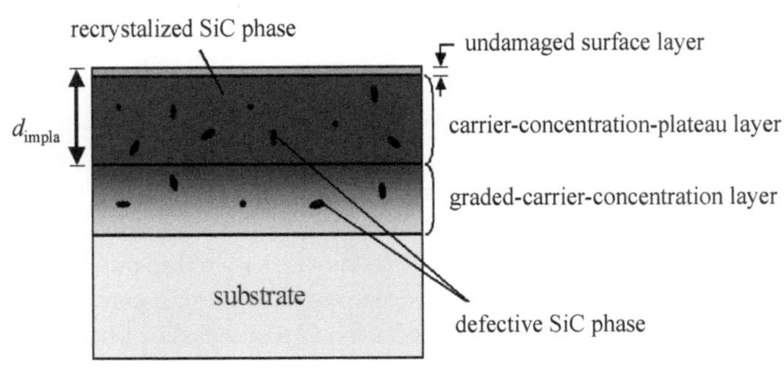

Figure 15. The structural model of the ion implanted SiC wafers used in the calculation of reflectance spectra [15].

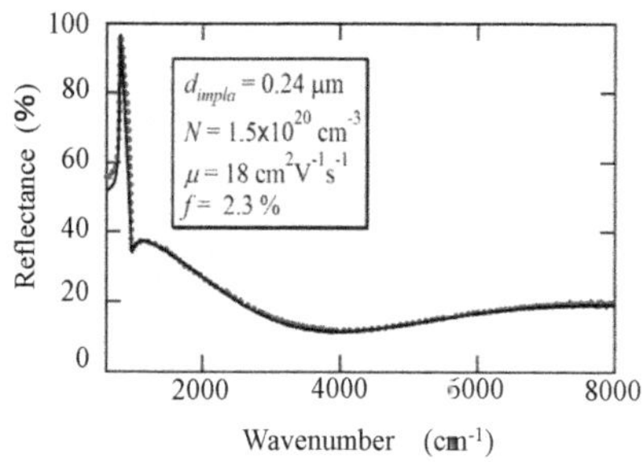

Figure 16. Observed (dotted line) and fitted IR reflectance spectra (solid line) for the samples annealed at 1400℃ for 30 min. The best fit parameters are described in the figure [15].

where ε_c and ε_d are the dielectric constants of re-crystallized and damaged phases, respectively, and f is the volume fraction of damaged phase We assumed that the frequency dependence of both the dielectric constants of re-crystallized phase and defective phase follows the MDF model given by eq.(1). Referring to the result of TEM observations, we employed the structural model that the ion implanted layer is composed of 3 layers: an undamaged surface layer, a carrier-concentration- plateau layer, and a graded-carrier-concentration layer as shown in Figure 15. Furthermore, we assumed that the volume fraction of defective phase in a graded-carrier-concentration layer is same as that in a carrier-concentration-plateau layer. For a graded-carrier- concentration layer, we used the multi-layer structure approximation assuming that the free carrier concentration decreases exponentially with depth and the mobility changes in inverse proportion to carrier concentration.

4.4. Annealing temperature dependences of electrical activity and re-crystallization [15]

As an example of curve fitting analysis, the spectrum of the sample annealed at 1400°C for 30 min and the fitted curve are show in Figure 16. We obtained a good fit in the whole spectral region measured. The best-fit parameters derived are also described in the figure. Figure 17 (a) shows the annealing temperature dependence of the volume fraction of the defective phase. By post implantation annealing, the volume fraction of defective SiC drastically decreases from 92 % (as implanted) to 2.9 % (1200°C annealed), and decreases a little with increasing of annealing temperature up to 1400°C. Figure 17 (b) shows the annealing temperature dependence of the carrier concentration (open circle) and the mobility (open triangle) in the re-crystallized phase. For comparison, the electrical properties derived from Hall effect measurements [35] are also plotted in the figure (filled symbols). We can see a good agreement in the electrical characteristics between IR reflectance spectroscopy and Hall effect measurements. The free carrier concentrations are almost constant in the temperature range studied, as in the case of the volume fraction of defective phase. In contrast, the carrier mobility becomes large with increasing the annealing temperature. These results show that the post implantation annealing at a temperature as low as 1200°C reduces the volume fraction of defective SiC drastically and put the impurities in substitutional lattice sites, but the crystalline recovery of re-crystallized phase is insufficient. In other words, the annealing temperature higher than 1400°C is necessary for improving the mobilities, as well as for activating the impurities.

Figure 18 shows the IR reflectance spectra for the samples annealed for various annealing periods. The spectrum for the sample annealed for 0.5 min is almost the same as that for the sample annealed at 1400°C for 30 min. There is little change with annealing period up to 10 min in the reflectance spectra except for the oscillation periods. Since the oscillation periods are concerned with the thickness of the implanted layer, these changes suggest that the thickness of the implanted layers is changed by evaporation or precipitation in the implanted SiC layer. From the analysis, the thickness of the implanted layer d_{impla} decreases from 0.25 μm (0.5 min annealed) to 0.19 μm (10 min annealed), and the thickness of a graded-carrier-concentration layer increases from 0.05 μm to 0.08 μm. The volume fraction of defective SiC phase decreases drastically down to 2.9 % by 0.5 min annealing and is almost constant up to 10min. The derived annealing period dependence of free carrier concentration and mobility also shows that the recovering of the crystallinity and the electrical activation are sufficient by the annealing even for 0.5 min. These results indicate that the high temperature annealing as high as 1700°C puts the impurities onto substitutional lattice sites and recovers the crystallinity of the implanted layers within 1 min.

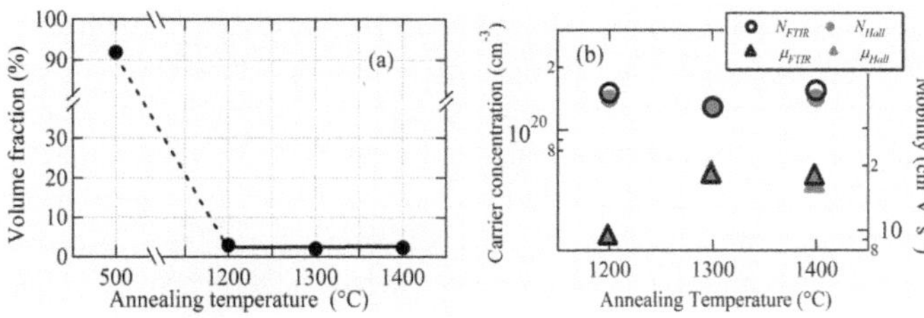

Figure 17. The annealing temperature dependences of (a) volume fraction of defective SiC phase, and (b) free carrier concentration and mobility in re-crystallized SiC phase. The values determined from Hall effect measurement also plotted in (b) for comparison [15].

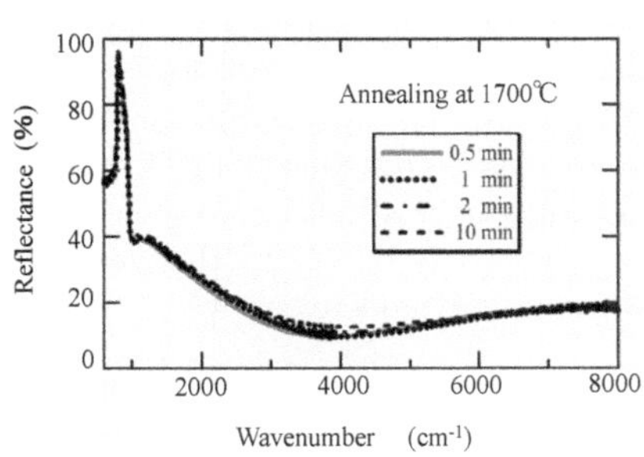

Figure 18. The IR reflectance spectra obtained from the samples annealed at 1700°C for various annealing periods [15].

5. Conclusion

We proposed the method for estimating the electrical properties, such as, carrier concentration and mobility of semiconductor wafers using IR reflectance spectroscopy. In the method, the observed spectra are fitted with the calculated ones, and the free carrier concentration and mobility are determined from the fitted parameters. In the calculation, we used the modified dielectric function (MDF) model for the dispersion relation of dielectric constants. We demonstrated the estimations of carrier concentrations and mobilities of commercially produced 6H-SiC wafers from observed IR reflectance spectra in the frequency range of 400–2000cm^{-1}. We showed that the free carrier concentration and mobility obtained from IR reflectance measurements agree well with the values obtained from Hall-effect measurements in the carrier concentration range of 10^{17}~10^{19} cm^{-3}, which suggests that we can estimate the carrier concentration and mobility accurately in a nondestructive and noncontact way. We demonstrated spatial mappings of carrier concentration and mobility in 2-inch 6H-SiC wafers using this method and showed its usefulness to characterize the spatial distribution of the carrier concentration and mobility in SiC wafers.

Next, we applied this method to the simultaneous determination of the carrier concentration, mobility and thickness of homo-epilayers, and the carrier concentration and mobility of substrates. IR reflectance spectra with the frequency range of 80–2000 cm^{-1} were measured for *n*-type 4H-SiC epilayers on *p*-type and *n*-type 4H-SiC substrates with different carrier concentrations. The obtained values of electrical properties for *n*-type epilayers on *p*-type substrates were compared with the values obtained from Hall-effect measurements, and those for *n*-type epilayers on *n*-type substrates were compared with the values from *C*–*V* measurements. Through these comparisons, we showed that the characterization method using IR reflectance measurements can determine the electrical property and the thickness of SiC homo-epilayers simultaneously and accurately. We also showed that the extension of the observation frequency range to Terahertz region (down to 20cm^{-1}) enables us to characterize the wafers and epilayers with carrier concentrations ranged from 10^{16} to 10^{19}cm^{-3} orders.

Finally, we performed the characterization of both the electrical properties and crystalline damage in high-dose phosphorous implanted and post implantation annealed 4H-SiC layers using IR reflectance spectroscopy. The characterization revealed that the impurities are activated by annealing at a temperature as low as 1200°C for 30 min, though the sufficient recovery of the crystallinity needs higher annealing temperatures than 1200°C. It is also found from the IR reflectance analyses that the annealing at 1700°C activates the impurities and recovers the crystallinity of implanted layer within 1 min. These results suggest that the method can give the information of, not only the electrical properties, but also the crystalline damages of ion-implanted SiC epilayers simultaneously.

In conclusion, the electrical characteristics of SiC wafers and the electrical properties and thickness of SiC epilayers can be obtained simultaneously from the analyses of IR reflectance spectroscopy in nondestructive and contactless manner, which makes possible to obtain the spatial mapping of the electrical characteristics and thickness of SiC epilayers by scanning a probing light beam. Therefore, the method we proposed is a useful technique as a monitoring tool of SiC device-process, i.e., the monitoring of the doping concentration, carrier mobility and thickness, and their uniformity over the wafers in homo-epitaxial growth process, and the recovery of crystallinity and electrical activation of impurities in post-implantation-annealing process.

Author details

Sadafumi Yoshida[1*], Yasuto Hijikata[2] and Hiroyuki Yaguchi[2]

*Address all correspondence to: s.yoshida@aist.go.jp

1 Advanced Power Electronics Research Center, National Institute of Advanced Industrial Science and Technology (AIST), Japan

2 Graduate School of Science and Engineering, Saitama University, Japan

References

[1] Stibal, R., Muller, S., Jantz, W., Pozina, G., Magnusson, B., & Ellison, A. (2003). *Phys. Status Solidi*, C0, 1013.

[2] Harima, H., Nakashima, S., & Uemura, T. (1995). Raman scattering from anisotropic LO-phonon-plasmon-coupled mode in n-type 4H- and 6H-SiC. *J. Appl. Phys.*, 78, 1996-2005.

[3] Burton, J. C., Sun, L., Pophristic, M., Lukas, S. J., Long, F. H., Feng, Z. C., & Ferguson, I. T. (1998). Spatial characterization of doped SiC wafers by Raman spectroscopy. *J. Appl. Phys.*, 84, 6268-6273.

[4] Chafai, M., Jaouhari, A., Terres, A., Anton, R., Martin, E., Jimenez, J., & Mitchel, W. C. (2001). Raman scattering from LO phonon-plasmon coupled modes and Hall-effect in n-type silicon carbide 4H-SiC. *J. Appl. Phys.*, 90, 5211-5215.

[5] Yugami, H., Nakashima, S., Mitsuishi, A., Uemoto, A., Furukawa, K., Suzuki, A., & Nakashima, S. (1987). Characterization of the free-carrier concentrations in doped β-SiC crystals by Raman scattering. *J. Appl. Phys.*, 61, 354-358.

[6] Tiwald, T. E., Woollam, J. A., Zoller, S., Christiansen, J., Gregory, R. B., Wetteroth, J. A., Wilson, S. R., & Powell, A. R. (1999). Carrier concentration and lattice absorption in bulk and epitaxial silicon carbide determined using infrared ellipometry. *Phys. Rev. B*, 60, 11464-11474.

[7] Weingärtner, R., Wellmann, P. J., Bickermann, M., Hofman, D., Straubinger, T. L., & Winnacker, A. (2002). Determination of charge carrier concentration in n- and p-doped SiC based on optical absorption measurements. *Appl. Phys. Lett.*, 80, 70-72.

[8] Holm, R. T., Gibson, J. W., & Palik, E. D. (1977). Infrared reflectance studies of bulk and epitaxial-film n-type GaAs. *J. Appl. Phys.*, 48, 212-223.

[9] Holm, R. T., Klein, P. H., & Nordquist, P. E. R., Jr. (1986). Infrared reflectance evaluation of chemically vapor deposited β-SiC films grown on Si substrates. *J. Appl. Phys.*, 60, 1479-1485.

[10] Macmillan, M. F., Henry, A., & Janzen, E. (1998). Thickness Determination of Low Doped SiC Epi-Films on Highly Doped SiC Substrates. *J. Electron. Mater.*, 27, 300-303.

[11] Yaguchi, H., Narita, K., Hijikata, Y., Yoshida, S., Nakashima, S., & Oyanagi, N. (2002). Spatial Mapping of the Carrier Concentration and Mobility in SiC Wafers by Micro Fourier-Transform Infrared Spectroscopy. *Mater. Sci. Forum*, 389-393, 621-624.

[12] Narita, K., Hijikata, Y., Yaguchi, H., Yoshida, S., & Nakashima, S. (2004). Characterization of Carrier Concentration and Mobility in n-type SiC Wafers Using Infrared Reflectance Spectroscopy. *Jpn. J. Appl. Phys.*, 43, 5156.

[13] Oishi, S., Hijikata, Y., Yaguchi, H., & Yoshida, S. (2006). Simultaneous Determination of Carrier Concentration, Mobility, and Thickness of SiC homoepilayers by Infrared Reflectance Spectroscopy. *Jpn. J. Appl. Phys.*, 45, L1226-L1229.

[14] Oishi, S., Hijikata, Y., Yaguchi, H., & Yoshida, S. (2007). Simultaneous Determination of the Carrier Concentration, Mobility and Thickness of SiC Homo-Epilayers Using Terahertz Reflectance Spectroscopy. *Mater. Sci. Forum*, 556-557, 423-426.

[15] Narita, K., Hijikata, Y., Yaguchi, H., Yoshida, S., Senzaki, J., & Nakashima, S. (2004). Characterization of electrical properties in high-dose implanted and post-implantation-annealed 4H-SiC wafers using infrared reflectance spectroscopy. *Mater. Sci. Forum*, 457-460, 905-908.

[16] Neyret, E., Ferro, G., Juillaguet, S., Bluet, J. M., Jaussaud, C., & Camassel, J. (1999). Optical investigation of residual doping species in 6H and 4H-SiC layers grown by chemical vapor deposition. *Materials Science and Engineering B*, 61-62, 253-257.

[17] Nakashima, S., & Harima, H. (2004). Spectroscopic analysis of electrical properties in polar semiconductors with over-damped plasmons. *J. Appl. Phys.*, 95, 3541-3546.

[18] Perkowitz, S. (1993). *Optical Characterization of Semiconductors*, Academic Press, London.

[19] Perkowitz, S., & Thorland, R. T. (1975). Generalized Dielectric Function and the Plasmon-Phonon Coupling in GaAs and CdTe. *Solid State Commun.*, 16, 1093-1096.

[20] Kukharskii, A. A. (1972). *Sov. Phys.-Solid State*, 14, 1501.

[21] Pensl, G., & Choyke, W. J. (1993). Electrical and optical characterization of SiC. *Physica B*, 185, 264-283.

[22] Chen, G. D., Lin, J. Y., & Jiang, H. X. (1975). Effects of electron mass anisotropy on Hall factors in 6H-SiC. *Appl. Phys. Lett.*, 68, 1341-1343.

[23] Karmann, S., Suttrop, W., Schöner, A., Schadt, M., Haberstroh, C., Engelbrecht, F., Helbig, R., Pensl, G., Stein, R., & Leibenzedar, S. (1992). Chemical vapor deposition and characterization of undoped and nitrogen-doped single crystalline 6H-SiC. *J. Appl. Phys.*, 72, 5437-5442.

[24] Iwata, H., & Itoh, K. M. (2001). Donor and acceptor concentration dependence of the electron Hall mobility and the Hall scattering factor in n-type 4H- and 6H-SiC. *J. Appl. Phys.*, 89, 6228-6234.

[25] Iwata, H., Itoh, K. M., & Pensl, G. (2000). Theory of the anisotropy of the electron Hall mobility in n-type 4H- and 6H-SiC. *J. Appl. Phys.*, 88, 1956-1961.

[26] Kinoshita, T., Itoh, K. M., Schadt, M., & Pensl, G. (1999). Theory of the electron mobility in n-type 6H-SiC. *J. Appl. Phys.*, 85, 8193-8198.

[27] Volm, D., Meyer, B. K., Hofman, D. M., Chen, W. M., Son, N. T., Persson, C., Lindefelt, U., Kordina, O., Sorman, E., Konstantinov, A. O., Monemar, B., & Janzen, E. (1996). Determination of the electron effective-mass tensor in 4H SiC. *Phys. Rev. B*, 53, 15409-15412.

[28] Kojima, K., Suzuki, T., Kuroda, S., Nishio, J., & Arai, K. (2003). Epitaxial Growth of High-Quality 4H-SiC Carbon-Face by Low-Pressure Hot-Wall Chemical Vapor Deposition. *Jpn. J. Appl. Phys.*, 42, L637-L639.

[29] Rutsch, G., Devaty, R. P., Choyke, W. J., Langer, D. W., & Rowland, L. B. (1998). Measurement of the Hall scattering factor in 4H and 6H SiC epilayers from 40 to 290 K and in magnetic fields up to 9 T. *J. Appl. Phys.*, 84, 2062-2064.

[30] Götz, W., Schöner, A., Pensl, G., Suttrop, W., Choyke, W. J., Stein, R., & Leibenzeder, S. (1993). Nitrogen donors in 4H-silicon carbide. *J. Appl. Phys.*, 73, 3332-3338.

[31] Capano, M. A., Cooper, J. A., Jr., Melloch, M. R., Saxler, A., & Mitchel, W. C. (2000). Ionization energies and electron mobilities in phosphorus- and nitrogen-implanted 4H-silicon carbide. *J. Appl. Phys.*, 87, 8773-8777.

[32] Nishio, J., Kushibe, M., Masahara, K., Kojima, K., Oono, T., Ishida, Y., Takahashi, T., Suzuki, T., Tanaka, T., Yoshida, S., & Arai, K. (2002). Investigation of Residual Impurities in 4H-SiC Epitaxial Layers Grown by Hot-Wall Chemical Vapor Deposition. *Mater. Sci. Forum*, 389-393, 215-218.

[33] Chang, W. Z., Feng, Z. C., Lin, J., Liu, R., Wee, A. T. S., Tone, L., & Zhao, J. H. (2002). Infrared reflection investigation of ion-implanted and post-implantation-annealed epitaxially grown 6H-SiC. *Surf. Interface Anal.*, 33, 500-505.

[34] Camassel, J., Wang, H. Y., Pernot, J., Godignon, P., Mestres, N., & Pascual, J. (2002). Infrared Investigation of Implantation Damage in 6H-SiC. *Mater. Sci. Forum*, 389-393, 859-862.

[35] Senzaki, J., Fukuda, K., & Arai, K. (2003). Influences of postimplantation annealing conditions on resistance loweing in high-phosphorus-implanted 4H-SiC. *J. Appl. Phys.*, 94, 2942-2947.

Chapter 3

Ion Synthesis of SiC and Its Instability at High Temperatures

Kair Kh. Nussupov and Nurzhan B. Beisenkhanov

Additional information is available at the end of the chapter

1. Introduction

As is known, such advantages of silicon carbide as a high hardness (4th place after diamond) [36, 37], high chemical and radiation resistance, high melting point, etc. became the basis of its wide application not only in microelectronics [1], but also as refractory and abrasive materials. Silicon carbide is included in the oxidation resistant composite materials [42] used in coating system for "Space Shuttle", capable of withstanding temperatures up to 1500°C at the entrance of the ship into the atmosphere. In many attempts to develop an effective oxidation-protection coating for carbon-carbon composites with excellent mechanical properties at elevated temperature, silicon carbide coating has shown the best performance for short periods of up to 1900K [39]. For longer periods and higher temperature applications, a challenging coating system should be developed.

Silicon carbide is regarded by researchers as a suitable material for the front wall structures of fusion reactors. The boers, cutting disks, grinding paper of SiC can be used for boring, drilling, surface grinding and cutting of steel, nonferrous metals, natural stone, concrete, wood and plastic.

The stability of silicon carbide to high temperature treatment is of special interest. As a special application, silicon carbide can be thermally oxidized in the form of SiO_2, and the devices which can be easily fabricated on Si substrate (Power MOSFET, IGBT, MOS controlled thyristor, etc.) can also be fabricated on SiC substrate [23]. In paper [23] the parabolic growth of thickness of thermal oxide versus oxidation time was observed, and the slope of the plots increases with increasing temperature. The thickness values of oxide films were about 23-77 nm (Si-face, wet oxidation), 18-63 nm (Si-face, dry oxidation), 210-810 nm (C-face, wet oxidation) and 125-260 nm (C-face, dry oxidation) for oxidation time 6 h and depend on temperature value (1000, 1050, 1110 or 1150°C).

Doped with different impurities, silicon carbide is used in semiconductor technology [63, 12].

Field-effect transistors, diodes and other electronic devices based on SiC have several advantages compared to similar silicon devices, for example, the opportunity to work at temperatures up to 600°C, high speed and high radiation resistance. A large number of polytypes of SiC makes it possible to create heteropolytype structures [31, 32, 33]. Currently, using the methods of vacuum sublimation [48], molecular beam epitaxy [15], the epitaxial and heteropolytype layers based on the cubic 3C-SiC and two hexagonal 6H-SiC, 4H-SiC on substrates of SiC, are grown. Heteroepitaxial layers of 3C-SiC on substrates of Si by chemical vapor deposition (CVD) [41] are grown. At the temperatures below 1200°C there are conditions for the growth of both poly- and nanocrystalline SiC with different degrees of crystallinity and structure of the cubic polytype 3C-SiC. Such conditions were realized in the magnetron sputtering [25, 56], laser ablation [53] and plasma deposition [36], plasma-enhanced chemical vapor deposition [19, 43], molecular beam epitaxy [16]. At temperatures below 1500°C in the direct deposition of carbon and silicon ions with energy of ~100 eV, the growth of nanocrystalline films with a consistent set of the polytypes 3C, 21R, 27R, 51R, 6H is possible [49, 50, 51].

In recent years there has been an intensification of studies on the synthesis of SiC by high-dose carbon ion implantation into Si [37, 35]. In addition, the synthesis of SiC by high dose implantation of carbon ions into silicon is also of fundamental scientific interest due to the wide practical application [9 - 11, 47], for example, to create a coating and insulating SiC layers in the manufacture of integrated circuits. High quality crystalline β-SiC film on SiO_2 can be obtained by multiple implantations of carbon ions into silicon and subsequent selective oxidation of the top layer of Si [52]. Intensively developing area is the formation by this technique in SiO_2 of nanostructured systems with inclusions of nanocrystals and clusters of Si, SiC and C, providing at the expense of size effects luminescence throughout the visible spectrum [57]. The study of the stability of these films to high temperature treatment is also of special interest.

The ion synthesis of silicon carbide and studies of crystallization process attract attention of researchers [12, 35, 37, 63]. The implantation of single energy carbon ions with a Gaussian distribution on the depth in silicon is of interest due to a presence of wide range of nanolayers with different concentrations of carbon and silicon atoms and, therefore, a presence of different clusters and nanocrystals of silicon, carbon and silicon carbide in the implanted layer after implantation and annealing. The properties of these layers have been investigated in detail. In previous investigations, the carbon ions with energy of 40 keV were used for considered purposes in a number of papers [8, 13, 20], and doses ranged 10^{16}-10^{18} cm^{-2} were used in almost all of investigations, when the ion synthesis of a silicon carbide film was carried out [2]. The IR absorption technique was widely applied for the investigation of these layers [26, 55]. It has been used, mainly, to confirm the formation of silicon carbide in the implanted layer and, to obtain the new information about the layer structure as well. In some papers, the dependence of both shift of the wave length of a minimum of IR transmission peak and the change of its half width versus the annealing temperature are used for interpretation of IR transmission spectra. In our opinion, it is necessary to investigate such important characteristics as a change of an area under a curve of IR transmission spectrum band and a change of peak amplitude at 800 cm^{-1} versus the annealing temperature which contains very valuable information about the structure changes in an implanted layer. If the thickness of ion-synthesized film is comparable or smaller than the wavelength of incident electromagnetic radiation, under certain geometric conditions of the experiment, one can observe not only the transverse optical oscillations of atoms (TO-phonons), as well as longitudinal optical lattice oscillations (LO-phonons) [3]. The detection of LO-phonon peak of SiC and its change after film annealing give additional information on the crystallization processes. It is necessary to carry out the circumstantial investigations devoted to an analysis in detail of change in a wider temperature interval of an half width of IR transmission peak which characterizes the degree of structure order of an ion implanted layer.

In the majority of analogous studies, the post-implantation isochronous annealing of samples was carried out at temperatures from 400 up to 1200°C [11, 14, 2, 27]. However, in several studies [28, 47] the temperature range was extended and, the ion implanted layers had been annealed at temperatures 1300 and 1405°C. We believe also, that the temperature interval 400-1200°C is not sufficient for a annealing of disordered layer and completion of crystallization processes. A more detailed investigation of processes at temperatures ranged from 20 up to 1405°C permits to observe a number interesting effects taking place in an implanted layer.

The authors of papers [11, 2, 5, 55] declare about a significant diffusion of carbon and, contrariwise, that is negatived in works [13, 26, 27]. The authors of papers [8, 4, 47, 34] show that a layer has the electron conduction after annealing. The data of [28] give of evidence about the p-type conduction.

In a number of studies [6, 2, 5] a synthesis of SiC on a (100) oriented silicon substrate is considered as preferable, but at the same time in papers [13] the orientation (111) of substrate is declared as a most suitable. [4] investigated the optical and photoelectric properties of the SiC-Si structure, formed by implantation into (100), (110) and (111) oriented n- and p-type silicon of ^{12}C ions with energies of 40 and 70 keV, and doses of 4.3×10^{17} and 5×10^{17} cm^{-2}. Analysis of the IR absorption spectra of silicon layers implanted by carbon ions with energy of 70 keV allowed finding a significant dependence of crystallinity of the SiC layer on the orientation of the substrate after annealing at temperatures of 700–900°C. Although the tetrahedral Si–C-bonds more intense formed at an orientation of the substrate (100), annealing at 1100°C all evens out differences in the absorption spectra for all three substrate orientations (100), (110) and (111). The photovoltage photoresponse was obtained in all implanted structures. Investigation of current-voltage characteristics showed improvement in the rectification effect of the structure after annealing. The possibility of creating of β-SiC—Si heterostructures by ion implantation technique (band gap of 2.39 eV and 1.11 eV, respectively) was shown.

Frangis et al. [17, 18] formed β-SiC in silicon by high-temperature implantation (850–950°C) of carbon ions with energies 200 keV and doses ranged within $(0.2–1)\times10^{18}$ cm^{-2}. Implantation was carried out into silicon wafers of orientation (100) and (111). In both cases, β-SiC was formed with the same orientation as the matrix. It also reported that implantation at a lower temperature (500°C), but at higher energy (300 keV) leads to the formation of good quality β-SiC.

Aleksandrov et al. [6] carried out the synthesis of single-crystal SiC layer with one-step technique of high current ion implantation of carbon atoms into silicon substrates with orientations (001) and (111). Single crystal layer of SiC, which contains a small number of twins, was synthesized by the implantation of carbon ions with dose of 6×10^{17} cm^{-2} into (001) oriented silicon wafer using a focused ion beam with current density of 300 A/cm^2. When the ion current density was 150 A/cm^2, a single crystal SiC layer with a high concentration of twins was formed at the interface with the substrate Si. On top of this layer is formed a layer of polycrystalline SiC. When the implantation of carbon ions was carried out into (111) oriented silicon, single crystal β-SiC layer is not formed even when implanted into substrate heated up to a temperature of 850°C. Polycrystalline SiC layer at the surface and single-crystal SiC layer with a high density of twins near the interface with the crystal Si matrix, are formed.

This chapter presents the study of silicon carbide and carbon layers on silicon synthesized by ion beam techniques. The investigations of silicon layers implanted by carbon ions with energy 40 keV and dose 3.56×10^{17} cm^{-2} after annealing over a wide temperature range from 20 up to 1400°C using the special IR analysis are described. The features of change of the SiC-peaks in the spectra of the infrared transmission due to the influence of the Gaussian profile of the distribution of carbon in silicon are shown. Experiments to observe the longitudinal optical oscillations (LO-phonons) associated with the silicon carbide were carried out. A type of a conduction of synthesized silicon carbide was studied. Definite information from a shape analysis of the IR transmission curve was obtained. A particular attention was attracted on some problems which were disputable in previous investigations. IR studies of high-temperature instability of homogeneous layers of silicon carbide on (100) and (111) oriented silicon substrates synthesized by multiple implantation of carbon ions with energies E = 40, 20, 10, 5 and 3 keV, are described. By IR spectroscopy, Auger electron spectroscopy and X-ray reflectometry the composition and the processes of structural adjustment of the layer during annealing are analyzed. By ion-beam sputtering and magnetron sputtering techniques the $SiC_{0.8}$ and C films on the silicon wafers were deposited. Characteristics of the films by X-ray reflectometry are analyzed.

2. Experimental

Carbon implantation was carried out under completely oil-free conditions using elaborated accelerator. A vacuum in the implantation chamber is created by help of the ceolite vacuum pumps (6.5×10^{-2} Pa) and titan magnet discharge pumps (1.3×10^{-4} Pa). These pumps permit to except completely the organic compounds in volume which could be to contaminate the surface of the implanting samples. Gas (dioxide of carbon) has been used to obtain the single-charged ions of $^{12}C^{+}$. The implantation dose was determined by integrating of the beam current registrated on the target with suppression of a secondary emission. In order to prevent a sample heating during implantation, the ion current density was kept at a level less than 3 mkA/cm^2. The temperature of the target during the implantation is controlled by a thermocouple and, it not exceeds 20–25°C. The implantation of carbon ions was carried out into single-crystal (100) and (111) oriented silicon wafers of sizes $7\times12\times0.4$ mm^3 with an electrical resistivity 4–5 and

10 Ohm cm, respectively. After cleaning and removing the native surface oxide in a chemical etch, the samples were mounted in the target chamber of the implanter.

A set of these silicon wafers were implanted by $^{12}C^{+}$ ions with energy 40 keV and dose 3.56×10^{17} cm^{-2} at room temperature. To observe the longitudinal optical oscillations (LO-phonons) of atoms in synthesized film, a rotating shaft was incorporated into work chamber of infrared spectrometer. A sample holder is attached on this shaft. This system permits to make the IR transmission measurements of an ion implanted layer versus an angle of incidence of electromagnetic radiation on sample surface over the range 0-360° with step of 5°. However, in practice, at measuring of spectra we change the angle of incidence from 0 up to ±75°. It was observed no differences in transmission spectra measured from samples sloped to the radiation at a rotation of shaft both clockwise and anti-clockwise from the normal. Isochronous annealing of ion implanted samples was carried out in vacuum over the temperature range 200-1400°C with steps of 50-200°C. The annealing was carried out in a low-inertia economical vacuum furnace especially elaborated and created for these purposes. It was carried out in conditions of completely oil-free pumping-out at a residual pressure ~ 1.3×10^{-4} Pa. The temperature was controlled by a help of tungsten-rhenium thermocouple.

To obtain a rectangular profile of the distribution of carbon atoms in the silicon, implantation of carbon ions of different energies and doses into second set of single-crystal silicon wafers of n- and p-type of conductivity was carried out sequentially in the following order: 1) E = 40 keV, D = 2.80×10^{17} cm^{-2}, 2) 20 keV and 0.96×10^{17} cm^{-2}, 3) 10 keV and 0.495×10^{17} cm^{-2}, 4) 5 keV and 0.165×10^{17} cm^{-2}, 5) 3 keV and 0.115×10^{17} cm^{-2}. The ratio of the concentrations of carbon and silicon atoms in the depth was about $N_C/N_{Si} = 0.7$. Post implantation annealing of the samples was performed in a vacuum in the temperature range 200-1200°C for 30 min with a step of 200°C. Then, the SiC films were subjected to prolonged isothermal annealing at the temperature of 1200°C for several hours in an atmosphere of inert gas (Ar) and, after specific time intervals infrared transmission spectra were recorded. The IR transmission spectra were recorded in differential regime on double-beam infrared spectrometer (400–5000 cm^{-1}). The spectra both at perpendicular incidence of infrared rays on the sample surface and at an angle of 73° with respect to the normal to the sample surface were measured. The composition of the layers was examined by Auger electron spectroscopy. The parameters were as follows: incident electron beam of diameter 1 μm, energy 10 keV, angle of incidence 45°, diameter of scanning region 300 μm, vacuum 1.33 $\times10^{-8}$ Pa, angle of Ar^{+} beam incidence 45°. Parameters of films were investigated using the X-ray reflectometry at small glancing angles by recording the angular dependence of the reflection coefficient for two spectral X-ray lines CuK_α (0.154 nm) and CuK_β (0.139 nm) at the facility "ComplexXRay C6". Selection of spectral lines CuK_α and CuK_β from polychromatic spectrum was carried out using thin semi-transparent and thick untransparent monochromators, respectively, made from the pyrolytic graphite with a mosaic angle of 0.5°.

SiC films on the silicon substrates (25°C) were also synthesized by ion-beam sputtering of a two-component target of graphite and silicon. The C films on the silicon substrate (75°C) by magnetron sputtering were synthesized. Parameters of SiC and C films on Si substrates were determined using the X-ray reflectometry.

3. Results and discussion

3.1. A model of an carbon implanted silicon layer and the mechanism of the low temperature formation of Si- and SiC crystallites

In Fig.1 the calculated profile of carbon atom distribution in Si constructed basing on data R_p and ΔR_p from [21] are presented. The Gaussian profile (Fig.1, curve 1) was calculated for the implantation of carbon ions with energy 40 keV and dose 3.534×10^{17} cm^{-2}, when the carbon concentration in the distribution peak is equal to stoichiometric composition of SiC, i.e. $N_C/N_{Si} = 1$, where N_C/N_{Si} is the ratio of the concentrations of C and Si atoms. The curve 2 in Fig.1 shows the calculated profile, corresponding to dose $D = 3.56\times10^{17}$ cm^{-2} of carbon ions used in this investigation.

3.1.1. LO-phonons and their applications to analysis of an implanted layer

As it is well known [45, 66], during interaction of electromagnetic waves with an infinite crystal lattice, the transversal optical oscillations (TO-phonons) of atoms are excited. In overwhelming majority of previous investigations, the synthesis of silicon carbide was identified by help of spectra of transversal optical phonons. No attempts was made to detect the longitudinal optical oscillations (LO-phonons) of atoms of lattice for a Gaussian concentration profiles of carbon in spite of that majority of studies in the field of ion synthesis of silicon carbide was carried out using these profiles. [3] found that an absorption at wave number 980 cm^{-1} is observed, if the angle of incidence of irradiation on the sample surface deviate from perpendicularity. Implantation of carbon ions with energies of 24 and 40 keV and a dose of 4.3×10^{17} cm^{-2} carried out at room temperature into a (111) oriented Si plate of p-type conductivity. Annealing was performed in a vacuum at temperatures of 900 and 1100°C for 30 minutes. The presence in the transmission spectrum of bands associated with the LO-and TO-phonons made possible to calculate such parameters of SiC, as high frequency, ε_∞, and the low-frequency dielectric constant, ε_0, attenuation coefficient of the phonons, the effective charge e^*/e and the force constant ϱ. So, the detection of LO-phonons may be used to obtain additional information about a structure of ion implanted layer. Morever, the frequency values of both the transversal- and longitudinal oscillations permit to determine the parameters of efficient charge which is a quantitative criterion of a compound polarity. The efficient charge value permits to calculate a mobility of free charge carriers.

If a thickness of ion synthesized film is less or comparable with a wave length of the electromagnetic radiation incident on film's surface, the limitations related with the condition of infinity of crystal lattice are lifted. As a result, one can to observe the longitudinal optical oscillations of lattice atoms at definite geometrical conditions of experiments [66]. In this relation, the special experiments to observe these phonons were carried out. For this purpose, the IR transmisssion spectra versus an angle of incidence of electromagnetic radiation on sample surface with step of 5° were measured.

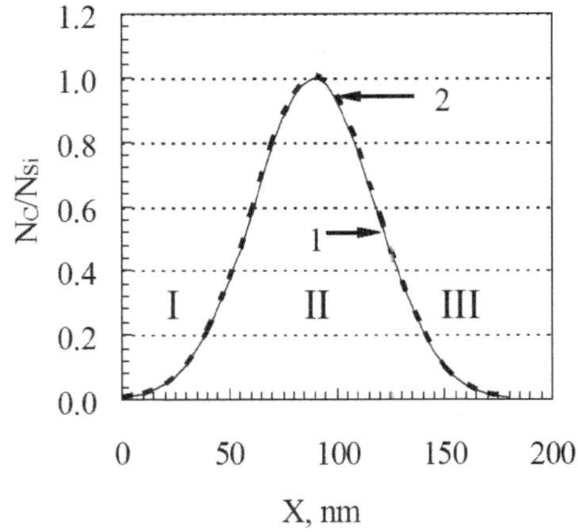

Figure 1. The calculated profiles of distribution of carbon atoms in Si constructed basing on data of R_p and ΔR_p from [21]. The energy of carbon ions is 40 keV, dose 3.534×10^{17} cm^{-2} (curve 1) and 3.56×10^{17} cm^{-2} (curve 2).

In Figs.2 and 3 the IR transmission spectra of both (100)- and (111) oriented silicon samples implanted by carbon ions with energy 40 kev and dose 3.56×10^{17} cm^{-2}, after isochronous annealing over the temperature range 200-1400°C, are presented. The spectra at both perpendicular incidence of infrared rays on the sample surface and at an angle of 73° with respect to the normal to the surface were measured. In Fig.4 the wave number values in maximum of IR transmission versus the annealing temperature are presented. The curves on this figure were constructed using the experimental data presented on Figs.2 and 3. The curves for TO-phonons were constructed basing on the infrared transmission spectra measured by using of perpendicular incidence of the infrared rays on the sample surface.

Figs. 2 and 3 show the appearance of an IR absorption peak at 965-970 cm^{-1} in the spectra from samples inclined to IR irradiation at an angle of 73° with respect to the normal to the sample surface. This absorption peak begins to be appeared after annealing at 1000°C for both types of substrate orientation together with the main peak at 797-800 cm^{-1} which corresponds to the transversal optical atomic oscillation of SiC. Basing on the values of wavelength of this peak, its amplitude and synchronous modification together with the peak for transversal optical oscillation of SiC during annealing as well, the peak at 965-970 cm^{-1} was associated with longitudinal optical phonons of silicon carbide. The temperature of LO-phonon appearance is about 300°C higher, than that for TO-phonons of SiC (700°C). It may be caused by smallness of the LO-phonon peak amplitude and, in consequence of this, by difficulties of their registration.

For (100) oriented substrates, the increase of the annealing temperature over the range 1000–1350°C leads a linearly increase of LO-phonon wavelength at minimum of amplitude from 930 to 965 cm^{-1} and, then wavelength increasing is saturated. As for (111) oriented substrates, unlinearly increase of LO-phonon wavelength in the range 955–970 cm^{-1} up to 1400°C is observed. Thus, the formation of SiC crystallites, i.e. the intensive formation of Si-C tetrahedral bonds of necessary length and bond angles, in the case of (111) oriented substrate is not completed up to the silicon melting point, as for (100) oriented substrate that is completed at 1350°C. The difference between the LO-phonon curves behaviour for (100) and (111) oriented substrates indicates on the differences in the crystallization mecanism. One can see an influence of silicon substrate orientation on SiC crystallization in the implanted layer from the TO-phonon curve also (Fig.4).

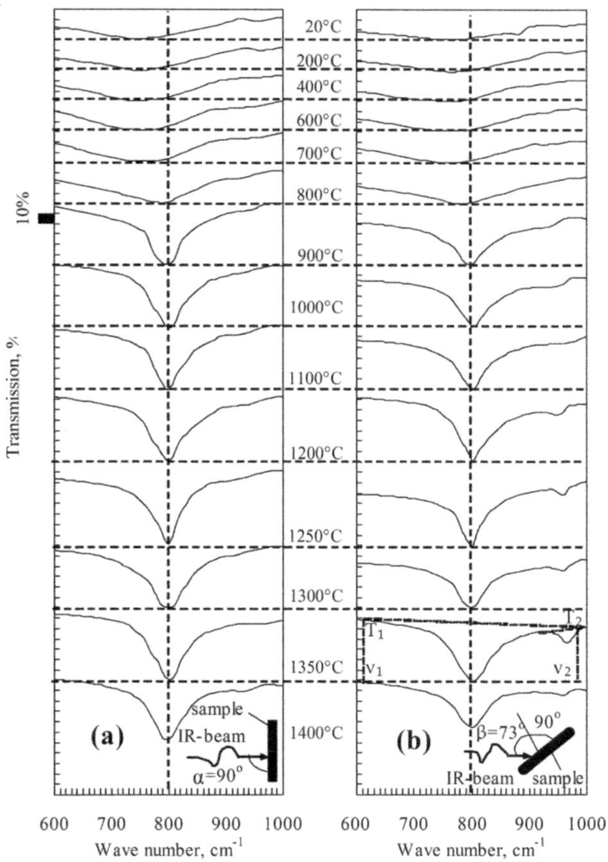

Figure 2. The IR transmission spectra of (100) oriented Si samples implanted by $^{+}C^{12}$ ions (E = 40 kev, D = 3.56×10^{17} cm^{-2}, after isochronous annealing over the temperature range 200-1400°C: a) α = 90°; b) β = 73°.

In some investigations no changes in the IR transmission spectra after annealing at 700°C [8], 850°C [14], 875°C [11], 900°C [26], 1100°C [2] were observed. That was explained by finishing of the β-SiC formation process. Really, that may be correct, if we are based on the analysis of TO-phonon curve only. However, as is seen from Fig.4, the TO-phonon curves show the saturated absorption and give no addititional information over the temperature range of 900–1400°C, as the LO-phonon curves undergo the substantial changes at these temperatures indicating on the structural changes in the ion implanted layer. Thus, one can conclude that the observation and measurement of LO-phonon peak are important for an analysis of crystallization process.

As it is known [38], the TO- and LO-phonons frequences are bounded with the equation:

$$\frac{\nu_{LO}}{\nu_{TO}} = \left(\frac{\varepsilon_0}{\varepsilon_\infty}\right)^{1/2} \tag{1}$$

Where ε_0 and ε_∞ are the low-frequency and the high-frequency dielectric constants, respectively. The effective charge e*/e is determined from the equation [54]:

$$\frac{e^*}{e} = \left(\frac{\varepsilon_0-\varepsilon_\infty}{4\pi}\right)^{1/2} \left(\frac{M_n}{N}\right)^{1/2} \left(\frac{3\omega_0}{\varepsilon_\infty+2}\right) \left(\frac{\pi(\varepsilon_0-\varepsilon_\infty)M_n\nu_{TO}^2}{Ne^2}\right)^{1/2} \left(\frac{3}{\varepsilon_\infty+2}\right) \tag{2}$$

where ω_0 is the resonance frequency, $N = 4.84\times10^{22}$ cm^{-3} is the concentration of ion pairs, $M_n = (M_+M_-)/(M_++M_-) = 1.396\times10^{-23}$ g is the reduced mass of ion pair, $\nu_{TO} = 2.395\times10^{-13}$ cm^{-1} is the frequency of TO-phonon irradiation.

The dimensionless parameter, ϱ, which is proportional to the absorption quantity, is determined from the equation:

$$\rho = \frac{\varepsilon_0-\varepsilon_\infty}{4\pi} \tag{3}$$

The values of ε_0, e*/e and ϱ determined from the equations (1)—(3) are equal to 9.82, 0.89, and 0.25, respectively. The value of ε_∞ has been chosen to be equal to 6.7, because according to [45] the sufficiently great dispersion of this value leads an insignificant changes. So, both the detection and the measuring of LO-phonons have been permited to determine some characteristics of synthesized film.

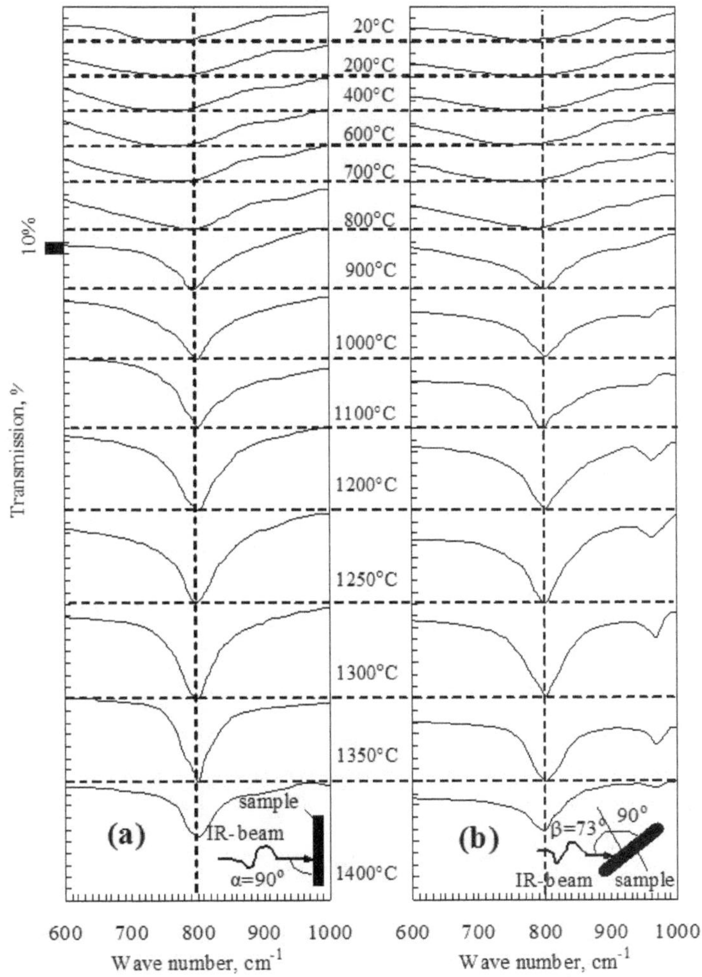

Figure 3. The IR transmission spectra of (111) oriented Si samples implanted by $^{+}C^{12}$ ions (E = 40 kev, D = 3.56x10^{17} cm^{-2}), after isochronous annealing over the temperature range 200–1400°C: a) α = 90°; b) β = 73°.

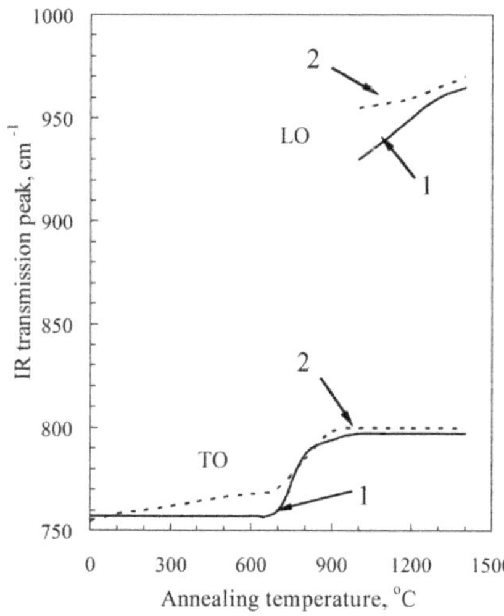

Figure 4. Wave numbers of IR transmission minimum for TO- and LO-phonon peaks of SiC versus an annealing temperature for carbon implanted silicon layers on Si substrates (E = 40 kev, D = 3.56×10^{17} cm^{-2}): 1 - Si(000); 2 - Si(111).

3.1.2. The IR transmission analysis of ion-implanted layer on (100) and (111) oriented silicon

In Figs.5a and b, the values of amplitude and halfwidth (FWHM) of the IR transmission peak, respectively, versus an incidence angle of the infrared radiation on the carbon implanted (100) oriented silicon samples, are presented. These data were obtained after isochronous annealing of samples over the range 200-1200°C for 30 min with the step of 200°C. Beginning from an angle α = 50° (Fig.5a), almost linear decreasing of TO-phonon peak amplitude and simultaneous increasing of LO-phonon peak amplitude are observed. The changes of LO- and TO-phonons peak amplitudes are correlated with one another.

The halfwidth (FWHM) changes of the peak have a more complicated dependence from the incidence angle of the infrared radiation on the sample surface (Fig.5b). This dependence has no correlation with the data in Fig.5a. The half-width of the peak is usually dependent on the quality of the crystal structure of the film and should not depend on the angle of incidence of IR radiation. Apparently, the decreasing of halfwidth of the TO-phonon peak can be explained by the presence of non-tetrahedral Si-C-bonds of a certain type, which are oriented in the space of the film in such a way that with increasing angle α above 35° they cease to absorb infrared radiation at frequencies near 800 cm^{-1}. This is equivalent to the effect of decay of these bonds, since it leads to a decrease in the amplitude and the narrowing of the TO-phonon peak, but can not testify about improving the structure of the layer. This is also accompanied by the appearance of deformed LO-phonon peak in the frequency range near 950 cm^{-1}. With the increase of the angle α up to 73°, the narrowing of LO-phonon peak and an increase in its amplitude are taken place. The interpretation of these results requires further investigation.

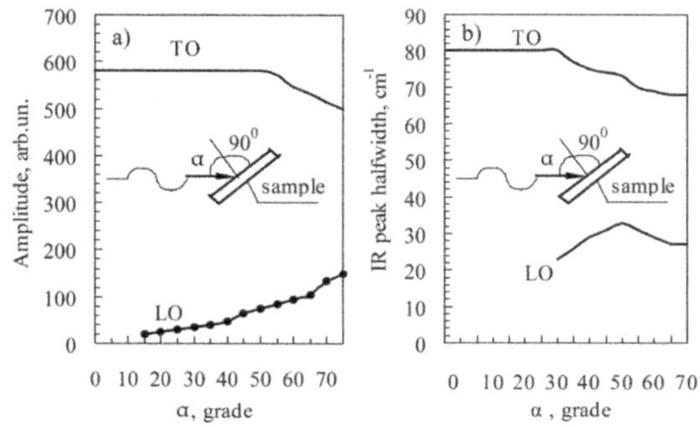

Figure 5. Amplitude (a) and halfwidth (FWHM) (b) of TO- and LO-phonons peaks of SiC of the IR transmission versus the incidence angle of IR radiation on the surface of the carbon implanted Si.

In previous IR investigations of implanted by $^{12}C^+$ ions silicon layers, a shift in frequency of an absorption maximum versus the annealing temperature, and also the changes of half-width and amplitude of peak were observed. Analyzing the IR transmission spectra presented in Figs.2 and 3, one can see not only the changes of these parameters. A base-line to each spectrum (Fig.2, 1350°C) was drawn. As is seen, the areas of obtained figures are changed, too (Fig.6). In our opinion, the area under the IR transmission curve is associated with the number of absorbing objects in the ion-implanted layer and better shows the transformation of these objects during isochronal annealing. As the object may be not only the crystallites of SiC, but also the another types of infrared active compounds of carbon atoms with carbon or silicon atoms, and silicon atoms one with another, which one can unify under one common appellation - clusters.

Basing on the mentioned above, the IR transmission spectra (Figs.2 and 3) were analyzed in detail accordingly to all listed points. In Fig.6 an area of the IR transmission peak (see Fig.2) associated with TO-phonons of SiC obtained both at perpendicular incidence of infrared rays on the sample surface (curve 1) and at an angle of 73° with respect to the normal to the sample surface (curve 2), versus the annealing temperature are presented. Area values can be determined by direct measurement or by using the expression (Fig.2, 1350°C):

$$A=\frac{1}{2}(T_1+T_2)(\nu_2-\nu_1)-\int\tau(\nu)d\nu\approx\frac{1}{2}(T_1+T_2)(\nu_2-\nu_1)-\sum\tau(\nu)\delta\nu \qquad (4)$$

where A – total absorption (or transmission) in relative units in the wave number range $\nu_1<\nu<\nu_2$, $\tau(\nu)$ – transmission at frequency ν, T_1 and T_2 – the values of IR transmission at wave numbers ν_1 and ν_2, respectively, $\delta\nu$ – step of measurements, equal to 2.5 or 5 cm^{-1}. The areas corresponded to LO-phonons have been not measured due to of their infinitesimal. Further the data obtained for a perpendicular incidence of IR-rays on sample surface will be discussed, as an analysis of the curve 2 is difficult due to the absence of the reflection data. As is seen from Fig. 6 (curve 1), the four peaks at 600, 1000, 1200 and 1350°C are evidently observed, which, seemingly, are related with four physical processes occuring in ion implanted layer in four temperature ranges. The same maxima are observed for curve 2 in approximately the same temperature ranges. On this curve a fifth maximum at 200°C is also observed.

It is necessary to note, that in most previous investigations, mainly, the informations corresponding to the peak in range 900-1000°C are obtained using the dependences of halfwidth and a shift in frequency of an absorption maximum from the annealing temperature. The physical processes corresponded to the peaks of both 600 and 1200-1400°C have not been studied in detail, partly, due to the hard access of the temperature range of 1200-1400°C and, partly, due to weak defined processes at 600°C. It follows from the Fig.6 that the change of area of SiC-peak takes place over the whole temperature range from 20°C up to 1400°C.

In Fig.7 the values of IR transmission amplitude for TO-phonons of SiC at wavenumber 800 cm^{-1} and for LO-phonons of SiC versus an annealing temperature for spectra presented in Figs.2 and 3 (curves 1, 1′ – for the perpendicular incidence of IR rays on sample, the curves 2, 2′, 3, 3′ – for an angle of 73°), are shown. When constructing these dependences we have believed that the IR transmission amplitude at 800 cm^{-1} is proportional to the concentration of tetrahedral oriented Si–C-bonds of atoms incorporated into crystallites of SiC. All factors which can affect on a broadening of peak corresponding to an infinitely thin ideal film of SiC have been neglected. This approximation is to some extent, may distort the true picture of the physical phenomena occurring in ion-implanted layer. However, this assumption is very important in a qualitative sense, since it allows understanding the general course of the process and separating a region of the IR transmission peak [30] due to the contribution of crystallites of SiC in the area value, from a region due to the optically active clusters. As seen in Fig.7, the overall shape of the curves 1 and 2 for the TO-phonons is about the same and has a number of features, in particular, at 1300°C, which is also found on the curve 3 for the LO-phonons. The changes of amplitude take place over the all temperature range from 20 up to 1400°C. There are differences in the crystallization processes of carbon implanted silicon layers for the orientation of the substrate (100) and (111).

In Fig.8 the half-width of the TO-phonon peak of SiC at perpendicular incidence of IR radiation on sample surface for spectra shown in Figs.2 (curve 1 – for substrate Si(100)) and 3 (curve 1' – for substrate Si(111)), versus an annealing temperature, is presented. The analogous dependences have been presented in previous papers [8, 47, 26]. However, the temperature range was narrower and, therefore, the maxima at 1200 and 1350°C were not found.

To explain these effects let us consider the ideal case again. We assume that the broadening of the absorption band due to different processes is absent and, therefore each frequency in the transmission spectrum corresponds to one or other bond between the atoms of the implanted layer. On this basis, we can see that, immediately after the implantation the contour of the transmission curve covers a wide range of frequencies, i. e. in ion-implanted layer there are many different bonds that absorb at different frequencies. If one attributes the frequency of 800 cm^{-1} to the tetrahedral oriented Si-C-bond of length of 0.194 nm (bond characteristic of the silicon carbide), so in the implanted layer there are the systems with the bond lengths of both larger and smaller than this. In general, the presence of different bond lengths between the atoms of the ion-implanted layer is completely natural because atoms can stop at different distances from each other in the process of implantation.

In our case, the most interesting bonds are the single-, double- and triple silicon-silicon (Si–Si, Si=Si, Si≡Si), silicon-carbon (Si–C, Si=C, Si≡C) and carbon-carbon (C–C, C=C, C≡C) bonds presented in Fig.9. Simple covalent bond C–C, formed by the overlap of two sp3-hybrid electron clouds along the line connecting the centers of atoms, is σ-bond. One of the electron pairs in the double C=C bond forms σ-bond, and the second bond is formed by p-electrons with the clouds in the form of "eight", which overlapping, form a π-bond. Triple C≡C bond is a combination of one σ-bond and two π-bonds [22]. The lengths of single bonds are shown in proportion to ones which are characteristic for these bonds in a tetrahedral orientation, although they can have various values in the ion implanted layer. And in the case of double and triple bonds they may be either higher or lower than the values given. The length of a single bond of the same type of atoms was taken equal to twice the covalent radius of atoms, and the length of the Si–C-bond was taken as half the sum of double the values of covalent radii of Si- and C without correction for their ionicity. The lengths of double and triple bonds were taken at 0.021 and 0.034 nm shorter than a single bond, respectively. Triple bond between two atoms can be represented by two tetrahedra sharing a common face, and for the double bond - by two tetrahedra sharing a common edge [44].

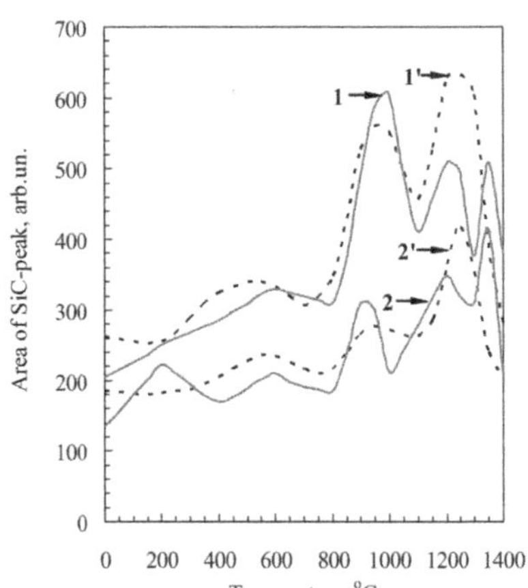

Figure 6. An area of the IR transmission peak for TO-phonons of SiC at perpendicular incidence of the radiation on the sample surface (curves 1, 1') and at 73° from a normal (curves 2, 2') for spectra presented in Figs.2 (curves 1, 2 - substrate Si(100)) and 3 (curves 1', 2' - substrate Si(111)), versus an annealing temperature.

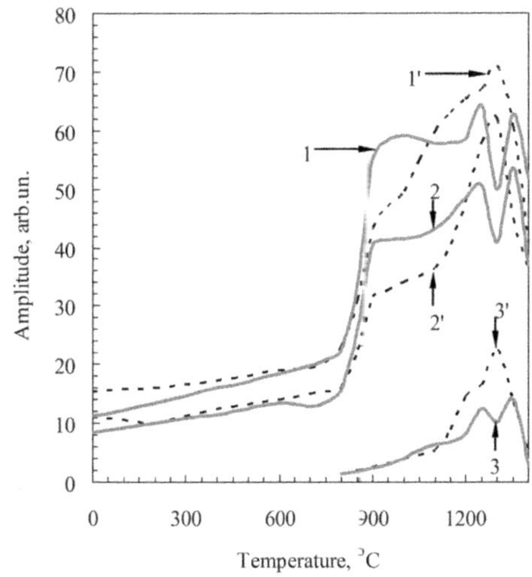

Figure 7. The IR transmission amplitude values for TO-phonons of SiC (curves 1, 1', 2, 2') at 800 cm^{-1} and for LO-phonons of SiC (curves 3, 3') at perpendicular incidence of the radiation on the sample surface (curves 1, 1') and at 73° from a normal (curves 2, 2', 3, 3') to surface for spectra presented in Figs.2 (curves 1, 2, 3 - substrate Si(100)) and 3 (curves 1', 2', 3' - substrate Si(111)), versus an annealing temperature.

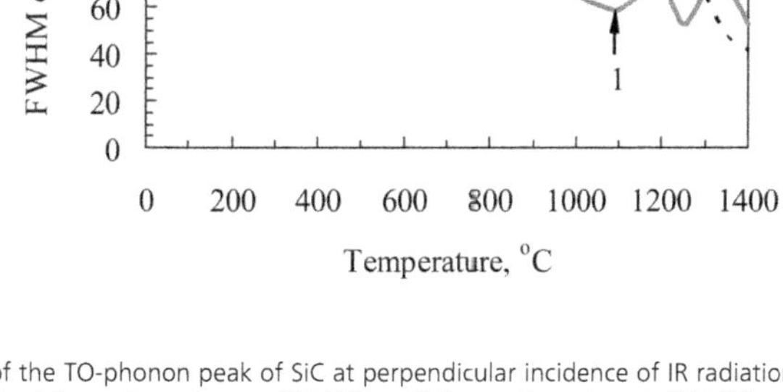

Figure 8. Half-width of the TO-phonon peak of SiC at perpendicular incidence of IR radiation on sample surface for spectra presented in Figs.2 (curve 1 – substrate Si(100)) and 3 (curve 1' – substrate Si(111)), versus an annealing temperature.

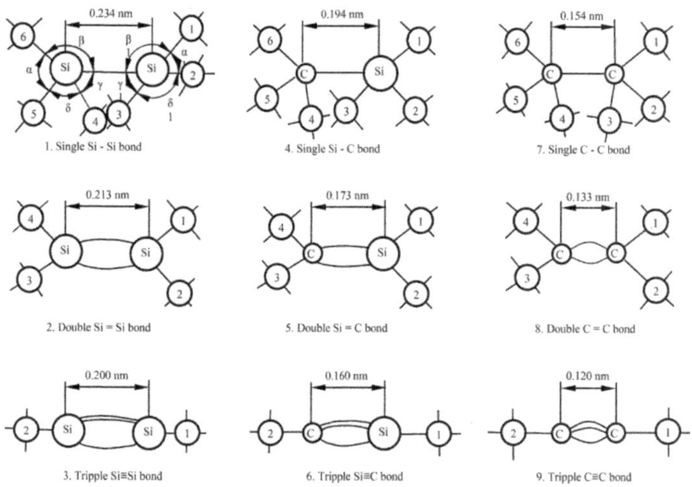

Figure 9. Various types of bonds between silicon, carbon atoms or their combination.

In the process of implantation of carbon into silicon vast majority of the covalent bonds of the substrate, starting with the amorphization threshold, are not covalent, because of violation of bond lengths and angles between them. However, among the formed Si–C-bonds there are tetrahedral bonds, the distances and angles between atoms of which correspond exactly to the crystallites of silicon carbide. This is confirmed by the presence of absorption at 800 cm^{-1} and by the results of the authors [26] who identified by electron diffraction the presence of silicon carbide crystallites immediately after the implantation of carbon into silicon.

We believe that the ion-implanted layer consists mainly of various combinations of the nine types of bonds, shown in Fig. 9. Moreover, it is possible the presence in the implanted layer of the single elongated bonds, sesqui, free ("dangling") and hybridized bonds, as well as resonances and other higher order interactions as well. By making these assumptions, we proceed not only from the contour of the spectrum of infrared transmission, covering a wide range of frequencies. We are basing also on the ability of carbon and silicon atoms to form besides single bonds also double and triple bonds [29, 44, 64, 46]. In paper [27], an aggregate of carbon atoms was named as a cluster. In this paper, as a cluster we have in mind all the nine types of bonds and combinations thereof, from which are formed during annealing a three-dimensional clusters and crystallites of Si and SiC.

Table 1 presents the values of the binding energy for all nine types of bonds, shown in Figure 9. The sum of the energies of two and three single C–C bonds are equal to 688 and 1032 kJ $mole^{-1}$, respectively, such that by 73 and 220 kJ $mole^{-1}$ higher than energy values of the C=C and C≡C bonds listed in Table 1. This suggests that the structures of the C=C and C≡C bonds for one bond has less energy than the structures with a single C–C bonds.

Types of bond	Binding energy, kJ mole^{-1}	Types of bond	Binding energy, kJ•mole^{-1}	Types of bond	Binding energy, kJ•mole^{-1}
Si–Si	187	Si–C	290	C–C	344
Si=Si	<374	Si=C	<580	C=C	615
Si≡Si	<561	Si≡C	<870	C≡C	812

Table 1. The values of binding energy for nine types of bond.

By analogy, reasonable to assume that clusters Si=Si, Si≡Si, Si=C, Si≡C per bond also has less energy than clusters with single Si–Si and Si–C bonds. Consequently, the energy of double and triple bonds Si=C, Si=Si, Si≡C and Si≡Si must be less than the sum of the energies of two or three single Si–Si and Si–C bonds, as is shown in Table 1. It is evident that the most strongly bonds are the carbon-carbon and then carbon-silicon and silicon-silicon clusters.

Let us consider the special features of the curves on Figs.6–8 basing on the assumptions presented above.

The temperature range 20-600°C

It is known [65] that a typical recrystallization temperature of amorphous silicon lies in the temperature range 500–600°C. When the dose of the implanted carbon ions is much higher than the amorphization threshold and, the implanted atoms combining with the silicon atoms can form the inclusions of new compounds of considerable volume, the crystallization of silicon, in the case of a Gaussian distribution profile of implanted atoms, starts at the surface and the interface "disturbed layer – substrate" and goes in the direction to the maximum of the carbon distribution with increasing annealing temperature. The silicon crystallites are formed in regions where the concentration of silicon atoms exceeds the concentration of carbon.

The values of concentration ratio N_C/N_{Si} (Fig.1) are small on the edges of distribution and, a significant number of silicon atoms falls at each implanted carbon atom. In this temperature range, such combinations of clusters in ion-implanted layer are decaying which consist mainly of bonds of Si–Si, Si=Si and elongated Si–C, as they have the lowest energy dissociation among the types of bonds listed above. As decay of the clusters, we mean such regrouping of the atoms in system and the change the lengths of chemical bonds and angles between them, which lead to the most energetically favorable state of system. In such state there is a system of atoms with tetrahedrally oriented bonds, which are the most stable and durable. All other types of bonds and their geometrical arrangements are energy unprofitable; they are insufficiently stable and can decay during annealing.

As it is seen from curves 1 and 1′ in the Figs. 6 and 7, the increasing of both an area of SiC-peak of IR transmission and its amplitude at 800 cm^{-1} takes place over the temperature range 20–600°C, i.e. the formation of SiC crystallites takes place at temperatures significantly less that it is necessary for the SiC formation by a thermal growth. The mechanism of this phenomenon is of interest.

In the system "implanted layer – substrate", the phonons are generated in the process of annealing. This system is a single entity and, it is reasonable to assume that between the implanted layer and the substrate there is a continuous interaction of phonons. Since the thickness of the substrate is much greater than the thickness of the implanted layer, then, with respect to the layer, the substrate can act as a huge reservoir of phonons, which able, due to its heat capacity, receive or deliver the phonons to the implanted layer. As the implanted layer is thermodynamically nonequilibrium system, the process of interaction of phonons with the atoms will go towards reducing the free energy of the system. During each collision of a phonon with cluster of the implanted layer, the phonon absorption will occur if the energy of the system will decrease.

Energy released during the decay of clusters, is transferred to the lattice, which can accumulate and transfer it to another cluster, followed by the formation of energetically favorable system, in this case, the crystallites of Si and SiC. It is also possible direct transfer of energy of the decaying Si–Si-cluster to other types of clusters. In this temperature range, mainly, weakly bound Si–Si-clusters can decay during the interaction with phonons. More energy requires for the formation of crystallites of Si and SiC. This energy is transfered to the reacting atoms by the lattice. The considered above mechanism of formation of crystallites of Si and SiC, the so-called over-barrier mechanism, when the interacting atoms overcome the energy barrier of height $E \cong E_n$, is shown in Fig.10. This mechanism of formation of tetrahedrally oriented bonds Si–Si and Si–C from an energy point of view is advantageous for the crystal lattice, since it thereby reduces its free energy.

All possible mechanisms of the tetrahedral oriented Si–Si- and Si–C-bonds formation should be energy profitable for the crystalline lattice to decrease free energy of system. The formation of crystallites of Si and SiC in the range 20–600°C occurs mainly due to the decay of clusters such as the longest Si–Si- and Si–C-bonds, but also partly due to the disintegration of other types of clusters. A number of studies have shown [5, 26], that immediately after the implantation, the presence of a minute quantity of SiC crystallites with hexagonal structure is observed in the ion implanted layer and, they are transformed into β-SiC at the temperature of 400°C and higher. It is impossible to identify the presence of Si crystallites due to their optical inactivity in this range of the infrared absorption. However, their presence is not in doubt, as much less energy is required to expend for their formation than for the formation of SiC crystallites at the implementation of over-barrier mechanism. As we showed earlier, in layers $SiC_{0.03}$ with a low carbon concentration the crystallites of Si increase their sizes from 2 to 3 nm in the range 20–700°C (Fig.11).

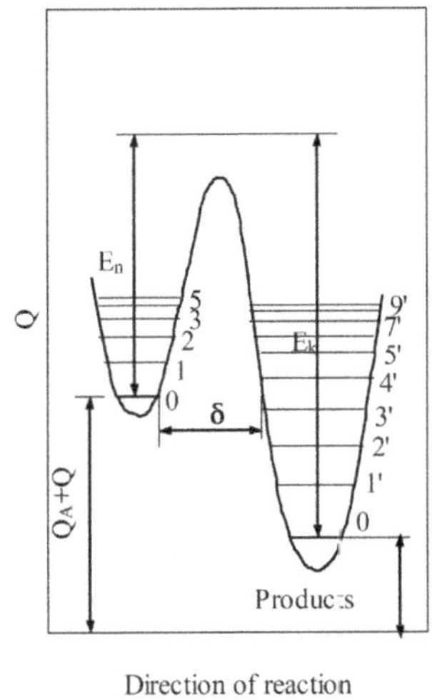

Figure 10. Illustration of the over-barrier mechanism of the formation of the tetrahedral oriented bonds

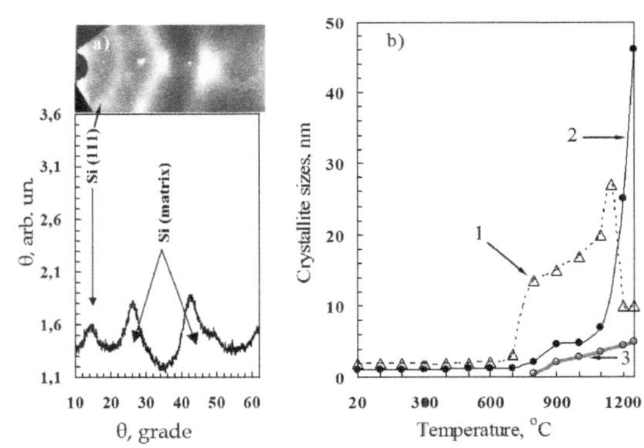

Figure 11. X-ray diffraction patterns of the $SiC_{0.12}$ layer after implantation (a) and average sizes of crystallites in the (111) plane after implantation and annealing (b): 1 – Si (for layer $SiC_{0.03}$), 2 – Si (for layer $SiC_{0.12}$), 3 – SiC (for layer $SiC_{0.12}$).

Peak area immediately after implantation is not zero, i. e. a part of the carbon atoms is included into composition of the optically active clusters (Fig.6). If we assume that, after annealing at 1000-1250°C almost all carbon atoms are optically active and optically inactive clusters broke up, we see that immediately after the implantation of carbon into the (100) and (111) oriented silicon at least 65% and 60% of carbon atoms were concentrated in optically inactive clusters, respectively, if the implantation was carried out by a dose sufficient to obtain the stoichiometric concentration (E = 40 keV, D = 3.56×10^{17} cm^{-2}). A number of carbon atoms is included into stable types of optically inactive clusters which stable up to melting point of silicon. It is known [44], that the optically inactive objects consist of the clusters and their chains that lie in one plane. The formation of clusters and chains of planar systems of nets may be due to energy considerations. For example, in [5] the formation of alternating layers of single crystal silicon with amorphous silicon precipitates enriched with carbon in the ion-implanted layer, attributed to the fact that the system in such a way reduces its free energy.

Spatial pattern of ion-implanted layer is difficult to model, since a Gaussian distribution profile of implanted atoms is characterized by the change by depth of the concentration of carbon atoms N_C/N_{Si} and, thus, the mechanism of physical processes from one layer to layer is changed. We can construct a flat infrared inactive net in the middle of layer where $N_C/N_{Si} = 1$. Flat optical inactive net consisting of C and Si atoms, linked by single, double and triple bonds, may also contain free ("dangling") bonds of the silicon and carbon atoms. These bonds may connect to atoms of the other flat net or on the association of the atoms which do not lie in one plane. The atoms which do not lie in one plane, can form an association of optically active clusters.

With increasing annealing temperature at first the decay of elongated single bonds at two atoms united by a triple bond, is taken place. These pairs of atoms inhibit the diffusion of atoms in the layer. Then, the energetically unfavorable single bonds of atoms are disintegrated in the planar nets of clusters. The subsequent increase in annealing temperature would lead to the disintegration of the planar nets forming a number free carbon and silicon atoms, as well as pairs of Si and C atoms, linked together by multiple bonds. Free carbon and silicon atoms can move on short distances and join to form the crystallite Si or SiC, which is profitable from an energy point of view and, as a result, the energy of system is decreased.

In addition to the planar nets of clusters, the ion-implanted layer may contain long chains of clusters, which are also optically inactive. The chains can be formed by alternating different types of bonds, and their degradation temperature may be different. There are also local clusters non-interacting with the surrounding atoms and consisting of three-, four- and more atoms linked together by double bonds. They are most stable clusters due to the full richness of their bonds. Perhaps these clusters are not disintegrated up to the melting point of layer.

In Fig.12 the IR transmission amplitude versus the annealing temperature for different frequencies is shown. It is evident that clusters absorbing at frequencies of 850 and 900 cm^{-1}, in the range 20–600°C did not disintegrated, as their amplitude remains unchanged. The position of minimum of IR transmission peak does not change and is located at 757 cm^{-1} (Fig.4). This indicates the dominant role of one type of clusters, which absorb at 757 cm^{-1}. The increase of the amplitude of the peak (Fig. 12, curve 3) indicates an increase in the concentration of these clusters with increasing annealing temperature. At the same time the concentration of clusters, which absorb at 700 cm^{-1} and correspond to the elongated single bond, is simultaneously increased. The energy of both the formation and decay of these bonds is a least one ($E = h\nu$), as they absorb the radiation of lowest frequencies among considered. It follows from Fig.7 (curve 1) and 12 (curve 1), that the bonds being very similar to tetrahedral bonds of SiC which absorbs at 800 cm^{-1}, are formed in the implanted layer. That follows also from decrease of peak halfwidth (Fig.8) in temperature range 20–400°C. Its increase in range 400–600°C may be associated with intensive restructuring of Si–Si bonds of amorphous silicon before recrystallization at the surface and near the substrate, where the concentration of carbon is low. The increase in peak area of the TO-phonon SiC in this range (Fig. 6, curves 1, 1 ') is caused by an increase in the number of tetrahedral bonds and close to them, which absorbs in the range 750-850 cm^{-1} (Fig.12, curves 1, 3, 5).

No significant differences between the properties of the films on the substrate orientation (100) and (111) in this temperature range were observed, except for the fact that the area of SiC-peak and amplitude at 800 cm^{-1} are slightly higher for the orientation (111), which indicates a higher number of tetrahedral bonds of SiC and a smaller number of optically inactive clusters.

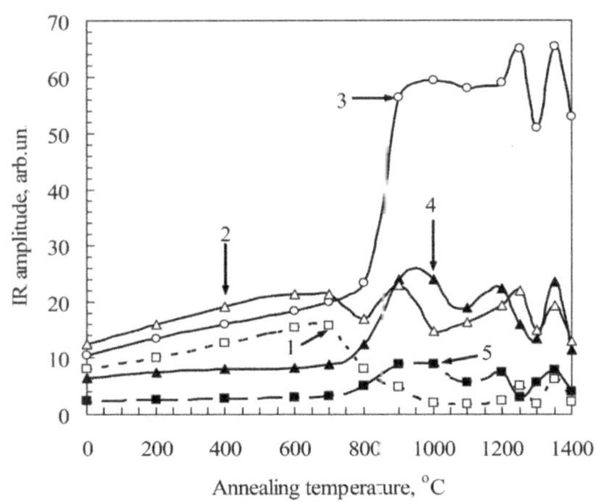

Figure 12. Amplitude of IR transmission for various wavenumber values versus an annealing temperature. (1 - □) 700 cm^{-1}, (2 - Δ) 750 cm^{-1}, (3 - ○) 800 cm^{-1}, (4 - ▲) 850 cm^{-1}, and (5 - ■) 900 cm^{-1}.

The temperature range 600-800°C

The area under IR spectrum curve is decreased in this temperature range (Fig.6) due to the decay of the optical active clusters absorbing at the frequencies close to 700 and 750 cm^{-1} (Fig.12). As shown in previous studies (Fig. 13d, e), this is due to the decay of elongated Si –C-bonds in the layers with a low concentration of carbon. An intensive process of disintegration of these bonds occurs in layers $SiC_{0.4}$ and $SiC_{0.12}$ between the surface and the maximum of the carbon distribution. In addition of the decay of the optical active non-tetrahedral Si–C-bonds, seemingly, deformed Si-Si bonds are disintegrated, too.

The intensive rearrangement of clusters is characterized for this temperature range. As a result of multiple collisions the atoms of clusters, successively passing from the initial- through the intermediate states to the most energy favorable end position, form the tetrahedrally oriented bonds of Si- and SiC crystallites. A significant change of halfwidth of the Si–C-peak of IR spectrum begins for film on (100) oriented substrate (Fig.8). In case of (111) oriented silicon substrate the same is taken place some later. A certain increase of the concentration of clusters absorbing on the wavenumbers ranged from 850 to 900 cm^{-1} (Fig. 12, curves 4, 5) is simultaneously taken place. The ordering of layer structure in region near the substrate is taken place, too.

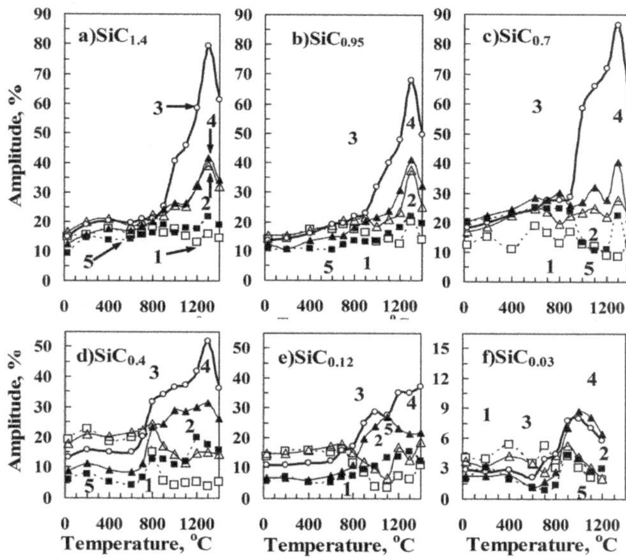

Figure 13. Effect of the annealing temperature on the IR transmission amplitude at wavenumbers of (1-□) 700 cm^{-1}, (2-Δ) 750 cm^{-1}, (3-○) 800 cm^{-1}, (4-▲) 850 cm^{-1}, and (5-■) 900 cm^{-1} under normal incidence of IR radiation on the sample surface: a) $SiC_{1.4}$; b) $SiC_{0.95}$; c) $SiC_{0.7}$, d) $SiC_{0.4}$, e) $SiC_{0.12}$; f) $SiC_{0.03}$.

The temperature range 800-1000°C

Accordingly to Figs. 6-8 and 12 almost whole ion implanted layer takes place in the process of crystallization of Si and SiC in this temperature range. The probability of over-barrier mechanism of the formation of Si- and SiC crystallites, seemingly, is increased as is seen from the significant increase of the IR transmission amplitude at 800 cm^{-1} (Fig.7, curves 1, 1′). The flat net of clusters and the chains of them in a great extent are disintegrated. The intensive decay of the infrared active non-tetrahedral Si–C-bonds is taken place and, the sesquialteral- and, partially, the double silicon-silicon bonds simultaneously with the infrared inactive single bonds can disintegrated as well. Seemingly, the energy of the phonons may be sufficient for the disintegration of the infrared inactive C–Si-, C–C- and even C=Si-bonds. Simultaneously, the formation of the most energy favorable tetrahedrally oriented bonds of Si- and SiC-crystallites is taken place (Figs. 6, 7, curves 1, 1′; Fig.12, curve 3). The continious process of the formation of the infrared active clusters absorbing on the frequencies close to

700–800 cm^{-1} is taken place in the layer and, the concentration of these clusters with the increasing of the annealing temperature is increased. The absorption at 800 cm^{-1} begins significantly predominate over the ones at another frequencies. That leads to the significant increase of the IR transmission amplitude at this frequency in comparison with the increase of amplitude at another ones (Fig.12, curve 3) and, that is perceived as a frequency shift of the IR absorption maximum (Fig.4). The increase of concentration of clusters absorbing on frequencies higher than 800 cm^{-1} is observed, too (Fig. 12, curves 4, 5).

So, the increase of the area under the IR transmission curve in the temperature range 800-1000°C is caused, mainly, by the absorption at frequency of 800 cm^{-1}, i.e. by bonds characteristic to the SiC-crystallites and, by bonds absorbing on frequencies both more and less than 800 cm^{-1} as well. In the case of the (100) oriented substrate the number of tetrahedral Si –C-bonds reached at 1000°C some maximum and does not change up to 1200°C, whereas in the case of orientation (111) it increases going smoothly in the range 900-1300°C. Comparison with the data in Fig.13 shows that such flat areas of curves at these temperatures are typical for the layers $SiC_{0.7}$, and especially for $SiC_{0.4}$. The total dose of implanted ions of carbon in the case of $SiC_{0.7}$ was $D(SiC_{0.7}) = 4.54\times10^{17}$ cm^{-2}, and $D(SiC_{0.4}) = 2.72\times10^{17}$ cm^{-2}, and is comparable to the dose of carbon ions with an energy of 40 keV for the considered Gaussian distribution of carbon: D (40 keV) = 3.56×10^{17} cm^{-2}. The halfwidth of IR-spectrum maximum (Fig.8) in the range 800-1000°C is rapidly decreased. That is an evidence of a significant ordering of the ion implanted layer structure caused by the formation of Si- and SiC-crystallites. It goes more intensively in case of (100) orientation of substrate.

The temperature range 1000-1100°C

In spite of the fact that the formation of new SiC crystallites in the implanted layer at these temperatures is not taken place (Fig.7, curve 1, 1'; Fig.12 and 16, curves 3), the decrease of area of SiC-peak of IR transmission curve is significant (Fig.6, curves 1, 1'). As it was shown earlier (Fig.11), the dimensions of Si- and SiC-crystallites are enlarged with the increase of the annealing temperature. Thereby, we believe that in this temperature range the uniting of small crystallites of Si and SiC in the larger ones is taken place, resulting the frequency shift of LO-phonon peak in IR spectrum to a higher frequency (Fig. 4), as well as the growth of its amplitude (Fig. 7, curve 3). When combining the crystallites, in the area of their union the decay of the both optically active and inactive clusters is taken place. This explains the decrease in the amplitude of the infrared transmittance for clusters absorbing at frequencies of 800-900 cm^{-1} (Fig. 12) and the corresponding decrease in the area (Fig. 6). Apparently, this temperature is insufficient for the decay of clusters C≡Si and C=C and, as a result the formation of new SiC crystallites is not observed. However, a volume of polycrystalline Si is continuously increased due to the disintegration of Si-Si, Si=Si bonds in regions with low concentration of carbon (Fig.1, regions I and II).

In the case of the substrate orientation Si(111), the SiC-peak area is reduced to a lesser extent, and is larger after annealing at 1100°C than the peak area for the substrate Si(100) due to the fact that the number of tetrahedral bonds and crystallites continues to grow.

The temperature range 1100-1200°C

Both the area of Si–C-peak (Fig. 6) and its half-width (Fig. 8) increase in this temperature interval, while the volume of the polycrystalline SiC is unchanged (Fig. 12, curve 3 and Fig. 7, curve 1), as the further growth of SiC crystallite size due to their association (Fig. 4, curves of LO-phonons) is taken place. The growth of the area and half-width of Si–C-peak are caused by the formation of new optically active clusters absorbing at frequencies of 750-900 cm^{-1} (Fig. 12) due to the decay of stable clusters. As a result, the ion-implanted layer degrades the structure (increase in half-width of the peak).

The temperature range 1200-1250°C

No significant changes in the formation of new infrared active clusters over this range are taken place. The decaying clusters with short Si–C-bonds, absorbing at frequencies of 800–900 cm^{-1} (Fig. 12, curves 1-3) are converted into clusters with long bonds, absorbing at 700 and 800 cm^{-1} (Fig. 12, curves 4 and 5). Therefore, the area of Si–C-peak is not changed (Fig.6) for both orientations of substrate (100) and (111). The annealing temperature 1250°C may be sufficient for the decay of sesqui- and double Si–C-bonds. As a result, the concentration of tetrahedrally oriented Si–C-bonds increases, the atoms are combined into crystallites of SiC, and the amplitude of the IR spectrum at 800 cm^{-1} increases (Fig. 12, curve 3). There is a further streamlining of the structure of the ion-implanted layer (Fig. 8), both due to the formation of new crystallites of SiC, as well as due to an increase in their size (Fig. 4, curve of LO-phonons).

The temperature range 1250-1300°C

Seemengly, there may be competing processes here. Firstly, in the case of the substrate Si(100) in this interval there is a decay of a large number of tetrahedral bonds (Fig. 7, curve 1), which is the main reason for reducing the area of SiC-peak at 1300°C (Fig. 6, curve 1). At the same time the amplitude of the LO-phonon is decreased (Fig. 7, curve 3) and the half-width of the peak is increased (Fig. 8, curve 1), indicating a deterioration of the structure. In contrast, in the case of the substrate Si(111) the peak area (Fig. 6, curve 1') and the amplitude at 800 cm^{-1} (Fig. 7, curve 1') are increased, and this is accompanied by a decrease in the half-width of the TO-phonon peak and an increase in amplitude of the LO-phonon peak (Fig. 7, curve 3'), i.e. by the improving of the layer structure. We assume that there may be two dominant mechanism of the influence of substrate orientation on the layer structure at high temperatures. During the recrystallization of the damaged layer in the interface "the SiC film – Si substrate", both a destruction of the silicon crystallites and the uniting of their atoms with the substrate are taken place. The difference of the recrystallization of the substrate Si(100) may be the appearance of forces and conditions for the destruction of the defective crystallites of silicon carbide. This leads to a decrease in amplitude at 800 cm^{-1} and increase the half-width of the peak due to the appearance of non-tetrahedral Si–C-bonds. The second mechanism may be associated with different concentrations of carbon near the surface of silicon. After implantation into (100) oriented Si substrate, the carbon-riched surface layer is much thicker than in case of Si(111) substrate, so the sublimation and desorption of carbon at high temperatures will lead to a significant decrease in the amplitude values of the SiC-peak at all frequencies. Ie, the experiments to study an influence of substrate orientation on the desorption of implanted carbon are necessary.

The temperature range 1300-1350°C

Despite the increase in the desorption of carbon, the area of SiC-peak (Fig. 6), as well as the amplitude at all frequencies of spectra are increased in case of the film on the substrate Si(100) (Fig. 12, curves 1-5). The atoms of clusters, which formed due to the destruction of defective crystallites of SiC in the previous temperature range, re-unite again in form of the crystallites of SiC, as well as in form of optically active clusters, which absorb at frequencies near 800 cm^{-1}. In addition, stable clusters with Si=C, Si≡C and C=C bonds are disintegrated (Fig. 9). Growth of SiC crystallite size leads to a frequency shift of LO-phonons peak in the short-wavelength region (Fig. 4). Since in an isolated system unacceptable the processes occurring with increasing free energy, the uniting of two crystallites occurs, if is accompanied by a gain in energy in comparison with the energy expended in their decay. In the case of the orientation of the substrate Si(111) a decrease of the area of SiC-peak (Fig. 6, curve 1'), as well as the amplitude at 800 cm^{-1} (Fig. 7, curve 1') occur due to increased desorption of carbon.

The temperature range 1350-1400°C

Although at these temperatures the decay of optically inactive clusters, formation of both new crystallites and optically active SiC-clusters should be the greatest, nevertheless the growth of area and the amplitude of the SiC-peak of IR spectrum is not observed. On the contrary, they decrease (Fig. 6 and 7), which can be explained to the dominant influence of sublimation and desorption of carbon. The volume of polycrystalline SiC is also reduced, which is accompanied by a decrease in the amplitude of LO-phonons (Fig. 7, curve 3). A further ordering of the structure of ion-implanted layer occurs, as evidenced by the decrease in the peak half-width of IR spectrum.

In conclusion, it is necessary to note that the quantity of absorbing Si–C-bonds in the silicon layer with Gaussian distribution of implanted carbon reaches a maximum at 1000°C for (100) oriented substrate and, at 1000 and 1250°C for (111) oriented substrate (Fig. 6). Most of the carbon atoms combine with atoms of silicon, forming the tetrahedrally oriented bonds of SiC (Fig. 12, curve 3). Seemingly, there are flat nets and chains of clusters (Fig. 9), which consist mainly of bonds Si–Si, Si = Si, C–Si, C–C and, this temperature is sufficient for their decay. Some significant part of the carbon atoms form bonds of higher order, which decay at temperatures of 1200-1400°C and above. At high temperatures, 1300-1400°C (Figs. 6 and 7) occur intense desorption processes of carbon.

A shape of IR transmission peak

All presented spectra (Fig.2) have shape different from simply dispersive spectrum (theoretically calculated). The transmission band both left and right from the transmission peak are perceptible asymmetric and, one can not describe it's shape by a simple analytical function. The shape asymmetry is decreased with the annealing temperature increasing and, it is minimal in the temperature range 1300–1350°C. The further increase of the annealing temperature leads to the increase of asymmetry, too. The contour of the transmission peak for TO-phonons at perpendicular incidence of electromagnetic radiation on sample surface after annealing at 1350°C is most close to the dispersive one.

Obviously, the asymmetry of IR transmission contour is related with the presence of the infrared active clusters in the ion implanted layer, and the concentration of clusters is minimal when the asymmetry is minimal, i.e. at 1350°C. In this relation, a largest area of SiC-peak corresponds to maximum amplitude of absorption at wavenumber 800 cm^{-1} (Figs.7 and 8). As is seen from amplitude values in Fig.12 (curves 1, 2, 4, 5), there is a certain quantity of the non-tetrahedral optical active clusters at 1350°C. Seemingly, a presence of very stable optical inactive clusters, which are not disintegrated even at the melting point of Si, is possible.

Measurement of a conduction type of carbon implanted silicon layer

The (100) oriented substrates of n- and p-Si of dimensions 7×5×0.3 mm^3 with resistivities 4–5 Ом см have been implanted by carbon ions with values of energy 40 keV and dose 3.56×10^{17} cm^{-2} to determine a type of conduction. After implantation, the samples have been isochronously annealed in vacuum over the temperature range from 200 up to 1200°C with step 200°C for 30 min. A surface layer of the annealed samples have been removed by etching in an acid mixture HF:HNO_3 in composition of 1:10. A type of conduction of the implanted surface has been determined using thermo-emf after each 0.5 mm along both horisontal and vertical directions. The thermo-emf have fixed with approximately equiprobability the both n- and p-type of conduction on n-Si substrates, while on the p-Si substrates the thermo-emf have shown the p-type of conduction only. We believe that p-type of conduction on the n-Si substrates is provided by the SiC crystallites. The conduction of the Si crystallites is similar to the conduction of the substrate. If the substrate is p-Si, so the both Si- and SiC-crystallites have the p-type of conduction. So, the synthesized SiC-crystallites have p-type of conduction independently from the type of substrate.

3.2. Investigation of high-temperature instability of solid SiC films synthesized by ion implantation

As stated in paragraph 2, for construction of a rectangular profile of the distribution of carbon atoms in the silicon, the implantation of carbon ions of different energies and doses in the second group of single-crystal silicon wafers of n- and p-type conductivity was carried out sequentially in the order according to Table 2. The doses of ions were chosen in such a way to obtain a layer $SiC_{0.7}$ with the ratio of the concentrations of carbon and silicon atoms through a depth of about N_C/N_{Si} = 0.7. Postimplantation annealing of the samples was performed in a vacuum in the temperature range 200-1200°C for 30 min with a step 200°C. In some cases, to compare also were analyzed the samples with films $SiC_{0.95}$.

E, keV		40	20	10	5	3
D($SiC_{0.7}$), 10^{17} cm^{-2}		2.80	0.96	0.495	0.165	0.115
D($SiC_{0.95}$), 10^{17} cm^{-2}		4.48	1.54	0.792	0.264	0.184
N_C(Gibbons) profile [21]	$R_p(E)$, nm	93.0	47.0	24.0	12.3	7.5
	$\Delta R_p(E)$, nm	34.0	21.0	13.0	7.0	4.3

Table 2. Values of energy, E, dose, D, projected range, $R_p(E)$, and straggling, $\Delta R_p(E)$, for $^{12}C^+$ ions in Si, used for constructing a rectangular distribution profiles $SiC_{0.7}$ and $SiC_{0.95}$.

3.2.1 Influences of annealing, sputtering and the film composition changes during high dose implantation on the thickness and shape of the distribution profile of carbon atoms

Fig. 14 shows the calculated profile N_C(Gibbons) of distribution of carbon atoms through the depth of silicon for the energies and doses of ions according to Table 2, which is the sum of Gaussian distributions constructed with the use of $R_p(E)$ and $\Delta R_p(E)$ by [21] (LSS) in accordance with the expression:

$$N(x)=\frac{D}{\Delta R_p(2\pi)^{1/2}}\exp[-\frac{(x-R_p)^2}{2\Delta R_p^{\ 2}}] \tag{5}$$

where x – the distance from the surface.

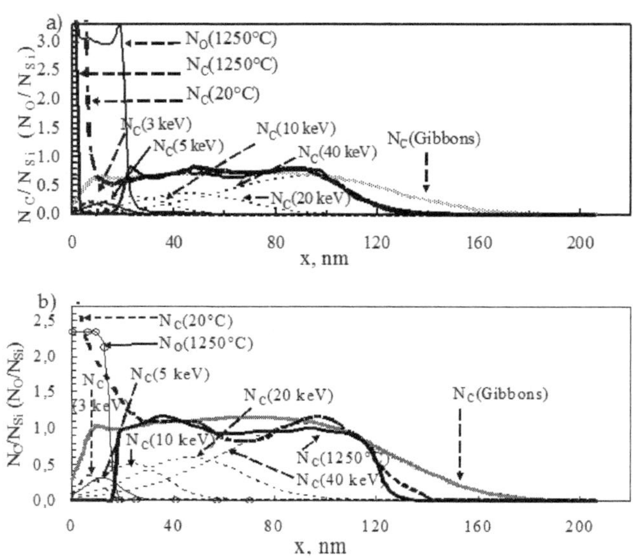

Figure 14. distribution profiles in Si produced by ion implantation (see Table 2). (a) $SiC_{0.7}$; (b) $SiC_{0.95}$; N_C(Gibbons) is the profiles calculated according to [21], where N_C(Gibbons) = N_C(40 keV) + N_C(20 keV) + N_C(10 keV) + N_C(5 keV) + N_C(3 keV). N_C(20°C), N_C(1250°C) and N_O(1250°C) are the Auger profiles of carbon and oxygen, respectively, in a layer after high-dose implantation and annealing at 1250°C for 30 min.

Fig. 14 also shows the experimental curves (Fig. 14, curves N_C(20°C), N_C(1250°C) and N_O(1250°C)), obtained by Auger electron spectroscopy, showing the ratio of the concentrations of carbon and oxygen atoms to silicon (N_C/N_{Si} и N_C/N_{Si}) through the depth of the sample after implantation (20°C) and annealing at 1250°C for 30 min in an argon atmosphere containing some oxygen. These distributions are constructed through a depth, taking into account the condition that the number of carbon atoms in silicon and, consequently, the integrals and the area under the curves N_C(Gibbons) and N_C(20°C) must be equal one another at a first approximation. The areas under the curves were equal to $S_G = S_{20°C} = \int (N_C / N_{Si}) dx = 90$ units (or 100%), and after annealing at 1250°C: $S_{1250°C} = 71$ units (or 79.4%) due to the formation of silicon oxide layer. When evaluating the number of carbon atoms in a thin surface region (8 nm), where N_C/N_{Si} is very great due to the low content of silicon atoms ($N_{Si} \ll 5\times10^{22}$ cm^{-3}), an approximation was made that the N_C/N_{Si} does not exceed 2.3 ($N_{graphite} = 11.6\times10^{22}$ cm^{-3} and $N_{silicon} = 5\times10^{22}$ cm^{-3}). At the same time, the area under the profile curve for the region x> 22.2 nm were estimated $S_G = 78$ units, $S_{20°C} = 66$ units and $S_{1250°C} = 65$ units. That is, the areas under the profile curves before and after annealing for x > 22.2 nm are almost equal, but less than calculated value, since part of the carbon atoms after implantation was concentrated near the surface (x <8 nm), and during annealing occur desorption of carbon from the layer (x<22.2 nm) and the formation of silicon oxide. The interface "the SiC film - Si substrate" in the experiment was more abrupt than it was expected. After annealing for 30 minutes, almost 20% of the total number of carbon atoms desorbed from the carbon-rich surface layer of the film. Fig. 14 shows that the average concentration of carbon and oxygen were: $N_C/N_{Si} = 0.7$ in the depth 22.2 < x < 110 nm and the $N_O/N_{Si} \approx 3.0$ at the surface layer x < 22.2 nm. In this case there is penetration of oxygen atoms into the layer up to 30 nm.

Some difference between the shape of the experimental and calculated curves of the profile is observed (Fig. 14). The distribution N_C(Gibbons) was made without taking into account the effects of sputtering and composition changes in the layer by high dose implantation. Accounting for the effect of surface sputtering during high dose of implantation of carbon ions (E = 40 keV, D = 2.8×10^{17} cm^{-2}) allows to assume the displacement of profile further into the layer with increasing dose, to some expansion of the profile and, consequently, to reduce the carbon concentration at the peak of the distribution in comparison with the calculated value. However, changing the composition of the single-crystal silicon substrate up to a mixture of C and Si atoms during the implantation suggests the formation of a significant amount of double and tripple Si–C- and C–C-bonds, which are more strong than the Si–Si-bonds, as well as the formation of stable carbon and carbon-silicon clusters. This results a decrease of $R_p(E)$ and $\Delta R_p(E)$ during implantation.

The decrease of $R_p(E)$ decreases the influence of surface sputtering on the position of the distribution maximum of carbon atoms, i.e., the maximum should remain nearly on the same depth. Moreover, the decrease of $\Delta R_p(E)$ will increase the carbon concentration at the maximum of peak and a more sharp decrease in concentration in the direction to the substrate and to the surface. This will cause a decrease in the depth of the interface "film SiC – the substrate Si», which becomes more sharp with increasing dose, as well as an occurrence of depression between peaks 40 and 20 keV, and possibly between 20 and 10 keV. The appearance of depression between peaks 40 and 20 keV in Fig. 14 for layers $SiC_{0.7}$ and $SiC_{0.95}$ may be due to these reasons.

The surface sputtering during the implantation of carbon ions with energies of 10, 5 and 3 keV is more intense with decreasing ion energy. This should lead to an increase in carbon concentration near the surface due to shift of the distribution maxima of these ions one to another (N_C(3 keV) and N_C(5 keV) in the direction of N_C(10 keV)). As a result, significant increase in the concentration of carbon in the surface layer is observed.

It is seen in Fig.14b that the average concentration of carbon and oxygen were: $N_C/N_{Si} = 0.95$ in the depth of the layer from 20 to 110 nm and the $N_O/N_{Si} \approx 2.33$ at the surface layer up to a depth of about 20 nm. At the same time observed the penetration of oxygen up to 80 nm to the depth. It was found for a layer $SiC_{0.95}$ that $S_G = S_{20°C} = \int (N_C / N_{Si}) dx = 144$ units (or 100%), and after annealing at 1250°C: $S_{1250°C} = 103$ units (or 71.3%). Thus, it appears that after annealing for 30 minutes, almost 30% of the carbon atoms desorbed from the surface layer of the film. At the same time the appearance of a layer of oxygen atoms at the surface is revealed. The area under the profile curve for the region x > 25 nm were estimated $S_G = 121$ units, $S_{20°C} = 99$ units and $S_{1250°C} = 95$ units. i.e., the area under the curves of the profile before and after annealing for region x > 25 nm are similar in magnitude, but again less than calculated one, since part of the carbon atoms after implantation was concentrated near the surface (x < 19 nm), and after annealing, there was desorption of carbon from layer (x <25 nm) and the formation of silicon oxide. The interface "the SiC film – Si substrate" in the experiment also was sharper than the expected one.

The presence of a sharp interface "the SiC film – Si substrate" permits to suppose that is possible to obtain promising results on the measurement of film thickness by X-ray reflectometry, although this method is typically used for films deposited with a very sharp interface "a film - substrate" and for ion-implanted layers usually does not apply. The parameters of the $SiC_{0.7}$ film by this method were investigated at small grazing angles θ by recording the angular dependence of the reflection coefficient using two spectral lines CuK_α (0.154 nm) and CuK_β (0.139 nm) on the installation "ComplexXRay C6" [61]. The oscillations of intensity were observed, assigned to the interference of X-ray reflections in the layers $SiC_{0.7}$ and SiO_2 (Fig. 15).

The first maximum of reflection with intensity $I_1 = 93207$ pulses at an angle of $2\theta = 0.418°$ is observed (Fig. 15). The angle of total external reflection is evaluated as an angle where the intensity is equal to a half of the first maximum ($I = I_1/2 = 46603$ pulses), ie $2\theta_c = 0.449°$, or $\theta_c = 0.2245° = 3.918$ mrad. Using the Henke program is determined that this value of θ_c corresponds to the value of film density 2.37 g/cm^3, which is close to the density of cristobalite (SiO_2) 2.32 g/cm^3. Further, with increasing of the incidence angle, the intensity of reflection increases again up to $I_2 = 76831$ pulses and that indicates the presence of a more dense structure. If the intensity falls up to the value $I = I_2/2 = 38415$ pulses, the value $2\theta_c = 0.486°$, the critical angle is equal to $\theta_c = 0.243° = 4.241$ mrad, which corresponds to a density 2.77 g/cm^3 and is close to the density of quartz (SiO_2) 2.65 g/cm^3. As shown in Fig. 15, then there is a second increase in intensity up to $I_3 = 34416$ pulse which corresponds to a denser structure. If the intensity falls up to the value $I = I_3/2 = 17208$ pulse, the value $2\theta_c = 0.526°$, and $\theta_c = 0.263° = 4.590$ mrad. This corresponds to a density $\rho = 3.25$ g/cm^3, which is close to the density of silicon carbide - 3.2 g/cm^3.

The layer thickness is determined by the formula $2d \sin\theta = \lambda$, or taking into account the small values of θ: $d = \lambda/2\theta$ nm, where λ - the wavelength of CuK_α (0.154 nm) or CuK_β (0.139 nm) radiation, and 2θ $_{av}$ was determined as an average from several (j - i) peaks (Table 3). To determine the thickness, four narrow peak of SiC, and two broad bands of SiO_2, probably from two different phases - cristobalite and quartz (Fig. 15), were used. The second broad band consists of 3 bands. The thickness of the resulting system (SiO_2 – $SiC_{0.7}$ – Si) was about 100 nm.

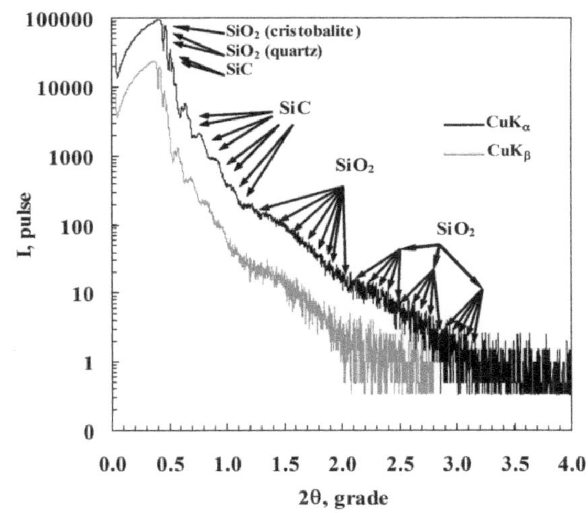

Figure 15. X-ray reflectometry using two spectral lines CuK_α (0.154 nm) and CuK_β (0.139 nm) ("ComplexXRay C6") of parameters of the $SiC_{0.7}$ films, synthesized by multiple implantation of carbon ions with energies of 40, 20, 10, 5 and 3 keV into silicon, after annealing at 1250°C.

Layer	$(2\theta)_j$	$(2\theta)_i$	j – i	$2\theta_{av}=[(2\theta)_j-(2\theta)_i]/(j-i)$	λ	d = λ/2θ, nm
SiC	1.138	0.598	4	0.135	0.15405	65.4
SiO_2	3.154	2.012	3	0.38	0.15405	23.2
SiO_2	2.012	1.178	1	0.83	0.15405	10.6

Table 3. Determination of the thickness of the layers in the system (SiO_3 – $SiC_{0.7}$ – Si) by X-ray reflectometry according equation 2 d• sin θ = λ.

Similar measurements for the $SiC_{0.95}$ layer also led to the observation of the intensity oscillations of X-ray reflections. The first maximum of reflection with intensity I_1 = 98703 pulses at an angle of 2θ = 0.396° is observed (Fig. 16). The critical angle of total external reflection is evaluated as an angle where I = I_1/2 = 49352 pulses, but in this case in the position of a minimum I = 53961 pulse, i.e. $2\theta_c$ = 0.458°, and θ_c = 0.229° = 3.979 mrad. This angle corresponds to the film density 2.46 g/cm³, which is close to the density of optical glass 2.51 g/cm³. Further, with increasing of incidence angle, the intensity of reflection is again increased up to I_2 =

57255 pulses, which indicates the occurrence of a more dense structure. If the intensity falls up to the value $I = I_2/2 = 28627$ pulses, the value $2\theta_c = 0.510°$, the critical angle is equal to $\theta_c = 0.255° = 4.451$ mrad. This corresponds to a density 3.06 g/cm³, which is close to the density of silicon carbide - 3.2 g/cm³.

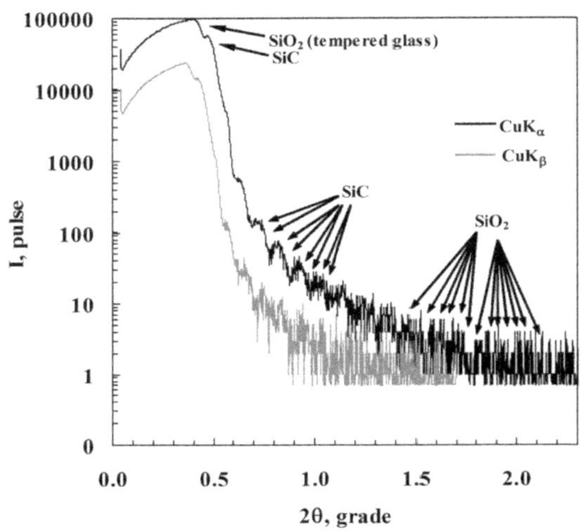

Figure 16. X-ray reflectometry of parameters of the films $SiC_{0.95}$, synthesized by multiple implantation of carbon ions with energies of 40, 20, 10, 5 and 3 keV into silicon after annealing at 1150°C.

To determine the thickness of the layers, four narrow peaks of SiC, and two broad bands of SiO_2 (Fig. 16) were used.

Layer	$(2\theta)_j$	$(2\theta)_i$	j - i	$2\theta_{av} = [(2\theta)_j - (2\theta)_i]/(j-i)$	λ	d = λ/2θ, nm
SiC	1.066	0.690	4	0.094	0.15405	93.9
SiO_2	2.19	1.426	2	0.382	0.15405	23.1

Table 4. Determination of the thickness of the layers in the system ($SiO_{2.33}$ – $S^{\cdot}C_{0.95}$ – Si) by X-ray reflectometry.

The thickness of the system ($SiO_{2.33}$ – $SiC_{0.95}$ – Si) was 117 nm, which was comparable to the estimated thickness of the film. SiO_2 peaks are visible not clearly. Perhaps this is because the annealing temperature was taken at 100°C lower (1150°C).

3.2.2 The study of the high-temperature instability of solid $SiC_{0.7}$ films synthesized by ion implantation

In Fig. 17, the IR transmission spectra of the homogeneous $SiC_{0.7}$ films, synthesized on substrates of Si(100) with resistivity 4–5 Ohm cm (a) and Si(111) with resistivity 10 Ohm cm (b), subjected to isothermal annealing at the temperature 1200°C for several hours in an atmosphere of inert gas (Ar), are presented. Comparing these two figures (a) and (b), one can see that the nature of the SiC films formed on substrates with different crystallographic orientations, are different. This difference manifests itself in the amplitudes and half-widths of the peaks corresponding to the excitation of both transverse and longitudinal optical lattice oscillations of SiC (TO- and LO-phonons).

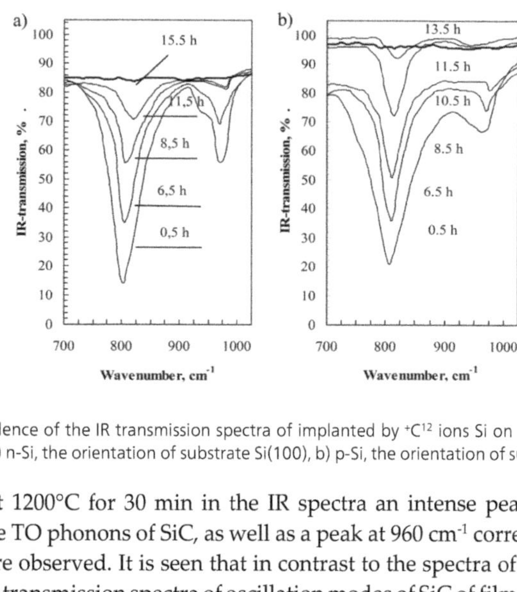

Figure 17. The dependence of the IR transmission spectra of implanted by $^{+}C^{12}$ ions Si on the annealing time at the temperature 1200°C: a) n-Si, the orientation of substrate Si(100), b) p-Si, the orientation of substrate Si(111).

After annealing at 1200°C for 30 min in the IR spectra an intense peak at 800 cm^{-1} which associated with the TO phonons of SiC, as well as a peak at 960 cm^{-1} corresponding to the LO-phonons of SiC, are observed. It is seen that in contrast to the spectra of film on (100) oriented Si substrate, the transmission spectra of oscillation modes of SiC of film on the (111) oriented silicon substrate are more blurred and the level of the transmission spectra of the two modes are superimposed on each other, and does not achieve the initial zero level in the wave number, equal to 915 cm^{-1}. This is caused by the half-width of these peaks (Fig. 17 and 18).

The narrowing of the peak (Fig. 18) up to 40 cm^{-1} occurs as a result of intensive formation of the tetrahedrally oriented Si–C-bonds absorbing at 800 cm^{-1}, as well as the decay of bonds which absorb at frequencies different from the value of 800 cm^{-1}. Since the tetrahedral bond corresponds to the crystalline phase of silicon carbide, the narrowing of the peak of the IR spectrum with a minimum at 800 cm^{-1} is associated with the process of ordering of the implanted layer. As is seen from Fig. 18, for the $SiC_{0.7}$ layer the narrowing of the peak is more intense with increasing time of isothermal annealing up to 6.5 hours in the case of the (111) oriented substrate in comparison with (100) orientation. After annealing for 8.5 hours or more a further narrowing of the peak is slowing, indicating a complete processes of SiC lattice ordering. Thus, it was established that the annealing duration of the less than 6.5 hours at 1200°C is insufficient to form the structure of silicon carbide.

As is seen from Figs. 17 and 19, the amplitude values of the peaks with increasing of annealing time at 1200°C are reduced. This indicates a decrease in the total volume of silicon carbide due to disintegration of SiC and desorption of carbon. Since the amplitude of the SiC-peak of infrared transmission is proportional to the concentration of Si–C-bonds, the measurements of its value were made in the spectra after isothermal annealing at the temperature of 1200°C (Fig. 19). For (100) oriented silicon substrate of n-type conductivity, the amplitude of the TO- and LO-phonon peaks of the infrared transmission (Fig. 19, curves 2 and 4) after annealing for 0.5–6.5 hours were higher than the same for (111) oriented silicon, and then the decay of SiC in this layer becomes more intense. However, as is seen from Figs. 17 and 19, after annealing for 11.5 and 13.5 hours the disintegration of silicon carbide is almost finished for the $SiC_{0.7}$ layer on the (111) substrate, while for the (100) orientation is observed after annealing during up to 15.5 hours.

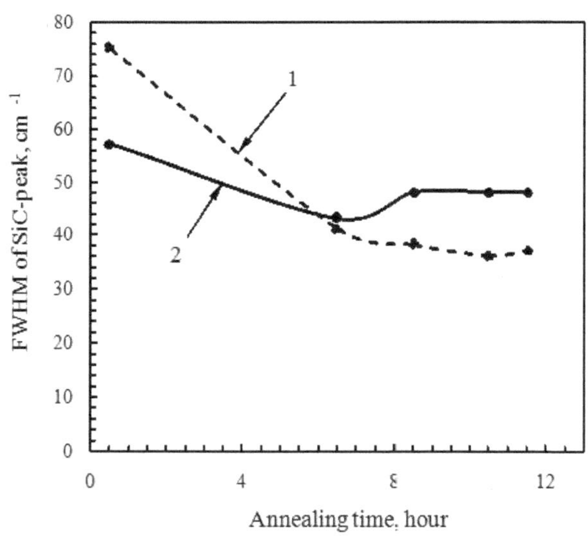

Figure 18. Dependence of the half-width of the TO-phonons peak of SiC of IR spectra on the annealing time at the temperature of 1200°C for $SiC_{0.7}$ layers: 1 - Si(111) substrate, 2 - Si(100) substrate.

It should be also noted that the signal from the LO-phonons in the spectra of both types of substrate disappears earlier (Fig. 19) than the signal from the TO-phonons, and in particular, at (111) orientation of substrate. Thus, a gradual decrease in the amplitudes of the TO-and LO-phonon peaks of SiC in the IR transmission spectra of ion-synthesized SiC films at the increasing time of high-temperature annealing indicates the decay of the SiC structure, i.e. the instability of these films to such regime of treatment.

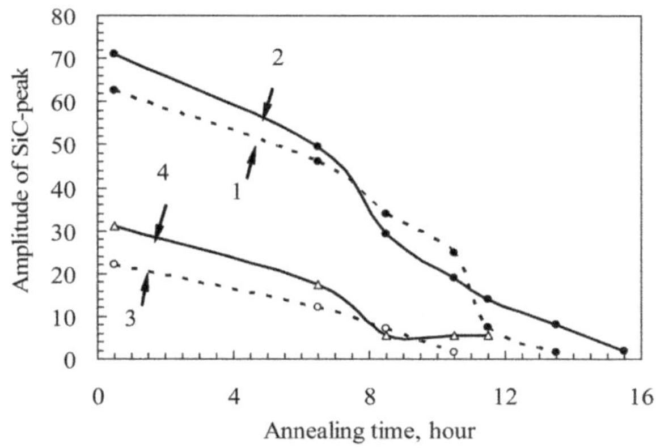

Figure 19. Amplitude of TO- and LO-phonon peaks of SiC of IR transmission versus the annealing time at the temperature of 1200°C for $SiC_{0.7}$ layers on silicon substrates of (100) and (111) orientation: 1 – Si(111), TO-phonon peak; 2 - Si(100), TO-phonon peak; 3 - Si(111), LO-phonon peak; 4 - Si(100), LO-phonon peak.

Since the amplitude of the infrared transmission at the wavenumber 800 cm^{-1} is proportional to the concentration of the tetrahedrally oriented Si–C-bonds, its magnitudes were measured in the spectra after isothermal annealing at the temperature of 1200°C (Fig. 20, curve 3). Assuming that the amplitude at any frequency is proportional to the number of the Si–C-bonds which absorb at this frequency, the amplitudes for the TO-phonons with wavenumbers 700, 750, 850 and 900 cm^{-1} (Fig. 20) in the case of incidence of IR radiation to the sample surface at an angle of 73° to the normal were also measured.

It is seen in Fig. 20a, b (curves 3) that after annealing at 1200°C for 0.5 hour of the $SiC_{0.7}$ film on the Si (100) substrate, the amplitude at wavenumber 800 cm^{-1} is higher than the same for the Si(111) substrate (70 and 58%), indicating a higher content of the tetrahedrally oriented SiC-bonds. It is also seen that the number of SiC-bonds which are close to tetrahedral orientation and absorb at 750 and 850 cm^{-1}, in the case of the substrate Si (100) is lower after annealing for 0.5 h (Fig. 20, curves 2 and 4) in comparison with the substrate Si(111) due to their more intense transformation into the tetrahedral SiC-bonds. The nonlinear nature of the curve 3 in the region 0.5 – 8.5 hours may be due to the formation of tetrahedrally oriented SiC-bonds in the layer simultaneously with their decay at the surface. The saturation of this process after annealing during 6.5 hours results in a faster decrease in the number of bonds during further annealing.

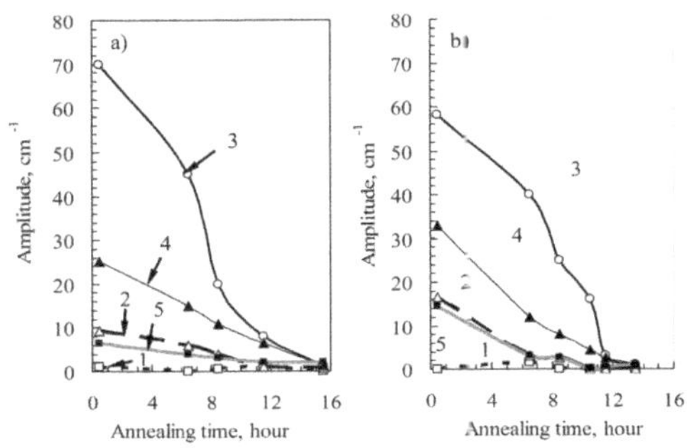

Figure 20. The amplitude of the infrared transmittance at fixed wavenumbers versus the duration of isothermal annealing of the $SiC_{0.7}$ layer (angle of incidence of infrared rays on the sample is 73° from the normal): 1 – 700 cm^{-1}, 2 – 750 cm^{-1}, 3 – 800 cm^{-1}, 4 – 850 cm^{-1}, 5 – 900 cm^{-1}; a) the substrate orientation Si (100), b) the substrate orientation Si (111).

This process is clearly demonstrated on the time dependence of the area of SiC-peak, which is proportional to the total number of optically active Si–C-bonds (Fig. 21). Although the peak amplitude at the minimum of IR transmission (Fig.19, curve 2) and at 800 cm^{-1} (Fig.20, curve 3) for the Si substrates with (100) orientation are higher than in the case of (111) orientation, the value of area of SiC-peak for (111) was higher after annealing for 0.5 hours. This is due to a greater half-

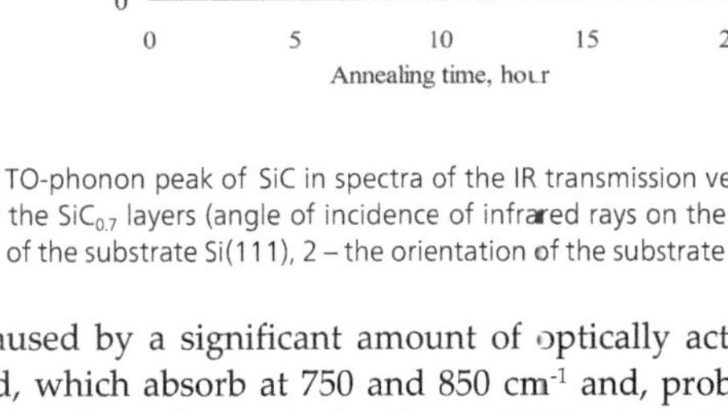

Figure 21. The area of the TO-phonon peak of SiC in spectra of the IR transmission versus the annealing time at the temperature of 1200°C for the $SiC_{0.7}$ layers (angle of incidence of infrared rays on the sample surface - 73° from the normal): 1 - the orientation of the substrate Si(111), 2 – the orientation of the substrate Si(100).

width of the peak caused by a significant amount of optically active Si–C-bonds close to tetrahedrally oriented, which absorb at 750 and 850 cm^{-1} and, probably due to the smaller amount of stable carbon silicon clusters in the film on (111) oriented silicon substrate. It is not contradict to the data for the SiC layer with Gaussian distribution of carbon in silicon, for which was shown that immediately after the implantation of carbon into the (100) and (111) oriented silicon at least 65% and 60% of carbon atoms are concentrated in optically inactive clusters, respectively (see 3.1.2, Fig. 6 "The temperature range 20-600°C"). The strong carbon clusters prevent the crystallization of SiC, and they are less susceptible to oxidation and prevent the penetration of oxygen into the SiC-layer. The SiC film on (100) substrate has more quantity of stable clusters after implantation and, as a result the smaller value of area of SiC-peak and smaller amount of optically active Si–C-bonds after annealing for 0.5 hours and, it less susceptible to oxidation at 1200°C – 15.5 h (instead of 13.5 h for (111) Si substrate). The relatively rapid decay of close to tetrahedral Si-C bonds (Fig. 20b, curves 2 and 4) led to the decrease with higher speed of the number of optically active Si–C-bonds in the case of (111) orientation of Si substrate. In general the dependence of the reduction of optically active Si–C-bonds on the annealing time is linear. This implies that the rapid decay of close to tetrahedral Si–C-bonds was occurring, mainly, due to their transformation into a tetrahedral bond. The linear dependence indicates the homogeneity of the layer, and a rectangular profile of the distribution of carbon atoms in silicon, as well as the fact that the decay rate of silicon carbide does not depend on the depth of the oxidation front. Some decrease in the slope of angle of the curves at the end of the annealing is occurring due to oxidation of the interface "the SiC film – Si substrate".

The value of the wave number of the minimum of SiC-peak of the infrared transmission (the position of the minimum of peak) defines a prevailing kind of optically active bonds which absorb at this wavenumber at this temperature. For the considered layers after annealing at 1200°C for 0.5 h, the transmission peaks with values of minimum at wavenumbers 803 and 806 cm^{-1}, characteristic of crystalline silicon carbide, are observed (Fig. 22). It was found that with increasing annealing time, the position of the peak minimum varies smoothly and moved in the direction of the wave number increase (Fig. 17 and 22).

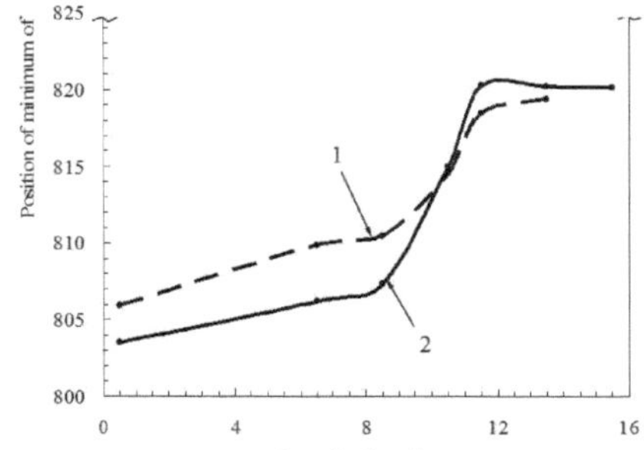

Figure 22. Wave number of the minimum of SiC-peak of IR transmission versus the annealing time at 1200°C for the $SiC_{0.7}$ layers on silicon substrates of (100) and (111) orientations: 1 - Si (111), TO-phonons, 2 - Si(100), TO-phonons, 3 - Si(111), LO-phonons, 4 - Si(100), the LO-phonons.

In our opinion, the frequency shifts of SiC-peak upward indicate a decrease in crystallite size of SiC. We have previously identified size effects, which manifested in the influence of the crystallite sizes of silicon carbide on its optical properties. It was shown (Fig.23) that the differences of the $SiC_{0.03}$, $SiC_{0.12}$ and $SiC_{0.4}$ layers with low carbon concentration from the $SiC_{1.4}$, $SiC_{0.95}$ and $SiC_{0.7}$ layers with high carbon concentration are manifested in the absence of LO-phonon peak of SiC in the IR transmission spectra and in a shift at 1000°C of minimum SiC-peak for TO-phonons in the region of wave numbers higher than 800 cm^{-1}, characteristic for the tetrahedral bonds of crystalline SiC, which is caused by small sizes of SiC crystallites (≤ 3 nm) and by an increase of contribution in the IR absorption of their surfaces, and the surfaces of Si crystallites containing strong short Si–C-bonds as well.

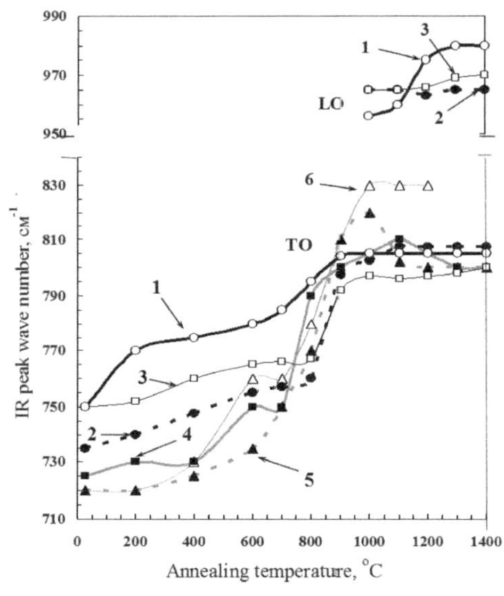

Figure 23. Wavenumber of the IR transmision peak for TO- and LO-phonons SiC as a function of the annealing temperature: 1 - $SiC_{1.4}$, 2 - $SiC_{0.95}$, 3 - $SiC_{0.7}$, 4 - $SiC_{0.4}$, 5 - $SiC_{0.12}$, 6 - $SiC_{0.03}$.

In this case (Fig. 22), the increase of annealing duration of $SiC_{0,7}$ layer leads to both the shift of the minimum of the IR transmission peak up to 820 cm^{-1}, and the reduction of the amplitude of the LO-phonon peaks and their subsequent disappearance, although $SiC_{0.7}$ is considered as the layer with a high concentration of carbon. At the same time more intense process of the shift of the minimum of SiC-peak occurs after annealing for longer than 8.5 hours, which leads to the disappearance of the peak of LO-phonons. This can occur when the penetration of oxygen deep into the layers, their interaction with the carbon atoms on the surface of the crystallites of silicon carbide with the formation of molecules of CO/CO_2. Desorption of carbon atoms causes a decrease in the size of the SiC crystallites and their disintegration. With increasing duration of annealing, the homogeneous $SiC_{0.7}$ layer entirely transforms into

SiO_2, and then goes the oxidation of the interface "SiC film – Si substrate", in which the carbon concentration decreases uniformly with depth according to a Gaussian law. Thus, the concentration of carbon in the remaining layer begins to decrease. This leads to the appearance of phenomenon which is characteristic for the $SiC_{0.4}$, $SiC_{0.12}$, $SiC_{0.03}$ layers, namely, to shift of the minimum of the IR transmission peak up to 820 cm^{-1}, and to a decrease of the amplitude of the LO-phonon peak and their subsequent disappearance. Thus, size effects are confirmed, published by us in 2011.

3.3. Parameters of SiC and C films on Si substrates synthesized by magnetron and ion-beam sputtering

3.3.1. Parameters of C films on Si substrates synthesized by by magnetron sputtering

Carbon thin films were obtained by reactive magnetron sputtering using an ARC 2000 system. A graphite target with a diameter ~50 mm and a thickness of 3 mm was used. The magnetron sputtering mode parameters were: cathode voltage U_c = 470 V, the ion beam current I_{ion} = 35 mA and the argon pressure inside the chamber ~1 Pa. The carbon films were deposited on a set of cleaned silicon substrates. The temperature of the substrate was 75°C.

The presence of a sharp interface "C film - Si substrate" permits to investigate the thickness and density of the film by X-ray reflectometry (CompleXRay C6) by recording the angular dependence of the reflection coefficient using two spectral lines CuK_α (0.154 nm) and CuK_β (0.139 nm). The oscillations of intensity were observed, assigned to the interference of X-ray reflections from the boundaries of carbon layer (Fig. 24).

It is known that the density of graphite is 2.2 g/cm^3 and the density of diamond - 3.51 g/cm^3. Since the density of the resulting film was 3.32 g/cm^3 (Table 5 and Fig.24), we concluded that the diamond-like carbon film was synthesized.

Layer	I_{max}	$I_{max}/2$	$2\theta_c$	θ_c, grade	θ_c, rad	ρ, g/cm^3
C	962849	481425	0.529	0.2645	4.616	3.32

Table 5. Determination of the density of the carbon layer by X-ray reflectometry and using the Henke program.

To determine the thickness, five narrow peaks of C, and a broad band of C were used (Fig. 24, table 5).

Layer	$(2\theta)_j$	$(2\theta)_i$	j - i	$2\theta_{av}=[(2\theta)_j-(2\theta)_i]/(j-i)$	λ, nm	d = λ/2θ, nm
C	1.684	1.164	5	0.104	0.15405	84.9
C	2.732	2.080	1	0.652	0.15405	13.5

Table 6. Determination of the thickness of the layers in the system (C - C - Si) by X-ray reflectometry.

Simulation using the Henke program (http://henke.lbl.gov/optical_constants/) [24] allows to obtain a theoretical curve, which is close to the experimental (Fig.25). The main parameters of the layer system, which allow to obtain an acceptable agreement of experimental and theoretical curves, were:

1. the diamond-like carbon film of thickness d = 84 nm, density ϱ = 3.3 g/cm^3, and surface roughness σ = 1.5 nm;

2. a thin graphite layer of thickness d = 5 nm with the density ϱ = 2.206 g/cm^3, and the roughness of (C– C) interface σ = 0.13 nm;

3. the silicon substrate with density ϱ=2.33 g/cm^3 and the roughness of interface (C–Si) σ=0 nm.

Thus, diamond-like carbon film of thickness d = 84 nm, density ϱ = 3.3 g/cm^3, and surface roughness of σ = 1.5 nm on a silicon surface by magnetron sputtering was synthesized.

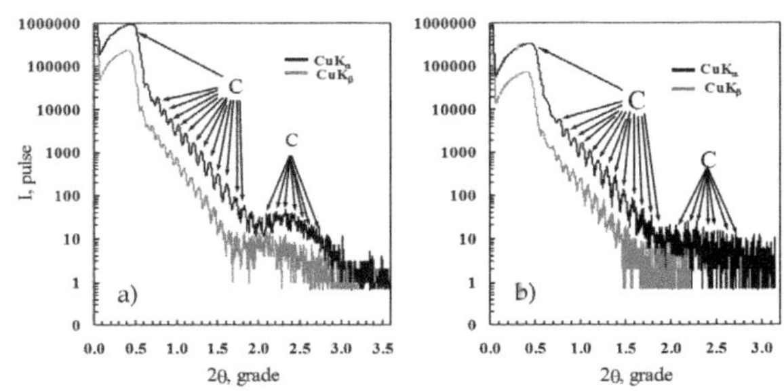

Figure 24. X-ray reflectometry using two spectral lines CuK$_\alpha$ (0.154 nm) and CuK$_\beta$ (0.139 nm) (CompleXRay C6) of parameters of carbon films synthesized by magnetron sputtering: a) sample A; b) sample B.

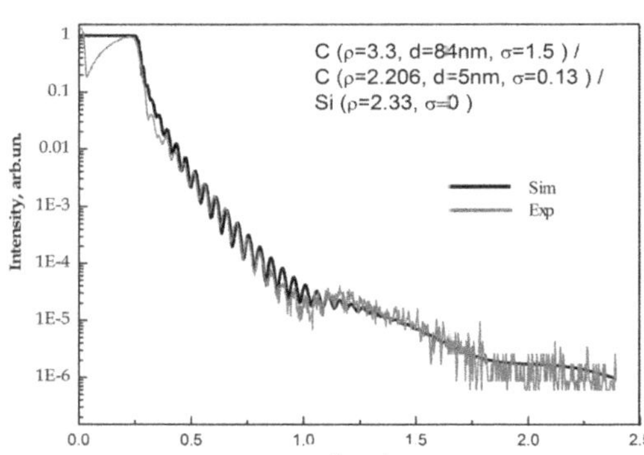

Figure 25. Simulation using the Henke program [24] of experimental results obtained by X-ray reflectometry of parameters of carbon films synthesized by magnetron sputtering.

3.3.2. Parameters of SiC films on Si substrates synthesized by by ion-beam sputtering

SiC films were prepared by ion-beam sputtering. For the simultaneous deposition on silicon substrates of C and Si atoms, a two-component target consisting of the overlapping wafers of silicon and graphite was used. Sputtering of the target was carried out in an argon atmosphere. The formation of Ar ion beam was happening in the ring electrode system (a hollow cathode and an anode), and magnets with crossed electrical and magnetic fields. Discharge power was 108 W (2.7 kV, 40 mA), argon pressure in the chamber 5.9×10^{-2} Pa, substrate temperature - 20°C. Samples with SiC films were annealed at 1250°C in an argon atmosphere for 30 min.

The presence of a sharp interface "the SiC film - Si substrate" permits to investigate the thickness and density of the film by X-ray reflectometry ("ComplexXRay C6"). The oscillations of intensity were observed, assigned to the interference of X-ray reflections from the boundaries of silicon carbide layer (Fig. 26). It is known that the density of graphite is 2.2 g/cm^3, and of silicon - 2.33 g/cm^3, of silicon carbide - 3.2 g/cm^3 and of diamond - 3.51 g/cm^3. Since the density of the obtained film was 3.03 g/cm^3 (Table 6 and Fig.26), we concluded, that the film close to silicon carbide, was synthesized. The film contains approximately [(3.03-2.33)/(3.2-2.33)] × 100% = 80.5% SiC, and [(3.2-3.03)/(3.2-2.33)] × 100% = 19.5% Si, i.e., about 80 atoms of C refer per 100 atoms of Si and the $SiC_{0.8}$ layer was formed.

Layer	I_{max}	$I_{max}/2$	$2\theta_c$	θ_c, grade	θ_c, rad	ρ, g/cm^3
SiC	101255	50628	0.508	0.254	4.433	3.03

Table 7. Determination of the density of the SiC layer by X-ray reflectometry and using the Henke program.

To determine the film thickness the position of maxima of five narrow peaks of SiC was used (Fig.26).

Layer	$(2\theta)_j$	$(2\theta)_i$	j - i	$2\theta_{av} = [(2\theta)_j - (2\theta)_i]/(j-i)$	λ, nm	$d = \lambda/2\theta$, nm
SiC	1.230	1.034	4	0.0490	0.15405	180.1

Table 8. Determination of the thickness of the layer in the system (SiC – Si) by X-ray reflectometry.

Simulation using the Henke program (http://henke.lbl.gov/optical_constants/) [24] allows obtaining a theoretical curve, which is close to the experimental (Fig.27). The main parameters of the layer system, which allow to obtain an acceptable agreement of experimental and theoretical curves, were:

1. the silicon carbide film of thickness d = 160 nm, the density $\varrho = 3.03$ g/cm^3, and surface roughness of $\sigma = 0.4$ nm;

2. the silicon substrate with density $\varrho = 2.33$ g/cm^3 and the roughness of interface (SiC-Si) $\sigma = 1.5$ nm.

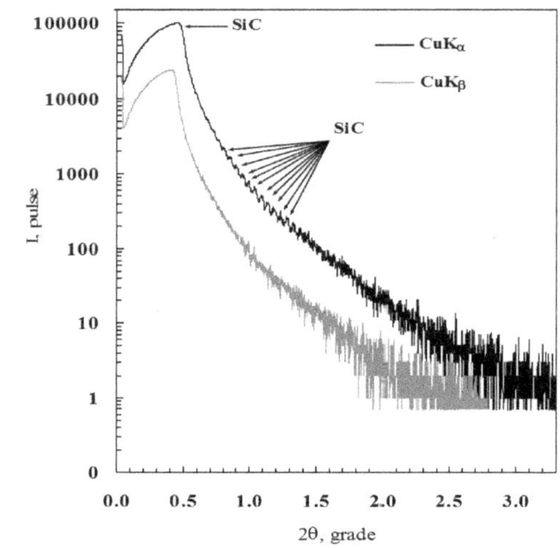

Figure 26. X-ray reflectometry using two spectral lines CuK_{α} (0.154 nm) and CuK_{β} (0.139 nm) (CompleXRay C6) parameters of silicon carbide films synthesized by ion-beam sputtering of the two-component target of silicon and graphite.

The film thickness d = 160 nm is different from the values of 180 nm, obtained from the average distance between the peaks $2\theta_{av}$. The reasons for the differences require further studies.

Thus, the silicon carbide film of thickness d = 160 nm, the density $\varrho = 3.03$ g/cm^3, and surface roughness of $\sigma = 0.4$ nm on the silicon surface by ion-beam sputtering of the two-component target of silicon and graphite was synthesized.

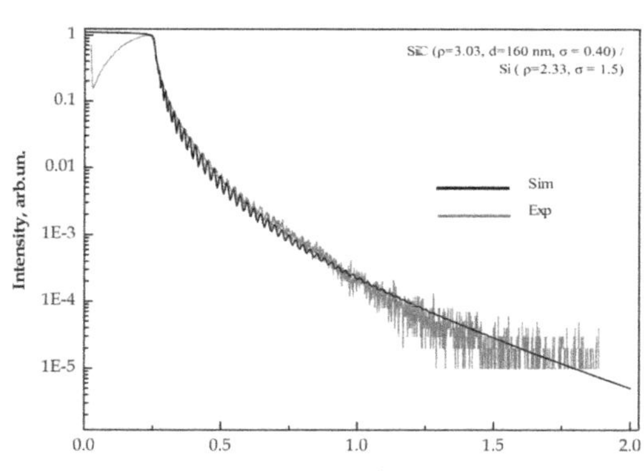

Figure 27. Simulation using the Henke program [24] of experimental results obtained by X-ray reflectometry of parameters of SiC films synthesized by ion-beam sputtering of the two-component target of silicon and graphite.

Conclusion

1. For silicon layer with the Gaussian profile of implanted carbon atoms, the peak of longitudinal optical oscillations (LO-phonons) of SiC at 965-970 cm^{-1} is found and, it permits to calculate a number of optical parameters of film. The values of the low-frequency dielectric constant, ε_0, the effective charge, e*/e, and the dimensionless parameter, ϱ, are equal to 9.82, 0.89, and 0.25, respectively. The dependences of LO-phonon peak amplitude, halfwidth and peak position from both an angle of incidence of the electromagneticradiationonsamplesurfaceandtheannealingtemperaturearedetermined.Thedifference between the LO-phonon curves behaviour indicates that the formation of SiC crystallites, i.e. the intensive formation of tetrahedral Si–C-bonds of necessary length and bond angles, in the case of (111) oriented substrate is not completed up to the silicon melting point, as for (100) oriented substrate that is completed at 1350°C.
2. The temperature dependence of an area under contour of IR spectrum curve in wide temperature range 200 – 1400°C was used to obtain valueable new information about the structure transformation in the carbon implanted silicon layer. The high sensitivity of temperature dependence of area to a change of substrate orientation was found.
3. A model of the ion implanted layer in the form of the system of the carbon-silicon clusters consisting of the carbon- and silicon atoms linked one with another by single-, double- and tripple bonds, and by single elongated-, sesqui, free ("dangling") and hybridized bonds, as well as resonances, is discussed. The infrared inactive flat nets and chains tied together by the infrared active clusters and, by tetrahedrally oriented bonds which are characteristic for silicon and silicon carbide, are described. Immediately after the implantation of carbon into the (100) and (111) oriented silicon at least 65% and 60% of carbon atoms are concentrated in these optically inactive clusters, respectively, if the implantation was carried out by a dose to obtain the stoichiometric concentration (E = 40 keV, D = 3.56×10^{17} cm^{-2}) of carbon and silicon. An over-barrier mechanism of the formation of Si and SiC crystallites is supposed and it explained by tendency of an isolated system to a minimum of free energy.
4. For carbon implanted silicon layer (E = 40 keV, D = 3.56×10^{17} cm^{-2}) on (100) and (111) substrate the crystallization process in several temperature ranges was studied. For (100) oriented Si substrate it was shown:
   - 20-600°C – the formation of tetrahedrally oriented Si–C-bonds due to disintegration of clusters which consist mainly of the Si–Si, Si=Si and elongated Si–C-bonds;
   - 600–800°C – the formation of Si crystallites and tetrahedrally oriented Si–C-bonds due to disintegration of strained Si-Si bonds and Si–C-bonds, which absorb at frequencies close to 700 and 750 cm^{-1}, recrystallization starts near the substrate and the surface and goes to the middle of layer;
   - 800–1000°C – begins to dominate the absorption at 800 cm^{-1}, and near it, due to formation of tetrahedrally oriented bonds of the SiC crystallites, as well as bonds close to tetrahedral;
   - 1000–1100°C - the growth of the crystallite size through consolidation of small crystallites of Si and SiC; decay of Si–C-bonds absorbing at 850 and 900 cm^{-1};

- 1100–1200°C – formation of new optically active clusters absorbing at frequencies of 750, 850 and 900 cm^{-1} due to the decay of optically inactive clusters;
- 1200–1250°C – the transformation of clusters absorbing at frequencies of 800-900 cm^{-1}, into clusters, which absorb at 700 and 800 cm^{-1}. and the growth of SiC crystallite size;
- 1250–1300°C – intensive decay of the tetrahedral Si–C-bonds, caused by the desorption of carbon and the destruction of small crystallites;
- 1300–1350°C – the decay of stable clusters with multiple bonds and the increase in the number and size of crystallites of SiC, and Si–C-bonds, close to tetrahedral;
- 1350–1400°C – a significant decrease of amount of the polycrystalline SiC due to desorption of carbon atoms.

5. An influence of substrate orientation on β-SiC formation has been studied. It was shown that the SiC-synthesis in the temperature range 900–1000°C is more preferable on (100) oriented silicon substrates and at 1200–1300°C – on (111) oriented silicon substrates. In the case of the (100) oriented substrate the number of tetrahedrally oriented Si–C-bonds reached at 1000°C some maximum and does not change up to 1200°C, whereas in the case of orientation (111) the number of bonds increases smoothly in the range 900–1300°C.
6. Using thermo-emf, it was determined a type of conduction of Si- and SiC-crystallites. The SiC-crystallites have p-type of conduction independently from the type of conduction of substrate, while the Si-crystallites have the same conduction as a substrate.
7. During the prolonged high-temperature isothermal annealing (1200°C) a gradual decrease in the amplitude of the TO- and LO-phonon peaks of IR transmission, characteristic of ion-synthesized SiC, indicates the disintegration of the structure of homogeneous SiC film, ie the instability of these films at these temperatures.
8. It is shown that the deformation of the rectangular Auger profile of C^{12} distribution in Si in comparison with the calculated profile, namely the thinning of the interface "the SiC film – Si substrate", and the increase of the carbon concentration at the surface and in regions near the maxima of the distributions for individual carbon ion energy (40, 20 keV), is caused by both the surface sputtering and the change in the composition of the silicon layer during high dose implantation of carbon.
9. The presence of a sharp interface "the SiC film – Si substrate" permits to study the composition, the density and the thickness of the $SiC_{0.7}$ and $SiC_{0.95}$ films by X-ray reflectometry (CompleXRay C6) at small grazing angles θ by recording the angular dependence of the reflection coefficient using two spectral lines CuK_α (0.154 nm) and CuK_β (0.139 nm), although this method for ion-implanted layers usually does not apply. Using the Henke program for $SiC_{0.7}$ film was determined a surface layer with density 2.37 g/cm^3, which is close to cristobalite (SiO_2) 2.32 g/cm^3. Further located the film with a density 2.77 g/cm^3 and is close to quartz (SiO_2) 2.65 g/cm^3. A structure of thickness 65 nm and density $\varrho = 3.25$ g/cm^3, is close to the silicon carbide - 3.2 g/cm^3. The study of $SiC_{0.95}$ film show the presence of surface layer with density 2.51 g/cm^3 (optical glass) and the silicon carbide layer of thickness 94 nm and density 3.06 g/cm^3.

10. It is shown that the decomposition of optically active Si–C-bonds is more intense in the case of the (111) orientation of the silicon substrate. Based on the linear nature of reducing the number of Si–C-bonds in a homogeneous $SiC_{0.7}$ layer with increasing annealing time, it was concluded that the decay rate of silicon carbide does not depend on the depth of the oxidation front.

11. It was confirmed the size effect caused by the small size of nanocrystals, which are manifested in the shift of the minimum of the peak of IR transmission up to to 820 cm^{-1}, the decrease of the amplitude of the LO-phonon peak and its subsequent disappearance during the oxidation of the interface "SiC film - Si substrate ", where the concentration of carbon atoms decreases.

12. Diamond-like carbon film of thickness d = 84 nm, density ϱ = 3.3 g/cm^3, and surface roughness of σ = 1.5 nm on a silicon surface by magnetron sputtering was synthesized. The silicon carbide film of thickness d = 160 nm, the density ϱ = 3.03 g/cm^3, and surface roughness of σ = 0.4 nm on the silicon surface by ion-beam sputtering of the two-component target of silicon and graphite was synthesized. Simulation using the Henke program allows obtain theoretical curves, which are close to the experimental curves.

Author details

Kair Kh. Nussupov* and Nurzhan B. Beisenkhanov

*Address all correspondence to: rich-famouskair@mail.ru

Kazakh-British Technical University, Kazakhstan

References

[1] Afanas'ev, A. V., Il'in, V. A., Korlyakov, A. V., Lebedev, A. O., Luchinin, V. V., & Tairov, Yu. M. (2011). Karbid kremniya. Vklad SPbGETU « LETI». Priznanie i perspektivy. V kn.: Fizika i tekhnologiya mikro- i nanosistem. S. 50-86. (Silicon carbide. The contribution of SPbGETU "LETI". Recognition and prospects. *In the book. Physics and technology of micro-and nanosystems. P.50-86). Edited by Luchinin V.V. and Malinovski V.V. Publisher "Russian Collection". St. Petersburg,* 239, in Russian.

[2] Akimchenko, I. P., Kisseleva, K. V., Krasnopevtsev, V. V., Milyutin, Yu. V., Touryanski, A. G., & Vavilov, V. S. (1977a). The structure of silicon carbide synthesized in diamond and silicon by ion implantation. *Radiation Effects,* 33, 75-80.

[3] Akimchenko, I. P., Kazdaev, Kh.R, & Krasopevtsev, V. V. (1977b). IK-pogloshenie β-SiC, sintezirovannogo pri ionnoi implantacii C v Si. (IR absorption of β-SiC, synthesized by ion implantation of C into Si). *Fizika i Tekhnika Poluprovodnikov, T.11, vyp. 10,* 1964-1966, in Russian.

[4] Akimchenko, I. P., Kazdaev, H. R., Kamenskikh, I. A., & Krasnopevtsev, V. V. (1979). Opticheskie i fotoelektricheskie svoistva struktury SiC–Si, poluchennoi pri implantacii ionov ugleroda v kremnii. (Optical and photoelectric properties of the SiC-Si structure, obtained by implantation of carbon ions into silicon). *Fizika i Tekhnika Poluprovodnikov, T.13, vyp.2*, 375-378, in Russian.

[5] Akimchenko, I. P., Kisseleva, K. V., Krasnopevtsev, V. V., Touryanski, A. G., & Vavilov, V. S. (1980). Structure and optical properties of silicon implanted by high doses of 70 and 310 keV carbon ions. *Radiation Effects*, 48, 7-12.

[6] Aleksandrov, P. A., Baranova, E. K., Demakov, K. D., Komarov, F. F., Novikov, A. P., & Shiryaev, S. Y. (1986). Sintez monokristallicheskogo karbida kremniya s pomoshyu odnoshagovoi tehniki vysokointensivnogo ionnogo legirovaniya. (Synthesis of single-crystal silicon carbide by using one-step technique of high-intensity ion implantation). *Fizika i Tekhnika Poluprovodnikov, T.20*, 1, 149-152, in Russian.

[7] Aleksandrov, P. A., Baranova, E. K., Gorodetski, A. E., Demakov, K. D., Kutukova, O. G., & Shemardov, S. G. (1988). Issledovanie raspredeleniya amorfnoi i kristallicheskoi fazy ionno-sintezirovannogo SiC v Si. (The investigation of the distribution of the amorphous and crystalline phases of ion-synthesized SiC in Si). *Fizika i Tekhnika Poluprovodnikov, T.22*, 4, 131-132, in Russian.

[8] Baranova, E. K., Demakov, K. D., Starinin, K. B., Strelcov, L. N., & Haibullin, I. B. (1971). Issledovanie monokristallicheskih plenok SiC, poluchennykh pri bombardirovke ionami C^+ monokristallov Si. (The study of single-crystal films of SiC, obtained by bombardment by C^+ ions of Si single crystals). *Doklady AN SSSR*, 200, 869-870, in Russian.

[9] Bayazitov, R. M., Haibullin, I. B., Batalov, R. I., Nurutdinov, R. M., Antonova, L. Kh, Aksenov, V. P., & Mikhailova, G. N. (2003). Structure and photoluminescent properties of SiC layers on Si, synthesized by pulsed ion-beam treatment. *Nucl. Instrum. and Meth. in Phys. Res., B.*, 206, 984-988.

[10] Belov, A. I., Mihailov, A. N., Nikolichev, D. E., Boryakov, A. V., Sidorin, A. P., Grachev, A. P., Ershov, A. V., & Tetelbaum, D. I. (2010). Formation and "white" photoluminescence of nanoclusters in SiO_x films implanted with carbon ions. *Semiconductors.*, 44(11), 1450-1456.

[11] Borders, J. A., Picraux, S. T., & Beezhold, W. (1971). Formation of SiC in silicon by ion implantation. *Appl Phys. Lett*, 18(11), 509-511.

[12] Chen, D., Wong, S. P., Yang, Sh., & Mo, D. (2003). Composition, structure and optical properties of SiC buried layer formed by high dose carbon implantation into Si using metal vapor vacuum arc ion source. *Thin Solid Films*, 426, 1-7.

[13] Durupt, P., Canut, B., Gauthier, J. P., Roger, J. A., & Pivot, J. (1980). RBS, Infrared and diffraction compared analysis of SiC Synthesis in C implanted silicon. *Mater. Res. Bull*, 15, 1557-1565.

[14] Edelman, F. L., Kuznetsov, O. N., Lezheiko, L. V., & Lubopytova, E. V. (1976). Formation of SiC and Si_3N_4 in silicon by ion implantation. *Radiation Effects*, 29, 13-15.

[15] Fissel, A., Kaizer, U., Pfennighaus, K., Schroter, B., & Richter, W. (1996). Growth of 6H-SiC on 6H-SiC(0001) by migration enhanced epitaxy controlled to an atomic level using surface superstructures. *Applied Physics Letters*, 68, 1204-1206.

[16] Fissel, A., Schroter, B., Kaiser, U., & Richter, W. (2000). Advances in the molecular-beam epitaxial growth of artificially layered heteropolytypic structures of SiC. *Appl. Phys. Lett*, 77, 2418-2420.

[17] Frangis, N., Nejim, F., Hemment, P. L. F., Stoemenos, J., & Van Landuyt, J. (1996). Ion beam synthesis of β-SiC at 950°C and structural characterization. *Nuclear Instruments and Methods in Phys. Res. B*, 112, 325-329.

[18] Frangis, N., Stoemenos, J., Van Landuyt, J., Nejim, F., & Hemment, P. L. F. (1997). The formation of 3C-SiC in crystalline Si by carbon implantation at 950°C and annealing- a structural study. *Journal of Crystal Growth*, 181, 218-228.

[19] George, V. C., Das, A., Roy, M., Dua, A. K., Raj, P., & Zahn, D. R. T. (2002). Bias enhanced deposition of highly oriented β-SiC thin films using low pressure hot filament chemical vapour deposition technique. *Thin Solid Films*, 419, 114-117.

[20] Gerasimenko, N. N., Kuznetsov, O. N., Lezheyko, L. V., Lyubopytova, E. V., Smirnov, L. S., & Edelman, F. L. (1974). Nekotorye svoistva plenok SiC, poluchennykh ionnym vnedreniem v strukture Ai-SiC-Si. (Some properties of the SiC films, obtained by ion implantation in the structure of Al-SiC-Si.). *Mikroelektronika, T.3., Vyp.5*, 467-468, in Russian.

[21] Gibbons, J. F., Johnson, W. S., & Hylroic, S. W. (1975). Projected Range Statistics. *Part 1. 93p. 2nd ed. Dowden, Stroudsburg. PA.*

[22] Glinka, N.L. (1985). Obshaya khimiya (General chemistry). *24-e izd. Pod red. Rabinovich V.A. L: Khimiya.*, 702, in Russian.

[23] Gupta, S. K., & Akhtar, J. (2011). Thermal Oxidation of Silicon Carbide (SiC)- Experimentally Observed Facts. *In the book "Silicon Carbide- Materials, Processing and Applications in Electronic Devices", edited by Moumita Mukherjee, InTech, Chapter 9.*, 207-230.

[24] Henke, B. L., Gullikson, E. M., & Davis, J. C. (1993). X-ray interactions: photoabsorption, scattering, transmission, and reflection at E=50-30000 eV, Z=1-92. *Atomic Data and Nuclear Data Tables.* [2], 181-342.

[25] Kerdiles, S., Rizk, R., Gourbilleau, F., Perez-Rodriguez, A., Garrido, B., Gonzalez-Varona, O., & Morante, J. R. (2000). Low temperature direct growth of nanocrystalline silicon carbide films. *Mater.Sci.Eng., B*, 69, 530-535.

[26] Kimura, T., Kagiyama, Sh., & Yugo, Sh. (1981). Structure and annealing properties of silicon carbide thin layers formed by ion implantation of carbon ions in silicon. *Thin Solid Films*, 319-327.

[27] Kimura, T., Kagiyama, Sh., & Yugo, Sh. (1982). Characteristics of the synthesis of β-SiC by the implantation of carbon ions into silicon. *Thin Solid Films*, 94, 191-198.

[28] Kimura, T., Yugo, Sh., Bao, Zh. S., & Adachi, Y. (1989). Formation of silicon carbide layers by the ion beam technique and their electrical properties. *Nuclear Instruments and Methods in Phys.Res., B.*, 39, 238-241.

[29] Khokhlov, A. F., Pavlov, D. A., Mashin, A. I., & Khokhlov, D. A. (1994). Vozniknovenie dvoiinykh svyazyei kremnii-kremnii v plenkakh a-Si:H, obluchennykh nyeonom i uglerodom pri otzhige. (The appearance of double silicon-silicon bonds in the a-Si:H films, exposed by neon and carbon during annealing). *Fizika i tekhnika poluprovodnikov. T.28, vyp.10. S.*, 1750-1754, in Russian.

[30] Khohlov, A. F., Pavlov, D. A., Mashin, A. I., & Mordvinova, Y. A. (1987). Izovalentnoe legirovanie amorphnogo kremniya uglerodom. (Isovalent doping of amorphous silicon by carbon). *Fizika I tehnika poluprovodnikov, T.21, vyp.3*, 531-535, in Russian.

[31] Lebedev, A. A., Mosina, G. N., Nikitina, I. P., Savkina, N. S., Sorokin, L. M., & Tregubova, A. C. (2001). Investigation of the structure of (p)3C - SiC-(n)6H-SiC. *Tech. Phys. Lett*, 27, 1052-1054.

[32] Lebedev, A. A., Strel'chuk, A. M., Savkina, N. S., Bogdanova, E. V., Tregubova, A. S., Kuznetsov, A. N., & Sorokin, L. M. (2002). Investigation of the p^-3Cn-SiC/n^+-6H-SiC heterostructures with modulated doping. *Tech. Phys. Lett*, 28(12), 1011-1014.

[33] Lebedev, A. A., Nelson, D. K., Razbirin, B. S., Saidashev, I. I., Kuznetsov, A. N., & Cherenkov, A. E. (2005). Study of the properties of two-dimensional electron gas in p^-3Cn-SiC/n^+-6H-SiC heterostructures at low temperatures. *Semiconductors*, 39(10), 194-996.

[34] Lezhyeiko, L. V., & Lyubopytova, Ye. V. (1976). Svoistva p-n i n-n-geteroperekhodov Si–β-SiC, poluchennykh ionnym vnedreniem. (Properties of p-n and n-n-heterojunctions Si-β-SiC, obtained by ion implantation). *Fizika i tekhnika poluprovodnikov, T. 10*, 6, 1749, in Russian.

[35] Liangdeng, Y., Intarasiri, S., Kamwanna, T., & Singkarat, S. (2008). Ion beam synthesis and modification of silicon carbide. *In book "Ion beam applications in surface and bulk modification of insulators".- IAEA, Austria, Vienna, IAEA-TECDOC-1607*, 63-92.

[36] Liao, F., Girshick, S. L., Mook, W. M., Gerberich, W. W., & Zachariah, M. R. (2005). Superhard nanocrystalline silicon carbide films. *Appl. Phys. Lett*, 86, 171913-171915.

[37] Lindner, J. K. N. (2003). High-dose carbon implantations into silicon: fundamental studies for new technological tricks. *Appl.Phys. A*, 77, 27-38.

[38] Lyddane, R. H., Sachs, R. G., & Teller, E. (1941). On the polar vibration of alkali halides. *Phys.Rev*, 59, 673-676.

[39] Morimoto, T., Ogura, Y., Kondo, M., & Ueda, T. (1995). Multilayer coating for carbon-composites. *Carbon*, 33(4), 351-357.

[40] Mukhamedshina, D. M., & Beisenkhanov, N. B. (2012). Influence of Crystallization on the Properties of SnO_2 Thin Films. *In book: "Advances in Crystallization Processes". Chapter 9. InTech. Dr. Yitzhak Mastai (Ed.).*, 219-258.

[41] Nishino, S., Jacob, C., & Okui, Y. (2002). Lateral over-growth of 3C-SiC on patterned Si(111) substrates. *Journal of Crystal Growth, 237-239*, 1250-1253.

[42] Oguri, K., & Sekigawa, T. (2004). Heat resistant material and hot structure member both space shuttle, space shuttle, and method for producing heat resistant material for space shuttle. *United State Patent* [US2004/0180242 A1], Sep.16.

[43] Pajagopalan, T., Wang, X., Lanthouh, B., & Ramkumar, C. (2003). Low temperature deposition of nanocrystalline silicon carbide films by plasma enhanced chemical vapor deposition and their structural and optical characterization. *J. Appl. Phys*, 94(3), 5252-5260.

[44] Pauling, L., & Pauling, P. (1975). Chemistry. *Freeman W.H. and company. San Francisco.*

[45] Peierls, R. E. (1956). Quantum theory of solids. *Oxford: Clarendon Press*, 54-58.

[46] Raabe, G., & Michl, J. (1985). Multiple Bonding to Silicon. *Chem. Rev.*, 85(5), 419-509.

[47] Rothemund, W., & Fritzsche, C. R. (1974). Optical absorption and electrical conductivity of SiC films produced by ion implantation. *J.Electrochem. Soc.: Solid-state science and technology*, 121(4), 586-588.

[48] Savkina, N. S., Lebedev, A. A., Davydov, D. V., Strelchuk, A. M., Tregubova, A. S., Raynaud, C., Chante, J. P., Locatelli, M. L., Planson, D., Nilan, J., Godignon, P., Campos, F. J., Nestres, N., Pascual, J., Brezeanu, G., & Batila, M. (2000). Low-doped 6H-SiC n-type epilayers grown by sublimation epitaxy. *Materials Science and Engineering*, B77, 50-54.

[49] Semenov, A. V., Puzikov, V. M., Dobrotvorskaya, M. V., Fedorov, A. G., & Lopin, A. V. (2008). Nanocrystalline SiC films prepared by direct deposition of carbon and silicon ions. *Thin Solid Films*, 516(10), 2899-2903.

[50] Semenov, A. V., Puzikov, V. M., Golubova, E. P., Baumer, V. N., & Dobrotvorskaya, M. V. (2009). Low-temperature production of silicon carbide films of different polytypes. *Semiconductors*, 43(5), 685-689.

[51] Semenov, A. V., Lopin, A. V., Puzikov, V. M., Baumer, V. N., & Dmitruk, I. N. (2010). Fabrication of heterostructures based on layered nanocrystalline silicon carbide polytypes. *Semiconductors*, 44(6), 816-823.

[52] Serre, C., Romano-Rodríguez, A., Pérez-Rodríguez, A., Morante, J. R., Fonseca, L., Acero, M. C., Kögler, R., & Skorupa, W. (1999). SiC on SiO_2 formed by ion implantation and bonding for micromechanics applications. *Sensors and Actuators A: Physical*, 74(1-3), 169-173.

[53] Spillman, H., & Wilmott, P. R. (2000). Growth of SiN_x and SiC_x thin films by pulsed reactive crossed-beam laser ablation. *Applied Physics. A.- Materials Science & Processing*, 70(3), 323-327.

[54] Spitzer, W. G., Kleinman, D., & Walsh, D. (1959). Infrared properties of hexagonal silicon carbide. *Phys. Rev.*, 113(1), 127-132.

[55] Srikanth, K., Chu, M., Ashok, S., Nguyen, N., & Vedam, K. (1988). High-dose carbon ion implantation studies in silicon. *Thin Solid Films*, 323-329.

[56] Sun, Y., & Miyasato, T. J. (1998). Characterization of cubic SiC films grown on thermally oxidized Si substrate. *J. Appl. Phys*, 84(5), 2602-2611.

[57] Tetelbaum, D. I., Mikhaylov, A. N., Belov, A. I., Vasiliev, V. K., Kovalev, A. I., Wainshtein, D. L., Golan, Y., & Osherov, A. (2009). Luminescence and structure of nanosized inclusions formed in SiO_2 layers under double implantation of silicon and carbon ions. *Journal of Surface Investigation. X-ray, Synchrotron and Neutron Techniques*, 3(5), 702-708.

[58] Touriyanski, A. G., Aprelov, S. A., Gerasimenko, N. N., Pirshin, I. V., & Senkov, V. M. (2007). Otnositelnaya rentgenovskaya reflektometriya diskretnykh sloistykh struktur (Relative X-ray reflectometry of discrete layered structures). *Pis'ma v ZHTF (Technical Physics Letters)*, 33(5), 87-94.

[59] Touriyanski, A. G., Gerasimenko, N. N., Aprelov, S. A., Pirshin, I. V., Poprygo, A. I., & Senkov, V. M. (2005). Investigation of ion-implanted layers by X-ray reflectometry. *Proc. SPIE. X-Ray and Neutron Capillary Optics II.*, 5943, 143-149.

[60] Touryanski, A., Gerasimenko, N., Pirshin, I., & Senkov, V. (2009). Mnogofunktsionalnyi rentgenovskii reflektometr dlya issledovaniya nanostruktur (Multipurpose X-ray reflectometer for the study of nanostructures). *Nanoindustriya (Nanoindustry)*, 5, 40-45, in Russian.

[61] Touryanski, A. G., Vinogradov, A. V., & Pirshin, I. V. (2000). X-ray reflectometer. *Patent no.6041098,US Cl. 378-70. Official Gazette March 21,2000*, 2960.

[62] Touryanski, A. G., Vinogradov, A. V., & Pirshin, I. V. (2000). Two-channel X-ray reflectometer. *Nucl. Instrum. and Meth. in Phys. Res.*, A 448, 184-187.

[63] Yan, H., Wang, B., Song, X.M., Tan, L. W., Zhang, S. J., Chen, G. H., Wong, S. P., Kwok, R. W. M., & Leo, W. M. L. (2000). Study on SiC layers synthesized with carbon ion beam at low substrate temperature. *Diamond and related materials*, 9, 1795-1798.

[64] West, R., Fink, M. J., & Michl, J. (1981). Tetramesityldisilene, a Stable Compound Containing a Silicon-Silicon Double Bond. *Science.*, 214, 1343-1344.

[65] Williams, J. S., & Poate, J. M. (1984). *Ion Implantation and Beam Processing. Academic Press, Melbourne, 1984. Edited by Williams J.S., Poate J.M. 419 p.*, 30-31.

[66] Ziman, J. M. (1960). Electrons and phonons. *Oxford: Clarendon Press.*, 209.

Section 2

Device Processing

Chapter 4

Etching of Silicon Carbide Using Chlorine Trifluoride Gas

Hitoshi Habuka

Additional information is available at the end of the chapter

1. Introduction

Silicon carbide (SiC) is known as an excellent material. Single-crystalline 4H-silicon carbide is a fascinating wide band-gap semiconductor material [1-3], suitable for high power and high temperature electronic devices [4] because of its suitable properties, such as high electron mobility, high thermal conductivity, high chemical stability, high mechanical hardness, high break down electric field and small dielectric constant [4, 5]. Additionally, many researchers have reported the stability of silicon carbide micro-electromechanical systems (MEMS) under corrosive conditions using acid and alkaline chemical reagents [6-9]. Polycrystalline 3C-silicon carbide is widely used for various purposes, such as dummy wafers and reactor parts, in silicon semiconductor device production processes.

In the semiconductor materials production technology [10], the electronics devices manufacturing process needs an easy and cost effective technique, such as wet and/or dry cleaning, for preparing the clean surface of the substrate materials. However, the suitable properties of silicon carbide often provided difficult Problems. The chemically and mechanically stable nature often makes it very difficult to prepare the entire surface in the wafer production process, such as surface polishing and removal of any damaged layer. Useful chemical reagents and processes should be developed for silicon carbide material production.

Wet and dry etching methods of silicon carbide have been studied by many researchers [5, 11-21], using various gases and various wet etchants. However, the largest etching rate reported was nearly 1 μm/min. Here, chlorine trifluoride (ClF_3) gas is very reactive even at low temperatures and has a very strong capability to etch various materials, such as silicon [22] without plasma assistance.

In Section 2, details of polycrystalline 3C-silicon carbide etching using chlorine trifluoride gas [23, 24] are reviewed, particularly focusing on the etching rate, gaseous products, surface chemical bonds and the surface morphology of the silicon carbide. In Section 3, the dry etching of single-crystalline 4H-silicon carbide using chlorine trifluoride gas [25-29] over the wide temperature range of 570-1570 K is reviewed, particularly about the etching rate, surface chemical reaction rate constant, surface morphology and etch pits.

2. Polycrystalline 3C-Silicon Carbide Etching Using Chlorine Trifluoride Gas

2.1. Reactor and processes using chlorine trifluoride gas

In order to etch silicon carbide by chlorine trifluoride gas, the horizontal cold-wall reactor shown in Figure 1. is used. This reactor consists of a gas supply system, a quartz chamber and infrared lamps. A 30 mm wide x 40 mm long x 0.2-1 mm thick 3C-silicon carbide substrate manufactured using chemical vapor deposition (CVD) (Admap Inc., Tokyo) is held horizontally on the bottom wall of the quartz chamber.

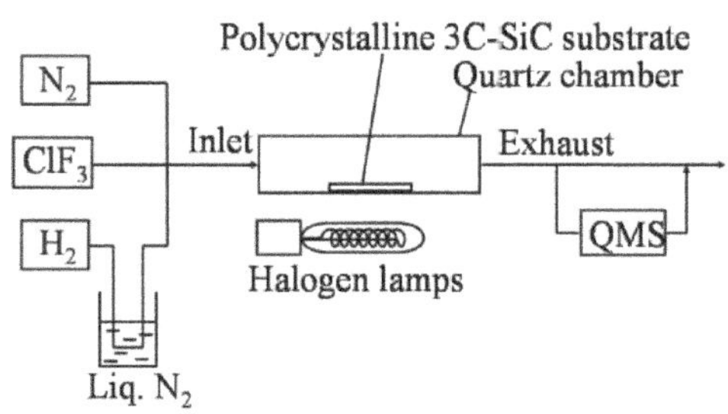

Figure 1. Horizontal cold-wall reactor used for etching polycrystalline 3C-silicon carbide substrate.

The gas supply system introduces chlorine trifluoride gas, nitrogen gas and hydrogen gas. Hydrogen gas is used to remove the silicon oxide film on the silicon carbide substrate surface, the same as those on the silicon surface [22]. The height and width of the quartz chamber are compactly designed to be 10 mm and 40 mm, respectively, similar to the chamber in our various studies [22, 30].

The etching using chlorine trifluoride gas is carried out following the process shown in Figure 2. There are mainly three steps.

(a) cleaning the silicon carbide substrate surface by baking in ambient hydrogen at 1370 K for 10 min,

(b) changing the gas from hydrogen to nitrogen, and

(c) etching the silicon carbide substrate surface using chlorine trifluoride gas.

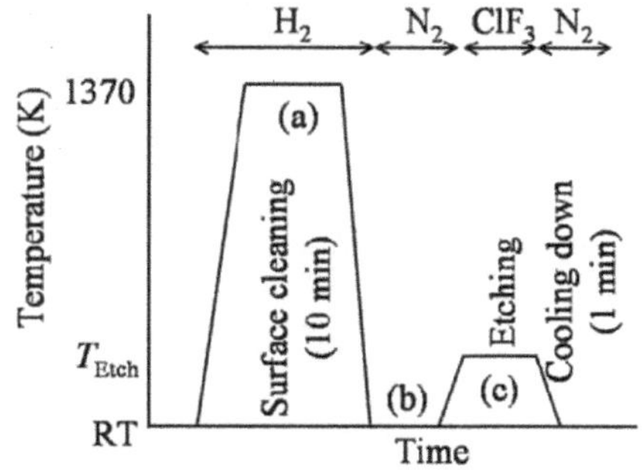

Figure 2. Process for cleaning and etching polycrystalline 3C-silicon carbide surface.

During step (a), hydrogen gas is introduced at atmospheric pressure into the reactor at a flow rate of 2 slm. Next, in step (b), the quartz chamber and the silicon carbide substrate are cooled to room temperature. The hydrogen gas present in the quartz chamber must be sufficiently purged with nitrogen gas to avoid an explosive reaction between hydrogen and chlorine trifluoride. During step (c), the silicon carbide substrate is heated and adjusted to temperatures between 670 K and 970 K. The silicon carbide substrate is etched by chlorine trifluoride (>99.9 %, Kanto Denka Kogyo Co., Ltd., Tokyo) at a flow rate of 0.1-0.25 slm without further purification and without dilution. In order to evaluate the gaseous products, part of the exhaust gas from the reactor is fed into a quadrupole mass spectra (QMS) analyzer, as shown in Figure 1. The silicon carbide etching rate using chlorine trifluoride gas is evaluated from the decrease in the weight of the silicon carbide substrate. The surface morphology of the polycrystalline 3C-silicon carbide substrate before and after the etching is observed using an optical microscope. The root-mean-square (RMS) surface roughness and the average roughness, R_a, are measured. In order to evaluate the condition of the chemical bonds of the silicon carbide surface before and after the etching, X-ray photoelectron spectra (XPS) are obtained.

2.2. Etching rate

The etching rate of the polycrystalline 3C-silicon carbide substrate surface is shown in Figure 3, which was obtained using 100% chlorine trifluoride gas at various gas flow rates in the temperature range of 670 to 970 K at atmospheric pressure. The squares, circles and triangles show the etching rate at the chlorine trifluoride gas flow rate of 0.2, 0.1 and 0.05 slm, respectively. As shown in Figure 3, the etching rate at the substrate temperature less than 670K is quite low; its value is less than 1 μm/min. However, with the increasing substrate temperature, the etching rate significantly increases particularly near 720 K. At the substrate temperature of 770 K, the etching rate at the flow rate of 0.2 slm becomes 20 μm/min; it remains constant at substrate temperatures greater than 770 K.

As shown in Figure 3, the etching rate changes with the flow rates. The etching rates are 25, 10 and 5 μm/min at the flow rates of 0.2, 0.1 and 0.05 slm, respectively. For each chlorine trifluoride gas flow rate, the trend in the flat etching rate at temperature greater than 770K is maintained.

In order to evaluate the influence of chlorine trifluoride gas concentration, the etching rate of the polycrystalline 3C-silicon carbide substrate surface using 10-100% chlorine trifluoride gas in ambient nitrogen was measured at the flow rate of 0.2 slm, atmospheric pressure and 670-970 K, as shown in Figure 4. In this figure, the square, reverse triangle, circle, diamond, and triangle show the substrate temperatures of 670, 730, 770, 870 and 970 K, respectively. Being consistent with Figure 3, the etching rate at the substrate temperature of 670K is very low. The etching rates at 730 K are significantly higher that those at 670 K.

At the substrate temperature of 770 K, the etching rate is proportional to the chlorine trifluoride gas concentration. At 870 and 970 K, the etching rate at each chlorine trifluoride gas concentration is the same as that at 770 K. Therefore, when the substrate temperature is higher than 770 K, the etching rate over a very wide chlorine trifluoride gas concentration range is not affected by the substrate temperature.

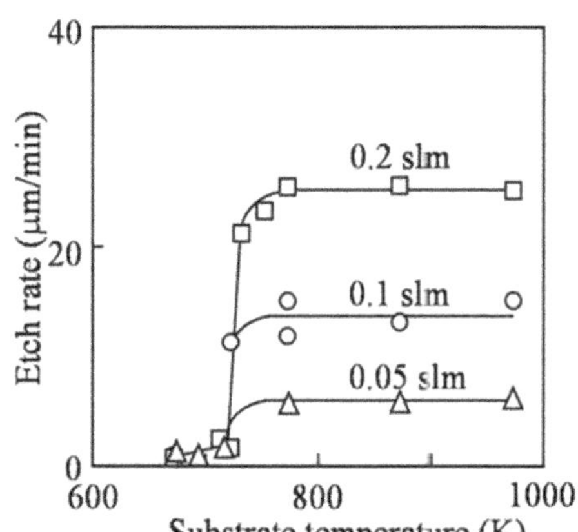

Figure 3. Etching rate of the polycrystalline 3C-silicon carbide substrate surface by chlorine trifluoride gas (100%) at atmospheric pressure in the temperature range between 670 and 970 K. Square: 0.2 slm, circle: 0.1 slm, and triangle: 0.05 slm.

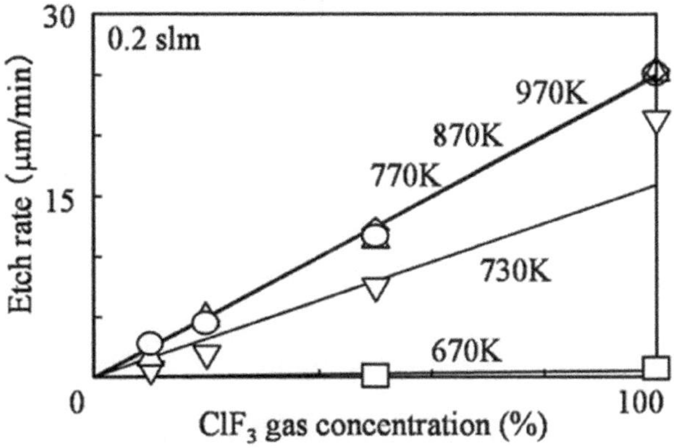

Figure 4. Etching rate of the polycrystalline 3C-silicon carbide substrate surface using chlorine trifluoride gas at 10-100%, 0.2 slm, atmospheric pressure, and 670-970K. Square: 670 K, reverse triangle: 730K, circle: 770K, diamond: 870 K, and triangle:970K.

2.3. Surface morphology and roughness

The change in the surface morphology of the 3C-silicon carbide is explained. Figure 5 shows photographs of the silicon carbide surface etched using chlorine trifluoride gas at the flow rate of 0.1 slm and atmospheric pressure at 670, 720 and 770 K for 15 min. The values in parentheses are the etch depth. Figure 5 (a) shows the initial surface which has the periodically line-shaped hills and valleys. As shown in Figure 5 (b), although the surface still has the line-shaped morphology at 670 K after 15 minutes, its pattern remains but is unclear. At the higher temperatures of 720 and 770 K, the line-shaped appearance does not remain, as shown in Figures 5 (c) and (d). In these figures, there are very small and very shallow pits having a round edge. This shows that the etching using chlorine trifluoride can smooth the large hills and valleys which existed on the silicon carbide surface.

In order to show the detail of surface smoothing effect, the change in the surface appearance is shown, in Figure 6, along etch period at a substrate temperature of 770 K and a flow rate of 0.1 slm of chlorine trifluoride. This figure shows photographs of the etched silicon carbide surface at (a) 0 min, (b) 5 min, (c) 10 min, (d) 15 min, and (e) 30 min. The values in parenthes are the etch depth. Figures 6 (a) and (d) are the same as Figures 5 (a) and (d), respectively. The relationship between the etch period and the etch depth is shown in Figure 7 (a). The line-shaped pattern in Figure 6 (a) is slightly rounded after 5 minutes. At 10 minutes, there is only a trace of the line-shaped appearance, as shown in Figure 6 (c). The 3C-silicon carbide surface has a round-shaped morphology after 15 minutes as shown in Figure 6 (d), since the line-shaped pattern is removed during the etch period between 10 and 15 minutes. The surface morphology in Figure 6 (d) is maintained at 30 minutes in Figure 6 (e), and the rounded edges of the very shallow pits do not become sharp during the last 15 minutes. Thus, the round-shaped morphology of silicon carbide surface will be maintained during etching for longer than 30 minutes.

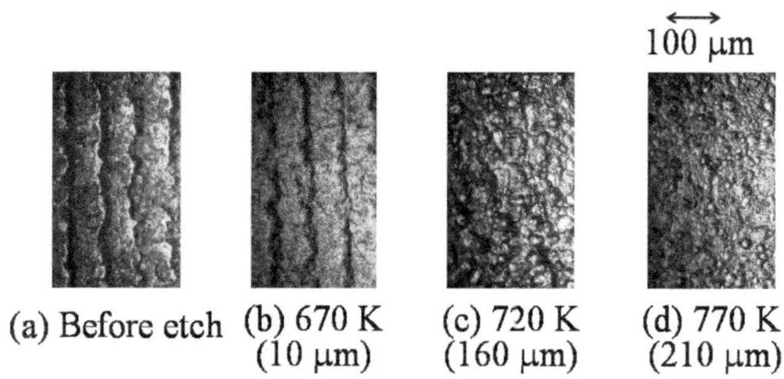

Figure 5. Photograph of the silicon carbide substrate surface etched using chlorine trifluoride at atmospheric pressure after 15 min at (b) 670 K, (c) 720 K, and (d) 770 K. (a) is the silicon carbide substrate surface before etching. The values in parentheses are the etch depth.

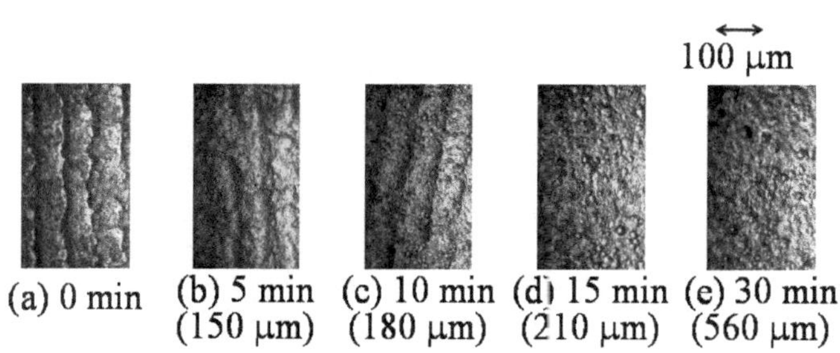

Figure 6. Photograph of the silicon carbide surface etched using chlorine trifluoride at atmospheric pressure and 770 K at (a) 0 min, (b) 5 min, (c) 10 min, (d) 15 min, and (e) 30 min. The values in parenthes are the etch depth.

In order to evaluate the surface smoothing effect of silicon carbide by chlorine trifluoride gas, the surface roughness is measured using the root-mean-square (RMS) roughness as shown in Figure 7 (b). Figure 7 (a) also shows the etch depth. The initial surface has an RMS roughness of 5 μm. The surface roughness decreases with increasing etch period. At 10 minutes, when 180 μm has been etched, the RMS roughness becomes a low value of 1 μm. Consistent with Figures 6 (d) and (e), the RMS roughness is maintained at nearly 1 μm at 30 minutes when the etch depth becomes greater than 500 μm. Solid line in Figure 7 (b) indicates the possibility that the chlorine trifluoride gas has a smoothing effect on the 3C-silicon carbide surface.

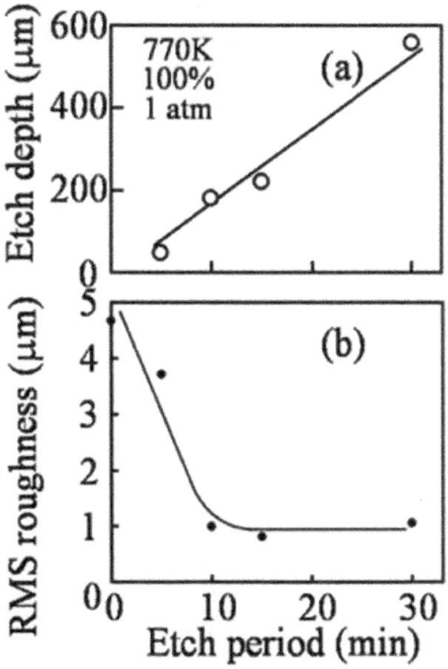

Figure 7. Etch depth and roughness of the silicon carbide surface etched by chlorine trifluoride gas at atmospheric pressure and 770 K within 30 min: (a) etch depth, and (b) root mean square (RMS) roughness.

Figure 8 shows photographs of the polycrystalline 3C-silicon carbide surface etched using a gas mixture of chlorine trifluoride and nitrogen at 670-870K for 15 min. The concentration and the flow rate of the gas mixture are 10-100% and 0.2 slm, respectively. Figure 9 shows the RMS roughness of the polycrystalline 3C-silicon carbide surface etched using chlorine trifluoride gas at atmospheric pressure and 670-870K for 15 min. The triangle, diamond and circle show the RMS roughness at the chlorine trifluoride gas concentrations of 10, 50 and 100 %, respectively.

The photograph indicated by 'Before etch' in Figure 8 shows the initial surface, which has very narrow, vague and shallow trenches formed by mechanical polishing. Using the chlorine trifluoride gas concentration of 10%, the change in the surface morphology is explained. The surface etched at 730K and 10% is recognized to have circular-shaped pits. Although the etching rate under this condition is very low, its surface shown in Figure 8 has an etched depth of 15 μm. This surface shows many circle-like pits, the edge of which is clearly shown. These may be the grain boundary or some disordered region which can be etched at a slightly higher etching rate.

At 770 K and 10 %, the shape of the circular-shaped pits still clearly exists, similar to that at 730 K and 10 %. The photograph at 870 K and 10% shows pits smaller than those at 770 K and 10%. Simultaneously, the conical shape of the pits still exists. This shows that chlorine trifluoride gas at the low concentration of 10% has a quite small role of smoothing the surface, but rather tends to roughen it. This trend is measured using the RMS roughness, shown by the triangles in Figure 9. The RMS roughness before etching is nearly 0.5 μm; it increases with the increasing substrate temperature at the chlorine trifluoride gas concentration of 10%.

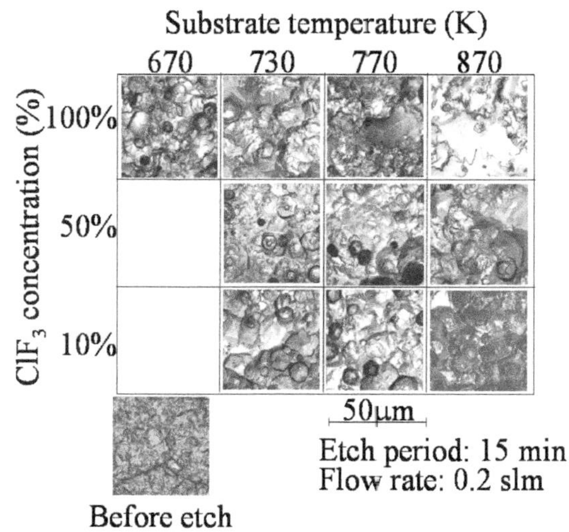

Figure 8. Photograph of the polycrystalline 3C-silicon carbide surface etched using chlorine trifluoride gas at atmospheric pressure for 15 min at 670-870 K, 10-100% and 0.2 slm.

Next, at the fixed chlorine trifluoride gas concentration of 50%, the change in the etched surface morphology is explained using Figure 8. The surface etched at 730 K and 50% shows the clear edge shape of the pits. The surface etched at 770 K and 50% still has a clear edge of the conical-shaped pits. Although some pits still have such the clearly observed edge shape, the rest of the surface has no clear edges. The surface morphology at 870 K and 50% shows both clear and vague edges. Since semi-smoothed and clear pits coexist there, the RMS roughness, indicated by the diamonds in Figure 9, still slightly increases with the increasing substrate temperature.

The change in the surface morphology etched at 100% is also shown in Figure 8. The surface etched at 670 K and 100% shows the clear edge of the pits. In contrast to this, the surface etched at 730 K and 100% shows the slightly vague edge of the pits. For the surface etched at 770 K and 100 %, the conical-shaped pits still remain, but are few. The edge of the conical-shaped pits disappears, when the surface is etched at 870 K. Since this trend in smoothing the surface appears in the RMS roughness behavior, the RMS roughness at 100%, indicated using circles in Figure 9, slightly decreases with the increasing substrate temperature.

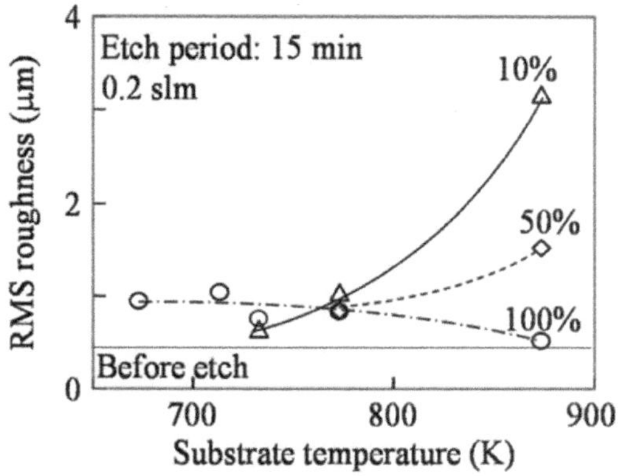

Figure 9. RMS roughness of the polycrystalline 3C-silicon carbide substrate surface etched using chlorine trifluoride gas at atmospheric pressure and 670-870 K for 15 min. Triangle: 10 %, diamond: 50 %, and circle: 100%.

From the view point of the effect of the etchant gas concentration at each temperature, the shape of the pits tends to become unclear with the increasing chlorine trifluoride gas concentration. Therefore, as the overall trend, the higher temperature and the higher chlorine trifluoride gas concentration produces a smoother surface of the polycrystalline 3C-silicon carbide.

The polycrystalline 3C-silicon carbide etching rate can be adjusted using the combination of gas flow rate, gas concentration and the substrate temperature, in order to obtain surfaces suitable for various purposes. This technique is expected to be used for various applications, such as the dry cleaning of the silicon carbide substrate surface instead of wet method, and the removal of the damaged layer formed during the chemical mechanical polishing using diamond slurry.

2.4. Surface chemical condition and etching rate

The fraction of silicon and carbon on the silicon carbide surface remaining after the etching is useful information for developing various processes which are performed after the etching. Thus, the chemical bonds at the silicon carbide surface are measured using X-ray photoelectron spectroscopy (XPS) before and after etching by chlorine trifluoride gas. Figure 10 shows the fraction of carbon, silicon, oxygen, chlorine and fluorine on the silicon carbide substrate before and after etching the depth greater than 150 μm using the chlorine trifluoride gas for 15 min at atmospheric pressure and at 720 K.

The silicon carbide surface before the etching has a carbon fraction nearly equal to that of silicon. However, the fraction of carbon significantly increases to 75 % after the etching. This indicates that the production of volatile carbon compound is slower than that of silicon compound at this temperature.

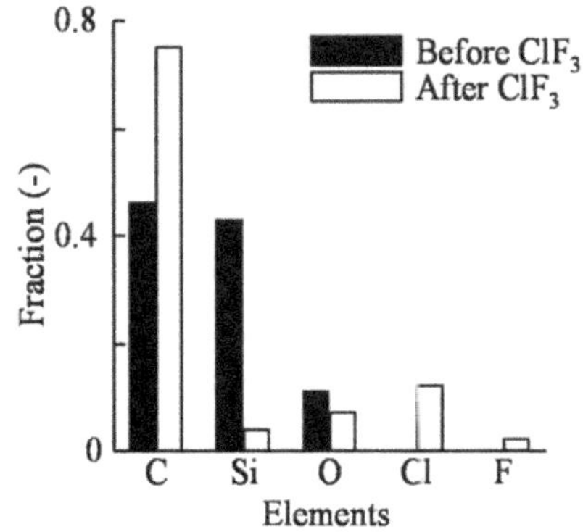

Figure 10. Fraction of carbon, silicon, oxygen, chlorine and fluorine on the silicon carbide substrate surface before and after etching using chlorine trifluoride for 15 min at atmospheric pressure in the reactor. The temperature of the silicon carbide substrate is 720 K.

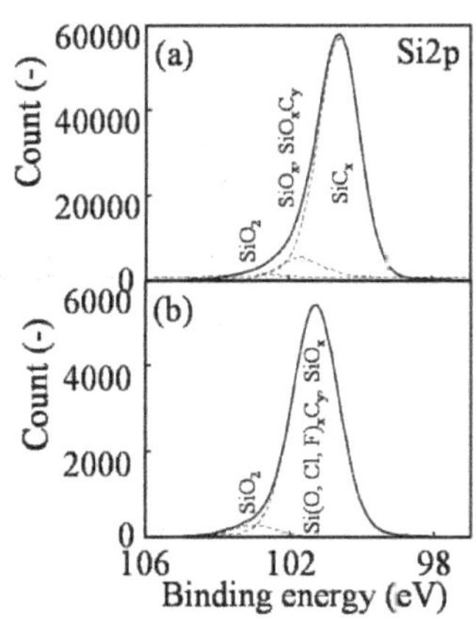

Figure 11. XPS spectra of Si 2p measured (a) before and (b) after etching using chlorine trifluoride gas for 15 min at atmospheric pressure. The temperature of the silicon carbide substrate is 720 K.

In order to evaluate the state of silicon and carbon, the XPS spectra of Si 2p and C 1s are shown in Figures 11 and 12, respectively. The conditions of etching are the same as in Figure 10.

An almost amount of silicon at the silicon carbide surface has chemical bond with carbon before etching as shown in Figure 11 (a). However, after the etching, a significant amount of silicon oxides and oxidized or halogenated silicon carbide are present as shown in Figure 11 (b). The chemical bonds of carbon simultaneously change. as same as those of silicon. Figure

12 (a) shows that chemical bonds of carbon with silicon dominate at the substrate surface before etching. After etching, an almost amount of carbon has chemical bonds with carbon as shown in Figure 12 (b). The surface after etching is covered with large amount of carbon having carbon-carbon bonds. This result is consistent with the dark appearance of the silicon carbide surface after the etching.

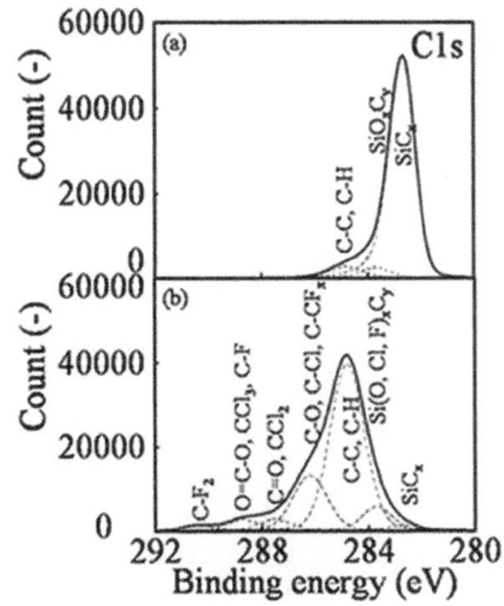

Figure 12. XPS spectra of C 1s measured (a) before and (b) after the etching using chlorine trifluoride gas for 15 min at atmospheric pressure. The temperature of the silicon carbide substrate is 720 K.

2.5. Chemical reactions

The gaseous products and the chemical reactions associated with silicon carbide etching are explained. Figure 13 shows the mass spectra of the gaseous species existing in the exhaust gas immediately after beginning silicon carbide substrate etching at 770 K using chlorine trifluoride gas at atmospheric pressure. The partial pressures are normalized using the pressure at the mass of 28 a.m.u., which is the largest partial pressure in this measurement and which can be assigned to silicon from the silicon tetrafluoride and nitrogen remaining in the QMS system.

Typical mass spectra, shown in Figure 13, are interpreted by taking into account the fragmentation in the QMS analyzer and the isotopic abundance of chlorine [31-33]. In this figure, the ion species at a mass of 14 a.m.u. is assigned to N^+, which is the fragment of nitrogen gas. The low partial pressures corresponding to chlorine trifluoride and its fragment, $ClF_3{}^+$, at masses of 92 and 94 a.m.u., and ClF^+ at 54 and 56 a.m.u., are detected. The partial pressures observed at masses of 70, 72 and 74 a.m.u. correspond to $Cl_2{}^+$, which is assigned to chlorine gas. Chlorine gas is produced due to the chemical reaction during silicon carbide etching, similar to that for the silicon etching [22, 34]. The partial pressures at masses of 35 and 37 a.m.u. can be assigned to Cl^+, which is a fragment from both chlorine trifluoride and chlorine (Cl_2). The partial pressures observed at masses of 19 and 20 a.m.u. correspond to F^+ and HF^+, respectively. F^+ and HF^+ are produced due to the fragmentation of chlorine trifluoride. Since the partial pressure of fluorine (F_2) at mass 38 a.m.u. did not appear in the mass spectra, the thermal dissociation of chlorine trifluoride gas [35] is negligible in the gas phase of the cold wall reactor used in this study.

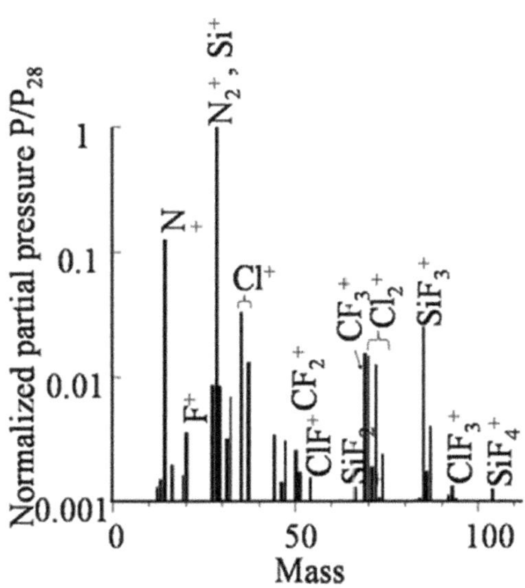

Figure 13. Mass spectra of gaseous species existing in the exhaust gas from the reactor during the etching of the silicon carbide surface using chlorine trifluoride gas at atmospheric pressure. The temperature of the silicon carbide substrate is 770 K. The ionization conditions are 70 eV and 1.73 mA.

The partial pressures at masses of 66, 85 and 104 a.m.u. can be assigned to $SiF_2{}^+$, $SiF_3{}^+$ and $SiF_4{}^+$, respectively, whose parent is silicon tetrafluoride, like silicon etching [34]. The gaseous carbon compound produced by etching is identified as carbon tetrafluoride (CF_4), because the partial pressures at masses of 50 and 69 a.m.u. correspond to $CF_2{}^+$ and $CF_3{}^+$, respectively, which can be assigned as fragments of carbon tetrafluoride.

Thus, silicon and carbon become the gaseous species of silicon tetrafluoride and carbon tetrafluoride, respectively. The overall chemical reaction between silicon carbide and chlorine trifluoride is as follows:

$$3SiC + 8ClF_3 \rightarrow 3SiF_4 + 3CF_4 + 4Cl_2 \quad (1)$$

3. Single-Crystalline 4H-Silicon Carbide Etching Using Chlorine Trifluoride Gas

3.1. Substrate, reactor and process

The substrate used is the n-type single-crystalline 4H-silicon carbide wafer having a (0001) surface, 8-degrees off-oriented to <11-20>. This substrate has nitrogen as the n-type dopant at the concentration of 3 - 5 x 10^{18} cm^{-3}. The 4H-silicon carbide substrate, having 5 mm wide x 5 mm long x 400 μm thick dimensions, is placed on the center of the polycrystalline 3C-silicon carbide susceptor, which has the dimension of 30 mm wide × 40 mm long × 0.2 mm thick produced by the chemical vapor deposition method (Admap Inc., Tokyo).

The reactor shown in Figure 1 is used following the process shown in Figure 2 except of hydrogen baking. The Si-face (0001) and C-face (000-1) of the 4H-silicon carbide substrates are etched using chlorine trifluoride gas. The etching is performed at the temperatures between 570 – 1570 K at the chlorine trifluoride gas flow rate of 0.1 – 0.3 slm. The average etching rate of 4H-silicon carbide is determined by the decrease in the substrate weight. Because the 3C-silicon carbide susceptor and the 4H-silicon carbide substrate are simultaneously etched in the reactor, the etching rate obtained is comparable to the average value for the wide 4H-silicon carbide substrate.

The X-ray topograph of Si-face and C-face of 4H-silicon carbide was taken at the beam-line BL15C of the Photon Factory of the High Energy Accelerator Research Organization (Proposal No. 2006G286), in order to evaluate the crystalline defects. The density and behavior of etch pit produced on the 4H-silicon carbide substrate using chlorine trifluoride gas at various temperatures were evaluated.

3.2. Numerical calculation of etching rate

The etching rate of single-crystalline 4H-silicon carbide is numerically calculated [29]. The geometry of horizontal cold-wall reactor, shown in the previous section (Figure 1), is taken into account for a series of calculations. In order to evaluate the silicon carbide etching rate and the overall rate constant in steady state in non-uniformly distributed temperature and gas flow fields, the two-dimensional equations of mass, momentum, energy, species transport and surface chemical reaction, linked with the ideal gas law, are solved. The discretized equations are coupled and solved using the SIMPLE algorithm [36] on a CFD software package, Fluent version 6 (Fluent, Inc., Lebanon, NH, USA).

The silicon carbide etching is assumed to follow the overall reaction in Eq. (2) [23, 24],

$$3SiC \cdot + \cdot 8ClF_3 \cdot \rightarrow \cdot 3SiF_4 \cdot + \cdot 3CF_4 \cdot + \cdot 4Cl_2 \quad (2)$$

Mass changes due to the chemical reaction of Eq. (2) are taken into account in the boundary conditions at the surface of silicon carbide. The overall reaction shown in Eq. (2) is assumed to be a first-order reaction.

$$\text{SiC etching rate} = 6 \times 10^7 M_{SiC} k_{SiC} [ClF_3] \cdot / \cdot \rho_{SiC} \quad (\mu mm\ min^{-1}), \tag{3}$$

where ρ_{SiC} is the density of solid silicon carbide (kg m^{-3}). M_{SiC} is the molecular weight of silicon carbide (kg mol^{-1}), respectively. The factor 6 x 10^7 is used for the unit conversion of m s^{-1} to μm min^{-1}. k_{SiC} is the overall rate constant for the reaction of Eq. (3). [ClF_3] is the concentration of chlorine trifluoride gas at the silicon carbide surface (mol m^{-3}). The concentration of each species at the surface is governed by a balance between the consumption due to the chemical reaction and the diffusion fluxes driven by the concentration.

The gas velocity and pressure at the inlet are 0.08 m s^{-1} and 1.0133 x 10^5 Pa, respectively. The heat capacities of chlorine trifluoride, nitrogen, tetrafluorosilane, tetrafluorocarbon and chlorine are taken from the literature [37]. The gas properties, such as the viscosity and the thermal conductivity of chlorine trifluoride, tetrafluorosilane, chlorine and nitrogen are estimated with the method described in the literature [38]. The Lennard-Jones parameters of σ and ε/k for chlorine trifluoride are 4.63 Angstroms and 355 K, respectively, which are obtained using a theoretical equation [38] taking the value of viscosity [39] into account. Each physical constant is expressed as a function of temperature. The properties of the mixed gas are estimated theoretically [40]. The binary diffusion coefficients of chlorine trifluoride, tetrafluorosilane and chlorine are estimated using the method described in the literature [38].

The overall rate constant, k_{SiC} in Eq. (3) is obtained so that the calculated etching rates agree with those measured at various conditions.

3.3. Etching rate

Figure 14 shows the etching rate of the Si-face and C-face of 4H-silicon carbide at the substrate temperatures between 570 K and 1570 K. The etching rate of the C-face of 4H-silicon carbide is slightly higher than that of the Si-face of 4H-silicon carbide. The etching rate of the Si-face and C-face of 4H-silicon carbide is near 5 μm min^{-1} and it is still flat at the temperatures between 770 K and 1570 K. This flat etching rate behavior is similar to that of polycrystalline 3C-silicon carbide, shown in Figure 3 in the previous section. The various surface morphologies at higher temperatures shown in the latter part of this section are obtained at nearly the same etching rate.

Figure 15 shows the typical behavior of the etching rate changing with the chlorine trifluoride gas concentration, obtained at 1370 K. The etching rate is proportional to the chlorine trifluoride gas concentration, similar to that of polycrystalline 3C-silicon carbide, shown in Figure 4. The fluctuation of the etching rate, shown in Figures 14 and 15, is entirely 20 %, which is due to the considerable distribution of the etching rate over the 3C-silicon carbide susceptor.

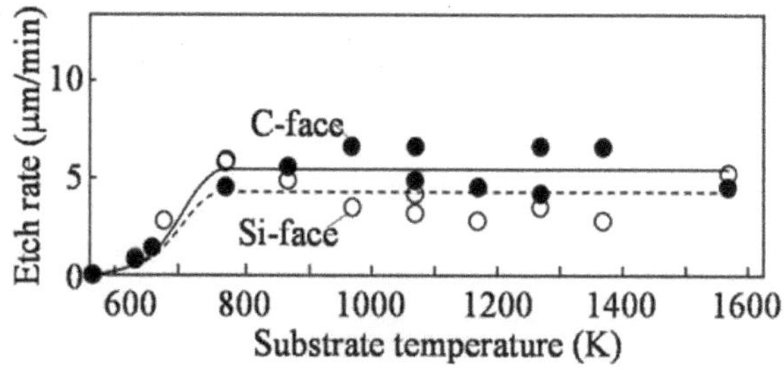

Figure 14. Etching rate of 4H-silicon carbide using chlorine trifluoride gas at 100 %, 0.1 slm and various temperatures. Dark circle: C-face, and white circle: Si-face.

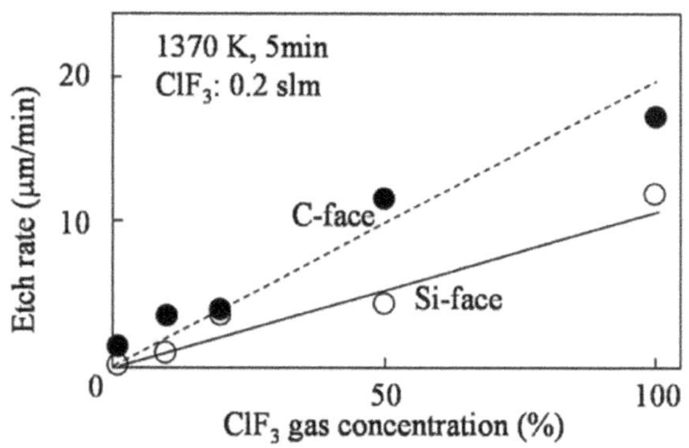

Figure 15. Etching rate of Si-face (white circle) and C-face (dark circle) 4H-silicon carbide changing with chlorine trifluoride gas concentration, at 1370 K and at the total gas flow rate of 0.2 slm for 5 min.

The relationship between the etching rate and the chlorine trifluoride gas flow rate is shown in Figures 16. Figures 16 (a) and 16 (b) show the etching rate of the Si-face and the C-face, respectively, of the 4H-silicon carbide substrate by chlorine trifluoride gas (100%) at atmospheric pressure, when changing with the chlorine trifluoride gas flow rate. The circle, square and triangle denote the etching rate at the substrate temperatures of 770, 870 and 970 K, respectively. As shown in these figures, the Si-face and C-face etching rates increase with the gas flow rate of the chlorine trifluoride gas. Additionally, the etching rate at 770, 870 and 970 K overlap each other for both the Si-face and the C-face, consistent with Figures 14. The etching rate of the Si-face is about 60 % of that of the C-face. The relationship between the Si-face etching rate and the C-face etching rate is similar to that of another empirically known etching technique, such as the potassium hydroxide method [41].

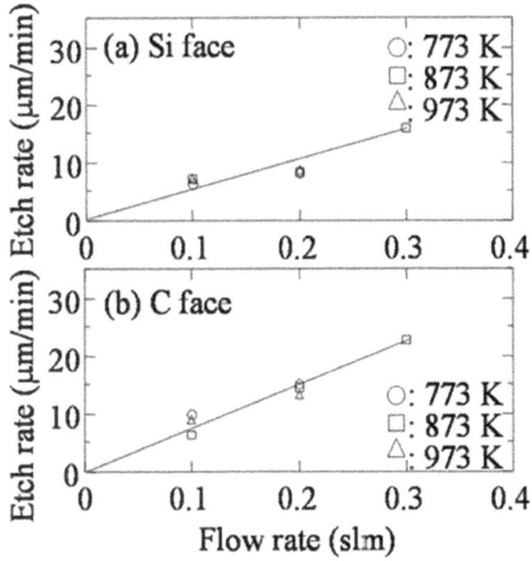

Figure 16. Etching rate of (a) Si-face and (b) C-face of 4H-silicon carbide substrate by chlorine trifluoride gas (100%) at atmospheric pressure in the flow rate range between 0.1 and 0.3 slm. Circle, square and triangle show the etching rates at the substrate temperatures of 770, 870 and 970 K, respectively.

3.4. Surface reaction rate constant

Figure 17 is the Arrhenius plot of the rate constants for etching of Si-face and C-face of 4H-silicon carbide. The rate constants is expressed in Eqs. (4) and (5).

$$k_{SiC}=4.1\ \exp\left(-6.6\times 10^{4}/RT\right) \quad (m/s) \quad \text{for Si-face} \tag{4}$$

and,

$$k_{SiC}=98\times \exp\left(-8.3\times 10^{4}/RT\right) \quad (m/s) \quad \text{for C-face} \tag{5}$$

where R is the gas constant (J mol^{-1} K^{-1}).

In order to show that the rate constant of Eq. (4) can reproduce the measured etching rate behavior, the measured and the calculated etching rate values of Si-face are shown in Figure 18, as the Arrhenius plot. The calculation shows that the etching rate at the temperatures near 670K is near 1 μm/min, and it becomes near 10 μm/min at 1000 K. The calculated etching rate tends to become flat at the higher temperatures at the chlorine trifluoride gas flow rate of 0.1 and 0.2 slm. Additionally, the etching rate obtained by the calculation increases with increasing the chlorine trifluoride gas flow rate. Because the great etchant flow rate can moderate the etchant depletion occurring in the downstream region, the average etching rate over the 4H-silicon carbide substrate and the 3C-silicon carbide susceptor can increase with the increasing flow rate of chlorine trifluoride gas. The etching rate for C-face calculated using Eq. (5) also showed the typical behavior of the measurement. Therefore, Eqs. (4) and (5) are applicable to reproduce the behavior of the 4H-silicon carbide (Si-face and C-face) etching rate.

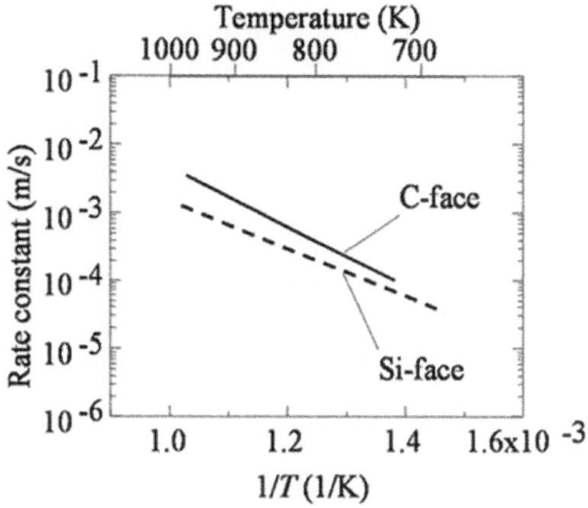

Figure 17. Rate constants for etching of Si-face (broken line) and C-face (solid line) of 4H-silicon carbide using chlorine trifluoride gas, obtained by numerical calculation.

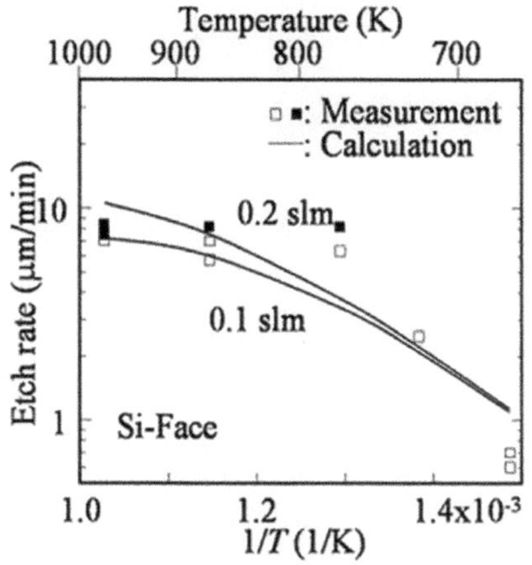

Figure 18. Arrhenius plot of 4H-silicon carbide Si-face etching rate at chlorine trifluoride flow rate of 0.1 and 0.2 slm. Square: measurement, solid line: calculation

3.5. Surface morphology

Figure 19. shows the surface morphology of the Si-face of 4H-silicon carbide before and after the etching at the chlorine trifluoride gas concentration of 100 % and at the flow rate of 0.1 slm. The etching was performed at the substrate temperatures of (b) 570, (c) 620, (d) 770, (e) 970, (f) 1270, (g) 1370 and (h) 1570 K. The etched depth was 5-18 μm.

Figure 19 (a) shows the surface morphology of the Si-face of 4H-silicon carbide substrate before the etching. Figure 19 (b) shows that there are many small pits after the etching at 570 K. At 620 K, the pits are very large, nearly 50 μm in diameter, as shown in Figure 19 (c). With the increasing temperature, the pits tends to become small and shallow, as shown in Figures 19 (d) - (h). The pit diameter after etching at 770 K, shown in Figure 19 (d), is nearly 10 % of that at 620 K, shown in Figure 19 (c). Figure 19 (e) shows that the pit diameter becomes even smaller and the pit density decreases at 970 K. This trend is very clear at the temperatures higher than 1270 K, as shown in Figures 19 (f), (g) and (h). Figure 19 (g) shows that the pit density significantly decreases at 1370 K. The sharp-shaped pits, presented at the lower temperatures, are not there, on the silicon carbide surface after the etching at 1570 K, as shown in Figure 19 (h).

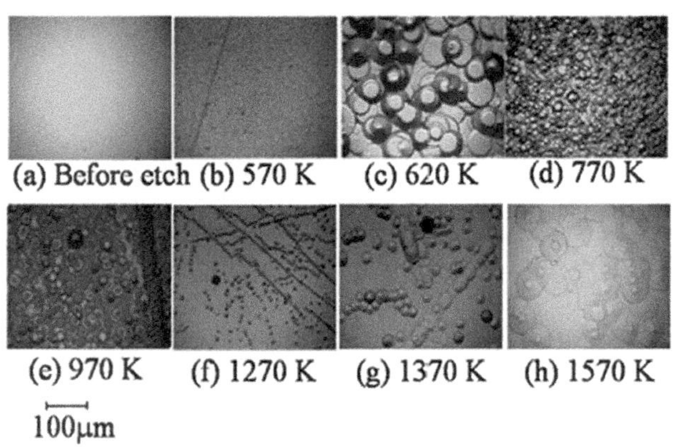

Figure 19. Surface morphology of Si-face of 4H-silicon carbide (a) before and after the etching using chlorine trifluoride gas at the concentration of 100 %, at the substrate temperature of (b) 570, (c) 620, (d) 770, (e) 970, (f) 1270, (g) 1370 and (h) 1570 K and at the flow rate of 0.1 slm. The etched depth is 5-18 μm.

Next, the influence of chlorine trifluoride gas concentration on the surface morphology is explained. Figure 20 shows the surface morphology of the Si-face 4H-silicon carbide surface taken by the optical microscope before and after the etching for 5 min at the total flow rate of 0.2 slm, at the substrate temperature of 1370 K and at various chlorine trifluoride gas concentrations. Figure 20 (a) shows the surface morphology of the Si-face of 4H-silicon carbide before the etching.

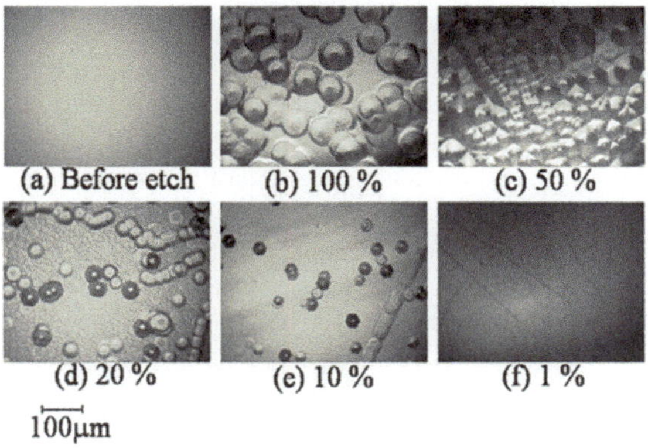

Figure 20. Surface morphology of Si-face 4H-silicon carbide surface (a) before and after etching at the total gas flow rate of 0.2 slm, 1370 K and various chlorine trifluoride gas concentrations of (b) 100, (c) 50, (d) 20, (e) 10 and (f) 1%, for 5 min.

Although there are many large pits at the chlorine trifluoride gas concentration of 100%, as shown in Figure 20 (b), they become significantly small and less at 20%, as shown in Figure 20 (d). Figure 20 (e) shows that the surface etched at 10% is flat with only a small number of pits. At 1%, most of the surface is flat, except for scratches, as shown in Figure 20 (f).

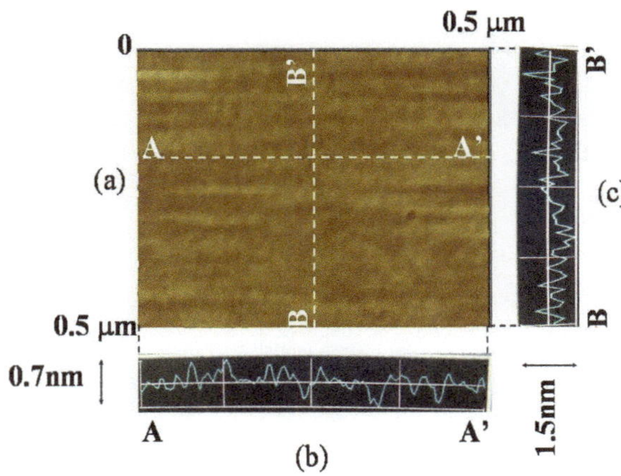

Figure 21. AFM photograph of Si-face of 4H-silicon carbide, etched for 0.5 min using the chlorine trifluoride gas concentration of 1% diluted in ambient nitrogen at the substrate temperature of 1570 K and at the total flow rate of 4 slm. (a): plan view, (b): A-A' cross section, and (c): B-B' cross section.

From Figures 19 and 20, the Si-face 4H-silicon carbide surface after etching tends to be flat with the increasing temperature and decreasing chlorine trifluoride gas concentration. Following this trend, the Si-face 4H-silicon carbide surface is etched at 1570 K at the chlorine trifluoride gas concentration of 1% for 0.5 min. Figure 21 shows the AFM photograph of the etched surface. Figures 21 (a), (b) and (c) are the plan view, A-A' cross section and B-B' cross section, respectively. Although the etched depth is only about 0.03 μm, it can reveal the trend of the surface, causing pit or not. As shown in Figure 21 (a), this surface does not show any etch pit; Figures 21 (b) and (c) shows no periodical shape reflecting the 4H-silicon carbide crystal step [42]. The root-mean-square (RMS) roughness are 0.1 and 0.2 nm on A-A′ line and B-B′ line, respectively, which are comparable to that of the polished 4H-silicon carbide substrate surface. Thus, the shallow etching for removing thin layer, such as damaged layer, is possible with maintaining the specular surface of the Si-face of 4H-silicon carbide.

Figure 22. shows the surface morphology of the C-face of 4H-silicon carbide before and after the etching at the chlorine trifluoride gas concentration of 100 %, at (b) 570, (c) 620, (d) 770, (e) 1070, (f) 1270, (g) 1370 and (h) 1570 K and at the flow rate of 0.1 slm. The etched depth is 10-30 μm. Figure 22 (a) is the C-face 4H-silicon carbide surface before the etching.

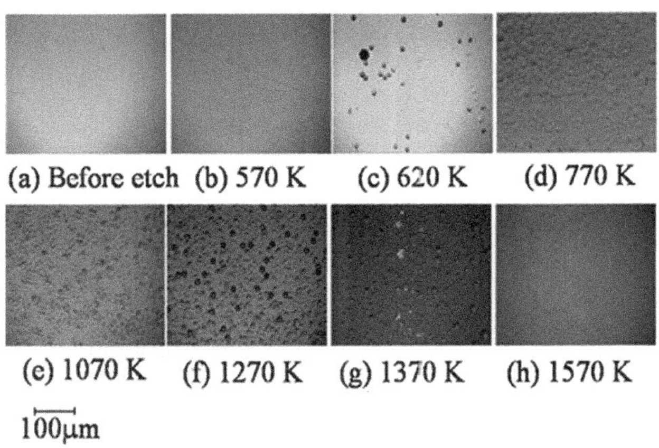

Figure 22. Surface morphology of C-face of 4H-silicon carbide (a) before and after the etching using chlorine trifluoride gas at the concentration of 100 %, at the substrate temperature of (b) 570, (c) 620, (d) 770, (e) 1070, (f) 1270, (g) 1370 and (h) 1570 K and at the flow rate of 0.1 slm. Etched depth is 10-30 μm.

There is the flat surface after the etching at 570 K, as shown in Figure 22 (b), because of significantly small etching rate. However, Figure 22 (c) shows that pits are produced at 620 K. The surface etched at the temperatures between 770 K and 1270 K have many small pits, as shown in Figures 22. (d), (e) and (f). Figures 22 (g) and (h) show that the pit diameter decreases at the temperatures higher than 1370 K. The surface etched at 1570 K shows a flat surface as shown in Figure 22 (h).

Next, the influence of the chlorine trifluoride gas concentration on the surface morphology of the C-face of 4H-silicon carbide is explained. Figure 23 shows the morphology of the C-face 4H-silicon carbide surface before and after the etching for 5 min at the various chlorine trifluoride gas concentrations of (b) 100, (c) 50, (d) 20, (e) 10 and (f) 1% at the total flow rate of 0.2 slm. The substrate temperature is fixed at 1370 K. The etched depth is 3-84 μm.

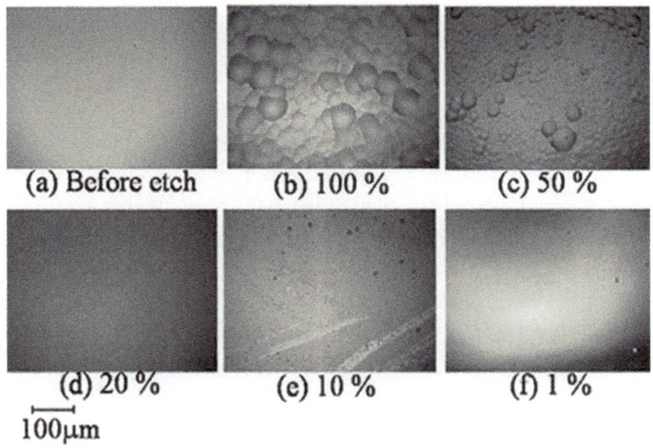

Figure 23. Surface morphology of C-face 4H-silicon carbide surface before and after etching at the total gas flow rate of 0.2 slm, 1370 K and various chlorine trifluoride gas concentrations for 5 min. (a) before etching, (b) 100, (c) 50, (d) 20, (e) 10 and (f) 1%. Etched depth is 3-84 μm.

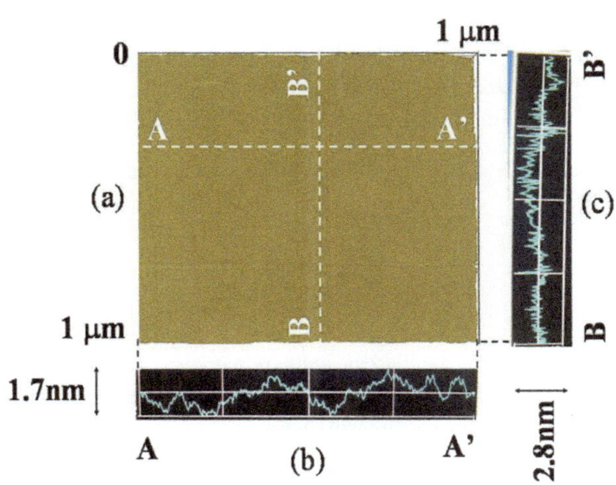

Figure 24. AFM photograph of C-face of 4H-silicon carbide, etched for 0.5 min using the chlorine trifluoride gas concentration of 1% diluted in ambient nitrogen at the substrate temperature of 1570 K and at the total flow rate of 4 slm. (a): plan view, (b): A-A' cross section, and (c): B-B' cross section.

Figure 23 (b) shows that the C-face 4H-silicon carbide surface has large and shallow pits after etching at the chlorine trifluoride gas concentration of 100%. As shown in Figure 23 (c), the etch pits become shallow at the chlorine trifluoride gas concentration of 50%. Figures 23 (d), (e) and (f) show that the etched surface is entirely flat at the chlorine trifluoride gas concentrations less than 20%. Particularly, the surface etched at 1% is flat, as shown in Figure 23 (f). Overall, the trend in the etched surface morphology of the C-face of 4H-silicon carbide is similar to that of the Si-face of 4H-silicon carbide, although the pit size of the C-face is smaller than that of Si-face.

Similar to the Si-face 4H-silicon carbide surface, the C-face 4H-silicon carbide surface is etched at 1570 K at the chlorine trifluoride gas concentration of 1% for 0.5 min. Figures 24 (a), (b) and (c) are the AFM photographs of the plan view, A-A' cross section and B-B' cross section, respectively. The etched depth is near 0.03 μm. Figure 24 (a) does not show any shape like an etch pit; Figures 24 (b) and (c) show no periodical shape reflecting the 4H-silicon carbide crystal step [42]. The RMS roughness is 0.4 and 0.3 nm on A-A′ line and B-B′ line, respectively, which are comparable to that of the polished 4H-silicon carbide substrate surface. Thus, the shallow etching without producing any trace of pit shape is possible, for the C-face of 4H-silicon carbide.

3.6. Surface morphology behavior and its rate process

Entire surface morphology behavior changing with the substrate temperature for Si-face and C-face of 4H silicon carbide is summarized as (i) very small change at very low temperatures (lower than 570 K), (ii) significant pit formation between 570 K and 1270 K, and (iii) pit formation reduced at high temperatures (higher than 1370 K).

The process of pit formation is described following the rate theory, assuming that the etch pit is formed due to the difference of the etching rate between the perfect crystal region and the weak spot having any kinds of damage and crystalline defect [10].

The rate constant of the etching in the perfect crystal region, k_P, is assumed to be expressed in Eq. (6).

$$k_P = A\exp\left(-\frac{E}{RT}\right) \tag{6}$$

where A is the pre-exponential factor, E is the activation energy, R is the gas constant, and T is the substrate temperature. In contrast to this, the weak spot, which has larger etching rate to cause pit, is assumed to have the slightly smaller activation energy than that in the perfect region. The rate constant at the weak spot, k_W, is assumed to be expressed in Eq. (7), using the difference of the activation energy from that in the perfect region, ΔE.

$$k_W = A\exp\left(-\frac{E-\Delta E}{RT}\right) \tag{7}$$

Assuming that the etchant gas concentration is the same in the perfect region and at the weak spot, the pit depth is expressed in Eq. (8), taking into account that $\Delta E/RT$ is very small.

$$Pitdepth = V_{\mathrm{E}}\left(\frac{k_{\mathrm{W}}-k_{\mathrm{P}}}{k_{\mathrm{P}}}\right) = V_{\mathrm{E}}\left(\exp\left(\frac{\Delta E}{RT}\right)-1\right) \cong V_{\mathrm{E}}\frac{\Delta E}{RT} \tag{8}$$

where V_{E} is the etching rate in the perfect region.

Here, assuming that V_{E} shown in Figure 14 is the etching rate in the perfect region, the normalized pit depth, h, is evaluated and shown in Figure 25. The h value is defined using the maximum value of the pit depth, in Eq. (9).

$$h = \frac{Pitdepth}{Pitdepth_{\mathrm{MAX}}} = \frac{\frac{V_{\mathrm{E}}}{T}}{\left(\frac{V_{\mathrm{E},}}{T}\right)_{Pitdepth_{\mathrm{MAX}}}} \tag{9}$$

In Figure 25, the h value at the temperatures lower than 500 K is very small; it significantly increases near 700 K. After showing its maximum, the h value gradually decreases with the increasing substrate temperature. Near 1600 K, the h value is significantly smaller than the maximum value. This trend qualitatively agrees with that of the 4H-silicon carbide surface etched using chlorine trifluoride gas. Thus, the surface morphology trend over wide temperature range can be understood mainly by the rate process.

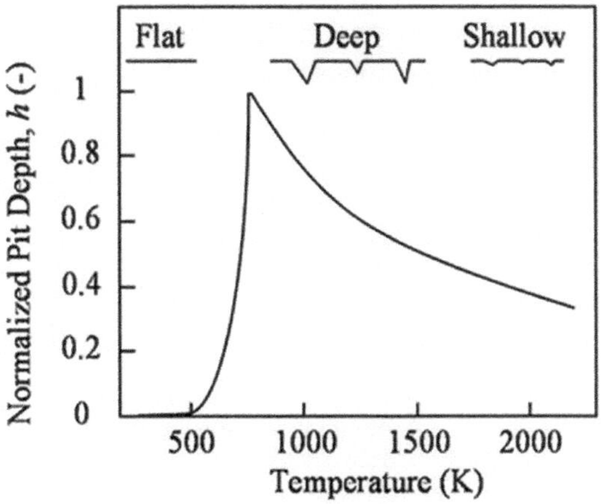

Figure 25. Normalized pit depth and temperature-dependent surface morphology behavior following the rate theory.

3.7. Etch pits and crystalline defect

AFM photographs of the pit shape formed by the chlorine trifluoride gas are shown in Figure 26. Figures 26 (a) and (b) show the pits on the Si-face and C-face, respectively, after the etching using the chlorine trifluoride gas (100%) at atmospheric pressure for 3 min at 870 K and 0.3 slm. Figure 26 (c) is the etch pit of Figure 26 (a), the edge of which is traced using dotted line.

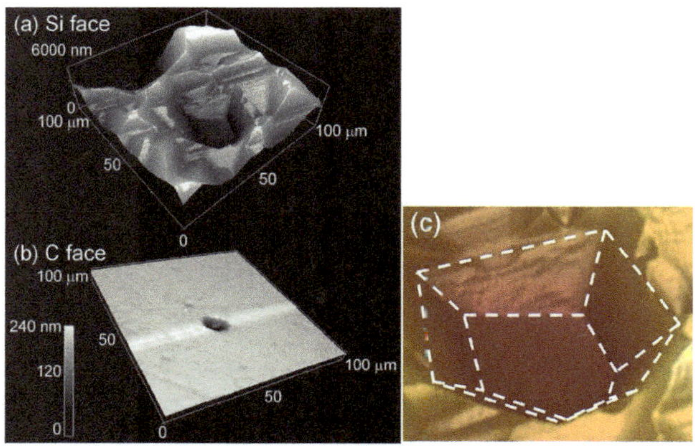

Figure 26. Etch pits on (a) Si-face and (b) C-face (870 K, chlorine trifluoride: 100 %, 0.3 slm). (c) is the magnification of the pit in (a), the edge of which is traced using dotted line.

Figures 26 (a) and (c) reveals a nearly hexagonal edge shape having a flat-shaped bottom. Its diameter and depth are about 0.03 mm and 2000-3000 nm, respectively. The many pits that exist over the entire surface are shown to have the same edge and bottom shape as that shown in Figure 26 (a). Figure 26 (b) shows the pit shape on the C-face. This surface is very flat and smooth having only a very small number of circular shaped pits, which are very shallow with a diameter of 0.01 mm and a depth of about 200 nm.

Many etch pits at the 4H-silicon carbide surface, produced by the etching using chlorine trifluoride gas, are expected to show a relationship with various crystalline defects, when the etching condition is appropriate. Thus, the Si-face and C-face 4H-silicon carbide surface are etched using chlorine trifluoride gas at 100% and at 700 K so that the pit depth become maximum value, as predicted by Figure 25. Additionally, the etch pits are compared with the X-ray topograph, because the X-ray topograph is suitable in order to evaluate an origin of etch pit [42, 43].

Figures 27 (a) and 28 (a) are the X-ray topograph of the Si-face and C-face 4H-silicon carbide surface, respectively. Figures 27 (b) and 28 (b) are the photograph of Si-face and C-face 4H-silicon carbide surface, respectively, etched using the chlorine trifluoride gas at 100% and at 700 K for 60 min. White arrows in these figures indicate the position of the spots in the X-ray topograph and the etch pits at the etched surface.

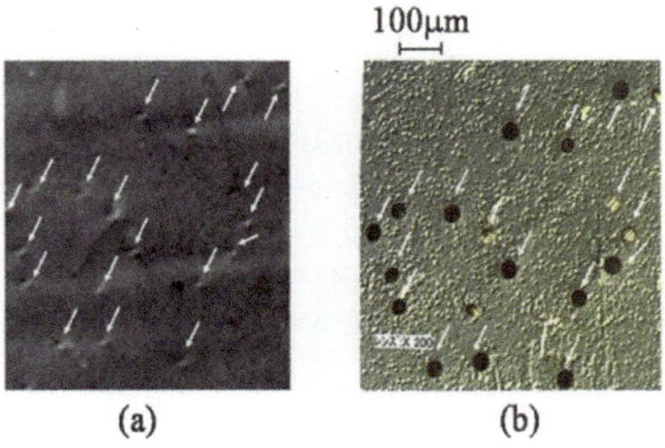

Figure 27. Comparison between the X-ray topograph and the etched Si-face 4H-silicon carbide surface. (a) X-ray topograph of the Si-face 4H silicon carbide surface, and (b) the Si-face 4H-silicon carbide surface etched using chlorine trifluoride gas at 100% and at 700 K for 60 min. Arrows in this figure indicate the spot in the X-ray topograph and the etch pit at the etched surface.

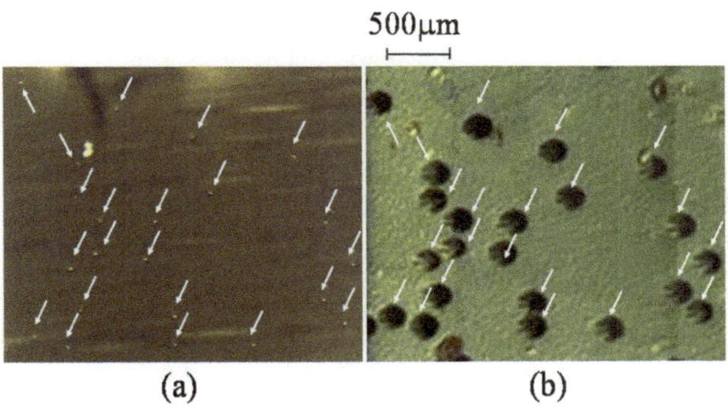

Figure 28. Comparison between the X-ray topograph and the etched C-face 4H-silicon carbide surface. (a) X-ray topograph of the Si-face 4H silicon carbide surface, and (b) the C-face 4H-silicon carbide surface etched using chlorine trifluoride gas at 100% and at 700 K for 60 min. Arrows in this figure indicate the spot in the X-ray topograph and the etch pit at the etched surface.

As indicated using white arrows, there are many etch pits, the positions of which correspond to those of the spots in the X-ray topograph. The dimension of spot in Figure 28 (a) is larger than that in Figure 27 (a); the diameter of etch pits in Figure 28 (b) is about 250 μm which is similarly larger than that in Figure 27 (b), about 40 μm. Taking into account the report [44] about the dimension of etch pits formed by KOH, Figures 28 (a) and (b) may show the screw dislocation; Figures 27 (a) and (b) may show the threading edge dislocation. Because the etching technique using the chlorine trifluoride gas may reveal the crystalline defects, like the KOH technique [44], the relationship between etch pits and various crystalline defects should be further studied.

Because of the functions to produce the specular surface and to reveal the crystalline defects, chlorine trifluoride gas is expected to be more useful than the other wet and dry techniques [41], for silicon carbide industrial process.

Next, the density and behavior of etch pit produced on the C-face of 4H-silicon carbide substrate using chlorine trifluoride gas at various temperatures were evaluated. The etch pit obtained using the chlorine trifluoride gas at 713 K for 10 min at 100% was observed and shown in Figure 29. The diameter of etch pits was near 10 μm, which was considered to be assigned to screw dislocations, following the previous study [28].

Because the etch pit density (EPD) changed with the substrate temperature [28], the etching was performed at various temperatures around 713 K The etch pit density obtained in an area of 500 x 500 μm^2 at various substrate temperatures is shown in Figure 30. At the temperatures below 673 K, the etch pit density was very small. At 683 K, the etch pit density increase to the value near 2 x 10^4 cm^{-2}. At 713K, the etch pit density showed the maximum value of 4 x 10^4 cm^{-2}. Although the etch pit density decreased at the temperatures higher than 723 K, its value maintained near 10^4 cm^{-2}.

The etch pit density obtained at 713 K coincided with the values widely accepted as the current dislocation density level of 4H silicon carbide. Thus, the etch pit density of C-face of 4H-silicon carbide obtained in this study is acceptable and is expected to show a relationship with the crystal quality. The etch pits obtained in this study were classified to the large circular-shaped and small oval-shaped pits, which ratio were 90% and 10%, respectively, at the etching temperature of 713 K. The former is considered to be assigned to the threading screw dislocation, and the latter can be the threading edge dislocation.

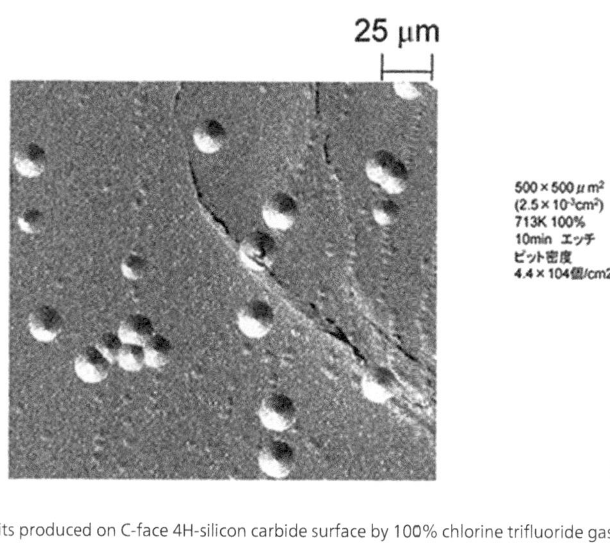

Figure 29. Etch pits produced on C-face 4H-silicon carbide surface by 100% chlorine trifluoride gas at 713 K for 10 min

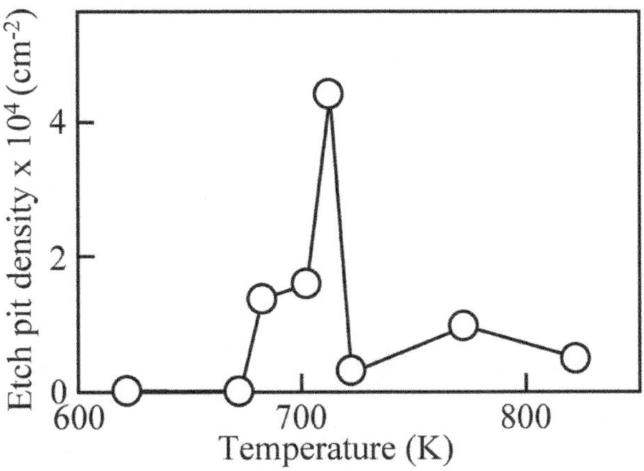

Figure 30. Etch pit density on C-face 4H-silicon carbide surface produced by chlorine trifluoride gas at various substrate temperatures.

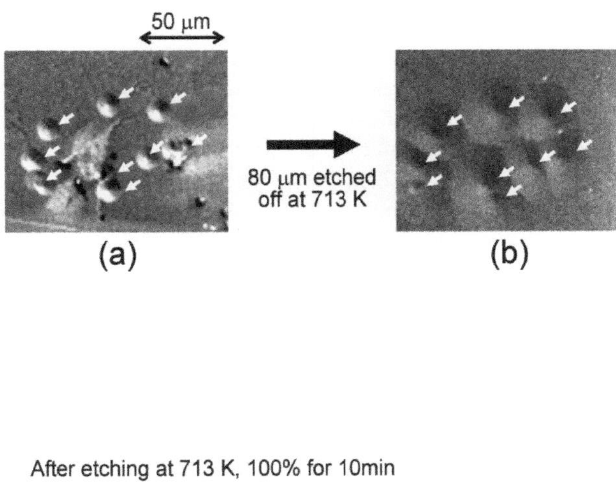

Figure 31. Comparison of etch pits (a) before and (b) after additional etching. White arrows incicate the etch pits.

In order to show that the origin of these etch pit was assigned to the crystalline imperfection, the etched surface was further etched using the same etching condition. Figure 31 (a) shows the nine etch pits observed on the C-face of 4H silicon carbide after the etching at 713 K for 10 min using the 100 % chlorine trifluoride gas. Their diameter was about 10 – 15 μm. This substrate was additionally etched at 713 K at the 100 % chlorine trifluoride gas. Figure 31 (b) shows the surface, 80 μm of which surface was etched off by the additional etching. Figure 31 (b) shows the large and shallow nine etch pits, which were overlapped with each other. The number and the position of etch pit center in Figure. 31 (b) corresponded to those in Figure 31 (a), respectively. Thus, over the depth of 80 μm, the origin to cause the etch pit was concluded to exist; the origin could be the dislocation, such as the threading dislocation, existing normal to the substrate surface.

The results obtained in this study indicate that the pits caused by the etching using chlorine trifluoride gas at around 713 K has the origin of crystalline imperfection, such as the threading dislocation. However, the interpretation of etch pits with respect to the underlying dislocation type should be further carefully performed by a comparative investigation [28]. Additionally, the density and shape of etch pit by this technique should be further clarified and verified through many characterization.

4. Summary

Silicon carbide etching using chlorine trifluoride gas with high etching rate occurs at the temperatures higher than 770 K. Its chemical reaction is as follows:

$$3SiC \cdot + \cdot 8ClF_3 \cdot \rightarrow \cdot 3SiF_4 \cdot + \cdot 3CF_4 \cdot + \cdot 4Cl_2$$

The etching rate is 10-20 μm/min and 5 μm/min, for polycrystalline 3C-silicon carbide and single-crystalline 4H-silicon carbide, respectively. The etching rate of Si-face of 4H-silicon carbide is slightly smaller than that of C-face. The etched surface tends to be carbon rich. The etched surface of Si-face of 4H-silicon carbide shows various kinds of morphology: pitted at low temperatures of 570 – 1270 K, and smooth at 1570 K. The C-face of 4H-silicon carbide shows the similar trend, and is entirely very smoother than that of Si-face. Most of the etch pits formed near 700 K at the Si-face and C-face of 4H-silicon carbide show relationship between dislocations revealed by the X-ray topograph. At the substrate temperature of 713 K, the etch pit density showed the maximum value of 4×10^4 cm^{-2}. The etch pit density obtained by this technique is considered to show the crystal quality, particularly, the dislocation density.

Acknowledgements

Various studies reviewed in this Chapter were performed with Ms. Satoko Oda, Mr. Yusuke Katsumi, Mr. Yu Kasahara, Ms. Keiko Tanaka, Mr. Kazuchika Furukawa, Mr. Yusuke Fukumoto, Dr. Yutaka Miura, Mr. Yoichi Negishi, Dr. Takashi Takeuchi, Prof. Masahiko Aihara, Prof. Minoru Takeda, Prof. Hironobu Kunieda and Prof. Kenji Aramaki of Yokohama National University, Mr. Yasushi Fukai, Mr. Katsuya Fukae, Mr. Naoto Takechi, Dr. Yuan Gao, and Mr. Shinji Iizuka of Kanto Denka Kogyo Co., Ltd., and Dr. Tomohisa Kato, Dr. Hajime Okumura and Dr. Kazuo Arai of National Institutes of Advanced Science and Technology. Mr. N. Okumura of Keyence Co., Ltd., is very much appreciated for the surface roughness evaluation. X-ray topography experiment has been performed under the approval of the Photon Factory Program Advisory Committee (Proposal No. 2006G286).

Author details

Hitoshi Habuka*

Address all correspondence to: habuka1@ynu.ac.jp

Department of Chemical and Energy Engineering, Yokohama National University, Japan

References

[1] Gogotsi, Y., Welz, S., Ersoy, D. A., & Mc Nallan, M. J. (2001). *Nature*, 411(6835), 283.

[2] Vyshnyakova, K., Yushin, G., Pereselentseva, L., & Gogotsi, Y. (2006). *Int. J. Appl. Ceramic Tech.*, 3, 485.

[3] Zinovev, A. V., Moore, J. F., Hryn, J., & Pellin, M. J. (2006). *Surf. Sci*, 600, 2242.

[4] Cooke, M. (2005, Dec). *III-Vs Review*, 18, 40.

[5] Chai, C., Yang, Y., Li, Y., & Jia, H. (2003). *Optical Materials*, 23, 103.

[6] Mehregany, M., Zorman, C. A., Roy, S., Fleischman, A. J., Wu, C. H., & Rajan, N. (2000). *International Materials Reviews*, 45, 85.

[7] Stoldt, C. R., Carraro, C., Ashurst, W. R., Gao, D., Howe, R. T., & Maboudian, R. (2002). *Sensors and Actuators a-Physical*, 97-98, 410.

[8] Rajan, N., Mehregany, M., Zorman, C. A., Stefanescu, S., & Kicher, T. P. (1999). *Journal of Microelectromechanical Systems*, 8, 251.

[9] Ashurst, W. R., Wijesundara, M. B. J., Carraro, C., & Maboudian, R. (2004). *Tribology Letters*, 17, 195.

[10] Shimura, F. (1989). Semiconductor Silicon Crystal Technology. 244, Academic Press, San Diego, USA.

[11] Kim, B., Kim, S., Ann, S., & Lee, B. (2003). *Thin Solid Film*, 434, 276.

[12] Schmid, U., Eickoff, M., Richter, C., Kroetz, G., & Schmitt-Landsiedel, D. (2001). *Sensors and Actuators A*, 94, 87.

[13] Wolf, R., & Helbig, R. (1996). *J. Electrochem. Soc.*, 143, 1037.

[14] Yih, P. H., & Steckl, A. J. (1995). J. Electrochem. Soc., ., 142, 312.

[15] Xie, Z. Y., Wei, C. H., Li, L. Y., Yu, Q. M., & Edgar, J. H. (2000). *J. Cryst. Growth*, 217, 115.

[16] Sanchez, E. K., Ha, S., Grim, J., Skowronski, M., Vetter, W. M., Dudley, M., Bertke, R., & Mitchel, W. C. (2002). *J. Electrochem. Soc.* 149, G 131.

[17] Syvajarvi, M., Yakimova, R., & Janzen, E. (2000). *J. Electrochem. Soc*, 147, 3519.

[18] Shor, J. S., Zhang, X. G., & Osgood, R. M. (1992). *J. Electrochem. Soc*, 139, 1213.

[19] Chabert, P. (2001). *J. Vac. Sci. Technol. B*,, 19, 1339.

[20] Khan, P. A., Roof, B., Zhou, L., & Asesida, I. (2001). *J. Electron. Mater.*, 30, 212.

[21] Jiang, L., Cheung, R., Brown, R., & Mount, A. (2003). *J. Appl. Phys.*, 93, 1376.

[22] Habuka, H., Koda, H., Saito, D., Suzuki, T., Nakamura, A., Takeuchi, T., & Aihara, M. (2003). *J. Electrochem. Soc.*, 150, G461-464.

[23] Habuka, H., Oda, S., Fukai, Y., Fukae, K., Sekiguchi, T., Takeuchi, T., & Aihara, M. (2005). *Jpn. J. Appl. Phys*, 44, 1376.

[24] Habuka, H., Oda, S., Fukai, Y., Fukae, K., Takeuchi, T., & Aihara, M. (2006). *Thin Solid Films*, 514, 193.

[25] Miura, Y., Katsumi, Y., Oda, S., Habuka, H., Fukai, Y., Fukae, K., Kato, T., Okumura, H., & Arai, K. (2007). *Jpn. J. Appl. Phys*,, 46, 7875.

[26] Habuka, H., Katsumi, Y., Miura, Y., Tanaka, K., Fukai, Y., Fukae, T., Gao, Y., Kato, T., Okumura, H., & Arai, K. (2008). *Materials Science Forum*,, 600-603, 655.

[27] Habuka, H., Tanaka, K., Katsumi, Y., Takechi, N., Fukae, K., & Kato, T. (2010). *Materials Science Forum*, 645-648, 787.

[28] Habuka, H., Tanaka, K., Katsumi, Y., Takechi, N., Fukae, K., & Kato, T. (2009). *J. Electrochem. Soc*,, 156, H971.

[29] Miura, Y., Kasahara, Y., Habuka, H., Takechi, N., & Fukae, K. (2009). *Jpn. J. Appl. Phys.*, 48, 026504.

[30] Chinone, Y., Esaki, S., & Fujita, F. (1996). *Shirikon no Kagaku*, Realize,, Tokyo, 890.

[31] Habuka, H., Sukenobu, T., Koda, H., Takeuchi, T., & Aihara, M. (2004). *J. Electrochem. Soc.*, 151, G783-787.

[32] Lide, D. R., & Frederikes, H. P. R. (1997). CRC Handbook of Chemistry and Physics. , ed., D. R. Lide (CRC Press, Boca Raton, FL, 78 th ed. , 1.

[33] Lee, T. E. (1998). A Beginner's Guide to Mass Spectral Interpretation. (Wiley, Chichester, 1998).

[34] Chapman, J. R. (1993). Practical Organic Mass Spectrometry. Wiley,, Chichester.

[35] Habuka, H., Otsuka, T., Qu, W. F., & Jpn, . (1999). *J. Appl. Phys.*, 38, 6466.

[36] Saito, Y., Hirabaru, M., & Yoshida, A. (1992). *J. Vac. Sci. Technol B*,, 10, 175.

[37] Patankar, S. V. (1980). Numerical Heat Transfer and Fluid Flow. ,McGraw-Hill, New York, USA

[38] Webbook. http://webbook.nist.gov/chemistry.

[39] Reid, R. C., Prausnitz, J. M., & Poling, B. E. (1987). The Properties of Gases and Liquids. McGraw-Hill, New York 4 th

[40] Gakujutsushinkokai, Nihon. , (1997). Fussokagaku Nyumon, Nikan Kogyo ShinbunshaTokyo, Japan [in Japanese], 86.

[41] Fluent (2001). FLUENT User's ManualVer. 5.5 (Fluent, Inc., Hanover, 2001)

[42] Matsunami, H. (2003). *Technology of Semiconductor SiC and Its Application*, Nikkan Kogyo, Tokyo, Japan.

[43] Ohno, T., Yamaguchi, H., Kuroda, S., Kojima, K., Suzuki, T., & Arai, K. (2004). *J. Cryst. Growth*, 260, 209.

[44] Ma, X., Dudley, M., Vetter, W., & Sudarshan, T. (2003). *Jpn. J. Appl. Phys.*,, 42, L1077.

[45] Takanashi, J., Kanaya, M., & Fujiwara, Y. (1994). *J. Cryst. Growth*, 135, 61.

Chapter 5

Physics Behind the Ohmic Nature in Silicon Carbide Contacts

Zhongchang Wang

Additional information is available at the end of the chapter

1. Introduction

One of the most active fields in semiconductor research is the development of electronic devices capable of function at high powder and high frequency levels, high temperatures, and caustic circumstances. This surge of activity is strongly driven by the urgent desire for replacing the current Si- and GaAs-based electronics because they are unable to operate properly under harsh environmental conditions. As a promising substitute, the wide-band-gap semiconductor, silicon carbide (SiC), has captured considerable attention recently due to its excellent intrinsic properties, which involve large breakdown electric field, high electron saturation drift velocity, strong hardness, and good thermal conductivity. On the other hand, current significant improvements in the epitaxial and bulk crystal growth of SiC have paved the way for fabricating its electronic devices, which stimulates further interest in developing device processing techniques so as to take full advantage of its superior inherent properties.

One of the most critical issues currently limiting the device processing is the manufacturing of reliable and low-resistance Ohmic contacts especially contacts to *p*-type 4H-SiC [1]. The Ohmic contacts are primarily important in SiC devices because a Schottky barrier of high energy is inclined to form at an interface between metal and wide-band-gap semiconductor, which consequently results in low current driving, slow switching speed, and increased power dissipation. Much of effort expended to date to realize the Ohmic contact has mainly focused on two techniques. One is the high-dose ion implantation approach [2], which can increase carrier density in SiC noticeably while reducing its depletion width significantly so that increasing tunneling current is able to flow across the barrier region. The key problem of this method is the easy formation of lattice defects or amorphization during the ion implantation. These defects are unfortunately very stable and need to be recovered via annealing at an extremely high temperature of about 2000 K, thereby complicating the mass production of SiC devices.

The other alternative is to generate an intermediate semiconductor layer with narrower band gap or higher carrier density at the contacts/SiC interface via depositing and annealing technique [3]. To form such layers, a wide range of materials have been examined in a trial-and-error designing fashion, including metals, silicides, carbides, nitrides, and graphite. Of all these materials, the metallic alloys have been investigated extensively, largely because their fabrication process is simple, standard, and requires no exotic materials. In particular, most of research activities have been focused on TiAl-based alloys, the only currently available materials that yield significantly low contact resistance (Ohmic contact) to *p*-SiC [4]. Moreover, they demonstrate high thermal stability. Although a lot of intriguing results have been obtained regarding the TiAl-based contact systems, the mechanism whereby the Schottky becomes Ohmic after annealing has not been well clarified yet. In other words, the key factor to understanding the formation origin of Ohmic contact remains controversial. Mohney et al. [5] proposed that a high density of surface pits and spikes underneath the contacts contributes to the formation of Ohmic behavior based on their observations using scanning electron microscopy and atomic force microscopy. Nakatsuka et al. [6], however, concluded that the Al concentration in the TiAl alloy is important for the contact formation. Using the liquid etch and ion milling techniques, John and Capano [7] ruled out these possibilities and claimed that what matters in realizing the Ohmic character is the generation of carbides, Ti_3SiC_2 and Al_4C_3, between the metals and semiconductor. This, however, differs, to some extent, from the X-ray diffraction (XRD) findings of Chang et al. [8] showing that the compounds formed at the metal/SiC interface are silicides, $TiSi_2$, TiSi, and Ti_3SiC_2. In addition, Ohyanagi et al. [9] argued that carbon exists at the contacts/SiC interface and might play a crucial role in lowering Schottky barrier. These are just a few representative examples illustrating the obvious discrepancies in clarifying the formation mechanism of the Ohmic contact. Taking the amount of speculations on the mechanism and the increasing needs for better device design and performance control, understanding the underlying formation origin is timely and relevant.

To develop an understanding of the origin in such a complex system, it is important to focus first on microstructure characterization. Transmission electron microscopy (TEM) studies by Tsukimoto et al. [10] have provided useful information in this aspect. They have found that the majority of compounds generated on the surface of 4H-SiC substrate after annealing consist of Ti_3SiC_2 and hence proposed that the SiC/Ti_3SiC_2 interface is responsible for the lowering of Schottky barrier in the TiAl-based contact system. However, the role of this interface in realizing the Ohmic nature is still unclear. It is not even clear how the two materials atomically bond together from their experiments, which is very important because it may strongly affect physical properties of the system. Theoretically, we have calculated the atomic structures, adhesive energies, and bonding nature of the SiC/Ti_3SiC_2 interface [11]. However, this calculation does not actually reveal the formation mechanism of Ohmic contact because it only addresses the interface structure. Furthermore, lacking essential experimental information about the interfacial atomic-scale structure, such calculations have been incomplete.

Recent advances in the high-angle annular-dark-field (HAADF) microscopy [12,13], the highest resolution, have enabled atomic-scale imaging of a buried interface. However, direct interpretation of the observed HAADF images is not always straightforward because there might be abrupt structural discontinuity, mixing of several species of elements on individual atomic columns, or missing contrasts of light elements. One possible way out to complement the microscopic data is through atomistic calculation, especially the first-principles calculation. As well known, the atomistic first-principles simulations have long been confirmed to be able to suggest plausible structures, elucidate the reason behind the observed images, and even provide a *quantitative* insight into how interface governs properties of materials [14,15]. Consequently, a combination of the state-of-the-art microcopy and accurate atomistic modeling [16] is an important advance for determining interface atomic-scale structure and relating it to device properties, revealing, in this way, physics origin of the contact issues in SiC electronics.

In addition to determining atomic structure of the 4H-SiC/Ti_3SiC_2 interface, the goal of this work is to clarify the formation mechanism of the TiAl-based Ohmic contacts so as to provide suggestions for further improvement of the contacts. 4H-SiC will hereafter be referred to as SiC. In this Chapter, we will first attribute qualitatively the formation of ohmic contacts in the TiAl-deposited SiC system to an epitaxial and atomically abrupt interface between the SiC and Ti_3SiC_2 generated *via* annealing. The interface presumably serves as a primary current-transport pathway to lower the Schottky barrier formed at the interface between the deposited metals and SiC. Further quantitative studies reveal that the barrier mitigation arises from trapping of an atomic layer of carbon at the SiC/Ti_3SiC_2 interface, which assists the electron transport across the SiC [17,18]. The considerations on the role played by interface do not, however, exclude another possibility that the Ti_3SiC_2 atomic layers can be generated inside the SiC bulk interiors, which presents a behavior that may differ from that of their bulk [19]. Combining the state-of-the-art TEM with atomistic first-principles calculations, we demonstrate the presence of an atomic-scale Ti_3SiC_2-like bilayer embedded in the SiC, forming an atomically ordered multilayer that exhibits an unexpected electronic state with point Fermi surface. The valence charge is confined to a large extent to within the bilayer in a spatially connected manner, serving possibly as a conducting channel to enhance the current flow over the semiconductor. Further investigation into the contact regions unveils another new opportunity to allow the electron transport across the semiconductor, namely, *via* the terraces formed at the SiC/Ti_3SiC_2 interface [20]. Experimentally, the formed carbide Ti_3SiC_2 is demonstrated to bond directly to the silicon carbide at the terraces in an epitaxial and atomically ordered fashion, regardless of dimension of the terraces. There appear pronounced gap states at Fermi level in the semiconductor layers around the terraces, and charges are accumulated heavily around the terraces in a connected and broadly distributed manner. The presence of the metallicity and the likelihood to act as electron conduction channels to enable the current flow over the semiconductor make the terraces at the interface one of the origins underlying the ohmic contact in silicon carbide electronics. We therefore demonstrate in this chapter that origin of the long-standing contact issue in SiC devices can be understood and technologically manipulated at the atomic level, and suggest the key physical factors for establishing the ohmic nature.

2. Role of coherent SiC/Ti_3SiC_2 interface

The *p*-type 4H-SiC epitaxial layers (5-μm thick) doped with aluminum ($N_A = 4.5 \times 10^{18}$ cm^{-3}) which were grown on undoped 4H-SiC wafers by chemical vapor deposition (manufactured by Cree Research, Inc.) were used as substrates. The 4H-SiC substrates had 8°-off Si-terminated (0001) surfaces inclined toward a [$\bar{2}$110] direction because only 4H-type structure of SiC with polymorph (e.g. 3C, 4H, 6H, 15R etc.) was controllable by lateral growth of the epitaxial layers parallel to (0001)-oriented surface. After chemical cleaning of the substrate surface, a 10 nm-thick sacrificial oxide (SiO_x) layer was grown on the SiC substrate by dry-oxidation at 1423 K for 60 min. The electrode patterns were made by removing the SiO_x layers, where contact metals were deposited by dipping in 5 % diluted hydrofluoric acid solution for 1 min using a photolithography technique. Prior to the deposition of contact materials, the substrates were cleaned by deionized water. Then, Ti and Al stacking layers with high purities were deposited sequentially on the substrate in a high vacuum chamber where the base pressure was below 5×10^{-6} Pa. The thicknesses of the Ti and Al layers investigated in this study are 100 nm and 380 nm, respectively, and these layer thicknesses were chosen to give the average composition of the Ti(20 at%) and Al(80 at%), where the layer thicknesses were measured by a quartz oscillator during deposition. The reasons to choose this average composition was that aluminum rich (more than 75 at%) in TiAl contacts were empirically found to be essential to yield low contact resistance, resulting from formation of the Ti_3SiC_2 compound layers. After depositing, the binary TiAl contact layers were annealed at 1273 K for a storage time of 2 min in an ultra-high vacuum chamber where the vacuum pressure was below 1×10^{-7} Pa.

The surface morphology of the TiAl contact layers on 4H-SiC after annealing was observed using a JEOL JSM-6060 scanning electron microscope (SEM). Microstructural analysis and identification of the Ti_3SiC_2 layers at the contact layers/4H-SiC interfaces after annealing was performed using X-ray diffraction (XRD) and cross-sectional TEM. For XRD analysis, Rigaku RINT-2500 with Cu $K\alpha$ radiation operated at 30 kV and 100 mA was used. In particular, the interfacial structures and an orientation relationship between the contact layers and the 4H-SiC substrates were characterized by cross-sectional high-resolution TEM observations and selected area diffraction pattern (SADP) analysis, respectively, using a JEOL JEM-4000EX electron microscope operated at an accelerating voltage of 400 kV, where the point-to-point resolution of this microscope was approximately 0.17 nm. Z-contrast images were obtained using a spherical aberration (C_s) corrected scanning transmission electron microscope (STEM) (JEOL 2100F), which provides an unprecedented opportunity to investigated atomic-scale structure with a sub-Å electron probe. Thin foil specimens for the TEM and STEM observations were prepared by the standard procedures: cutting, gluing, mechanical grinding, dimple polishing, and argon ion sputter thinning techniques.

Calculations of electronic structure and total energy were carried out using the Vienna *ab initio* simulation package (VASP) within the framework of density functional theory (DFT) [21]. The projector augmented wave method was used for electron-ion interactions, and the generalized gradient approximation of Perdew et al. (PW91) was employed to describe the exchange-correlation functional. The single-particle Kohn-Sham wave function was expanded using plane waves with different cutoff energies depending on the calculated systems. Sampling of irreducible Brillion zone was performed with a regular Monkhorst-Pack grid of special *k* points, and electronic occupancies were determined according to the Methfessel-Paxton scheme. Independent *k* point convergence tests were conducted for distinct supercells. Ground state charge densities were calculated self-consistently using a Pulay-like mixing scheme and the stable blocked Davidson minimization algorithm. Total energies were calculated using the linear tetrahedron method with Blöchl corrections, which eliminates broadening-related uncertainties. All atoms were fully relaxed using the conjugate gradient algorithm until the magnitude of the Hellmann-Feynman force on each atom converged to less than 0.05 eV/Å, yielding optimized structures.

To determine the most stable interface theoretically, one first has to establish feasible models on the basis of the distinct terminations and contact sites and then compare them. However, a direct comparison of total energies of such candidate models is not physically meaningful since interfaces might have a different number of atoms. On the other hand, the adhesion energy (W_{ad}), which is key to predicting the mechanical properties of an interface, is physically comparable. Generally, the W_{ad} is defined as the energy required to reversibly separate an interface into two free surfaces, neglecting plastic and diffusion degrees of freedom. The energy needed in actual cleavage experiments is always greater than the W_{ad} owing to plastic deformation, but the extent of plastic deformation relies on the W_{ad}. Formally, the W_{ad} can be expressed in terms of surface and interfacial energies or by the difference in total energy between the interface and isolated slabs [22,23]:

$$W_{ad} \equiv \sigma_1 + \sigma_2 - \sigma_{12} = (E_1 + E_2 - E_{IF}) / A \quad (1)$$

Here σ_i is the surface energy of slab *i*, σ_{12} is the interface energy, E_i is the total energy of isolated slab *i*, E_{IF} is the total energy of the interface system, and *A* is the total interface area. In general, two steps can be taken to estimate the W_{ad}. First, the total energies were calculated for a series of separations as two rigid slabs were brought increasingly closer from a large initial separation. As a consequence, the calculated total energies were found to behave like a parabola, passing through a minimum at the equilibrium separation. The unrelaxed W_{ad} was obtained by computing the energy difference between the interface at the equilibrium state and the unrelaxed isolated slab. Next, each isolated slab as well as interfacial slab was allowed to optimize fully, yielding an estimation of relaxed W_{ad}.

2.1. Atomic-scale structures of the Ohmic contacts

The electric properties for the TiAl contact systems before and after annealing are meaured first to verify the formation of Ohmic contact. The current almost keeps zero before the annealing, while increasing nearly linear with the rise of applied bias, which unambiguously confirms the formation of the Ohmic contacts after annealing. Further XRD analyses demonstrate that a new reaction product, the ternary Ti_3SiC_2, is generated after annealing, which shows a strongly (0001)-oriented texture. The SiC retains (0001)-oriented texture after annealing, thereby facilitating development of hetero-epitaxy between reaction products and substrates. The TEM imaging reveals that no any other compounds contact directly the SiC surface, thereby ensuring an exclusive contact of Ti_3SiC_2 to SiC. Since the carbide itself is metallic in nature, the lowering in Schottky barrier in the TiAl-based contacts is hence attributed qualitatively to the epitaxial and atomically sharp SiC/Ti_3SiC_2 interface. A careful indexing of the selected area diffraction pattern (SADP) at the contacts/SiC interface reveals that the formed Ti_3SiC_2 layers have epitaxial orientation relationships, that is, (0001)Ti_3SiC_2//(0001)SiC and [0¯110]Ti_3SiC_2// [0¯110]SiC, with the SiC substrate. These orientation relationships are believed to be beneficial for forming a coherent and well matched interface between SiC and Ti_3SiC_2, since they both belong to the hexagonal space group with lattice constants of $a = 3.081$ Å and $c = 10.085$ Å for the SiC and $a = 3.068$ Å and $c = 17.669$ Å for the Ti_3SiC_2.

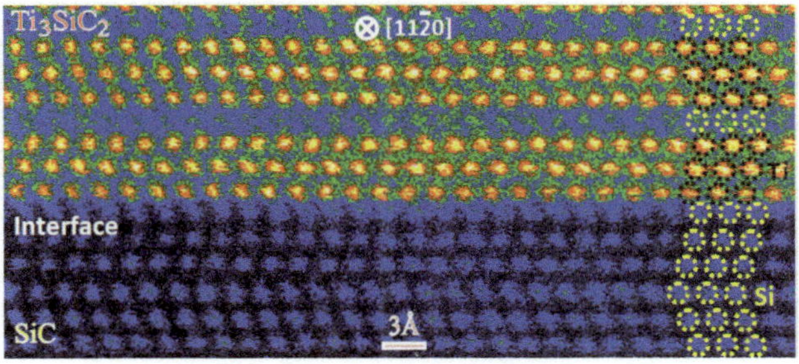

Figure 1. A typical HAADF-STEM image of the SiC/Ti_3SiC_2 interface in the annealed TiAl contact system observed from [11-20] direction. The points at which the phase contrast is no longer periodic in either the Ti_3SiC_2 or SiC define the interfacial region [23].

A representative HAADF image of the SiC/Ti_3SiC_2 interface is shown in Fig. 1, which confirms a clean and atomically sharp contact between the two materials. Since intensity of an atomic column in the Scanning TEM, to good approximation, is directly proportional to the square of atomic number (Z) [24], brighter spots in image represent atomic columns of Ti, while the comparatively darker ones are Si. Not surprisingly, the columns of C are not scattered strongly enough to be visualized owing to its small Z, thereby making this image incomplete. It should be noted that obtaining a signal of pure interfacial carbon is technically very difficult because the specimen can be easily affected by the environmental carbon, thereby precluding the element-selective imaging of carbon. We therefore rely on the first-principles calculations instead to discuss the possibility in the presence of C at the interface, as will be described later. To see the interface clearer, we magnify the cross-sectional HAADF image in Fig. 2(a) and further filter it to reduce noise, as shown in Fig. 2(b). The Si-terminated Ti_3SiC_2 is observed intuitively to make a direct contact with the Si-terminated SiC substrate with interfacial Si atoms of Ti_3SiC_2 sitting above hollow sites of interfacial Si plane of SiC. However, this straightforward interpretation is premature, as will be described later. Since there are no pits, spikes, or dislocations that might act as pathways for current transport, we conclude that this clean and coherent SiC/Ti_3SiC_2 interface should be critical for the formation of Ohmic contact.

To clarify the mechanism, it is prerequisite to determine the atomic structure of the SiC/Ti_3SiC_2 interface via complementing the obtained HAADF image (Fig. 1). We have considered a total of 96 candidate interfacial geometries using bulklike slabs, taking into account termination effect, stacking sequence, and full optimization. From the bulk 4H-SiC and Ti_3SiC_2 structures and the relative stacking order of Ti and Si, the observed image in Fig. 2(a) can be intuitively fitted by a SiSi model shown in Fig. 2(c). In this model, the interfacial Si atoms of Ti_3SiC_2 sit above the hollow sites of interfacial Si plane of SiC, where the optimal distance between interfacial Si-Si planes (denoted as *d*1 in Fig. 2(b)) and that between interfacial Si-Si atoms projected onto paper plane (denoted as *d*2 in Fig. 2(c)) are calculated to be 2.13 and 2.53 Å, respectively. These distances, however, deviate severely from their average experimental values, 2.5 Å and 2.8 Å, which are obtained by characterizing quantitatively the HAADF image (Fig. 2(a)). In addition, an examination of interface stability by calculating the W_{ad} indicates that this SiSi model is not favored (1.62 J/m^2). It is even less stable than the model with interfacial Si of Ti_3SiC_2 resting straight atop the interfacial Si of SiC (2.58 J/m^2), which contravenes again the observed image.

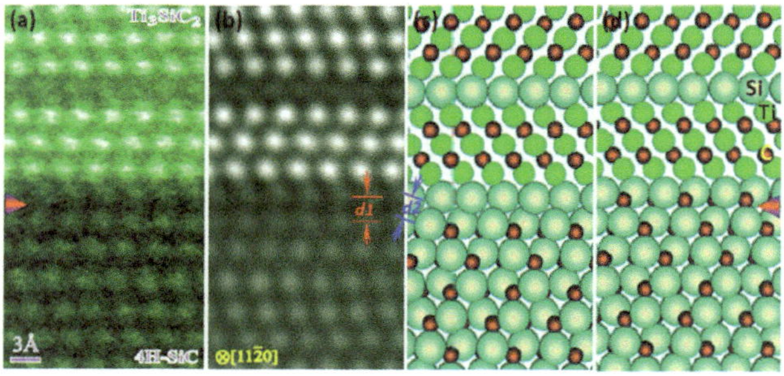

Figure 2. a) Magnified HAADF image of the SiC/Ti_3SiC_2 interface. (b) The same image as in (a) but has been low-pass filtered to reduce noise. Relaxed $SiC(0001)/Ti_3SiC_2(0001)$ interface models (c) without (SiSi) and (d) with (SiCSi) interfacial C atoms. The distance between interfacial Si-Si layers is represented by *d*1 and that between interfacial Si-Si atoms projected onto the paper plane by *d*2. The interfaces are represented by an arrow [23].

To resolve these paradoxes, we notice that a possibility might be ignored, that is, the unseen C might be trapped at the interface, altering local environment there. To test this scenario, we established a new model (named SiCSi) by introducing C into the interfacial layer from the consideration of crystal extension and stacking sequences. The calculated W_{ad} of this SiCSi model is 6.81 J/m^2, which indicates that interface is indeed strengthened substantially after incorporating C. Further examination of the relaxed atomic geometry (Fig. 2(d)) reveals that the incorporation of C does not induce a significant structural transformation. Namely, the two Si layers proximal to the interface maintain the stacking seen in Fig. 2(c), thus matching the HAADF image geometrically. Quantitatively, the *d*1 and *d*2 distances are now 2.53 and 2.81 Å, respectively, very close to the experimental values. Therefore, the introduction of interfacial C monolayer resolves the inconsistencies between simulations and experiments.

2.2. Electronic structure and bonding

Calculations of *p*-type Schottky barrier height (SBH) reveal that the interface with C (0.60 eV) has much lower SBH than the interface without C (1.05 eV), suggesting that the trapped C assist the lowering of SBH. To shed light on origin of the decrease in SBH and junction strengthening in the SiCSi interface, we characterized thoroughly interfacial electronic states and bonding nature. Figure 3(a) shows a planar-averaged charge-density difference along interface normal, where there appears a more dramatic accumulation of charge within the interfacial region for the interface with C. This indicates that the covalent bonding is strengthened in the SiCSi case. In addition, we note that the planar-averaged density difference for the SiCSi more prominently deviates from zero around interface, reflecting more significant charge transfer between the SiC and Ti_3SiC_2 slabs. Moreover, charge is observed to be depleted noticeably in both the sub-interfacial SiC and Ti_3SiC_2 region for the SiCSi, suggesting that the atoms second nearest to the interface contribute to interfacial bonding. These missing charges, to a large extent, make their way onto the more electronegative C ions, indicative of the formation of ionic bonding.

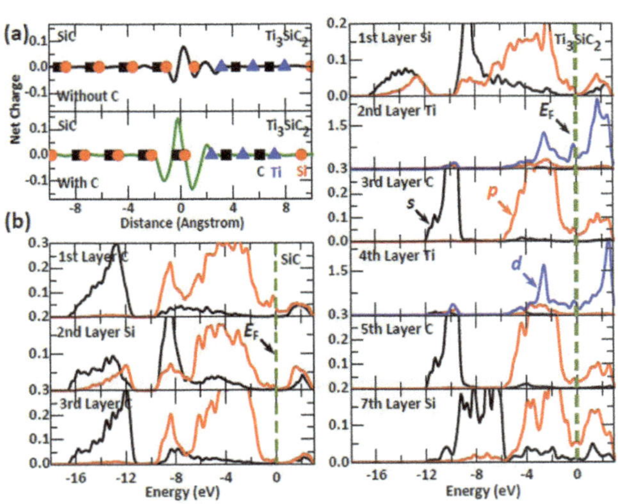

Figure 3. a) Planar-averaged charge-density difference for the relaxed interfaces with and without C along [0001] direction. (b) Density of states projected onto the atomic layers close to the relaxed SiCSi interface. The left bottom panel shows the PDOSs of SiC layers and the right one those of Ti_3SiC_2 layers. The first layer is the atomic layer proximal to interface. The E_F is set to zero [23].

We then presented in Fig. 3(b) DOS projected on selected atomic layers of the SiCSi interface. A key feature in this figure is that a strong interaction is observed between the sub-interfacial Ti *d* and Si *p* states below Fermi level (E_F), which continues well into SiC surface, inducing noticeable gap states in the interfacial C at E_F. This means that the interfacial C layer is metalized, indicative of possible electric conductivity. In fact, the gap states can extend as far as they can into deeper layers of SiC, as there appear weak but visible peaks at E_F in the PDOSs of the 2nd and 3rd layers of SiC. Therefore, a local weak metallicity might occur at top few layers of the semiconductor surface, which could enable current flow through the SiC. We also note significant hybridization between the interfacial C *sp* and Si *sp* states, suggestive of covalent bonding at interface.

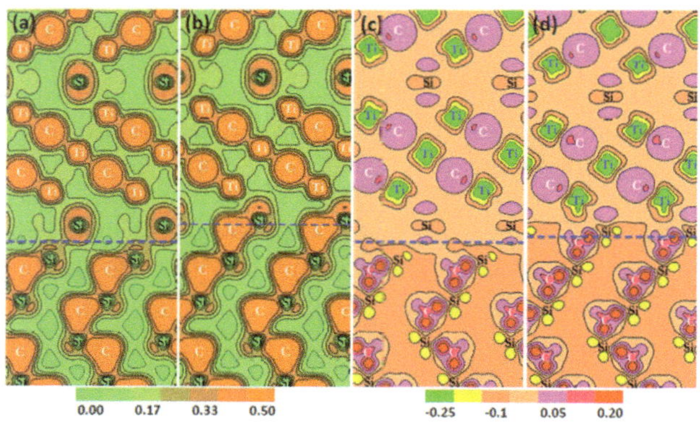

Figure 4. Contour plots of charge densities for (a) SiSi and (b) SiCSi interfaces taken along (11-20) plane. The interface is represented by a horizontal line and the atoms that intersect the contour plane are labeled. Corresponding contour plots of charge-density differences for (c) SiSi and (d) SiCSi interfaces [23].

Figure 4 shows contour plots of charge densities and their differences along (11-20) plane for the optimized SiSi and SiCSi interfaces. We notice in Fig. 4(b) that the bonding interaction between interfacial Si and C for the SiCSi interface is remarkably similar to the Si-C interaction deeper into SiC: the majority of charge is localized on C with humps directed towards their neighboring Si. We thus conclude that the interfacial bonding for the SiCSi is of mixed covalent-ionic nature. The interfacial bonds for the SiSi interface, however, have covalent nature with a small amount of charge accumulated within the interfacial region (Fig. 4(a)). In addition, the amount of charge accumulated on the interfacial Si-Si bonds of SiSi (Fig. 4(c)) is far less significant than that on the interfacial Si-C bonds of SiCSi (Fig. 4(d)). This heaver charge accumulation in the case of SiCSi, together with its mixed covalent-ionic character at interface, accounts for the largest W_{ad} associated with the SiCSi interface.

2.3. Quantum electron transport properties

Although the charge-distribution analysis can reveal valuable information on interfacial bonding, it provides restrained insight into how electrons distribute around E_F, which matters because density around E_F directly determines the current transmission. Figure 5 illustrates an electron-density isosurface and its slice along the (11-20) plane for the optimized

SiCSi interface around E_F. From Fig. 5(a), one can see that charges surrounding interfacial Si are connected and broadly distributed in a sheet-like fashion, which suggests possible electrical conductivity through this region. In addition, there also appear heavily accumulated electrons within the interfacial area, which are connected along the interface and extended as far as several atomic layers into the SiC. These characters can also be confirmed from the slice plot in Fig. 5(b), meaning that current might flow over top few atomic layers of semiconductor, thereby causing Ohmic property. As expected, the electron density at E_F is extremely high for the Ti_3SiC_2 (*i.e.*, sea of electrons) but becomes nil for the SiC layers away from interface (Fig. 5(b)), which can be understood from their intrinsic metallic and semiconducting nature [25].

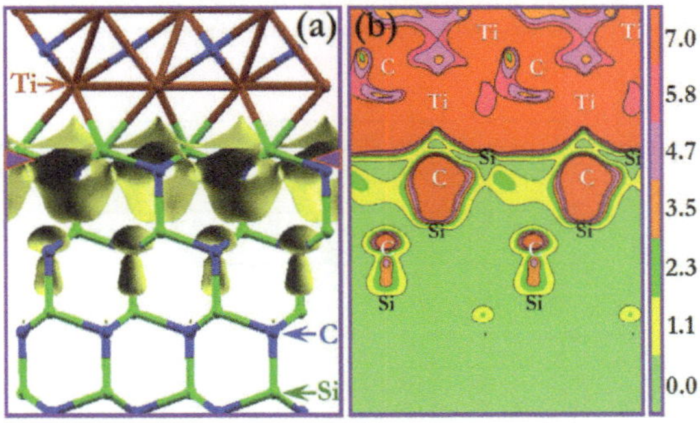

Figure 5. a) Isosurface and (b) electron density plot along the (11-20) plane in the energy window (E_F-0.5 eV, E_F) for the SiCSi interface. The interface is marked by two arrows [18].

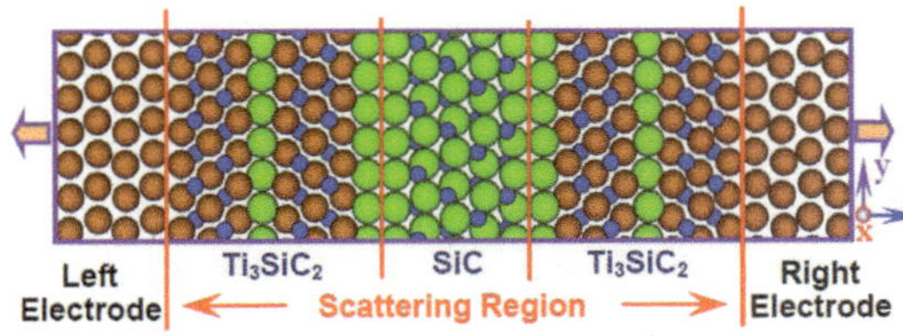

Figure 6. Schematic illustration of a two-probe $Ti/Ti_3SiC_2/SiC/Ti_3SiC_2/Ti$ quantum transport system. The system has infinite extent in the (*x*, *y*) direction and extends to ±∞ in the *z* direction. The SiC/Ti_3SiC_2 interfaces shown in Fig. 2 are adopted [18].

To examine electrical conductivity and gain insight into how the interface influences current transport, we devised a two-probe system [26], $Ti/Ti_3SiC_2/SiC/Ti_3SiC_2/Ti$, and investigated nonequilibrium quantum transport properties. Figure 6 schematically shows a model of the sandwich transport system, which can be divided into a left semi-infinite electrode, a scattering region, and a right semi-infinite electrode. The atomic and electronic structures of the semi-infinite Ti electrodes are assumed to be the same as those of bulk Ti. On the other hand, the electronic states of scattering region are calculated self-consistently. The scattering region consists of hexagonal SiC and Ti_3SiC_2 layers and the periodic boundary conditions are imposed along the directions parallel to the interface. The SiC/Ti_3SiC_2 interface could be either the SiSi or SiCSi, whereas other interfaces are maintained identical for the sandwich systems. In this sense, the difference between the two systems can be mainly attributed to their differing SiC/Ti_3SiC_2 interfaces. Furthermore, we also calculated the Ti/SiC/Ti system, wherein the SiTi model was taken as the SiC/Ti interface.

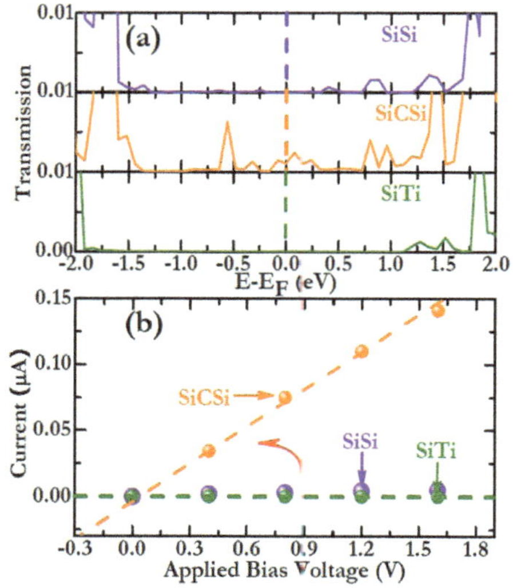

Figure 7. a) Transmission spectra under 0 V and (b) current-voltage characteristics for the sandwich systems involving the interfaces containing direct Si-Si bonding (SiSi), Si-C-Si bonding (SiCSi), and the direct contact of Ti to SiC (SiTi). Refer to Fig. 2 for their corresponding interfacial configurations [18].

Figure 7(a) shows transmission spectra for the relaxed SiSi, SiCSi, and SiTi systems, where one can see that the spectra differ from one another suggesting variations in electronic structures with interface geometries. The most interesting feature is the presence of transmission peaks at E_F for the SiCSi, which is attributable to the electrons distributed around the interface at E_F. Further calculations on electrical properties (*e.g.*, *I–V* curve) for the three systems reveal that the current in the SiCSi case increases much faster than either the SiSi or SiTi case as the applied bias voltage increases (Fig. 7(b)), which can be explained by its lowest SBH. We further examined how applied bias voltages vary from the interface to the SiC region by analyzing the difference in effective potential along the (11-20) plane between the bias voltage of 0.4 V and the one of 0.0 V for the relaxed system. The voltage is found to drop less intensively from the interface to the SiC in the SiCSi case, suggestive of less Schottky nature. Finally, in comparing the general trend of the calculated *I–V* with that of our experimental curve, we find that they agree qualitatively:

both the annealed specimen and the SiCSi model clearly show Ohmic behavior, thereby validating the application of the SiCSi model to describe the Ohmic contacts in the TiAl-based system. In addition to the role of this interface, we also found that an atomic-scale Ti_3SiC_2-like bilayer can be embedded in the SiC interior, forming an atomically ordered multilayer that exhibits an unexpected electronic state with point Fermi surface. The valence charge is confined largely to this bilayer in a spatially connected fashion, serving possibly as a conducting channel to enhance the current flow over the semiconductor.

3. Atomic-scale Ti_3SiC_2 bilayers embedded in SiC

3.1. Atomic structure of the embedded system

Figure 8(a) shows a HAADF image of the annealed TiAl/SiC system, where the SiC substrate is covered entirely by a layered compound, Ti_3SiC_2, as reported previously. In addition to the formation of this epitaxial and coherent SiC/Ti_3SiC_2 interface, another interesting feature is that an atomic-scale bilayer is generated in the SiC interior (marked by a square in Fig. 8(a)), which is located approximately 9.5 nm away from the interface. An enlarged image of the region surrounding the bilayer shows that it has a Ti_3SiC_2-like structure, as shown in Fig. 8(b), where brighter spots represent atomic columns of Ti (smaller circles), while dark ones those of Si (larger circles), since intensity of an atomic column in STEM is, to a good approximation, directly proportional to $Z^{1.7}$ (Z: atomic number) [24].

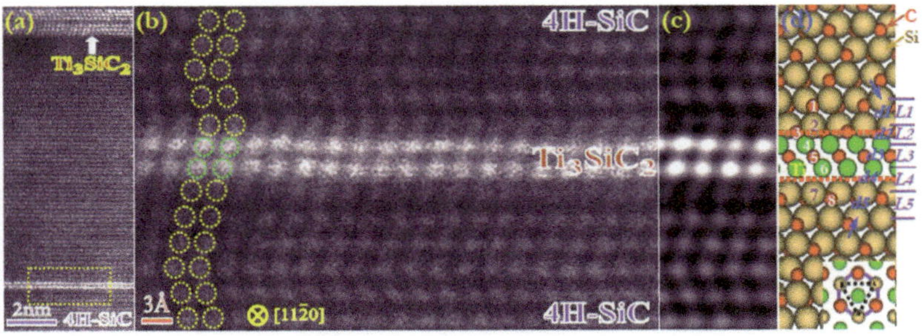

Figure 8. a) HAADF-STEM image for the Ti_3SiC_2-like bilayer embedded in 4H-SiC in the annealed TiAl contact system viewed along [11-20] direction. The bilayer sits about 9.5 nm below the SiC/Ti_3SiC_2 interface. (b) Magnified HAADF image of the region marked in (a) by a dotted square. An overlay is given, where the big circles indicate Si and small ones Ti. (c) The same image as in (b) but has been low-pass filtered to reduce noise. (d) The optimized SiC/Ti_3SiC_2/SiC multilayer model. The distances between layers around the bilayer are represented by *Lm* (*m* = 1 to 5) and those between neighboring atoms projected onto the paper plane by *dm*. The atomic layers are labeled 1 through 8. The inset shows top view of arrangement of Ti and its neighboring Si and C, where the Ti occupies atop the center of the hexagon composed of Si and C [19].

Evidently, the bilayer is embedded in the SiC in an atomically coherent and ordered fashion with no transitional or intermixing layers (see Fig. 8(c)). In view of bulk structures of 4H-SiC and Ti_3SiC_2 and the relative stacking sequence of Ti and Si, the image in Fig. 8(c) can be qualitatively fitted by an energetically stable model shown in Fig. 8(d). In this model, the optimal distances between layers around the bilayer (denoted as *L1* to *L5* in Fig. 8(d)) are calculated to be 2.52, 2.27, 2.28, 2.40, and 2.57 Å, respectively, very close to experimental values of 2.5, 2.2, 2.3, 2.3, and 2.7 Å estimated via quantitative characterization and averaging of different sites in the HAADF images. In addition, the calculated distances between neighboring atoms (labeled *d1* to *d5* in Fig. 8(d)) (2.67, 2.44, 2.44, 2.55, and 2.72 Å) also approach the experimental values (2.7, 2.3, 2.4, 2.6, and 2.7 Å). These mean that the model constructed (Fig. 8(d)) matches the HAADF image (Fig. 8(c)) both qualitatively and quantitatively in light of energetics and atomic distances.

3.2. Formation of point Fermi surface

To gain insight into how the embedded layer influences SiC electronically, we present in Fig. 9 band structure and density of states (DOS) of the multilayer system calculated using the optimal atomic geometry (Fig. 8(d)). Unexpectedly, several bands with a quadratic dispersion cross the Fermi level (E_F) at a single Γ point (Fig. 9(a)), rendering conduction bands (CB) and valence band (VB) touch at their tips and hence the multilayer become a gapless semiconductor. Inspection throughout the Brillion zone verifies that the Fermi surface crossing is a single point, which determines E_F, as in the bilayer graphene with no external stimuli (bands in graphene extend linearly both to lower and higher energy from point Fermi surfaces, as referred to as "massless Dirac"). This crossing of bands is confirmed in the DOS showing a curious vanishing of states at E_F for both spins (Fig. 9(b)) and further verified in a surface plot of the two bands proximal to E_F in a small *k* space presenting gapless character at the Γ point and a gap in the region away from the crossing point (Fig. 2(c)).

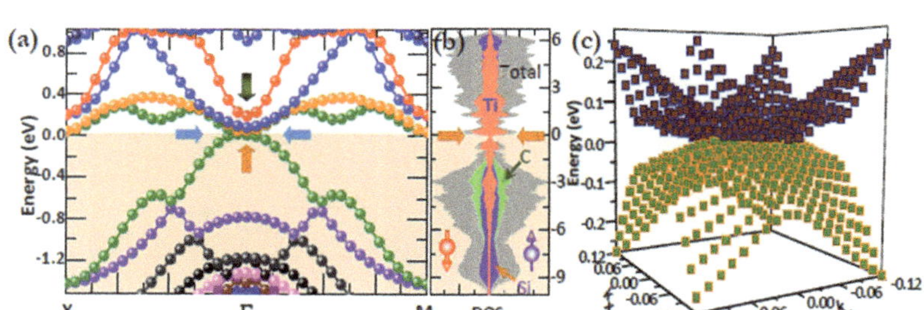

Figure 9. a) Blowup of band structure around E_F shown on the *xy* plane with $X = \pi/a(1, \bar{1}, 0)$ and $M = \pi/a(1,1,0)$, where the "*a*" is in-plane lattice constant. Note that the point Fermi surface is at Γ where bands cross precisely at E_F. (b) Total DOS and PDOS plots of C, Si, and Ti atom contributions for the optimized $SiC/Ti_3SiC_2/SiC$ multilayer, showing that bands surrounding E_F have characters of Ti in the bilayer. (c) "Surface" plot of the two bands that cross the E_F, centered surrounding the Γ point. The E_F position is aligned to zero [19].

In addition, extensive calculations using the LDA and PBE functional corroborate once again the peculiar crossing of the bands (band structure and DOS spectra are almost identical to those calculated using the PW91), which therefore indicates that the crossing at E_F is not an accidental degeneracy arising from the applied functional. This behavior is quite unusual, as it turns up neither in the SiC bulk showing E_F lying in a gap between states nor in the Ti_3SiC_2 bulk showing E_F lying in the middle of a band of electronic states, which can be attributed to the structural symmetry with Ti atoms sitting exactly atop center of the hexagon composed of Si and C (corresponding to the Γ point in *k* space), as shown in inset of Fig. 8(d). Further calculations reveal that this gapless nature appears in the multilayer consisting of a Ti_3SiC_2-like monolayer embedded in SiC as well, indicating that the unique state could be understood upon quantum confinement effect and the interface phenomenon accompanied by the polarity discontinuity. However, our additional calculations suggest that the point Fermi surface vanishes when a Ti_3SiC_2 trilayer is hypothetically embedded within SiC, which takes on metallic states with three bands (arise from the three Ti layers) crossing E_F, similar to what is seen in the band structure of Ti_3SiC_2 bulk. This transition from the zero-gap semiconductor to metal therefore highlight the importance of quantum confinement in inducing the point Fermi surface, since additional layer is believed to relieve the confinement effect and produce more states.

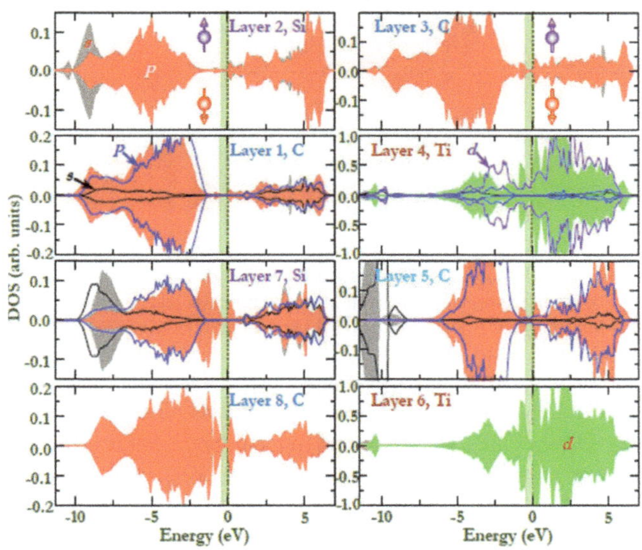

Figure 10. DOS projected on selected atomic layers in the multilayer system. Left panel gives the PDOSs of SiC layers and right panel those of Ti_3SiC_2 layers. The lines show PDOSs of the corresponding atoms in the bulk as a reference. The E_F is set to zero [19].

3.3. Electronic states

Further investigation of DOS projected on selected atomic layers provides evidence that the bands close to E_F involve dominantly Ti 3*d* states (layers 4 and 6 in Fig. 8(d)). These states undergo a sudden vanishing at E_F (Fig. 10), which is in stark difference from what is seen in the DOS projected (PDOS) on Ti in bulk Ti_3SiC_2 presenting continuous states at and around E_F (lines in Fig. 10). In addition, there are notable electronic states in the forbidden gap of bulk SiC for the C and Si in the multilayer, which is definitely attributable to the embedded bilayer. The presence of the induced gap states, in particular, those below the CB minimum of bulk SiC but close to E_F, can readily present a trap for CB electrons, which could modify electronic behaviors of the originally insulating SiC and thus be relevant for the current flow over semiconductor in the TiAl-deposited SiC system. Moreover, we note that overall features of PDOS for identical atom species may even differ from one another, which can be ascribed to their different bonding environments.

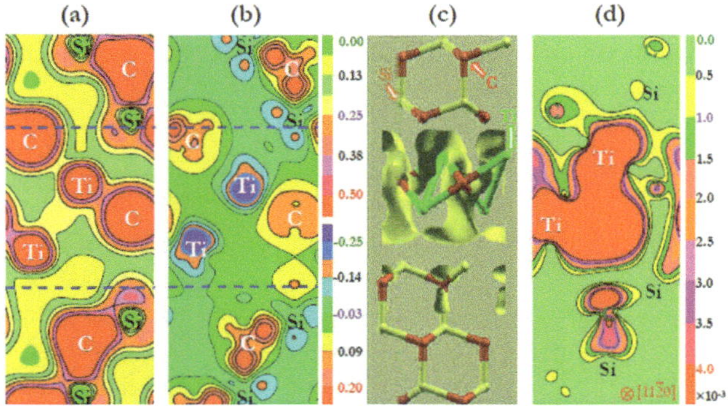

Figure 11. Contour plot of (a) charge density and (b) density difference for the multilayer system viewed along the (11-20) plane. The difference of charge density gives the redistribution of charge in the system relative to its isolated one. The upper scale denotes the magnitude of charge in (a) and the lower scale that of charge difference in (b). (c) Isosurface and (d) electron density plot along the same plane as in (a) in the energy window (E_F-0.5 eV, E_F) [19].

To shed further light on bonding nature and charge distribution in the multilayer, we present contour plots of charge density (Fig. 11(a)) and its difference (Fig. 11(b)) along the (11-20) plane. From the figures, we notice that (i) majority of charge is localized on C atoms with humps distorted toward their neighboring atoms, suggesting that bonds in both SiC and bilayer are of a mixed covalent-ionic nature, (ii) charge distribution on C in SiC exhibits more pronounced lobes than that on C in the bilayer (Fig. 11(a)), indicative of more covalent element for bonds in SiC, and (iii) ionicity originates from the large charge gain on C at an expense of charge loss on its neighboring cations (Fig. 11(b)). Further examination on electron distribution around E_F reveals that valence electrons are confined, to a large extent, to within the bilayer (Fig. 11(c)), in good agreement with the DOS analysis (Fig. 3). These charges are spatially connected and broadly distributed surrounding the bilayer with a small degree of leakage into as far as two atomic layers of SiC (Fig. 11(c)), as is further confirmed from a slice plot in Fig. 11(d). This implies that the bilayer buried in between the SiC may act as a conducting channel so as to enhance current flow over the semiconductor.

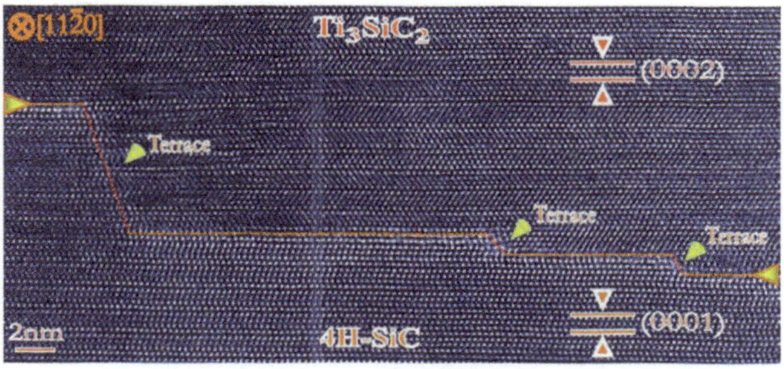

Figure 12. A cross-sectional HRTEM image of the contact of the formed Ti_3SiC_2 to the 4H-SiC substrate viewed from the [11-20] direction. The two materials are demarcated by a zigzag line. Terraces with varying dimensions are observed, as indicated by arrows [20].

4. Terraces at Ohmic contact in SiC electronics

Combining imaging with atomistic simulations, we determine the atomic-scale structures of terraces in between SiC and its contacts and relate the structures to their electronic states and bonding nature, aimed at revealing the impact of the terraces on the contacts of SiC electronics. The terraces were first characterized using the high-resolution TEM (HRTEM) and scanning TEM (STEM), upon which the first-principles calculations were performed. The combined study allows a deeper understanding of the role played by terraces in the ohmic contact formation on a quantum level. The terraces are structurally epitaxial, coherent and atomically ordered, and theoretically predicted to have electronic states at Fermi level (E_F) regardless of their dimension.

4.1. Atomic structures of terraces

Figure 12 shows a typical cross-sectional HRTEM image of the contact of Ti_3SiC_2 to SiC which includes terraces of various dimensions. Well arranged (000*l*)-oriented lattice fringes can be observed in both the SiC and Ti_3SiC_2 layers. The points at which the phase contrast is no longer periodic define the interfacial area, as indicated by a zigzag line. Evidently, the SiC substrate is covered entirely by the Ti_3SiC_2 layers even in the terrace region (arrows), meaning a direct contact of Ti_3SiC_2 to SiC at the terraces at atomic scale. The terraces are atomically abrupt and ordered, showing no amorphous layers, secondary-phase layers, contaminants, or transition areas. A small number of misfit dislocations are identified at the contact region (not shown here) due to the small lattice mismatch of SiC to Ti_3SiC_2 (less than 0.5%). Further analyses of selected area diffraction patterns reveal that the Ti_3SiC_2 layers have epitaxial orientation relationships, (0001)Ti_3SiC_2//(0001)SiC and [0-110]Ti_3SiC_2//[0-110]SiC, with the SiC substrate.

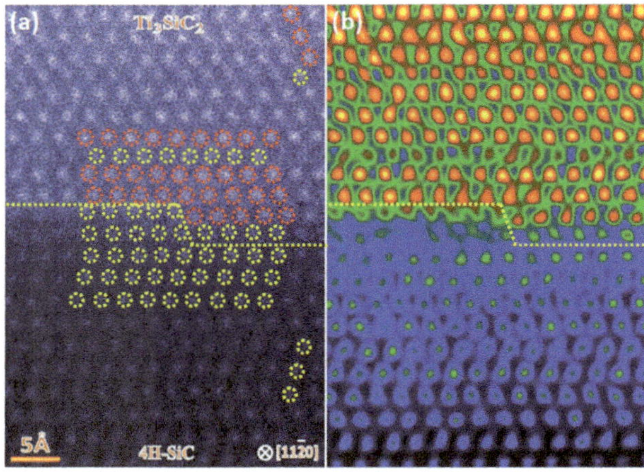

Figure 13. a) A typical HAADF-STEM image of a small terrace observed from the [11-20] direction. The terrace is indicated by a zigzag line. Bigger dotted circles denote Ti and the smaller ones Si. (b) The same image as in (a) but has been filtered to reduce noise [20].

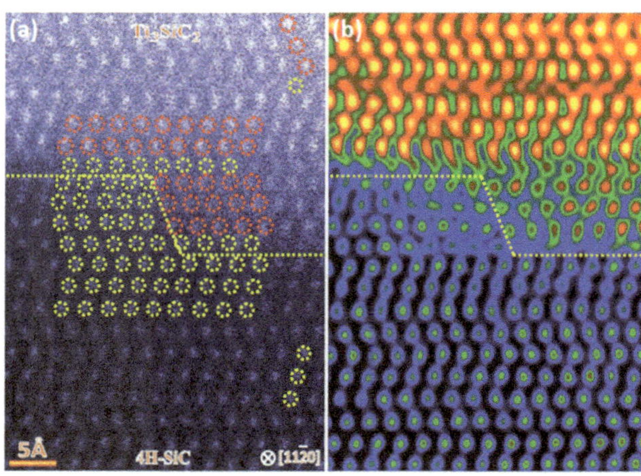

Figure 14. A typical HAADF image of an intermediate terrace observed from the [11-20] direction. The terrace is indicated by a zigzag line. (b) The same image as in (a) but has been filtered to reduce noise [20].

In general, this contact region contains terraces with a wide variety of dimensions that can be affected by numerous factors. However, to develop an understanding of such a complex contact, it is important to first focus on representative terraces. Here, we choose purposely three species of terraces based on the dimension: small, intermediate, and large terrace. The corresponding HAADF images are presented in Figs. 13–15, which confirm the atomically abrupt and ordered terraces. Moreover, heteroepitaxy is retained between the SiC and Ti_3SiC_2 for each terrace. Since the intensity of an atomic column in the HAADF image is proportional to $Z^{1.7}$ (Z: atomic number) [24], brighter spots in the images represent atomic columns of Ti while darker ones are Si. In view of the atomic arrangements in the SiC and Ti_3SiC_2 bulks, we define the terrace by a dashed line and identify the atoms in a few layers close to the line as in the terrace region hereafter. For the small terrace, the image can be intuitively interpreted as a bonding of Ti (Si) in Ti_3SiC_2 to Si in SiC (Fig. 13(b)). However, this is not the case for the intermediate terrace showing Si-Si bonding at the hollow site (Fig. 14(b)). Further interpretation of the HAADF mage of large terrace (Fig. 15(b)) reveals a Si-Si bonding as well. Not surprisingly, atomic columns of carbon are not scattered strongly enough to be visualized due to its small Z, rendering the image incomplete. Further complementing of these images so as to relate their atomic structures to their electronic properties requires a combination of imaging with the first-principles modeling.

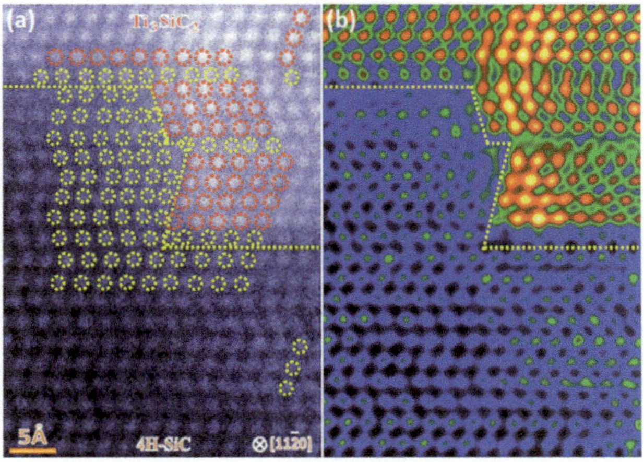

Figure 15. A typical HAADF image of a large terrace observed from the [11-20] direction. The terrace is indicated by a zigzag line. (b) The same image as in (a) but has been filtered to reduce noise [20].

4.2. Electronic states and bonding nature of the terraces

To gain insight into electronic properties of the terraces and the role they played in the ohmic contact formation, we perform first-principles calculations on the three typical terraces. Upon a consideration of bulk structures of SiC and Ti_3SiC_2, the aforementioned orientation relations, and relative stacking sequence near terraces shown in the HAADF (Figs. 13–15), atomic models of the three representative terraces were established (Fig. 16), taking into account full structural relaxation. It should be noted that these models may not exactly reflect the real terraces because it is the extreme difficult to interpret directly the HAADF images owing to the intricate atomic arrangements around the terraces and to the invisible C atoms.

However, these models are constructed upon a careful consideration of the space filling and local chemical environment, and importantly they exhibit the typical variations of the contact with terraces, which should be useful as an initial stage to look at the electronic states of terraces. It is also worthy of mentioning that neither the total nor the interface energies could be applicable to justify the models as an interpretation of the HAADF images because (i) the terrace models have a different number of atoms, (ii) it is unlikely to calculate the exact total terrace area, and (iii) there are numerous candidates for each type of terrace. In the optimized model of small terrace (Fig. 16(a)), one can notice a bonding of Ti (Si) atoms in Ti_3SiC_2 to Si atoms in SiC at the hollow site.

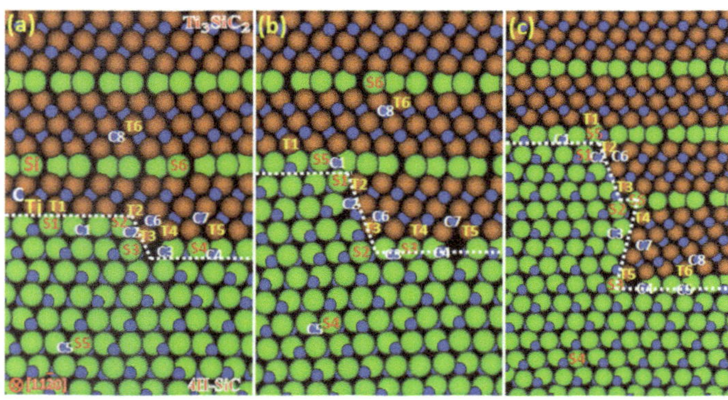

Figure 16. The relaxed atomic models of the three representative terraces with different dimensions: (a) small, (b) intermediate, and (c) large terrace. The representative atoms surrounding terrace are labeled for the atom-projected density of states analysis [20].

Figure 17 shows PDOS of several representative atoms at the small terrace (labeled in Fig. 17(a)), where a remarkable difference is seen between the atoms near and away from terrace. A key feature of this figure is that strong hybridization takes place between the Ti *d* (T1~T6) and Si *p* (S1~S6) levels below E_F, which continues well into the SiC region, inducing a pronounced gap state at E_F for the C atoms near the terrace (C1~C4, C6). However, such a gap state at E_F is vanished completely for the C deeper into the SiC (C5 in Fig. 17). Apart from the C, the Si atoms in SiC near terrace (S1~S3) also exhibit a weak yet visible peak at E_F that are totally absent in the bulk (S5 in Fig. 17). The presence of gap states can be attributed to the overlap of the Ti *d* with Si *p* levels. A notable hybridization is seen as well between the Si *sp* (S1~S4) and C *sp* (C1~C4, C6) levels, indicating the formation of covalency near terrace. Apart from the SiC, all of the atoms in Ti_3SiC_2 around terrace (C6, C7, S4, T1~T5) also exhibit a notable peak at E_F owing to the great degree of overlap between the Ti *d* and C (Si) *p* levels. Moreover, overall feature of PDOS for the Ti atoms around terrace (T1~T4) differs from that in the Ti_3SiC_2 bulk (T6), indicating that the terrace can have an impact on electronic states of Ti as well.

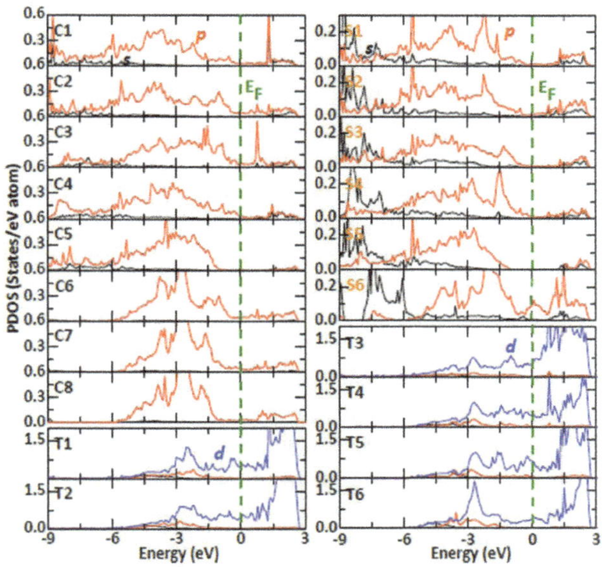

Figure 17. PDOS (states/eV atom) of several selected atoms near or far from the small terrace. All of atoms in the slab are fully relaxed. Refer to Fig. 16(a) for the sites of the selected atoms. The E_F is set to zero and indicated by a vertical dashed line [20].

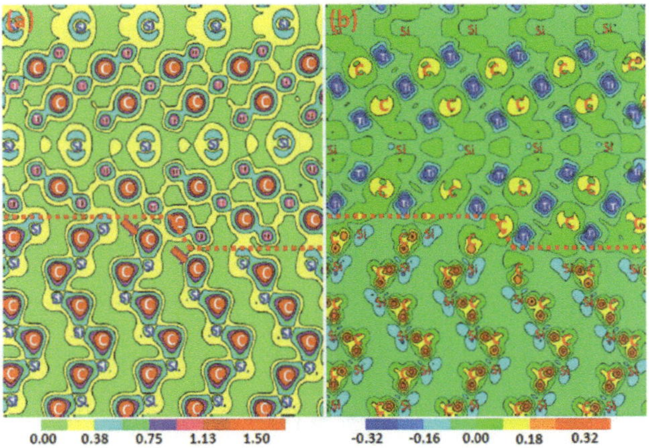

Figure 18. Contour plots of (a) charge density and (b) charge-density difference for the small terrace viewed along the (11-20) plane. Difference in charge density shows redistribution of charge near terrace relative to its isolated system. The position of terrace is indicated by a zigzag dashed line and the atoms intersecting the contour plane are labeled [20].

To identify the bonding nature directly, we further show the contour plots of charge density and density difference for the optimized small terrace viewed along the (11-20) plane (Fig. 18). From Fig. 18(a), a remarkable difference is observed in charge distribution on C: charge distribution around C in SiC (away from terrace) has humps directed toward neighboring Si, while that around C in Ti_3SiC_2 (away from terrace) is of almost spherical symmetry. However, the charge distribution on some C atoms near the terrace (indicated by arrows in Fig. 18(a)) shows a mixed character with a lobe on one side while a spherical outline on the other, which is reflected from their different PDOS (Fig. 17). Furthermore, the charge distribution on the Si-C bonds closet to the zigzag line (Fig. 18(a)) shares some features with that on the Si-C bonds away from the zigzag line (analog to the bonds in SiC bulk): (i) the majority of charges are distributed on all the C atoms, and (ii) there are visible distortions in the charge distribution on the C atoms directed toward their adjacent Si atoms. A certain level of covalency is seen on the atomic bonding along the zigzag line (which defines the terrace), which is due to the hybridization of Ti *d* with Si (C) *p* levels (Fig. 17). These imply a mixed covalent-ionic bonding for the small terrace. It is obvious from the density-difference plot (Fig. 18(b)) that the ionic nature arises from the charge transfer of Ti (Si) to C. In the Ti_3SiC_2 region away from the zigzag line, the Ti-C bonds are found to have more covalent element than the Ti-Si bonds, as more amount of charges are accumulated in between Ti and C, which indicates that the Ti-C is a stronger chemical bond than the Ti-Si.

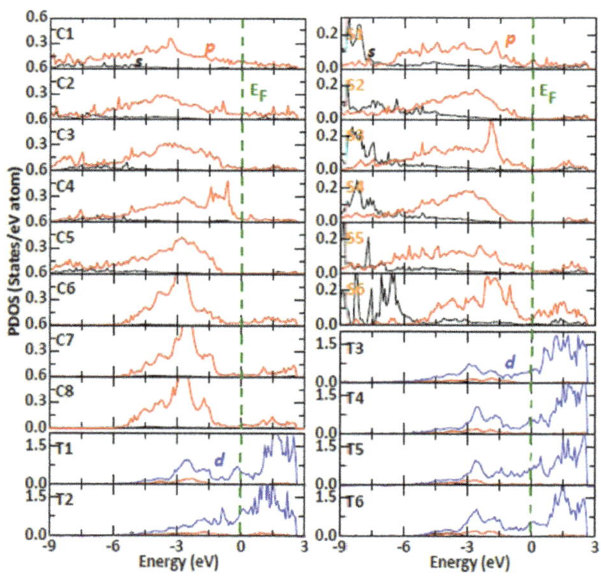

Figure 19. PDOS (states/eV atom) of several selected atoms at or far from the intermediate terrace. Refer to Fig. 16(b) for the sites of the selected atoms [20].

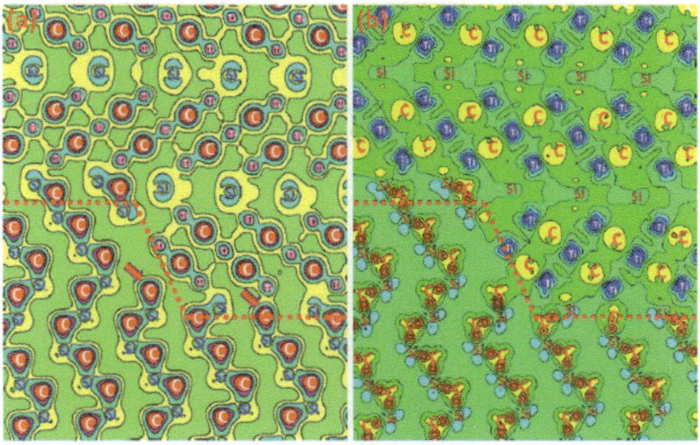

Figure 20. Contour plots of (a) charge density and (b) charge-density difference for the intermediate terrace viewed along the (11-20) plane. The scale is the same as in Fig. 18. The location of terrace is indicated by a zigzag line and the atoms intersecting the contour plane are labeled [20].

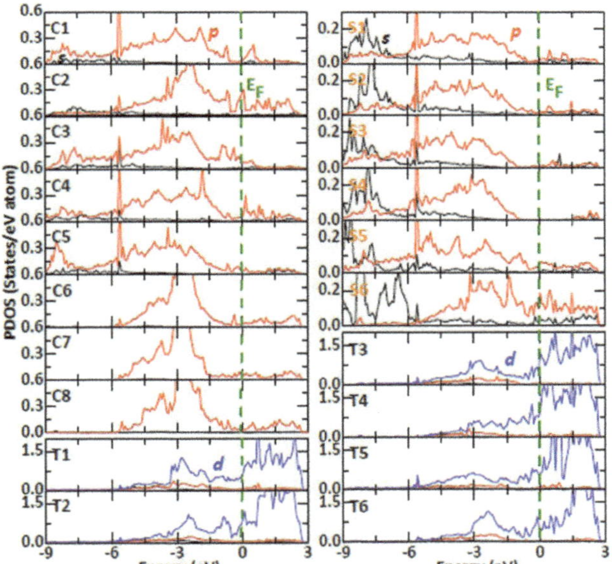

Figure 21. PDOS (states/eV atom) of several selected atoms at or far from the large terrace. Refer to Fig. 16(c) for the sites of the selected atoms [20].

The fully relaxed structure of the intermediate terrace is shown in Fig. 16(b), where one can see a Si-Si bonding at the hollow site (on two sides of the zigzag line). Figure 19 shows PDOS plot of several representative atoms on or near terrace (labeled in Fig. 16(b)), where one can note that (i) gap states at E_F appear for the C (C2~C4) and Si (S1,S2) atoms in SiC near terrace, (ii) a hybridization is observed between the Ti *d* (Si *p*) and O *p* levels just below E_F, an indication of the formation of covalency, and (iii) the impact of terrace on the electronic states of both SiC and Ti_3SiC_2 is confined, as the PDOS returns to its bulk value for the atoms away from terrace (*e.g.*, C5, C8, S4, S6, T6). Interestingly, overall feature of the PDOS for either the C (C1~C4) or Si (S1~S3) atoms around terrace differs from one another, suggesting a strong effect of terrace on electronic states of both SiC and Ti_3SiC_2.

Figure 20 shows contour plots of charge density and density difference for the relaxed intermediate terrace intersected along the same plane as in Fig. 18. Like what was seen in the small terrace, the majority of charges remain concentrated on C in two different ways: the charge distribution on the C in Ti_3SiC_2 (*e.g.*, C8) is of spherical symmetry, while that on the C in SiC (*e.g.*, C5) has notable lobes pointed toward their adjacent Si. However, some C atoms near terrace (indicated by arrows in Fig. 12(a)) have lobe and sphere outline simultaneously. Moreover, charges are distributed along the bonds near terrace (on two sides of zigzag line), which together with the charge localization on C atoms infers that the intermediate terrace has a mixed covalent and ionic bonding as well. Evidently, the charge gain on C is at expense of charge loss on their neighboring cations (Fig. 20(b)).

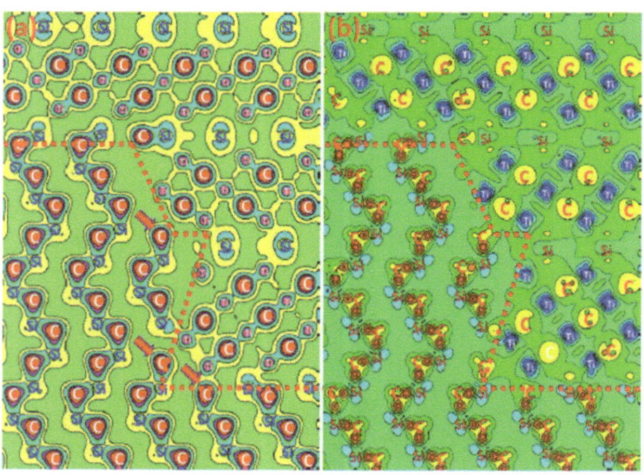

Figure 22. Contour plots of (a) charge density and (b) charge-density difference for the large terrace viewed along the (11-20) plane. The scale is the same as in Fig. 18. The site of terrace is indicated by a zigzag line and the atoms intersecting the contour plane are labeled [20].

Figure 16(c) illustrates optimized atomic geometry of the large terrace, where a Si-Si bonding is revealed. Figure 21 shows the PDOS of the large terrace, where one can notice that electronic structure is influenced remarkably by terrace. The key point is that there emerge notable peaks at E_F for both the C (C1~C5) and Si (S1~S3, S5, S6) surrounding terrace. However, such gap states are screened rapidly, since the atoms in SiC away from terrace (S4) show no peak at E_F at all, which implies that the effect of terrace is confined to within a couple of layers. Moreover, a substantial hybridization is seen between the Ti *d* and Si (C) *p* levels below E_F, which indicates the formation of covalent bonding. The covalency is also reflected in the charge-density plot (Fig. 22(a)) showing lobes for the C in SiC and charge accumulation along the bonds in Ti_3SiC_2. These, along with the charge distribution on C at expense of Ti and Si (Fig. 22(b)), indicate that bonding near terrace is again primarily ionic, yet maintains a certain degree of covalency. Finally, both the electron density and density difference for some atoms near terrace (indicated by arrows) deviate severely from those for their bulk counterparts, implying an essential electronic role of the Ti_3SiC_2 in the semiconductor.

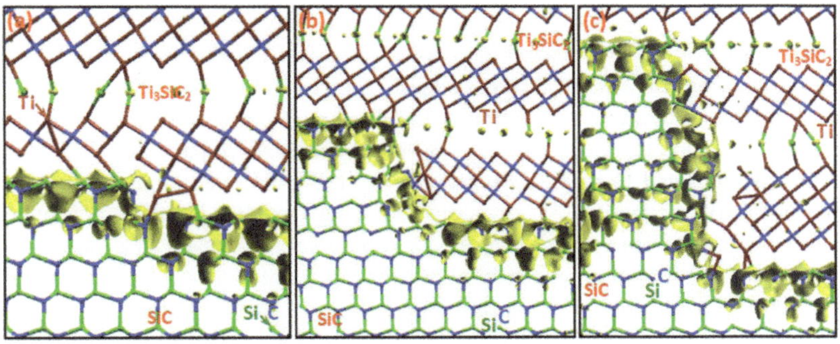

Figure 23. Isosurface plots for (a) small, (b) intermediate, and (c) large terrace along the (11-20) plane in an energy window (E_F–0.5 eV, E_F). The charges on Ti_3SiC_2 are omitted for clarity [20].

4.3. Electron distribution near Fermi energy

Although the PDOS and charge analyses can reveal valuable information on the bonding nature near terrace, they provide limited insights into matters regarding electron distribution around E_F, which is strongly related to the electronic conduction over terrace. Figure 23 shows electron-density isosurface near E_F along the (11-20) plane for the three terraces. The electrons are accumulated substantially around all the terraces in a spatially connected fashion, which are extended as far as several atomic layers into SiC, regardless of dimension of the terraces. Such a broad electron distribution provides a likely electron conduction channel to allow current transport across a few layers of the semiconductor, which indicates that terraces could also be one of the origins underlying the observed ohmic nature in the metal/SiC contact system. One can also note that the three terraces are comparable in the amount of accumulated charge at E_F, inferring that the dimension of terrace plays an insignificant role in affecting electron transport across the contact. It is worthy of mentioning that the electron density at E_F is extremely high in Ti_3SiC_2 (sea of electrons, not shown for clarity), yet turns almost nil for the SiC atoms away from terrace, which can be understood on the fact that the Ti_3SiC_2 is intrinsically metallic, whereas the SiC is semiconducting.

5. Discussion and conclusions

The current understanding of formation origin of Ohmic contact, which is based mainly on experimental studies of property measurement and structure characterization, can be summarized in three main points [27-30]: (1) the deposited Al (80 at%) might diffuse in part into the SiC and dope heavily the semiconductor because Al is well-known to act as a *p*-type dopant for SiC, (2) the high density of pits, spikes, or dislocations may be generated underneath the contacts after annealing so that current can transport primarily through these defects due to the possible enhancement of electric field at these features and semiconductor doping at these locations, (3) formation of intermediate semiconductor layer between the deposited metals and semiconductor, which consists of silicides or carbides, could divide the high barrier height into lower ones, thus reducing the effective barrier height.

The findings presented first demonstrate that no Al is clearly segregated around the interfacial region, in particular at the top few layers of SiC, which rules out the possibility of additional Al doping. Though a small amount of residual Al is found to be present, mostly in a form of Al_4C_3 compound, it may locate on the surface of annealed contacts rather than in the layer directly contacted to the SiC, thus playing a negligible role in Ohmic contact formation. The majority of deposited Al is evaporated during annealing because of its low melting point and high equilibrium vapor pressure. The dominant role played by Al in the TiAl system is to assist the formation of liquid alloy so as to facilitate chemical reaction. Furthermore, careful characterization of the interfacial region reveals that the substrate and the generated compound are epitaxially oriented and well matched at interface with no clear evidence of high density of defects. This suggests that the morphology might not be the key to understanding the contact formation. In support of this speculation, it has been observed previously that Ti Ohmic contacts can be possibly generated without any pitting and that pit-free Ohmic contacts can be fabricated.

One remaining theory is the alloy-assisted Ohmic contact formation. This alloy is determined to be ternary Ti_3SiC_2, which has also been corroborated by other expriments. Since the bulk Ti_3SiC_2 has already been found to be of metallic nature both in experiment and theory, the contact between Ti_3SiC_2 and its covered metals should show Ohmic character and thus the SiC/Ti_3SiC_2 interface should play a significant role in Ohmic contact formation. This idea is supported by the fact that the determined interface has a lowered SBH due to the large dipole shift at interface induced by the partial ionicity and the considerable charge transfer. In addition, the interfacial states, as indicated by the electron distribution at E_F, are also viewed as a contributing factor in reducing the SBH. These states might be further enhanced by the possible presence of point defects at interface, although these structural defects have not been detected by the TEM study.

Interestingly, our calculations predict that an atomic layer of carbon emerges as the first monolayer of Ohmic contacts, which eventually affects interface electronic structure. Such trapped carbon was previously studied in both other interfacial systems theoretically by DFT and the TiNi Ohmic contacts on 4H-SiC experimentally by Auger electron spectroscopy. It was proposed that the carbon could be segregated to the interfacial area, strengthening interface substantially and reducing the Schottky barrier dramatically. Further, it was reported that the Ohmic contact can be realized by depositing carbon films only onto the SiC substrate, indicative of the determinative role of carbon in the Ohmic contact formation. The important role played by carbon in our study can be traced to the two interfacial Si layers, which provide possible sites for carbon segregation due to the strong Si-C interaction. However, direct imaging of the trapped carbon is still difficult in present study and further characterization requires the high-voltage EM and/or other advanced microscopic techniques.

We then demontrate that atomic-scale Ti_3SiC_2-like bilayer can be embeded in the SiC interior, forming an atomically ordered multilayer that exhibits an unexpected electronic state with the point Fermi surface, in stark absence in repestive bulk constituents. The valence charge is found to be confined largely within the bilayer in a spatially connected way, which serves as a possible conducting channel to enhance the current flow over the semiconductor. Such a heterostructure with unusual properties is mechanically robust, rendering its patterning for technological applications likely. Finally, the atomic structures of terraces at the contacts in SiC devices are investigated and bridged to their electronic properties at an atomic scale. Experimentally, newly formed carbide Ti_3SiC_2 is demonstrated to bond directly to silicon carbide in the terrace region in an epitaxial and atomically ordered fashion, regardless of dimension of terraces. Further first-principles calculations reveal gap states in the semiconductor layers and a substantial charge accumulation around terraces in a connected and broadly distributed manner. The presence of gap states at Fermi energy and the likelihood to serve as electron conduction channels to allow current flow over the semiconductor identify the terraces as one of the origins underlying the ohmic contact in silicon carbide electronics. Such a combined experimental and theoretical investigation provides insight into the complex atomic and electronic structures of buried terraces, which should be applicable to addressing contact issues of interest in other electronic devices.

To summarize, we have determined in this chapter atomic-scale structures of Ohmic contacts on SiC and related them to their electronic structures and electric properties, aimed at understanding the formation mechanism of Ohmic contact in TiAl-based system. The combined HAADF-DFT study [31] represents an important advance in relating structures to device properties at an atomic scale and is not limited to the contacts in SiC electronics. Our results show that the main product generated by chemical reaction can be epitaxial and have atomic bonds to the substrate. The contact interface, which could trap an atomic layer of carbon, enables lowered Schottky barrier due to the large interfacial dipole shift associated with the considerable charge transfer. The atomic-scale Ti_3SiC_2-like bilayer is embedded well in SiC bulk interior in an epitaxial, coherent and atomically abrupt manner, which exhibits an unexpected state with a point Fermi surface. Moreover, the formed Ti_3SiC_2 can even be epitaxial and atomically ordered on SiC substrate near terrace, which inudces pronounced gap states at E_F in the semiconductor layers. Charges are accumulated heavily surrounding terrace in a spatially connected fashion, irrespective of dimension of the terraces, which suggests the possiblity of terraces as likely electron conduction channels to allow current transport across the semiconductor. The inducing of gap states and the capability to enable current flow over the semiconductor identify the terraces as one of the origins underlying the Ohmic nature in the metal/SiC contact system as well. These findings are relevant for technological improvement of contacts in SiC devices, and this chapter presents an important step towards addressing the current contact issues in wide-band-gap electronics.

Acknowledgements

The author acknowledges M. Saito, S. Tsukimoto, at the WPI Research Center, Advanced Institute for Materials Research, Tohoku University and Y. Ikuhara at the The University of Tokyo for their collaborations. The author thanks S. Watanabe at The University of Tokyo for allowing our use of computational resources. The present study was supported in part by a Grant-in-Aid for Scientific Research on Priority Area, "Atomic Scale Modification (474)" from the Ministry of Education, Culture, Sports, Science, and Technology of Japan. Z. W acknowledges financial supports from the Grant-in-Aid for Young Scientists (A) (Grant No. 24686069) and the Challenging Exploratory Research (Grant No. 24656376). The calculations were carried out on a parallel SR11000 supercomputer at the Institute for Solid State Physics, Univ. of Tokyo.

Author details

Zhongchang Wang*

Address all correspondence to: zcwang@wpi-aimr.tohoku.ac.jp

WPI Research Center, Advanced Institute for Materials Research, Tohoku University, Japan

References

[1] Perez-Wurfl, I., Krutsinger, R., Torvik, J. T., & Van Zeghbroeck, B. (2003). 4H-SiC Bipolar Junction Transistor with High Current and Power Density. *Solid State Electronics*, 47, 229-231.

[2] Lee, S. K., Zetterling, C. M., Danielsson, E., Östling, M., Palmquist, J. P., Högberg, H., & Jansson, U. (2000). Electrical Characterization of TiC Ohmic Contacts to Aluminum Ion Implanted 4H-Silicon Carbide. *Applied Physics Letters*, 77, 1478-1480.

[3] Pécz, B. (2001). Contact Formation in SiC Devices. *Applied Surface Science*, 184, 287-294.

[4] Crofton, J., Barnes, P. A., & Williams, J. R. (1993). Contact Resistance Measurements on p-type 6H-SiC. *Applied Physics Letters*, 62, 384-386.

[5] Mohney, S. E., Hull, B. A., Lin, J. Y., & Crofton, J. (2002). Morphological Study of the Al-Ti Ohmic Contact to p-type SiC. *Solid State Electronics*, 46, 689-693.

[6] Nakatsuka, O., Takei, T., Koide, Y., & Murakami, M. (2002). Low Resistance TiAl Ohmic Contacts with Multi-Layered Structure for p-Type 4H-SiC. *Materials Transactions*, 43, 1684-1688.

[7] Johnson, B. J., & Capano, M. A. (2004). Mechanism of Ohmic Behavior of Al/Ti Contacts to p-type 4H-SiC After Annealing. *Journal of Applied physics*, 95, 5616-5620.

[8] Chang, S. C., Wang, S. J., Uang, K. M., & Liou, B. W. (2005). Investigation of Au/Ti/Al Ohmic Contact to N-type 4H-SiC. *Solid State Electronics*, 49, 1937-1941.

[9] Ohyanagi, T., Onose, Y., & Watanabe, A. (2008). Ti/Ni Bilayer Ohmic Contact on 4H-SiC. *Journal of Vacuum Science and Technology B*, 26, 1395-1362.

[10] Tsukimoto, S., Nitta, K., Sakai, T., Moriyama, M., & Murakami, M. (2004). Correlation Between the Electrical Properties and the Interfacial Microstructurs of TiAl-Based Ohmic Contacts to p-type 4H-SiC. *Journal of Electronic Materials*, 33, 460-466.

[11] Wang, Z. C., Tsukimoto, S., Saito, M., & Ikuhara, Y. (2009). SiC/Ti_3SiC_2 Interface: Atomic Structure, Energetics, and Bonding. Physical Review B , 045318 , 79, 1-10.

[12] Nellist, P. D., Chisholm, M. F., Dellby, N., Krivanek, O. L., Murfitt, M. F., Szilagyi, Z. S., Lupini, A. R., Borisevich, A., Sides Jr, W. H., & Pennycook, S. J. (2004). Direct Sub-Angstronm Imaging of a Crystal Lattice. *Science*, 305, 1741.

[13] Wang, Z. C., Saito, M., Mc Kenna, K. P., Gu, L., Tsukimoto, S., Shluger, A. L., & Ikuhara, Y. (2011). Atom-Resolved Imaging of Ordered Defect Superstructures at Individual Grain Boundaries. *Nature*, 479, 380-383.

[14] Wang, Z. C., Tsukimoto, S., Saito, M., & Ikuhara, Y. (2009). Atomic and Electronic Structure of the YBa2Cu3O7/SrTiO3 Interface from First Principles. *Journal of Applied Physics*, 106, 093714-1-8.

[15] Wang, Z. C., Tsukimoto, S., Saito, M., & Ikuhara, Y. (2009). Individual Charge-Trapping Dislocations in an Ionic Insulator. *Applied Physics Letters*, 95, 184101-1-3.

[16] Wang, Z. C., Zeng, W., Gu, L., Saito, M., Tsukimoto, S., & Ikuhara, Y. (2010). Atomic-Scale Structure and Electronic Property of the LaAlO3/TiO_2 Interface. *Journal of Applied Physics*, 108, 113701-1-9.

[17] Wang, Z. C., Saito, M., Tsukimoto, S., & Ikuhara, Y. (2009). Interface Atomic-Scale Structure and its Impact on Quantum Electron Transport. . Advanced Materials , 21, 4966-4969.

[18] Wang, Z. C., Tsukimoto, S., Saito, M., Ito, K., Murakami, M., & Ikuhara, Y. (2009). Ohmic Contacts on Silicon Carbide: The First Monolayer and its Electronic Effect. *Physical Review B*, 80, 245303, 1-12.

[19] Wang, Z. C., Tsukimoto, S., Sun, R., Saito, M., & Ikuhara, Y. (2011). Atomic-Scale Ti_3SiC_2 Bilayers Embedded in SiC: Formation of Point Fermi Surface. *Applied Physics Letters*, 98, 104101-1-3.

[20] Wang, Z. C., Saito, M., Tsukimoto, S., & Ikuhara, Y. (2012). Terraces at Ohmic Contact in SiC Electronics: Structure and Electronic States. *Journal of Applied Physics*, 111, 113717-1-18.

[21] Kresse, G., & Hafner, J. (1993). Ab initio Molecular Dynamics for Liquid Metals. *Physical Review B*, 47, 558-561.

[22] Wang, X. G., Smith, J. R., & Evans, A. (2002). Fundamental Inflenence of C on Adhesion of the Al_2O_3/Al Interface. *Physical Review Letters*, 89, 286102-1-4.

[23] Wang, Z. C., Saito, M., Tsukimoto, S., & Ikuhara, Y. (2011). Heterointerfaces: Atomic Structures, Electronic States, and Related Properties. *Journal of the Ceramics Society of Japan*, 119, 783-793.

[24] Pennycook, S. J., & Boatner, L. A. (1988). Chemically Sensitive Structure-Imaging with a Scanning Transmission Electron Microscope. *Nature*, 336, 565-567.

[25] Ching, W. Y., Xu, Y. N., Rulis, P., & Ouyang, L. Z. (2006). The Electronic Structure and Spectroscopic Properties of 3C, 2H, 4H, 6H, 15R and 21R Polymorphs of SiC. *Materials Science and Engineering A*, 422, 147-156.

[26] Wang, Z. C., Kadohira, T., Tada, T., & Watanabe, S. (2007). Nonequilibrium Quantum Transport Properties of a Silver Atomic Switch. *Nano Letters*, 7, 2688-2692.

[27] Gao, M., Tsukimoto, S., Goss, S. H., Tumakha, S. P., Onishi, T., Murakami, M., & Brillson, L. J. (2007). Role of Interface Layers and Localized States in TiAl-Based Ohmic Contacts to p-Type 4H-SiC. *Journal of Electronic Materials*, 36, 277-284.

[28] Tanimoto, S., Kiritani, N., Hoshi, M., & Okushi, H. (2002). Ohmic Contact Structure and Fabrication Process Applicable to Practical SiC Devices. *Materials Science Forum*, 389-393, 879-884.

[29] Morkoc, H., Strite, S., Gao, G. B., Lin, M. E., Sverdlov, B., & Burns, M. (1994). Large-band-gap SiC, III-V Nitride, and II-VI ZnSe-based Semiconductor Device Technologies. *Journal of Applied physics*, 76, 1363-1398.

[30] Lu, W., Mitchel, W. C., Thornton, CA, , W. E., Landis, G. R., & Smith, S. R. (2003). Ohmic Contact Behavior of Carbon Films on SiC. *Journal of the Electrochemical Society*, 153, G177-G182.

[31] Wang, Z. C., Okude, M., Saito, M., Tsukimoto, S., Ohtomo, A., Tsukada, M., Kawasaki, M., & Ikuhara, Y. (2010). Dimensionality-Driven Insulator-Metal Transition in A-site Excess Non-stoichiometric Perovskites. *Nature Communications*, 1, 106.

Chapter 6

PECVD Amorphous Silicon Carbide (α-SiC) Layers for MEMS Applications

Ciprian Iliescu and Daniel P. Poenar

Additional information is available at the end of the chapter

1. Introduction

Silicon carbide (SiC) became an important material whose popularity has been constantly increasing in the last period due to its excellent mechanical, electrical, optical and chemical properties, which recommend it in difficult and demanding applications.

There are two main fields of applications of SiC. The first one is related to nano electronic [1] or even integrated circuits [2] which are using SiC (in monocrystalline or –sometimes-polycrystalline form) as basic structural material for high frequency [3], high power [4], high voltages [5], and/or high temperature devices [6] or combinations thereof [7]. In most of these applications, SiC act as are placement material for silicon which cannot be used under such extreme conditions.

The second area of applications is related to sensors [8] and actuators, i.e. structures, devices, and/or Microsystems realized (or at least embedding some elements of) micro- and nano electro mechanical systems (MEMS & NEMS). In this area, the usage of SiC (but mainly in polycrystalline and amorphous form) is mainly due to its compatibility and easy integrability with Si and Si-based microfabrication technology. In this direction, a lot of miniaturized devices, such as chemical sensors [9], UV detectors [10], MEMS devices [11-13], and even NEMS [14], are using SiC thin films or substrates (6H-SiC or 4H-SiC).

Polycrystalline-SiC (3C-SiC) thin films can be heteroepitaxially grown on Si substrates [14] due to the deposition temperature well below the Si melting point. [15] However, most of MEMS applications require thin films deposition methods at lower temperatures. This is necessary in order to ensure an overall low thermal budget for the entire fabrication of the SiC-based device(s), an essential prerequisite for postprocessing of MEMS structures on top of Si CMOS circuits, which ensures implementation of 'smart sensors'. Plasma enhanced chemical vapor deposition (PECVD) of SiC in an amorphouos state (α-SiC) can be such a solution. Early studies have been done on the structural, optical and electronic properties of this material [16, 17]. More specifically, one of the key challenges for PECVD of SiC for MEMS and NEMS applications is achieving alow residual stress together, if possible, with a high deposition rate and good uniformity [16], [13-20].

The main advantages of PECVD α-SiC deposition can be summarized as follows:

- Low temperature deposition, usually between 200-400°C (depending on the specific s of the machine and recipe employed for the deposition, as well as on the details of the device's fabrication process).
- As a direct consequence of the advantage mentioned previously, α-SiC is a highly suitable structural layer for surface micro machining applications using polyimide [21], amorphous silicon [22] or SiO_2 [23] as sacrificial layer materials.
- Stress control in a wide range (e.g. between −1200MPa and 400MPa [24] by tuning of deposition parameters [25], doping [26] or annealing [27].
- The ability of the fabricated device to operate at relatively high temperatures.
- Large mechanical strength of the deposited α-SiC layer.
- Wide bandgap for the deposited α-SiC layer, making it an almost ideal optoelectronic material, transparent for all visible wave lengths above 0.5μm and thus highly suitable for guiding light in the visible and infrared optical spectrum [28].
- A refractive index greater than 2.5 (significantly larger than that of SiO_2 and even that that of Si_3N_4) also make α-SiC an excellent candidate for optical waveguides [29].
- Capability to deposit conformal layers [30].
- The deposited α-SiC has a coefficient of thermal expansion (CTE) relatively similar with that of Si. This means that the risks of both developing very high internal stress within the film and, therefore, of subsequent delamination, are minimized [31] when the devices are operated even at high temperatures.
- The deposited α-SiC is highly suitable for applications requiring a high level of corrosion resistance and moderate operating temperatures (below 300°C) [32].

This chapter will focus on the PECVD deposition of α-SiC layers for MEMS/NEMS applications. The chapter is organized in three major parts:

- A detailed description of the typical PECVD reactor and of the important aspects necessary for a competitive α-SiC deposition process.
- The influence of main parameters on the deposition process and how one could achieve a low stress α-SiC film with anexcellent uniformity and a good deposition rate
- Post-deposition processing of the PECVD α-SiC layerfor various devices and applications.

2. PECVD reactors

The deposition of α-SiC layers in a Plasma Enhanced Chemical Vapor Deposition (PECVD) reactor is facilitated by the plasma generated between two electrodes (radio frequency-RF- or DC discharge) in the presence of reacting gases, the substrate being connected at one of these electrodes. There is a very large diversity of types of PECVD reactors on the market, either for industrial applications or specially designed for dedicated R&D. Usually the R&D equipments are more complex but also may allow much more of flexibility and degrees of freedom in controlling the deposition, thus enabling to obtain different layers with distinct properties using the same reactor.

The key elements in the selection of a PECVD reactor can be summarized as follows:

Deposition chamber. The operating temperature for most of the PECVD reactors is between 200-400°C. In order to achieve a uniform deposition special attention has to be paid to the inlet of the reactive gases, which can be of three types: several inlets around the bottom chuck (electrode), or one inlet through the top electrode, or multiple inlets (shower) from the top electrode. The last solution seems to ensure a better distribution of reactive gases between the working electrodes with positive effect on the film's uniformity. Meanwhile, preheating the gases (using a heated gas distribution system) before their actual introduction into the deposition chamber may also improve the deposition uniformity. Heating the deposition chamber itself (usuallyat 50-100°C) generates gradients of temperature that avoid particle deposition on the substrate during processing. Of course, adding all these elements into a standard system would finally be reflected in a higher cost of the tool.

Loadlock system. Two types of reactors can be distinguished, depending on whether a loading system is present or not: open systems (without lock load, i.e. relatively cheap reactors used only in research labs) and closed systems. The presence of a vacuumed loading system is also a critical element for good PECVD deposition, for a stable and repetitive process. There are two main aspects related to the presence of the load lock: one is related to safety while the other one is related tothe quality of the deposited layer. For the safety aspect, the presence of the loading system avoids the contact of the operator with the by-products resulted during processing, some of which are carcinogenic. As for the process quality, the fact that the chamber is kept permanently under vacuum results in excellent film quality with outstanding reproducibility.

Reconfiguration of the chamber. Cleaning of the chamber is also an important element in achieving good-quality layers. Most of the PECVD reactors also allow a „plasma cleaning" process, which is applied once a certain thickness of the deposited layer is achieved (usually 5-10μm). This cleaning process is designed to remove the products deposited on the chamber's walls or on electrodes, and it is performed mainly using CF_4/O_2 or C_4F_8/O_2 as reactive gasses. The process is followed by a short pre-deposition of the material desired to be deposited in the reactor. Mechanical cleaning must be also performed periodically.

Gas precursors. The PECVD systems frequently used in R&D are equipped with a large number of inlets for the reactive gases. In most of the cases, the equipment is used for multiple depositions such as SiO_2 (doped and undoped), Si_3N_4, α-Si or even TEOS (using a special

Liquid Delivery System –LDS). In our case, for the deposition of α-SiC layers, silane (SiH_4) and methane (CH_4) are the most often used gas precursors, although other precursors, e.g. methyltrichlorosilane (MTCS) [33] or SiH_4/acetylene (C_2H_2) [34], were also studied.

3. Influence of the deposition parameters

3.1. Materials and Methods

This section describes the influence of the main deposition parameters on the film properties, being a practical guide of parameter selection for a desired characteristic of the deposited α-SiC thin film. The experiments were performed on 4 inch diameter, *p* type, 500µm-thick silicon wafers with a (100) crystallographic orientation. The wafers were initially cleaned in piranha (H_2SO_4:H_2O_2 in the ratio of 2:1) at 120°C for 20 minutes and rinsed in DI water. The native silicon oxide layer was then removed by immersing the wafers in a classical BOE solution for 30 seconds. Stress measurement was performed using a KLA Tencor FLX-2320 system while the thickness and thickness uniformity of the thin filmswere measured with a refractometer (Filmetrics F50, USA). The deposition of the tested α-SiC layers was performed using a STS Multiplex Pro-CVD PECVD system described in detail elsewhere [35, 36]. This system enables two RF deposition modes: a low frequency (LF) mode at 380 kHz with a tuning power between 0 to 1 kW, and/or a high frequency (HF) mode at 13.56 MHz with a selected power in the range between 0 to 600 W. The depositions of the α-SiC layers were performed using pure SiH_4(pure) and CH_4 as precursors, with Ar as an overall dilution gas.

3.2. Pressure

The α-SiC deposition's uniformity is strongly dependent on the pressure in the reactor chamber (Fig. 1a).

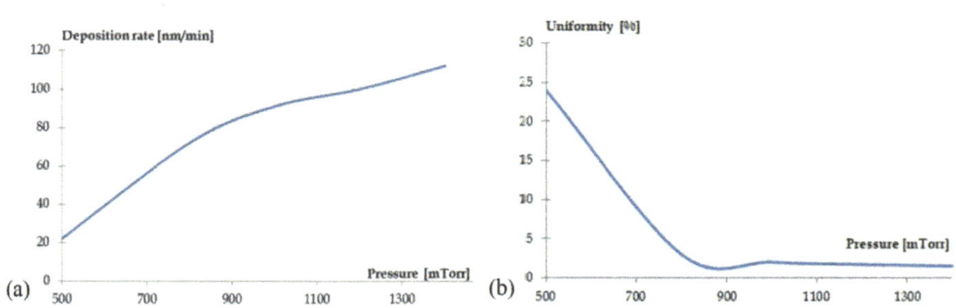

Figure 1. Variation of: (a)uniformity, and (b)deposition rate, with the pressure for α-SiC PECVD deposition in a STS Multiplex Pro-CVD PECVD system.

Below 800mTorr, large non-uniformity values were observed, but the deposition uniformity linearly improved as the pressure was varied between 500 to 800mTorr, finally settling to a constant valueof the uniformities under 2% for all the pressures in the range 900 to 1400mTorr. The thickness uniformity's map presented a "donut" shape, which means that the gas molecules present a high velocity, increasing the deposition rate at the edge of the wafer.

Another important aspect of the pressure is its influence on the deposition rate (Fig. 1b): low pressures reduce the concentration of reactive species thus resulting in a low deposition rate, which increases quasi-linearly with pressure.

3.3. RF Power

Both the RF power and the power deposition mode are key parameters for tuning the optical and mechanical properties of the deposited PECVD α-SiC layer. A low value of the residual stress is required in MEMS applications where free standing structure is fabricated. Meanwhile, a high deposition rate is desired mainly in industrial applications. Fig. 2 illustrates the variation of the deposition rate and residual stress versus the RF power for the HF mode. The linear dependence of the deposition rate on the HF power noticed for power levels below 300W can be explained by the proportional increase in the dissociation of reactant gases with increasing the power. After a 'threshold' value (300W) the dissociation into reactive species is no longer a crucial factor as probably all reactive species are easily and fully dissociated, hence further increasing the power has little effect on the deposition rate.

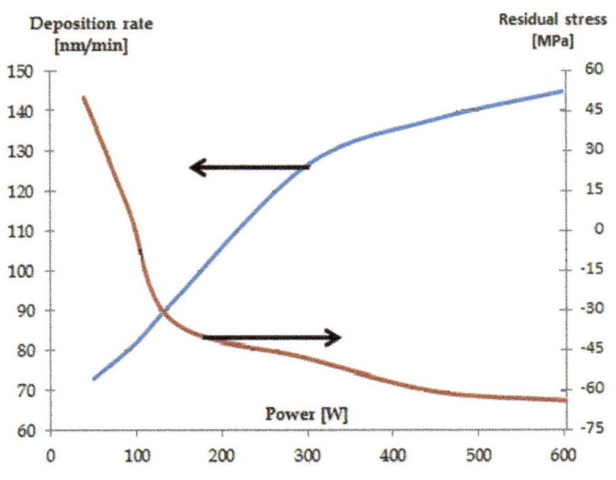

Figure 2. Variation of the deposition rate and residual stress with the HF power in the STS Multiplex Pro-CVD PECVD reactor. The deposition temperature was 300°C, the pressure 1100mTorr, and the gas flow rates of SiH_4, CH_4 and Ar were 45, 300 and 700sccm, respectively.

An interesting characteristic of PECVD α-SiC thin films deposited using the dual mode technique, particularly when compared to other PECVD deposition methods, is the very low value of the internal average stress, which can range between 50MPa and −70MPa. Fig. 2 shows the variation of this residual average with the RF power. Our results indicate that the average stress variation of a α-SiC film deposited in the HF deposition regime is not due to modifications of the film's chemical composition, but rather due to internal structural re-arrangement because the increasingly higher values of HF power quickly provide enough energy for the adsorbed species to migrate and find the more energetically favorable sites for the film growth to take place.

The refractive index was almost constant in the range between 2.5 and 2.6 while the uniformity of the deposition and uniformity of the refractive index was below 1.5%.

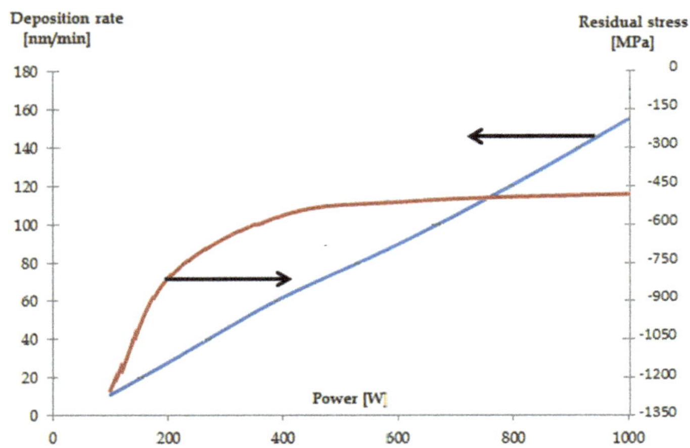

Figure 3. Variation of the deposition rate and residual stress with the LF power in the same PECVD reactor under the same conditions as in Fig. 2.

The variation of the deposition rate and residual stress of PECVDα-SiC films deposited in LF mode are presented in Fig. 3. The differences between HF mode and LF mode are quite evident if one compares the results shown in Fig. 3 with those shown in Fig. 2. The deposition rate depends linearly with the LF power. However, for low values of LF power, the stress is highly compressive (even below −1200MPa) but quickly reduces to about −500MPa and remains almost constant at this value for any LF power above 300W. This variation can be explained by the densification of the layer determined by the increasingly more energetic ion bombardment with increasing the LF power, which characterizes the LF deposition mode. At high frequencies, only the electrons are able to follow the RF field while the ions cannot follow the instantaneous variations of the electric field due to their heavier mass. The cross over frequency at which the ions start following the electric field is between 1 and 5MHz depending upon the mass of the ions. Consequently, below 1MHz, the ion bombardment is significantly higher, which not only enhances chemical reactions but also densifies the film. However, the refractive index of α-SiC film decreases with the increasing the LF power (from 2.9 to 2.5). It can be concluded that, for a low value of the residual stress in the α-SiC layers, the HF deposition mode is more suitable. Meanwhile, the HF mode assures a better dissociation of the gasses that is reflected in a higher deposition rate.The deposition in LF mode is more suitable when PECVD α-SiC is used as masking layer or in applications for harsh environments (due to the densification of the layer).

3.4. Temperature

A very interesting characteristic of the PECVD α-SiC deposition in the LF mode and which can be useful in many applications is that the deposition rate as well as the residual stress do not present relevant variations as a function of temperature [37]. In contrast, in the HF mode, temperature has an important effect on the stress value, although the deposition rate is not much affected by temperature in HF mode as well [37]. As is depicted in Figure 4, the internal average stress is compressive (around −110MPa) when the α-SiC layers are deposited at 200°C, but it becomes more and more tensile with increasing the deposition temperature,the stress (for temperature between 350 and 400°C). Most remarkable is the fact the stress is very low (around zero) when the deposition temperature is situated between 300 and 350°C.

The refractive index value remained almost constant at 2.6 for all the films deposited in the HF mode which proves that the stress variation is resulted from the film's structural re-arrangement, not from modifications of its chemical composition [37].

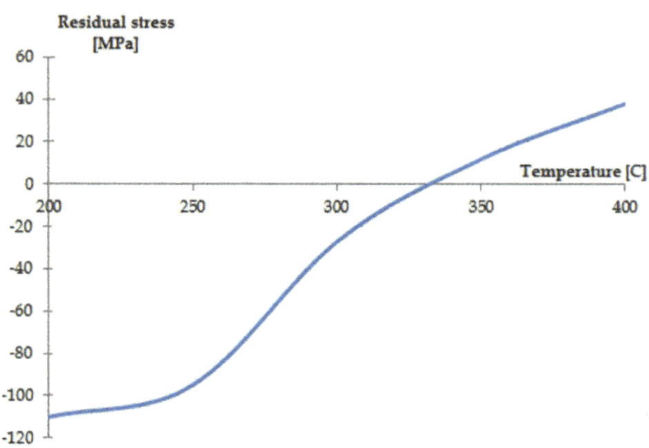

Figure 4. Variation of the average residual stress of PECVD α-SiC films with the deposition temperature. The films were deposited in the HF mode at a power of 150W, at a pressure of 1100mTorr, and with gas flow rates of SiH_4, CH_4 and Ar of 45, 300 and 700sccm, respectively.

3.5. SiH_4/CH_4 ratio

The influence of the SiH_4/CH_4 gas flow ratio, illustrated in Fig. 5, shows a linear dependence of the (compressive) residual stress on the SiH_4/CH_4 ratio for depositions in the HF mode. As expected, increasing the SiH_4 content from 1:10 to 2:10 in the gas flow makes the deposited α-SiC film more „Si-rich" and at the same time decreases the magnitude of the compressive stress from about −70MPa to −20MPa, while at the same time the refractive index increases from 2.6 to 2.8.

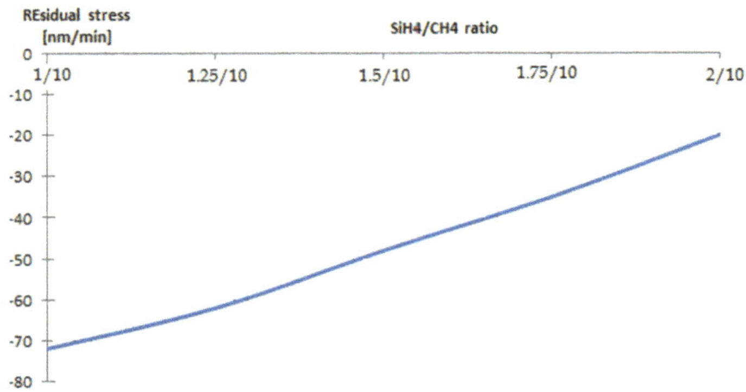

Figure 5. Variation of the average residual stress with the SiH_4/CH_4ratio for PECVD α-SiC films deposited at 300°C. The SiH_4 flow rate was kept constant at 45sccm, while all the other deposition conditions were the same as those indicated in Fig. 4.

To conclude, an optimized process recipe for the deposition of a low stress α-SiC film using a STSP PECVD reactor has the following parameters:

Process parameters/ Characteristic of the deposition process	Value
Pressure	1400mTorr
Power (HF mode)	600W
SiH_4, CH_4 and Ar gas flows	70, 500, and 700sccm
Deposition temperature	400°C
Refractive index	2.62
Residual Stress	± 5MPa
Deposition rate	180nm/min
Uniformity (refractive index & thickness)	<1%

Table 1.

4. MEMS application of PECVD α-SiC layers

4.1. Patterning of SiC layers

Micro patterning of thin α-SiC film layer deposited in PECVD reactor is similar to the patterning of crystalline or polycrystalline SiC layers. The patterning can be done in Reactive Ion Etching (RIE) or Inductive Coupled Plasma (ICP) deep RIE (DRIE) equipments by fluorine chemistry, i.e. using a fluorine-containing gas (CF_4,SF_6 or even CHF_3) and O_2 and using photoresist as masking layer. It must be mentioned that the selectivity of the etching process to Si (used as substrate), SiO_2, or even to photoresist is not outstanding (ranging from 1.2 up to 0.5) [16]. Metal masking layers have thus been initially been choosen [38, 39]. However, metal masks cause a major problem, namely the micromasking effect: as dry etching progresses, metallic particles are extracted from the mask and re-deposited onto the film/substrate where they can continue to have a protective role, resulting in a very uneven and rough final surface. For instance, aNF_3/O_2 chemistry was tested for etching a SiC layer [40] using to photoresist is mask, achieving etching rates of 0.135μm/min. Similarly, HBr/Cl_2 etching chemistries using SiO_2 etch masks have been developed [41]. This latter chemistry allowed a high selectivity (20:1) of etching SiC with respect to SiO_2, but the price to pay was a very low etching rate (0.02μm/min). In a more recent work, Senesky and Pisano reported the usage of AlN as masking layer for SiC structures in an ICP DRIE reactor using SF_6/O_2 chemistry [42]. The etching process yielded a SiC etch rate of 0.4μm/min, having a high selectivity (SiC/AlN) of 16:1. Moreover, the anisotropy of the process was very good, as features with a sidewall angle of 10° were reported. In another work,Cl_2 chemistry was used in an ICP DRIE reactor (by Pandraud et al.) for uniform patterning of a PECVD α-SiC layer for wave guide applications [43].

4.2. PECVD α-SiC as masking layer

The intrinsic chemical inertness of SiC makes PECVD α-SiC an interesting candidate as masking layer for harsh wet and dry etching.

4.2.1. Masking layer for orientation dependent etching of Si in alkaline solutions

A low etching rate of 78nm/h of low stress PECVDα-SiC in a 30%KOH solution was reported [19, 24, 37], while etching rates lower than 2nm/h in both 33%KOH at 85°C as well as in 25% TMAH are reported by Sarro in [16]. As expected, the etching rate depends on both the composition of the deposited α-SiC layer and its density. However, in general, the reported results are in the same range with the etch rate values of PECVD Si_3N_4 (13 nm/h) in KOH 20% at 85°C [36] but much better than the etch rate of thermal SiO_2 (462 nm/h)in the same etchant [44]. As the reported etch rates of PECVD α-SiC have such relatively low values, we can conclude that PECVD α-SiC can be successfully used as a mask in silicon bulk micromachining processes. Furthermore, we have practically tested this hypothesis and demonstrated the feasibility of using PECVD α-SiC as masking layer, as is detailed in the next section.

4.2.2. Masking layer for etching in HF based solutions

Deep wet etching of glass is an important technology for microfluidic applications [45]. The main etchant for glass materials is highly concentrated HF [46] (sometimes with a small amount of HCl [47] or H_3PO_4 added [48]). PECVD α-SiC is almost an inert material in these solutions, exhibiting etching rates lower than 10Å/h [37]. Such an extremely low etching rate in highly concentrated HF solutions combined with the reduced value of the stress are the main request of a good masking layer for deep wet etching of glasses [46, 49-51]. In our tests, a compound masking multilayer 'sandwich' made to low stress α-Si/low stress α-SiC/photoresist seemed to be the best solution in terms of depth of the etch achieved without any damage of the mask (through- etching of 1 mm thick Pyrex glass wafers, equivalent of a 2.5 hour exposure to 49% HF). The experimental results of processing Pyrex glass wafers (Corning 7740) in such a manner are presented in Fig. 6. It can be noticed that the mask is fully intact after the etching process while the shape of the etched hole describes a perfect isotropic process. The α-Si was used in this case as an adhesion layer. If only PECVD α-SiC is used as masking layer, a hugely isotropical etching process can still be observed [50].

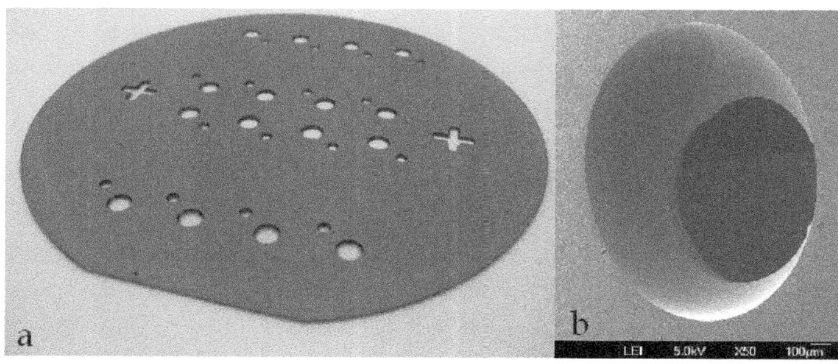

Figure 6. (a) Optical image of a glass wafer coated with PECVD α-Si/PECVD α-SiC/Photoresist after etching through in HF 49%, and (b) a SEM picture of one of the resulted etch-through holes in the glass wafer.

4.2.3. Masking layer for dry release structure in XeF_2

Dry release in XeF_2 is an emerging technology in surface and bulk micromachining of MEMS free-standing structures. The PECVD α-SiC films present a low etching rate in XeF_2 gas (around 7 Å/min) which makes the α-SiC very suitable as a structural layer for any dry-release processes using α-Si or polysilicon as a sacrificial layer. Fig. 7 presents SEM images with PECVD α-SiC cantilevers fabricated using dry released in XeF_2 [24].

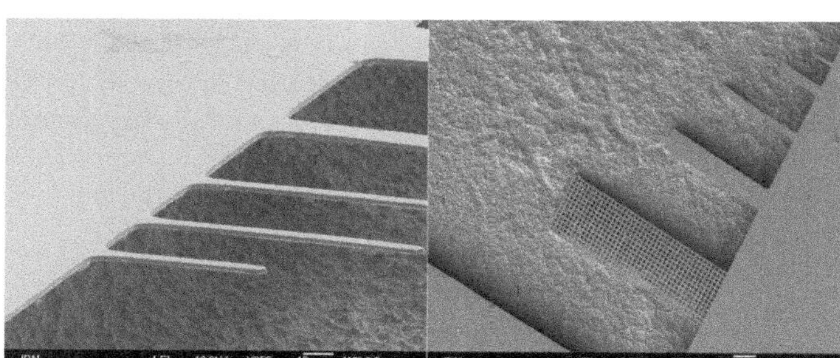

Figure 7. PECVD α-SiC cantilevers fabricated using dry-released process in XeF_2.

4.2.4. Protective layer for harsh environment

PECVD α-SiC films are also very suitable for structures intended to operate in harsh environments, due to α-SiC 's large hardness (2.48kg/m^2) [52], high fracture strength, high modulus [53], excellent wear resistance [54] and chemical inertness in acid or based solutions, low oxidation rate and strong covalent Si-C bonds [52]. Early work proved the potential of PECVD α-SiC as a potential material for encapsulation of micromachined transducers due to its good resistance in a large range of media such as piranha solution, HF and KOH [55].

The good mechanical strength and anti-stiction surface properties of PECVD α-SiC [56] as well as its inertness in corrosive environment recommends this material for diverse applications. An illustrating example is that of a 1μm-thick PECVD α-SiC combined together with Teflon like fluoro-polymer coatings to reduce the demolding energy by a factor of about 10, compared to a bare silicon mold [57]. In another similar application PECVD α-SiC was used for its hardness and for its good step-coverage (improving the wall roughness generated during the deep RIE process) while the Teflon layer acted as an anti-stiction layer [58]. Decreasing the demolding energy, in this application, is equivalent with the increasing of the life-time of the Si mold (used usually for rapid prototyping on hot embossing tools).

In another application [30], a 1μm-thick PECVD α-SiC layer was used as an anti-erosion coating layer of a piezoresistive pressure sensor. In order to reduce the residual stress in the α-SiC protection layer (initially evaluated at −450MPa), annealing was performed at 450°C for 1h. The resulting stress value was of only +60MPa. The α-SiC protective layer also showed a good coverage of the Al metallization layer. The erosion testing was performed in 45%KOH solution at 80°C. The final PECVD α-SiC-coated pressure sensor showed a small decrease in sensitivity, but exhibited a high erosion resistance and less temperature dependence. A similar application was also reported [59], in which low stress PECVD α-SiC layer was used for a capacitive pressure sensor fabricated using surface micromachining. Aluminum was used as material for electrodes while polyimide was selected as a sacrificial layer.

4.3. Fabrication of free standing structuresof PECVD α-SiC using to surface micromachining

The opportunity of using thick α-Si (amorphous silicon) layers [22] (up to 20μm thick, from our own experience) in surface micromachining gives rise to new processing opportunities, especially for microfluidic applications. The α-Si sacrificial layer can be easily removed by wet etching in an alkaline solution (TMAH or KOH) or by dry release in XeF_2. The PECVD α-SiC presents a high chemical inertness to all the etchants of the above mentioned processes, and, therefore, is a very attractive candidate as a structural layer for surface micromachining processes in which amorphous silicon is used as a sacrificial layer. An example of such a process used a 3μm-thick free standing structure fabricated from PECVD α-SiC on a glass substrate using 9 μm-thick PECVD α-Si as sacrificial layer [22]. The structure was released using wet etching in 30% KOH at 80°C. The structure (presented in Fig. 8) shows also that the PECVD deposition process ensured a very good step coverage, the thickness of the vertical wall being identical with that of the layer on the horizontal surfaces [22]. In another example of using PECVD α-SiC as structural material for surface micromachining, 1μm-thick self-sustaining microbridges and microtunnels were fabricated from PECVD α-SiC films (deposited at 320°C using CH_4 and SiH_4 as reactive gases), using a SiO_xN_yfilm as a sacrificial layer [60]. In a similar manner, self-sustained grids have also been fabricated [61].

Figure 8. SEM image of asuspended structure fabricated by surface micromachining using thick low-stress α-Si as sacrificial layer and PECVD α-SiC as a structural layer [22].

PECVD α-SiC was used in shunt capacitor RF MEMS microbridge-based switches [23]. In this application 300 nm and 500 nm thick PECVD α-SiC films were used as structural layers, while 2µm-thick PECVD SiO_2 was used as a sacrificial layer. The structures were released in BOE, followed by CO_2 supercritical drying in order to eliminate anystiction possibility. Polyimide was the choice for the sacrificial layer in an application which also used PECVD α-SiC as structural layer [21]. The main advantage of using polyimide is the opportunity of using a dry release process (in O_2 plasma), thus more easily overcoming stiction problems often encountered in wet sacrificial etching technology than by using XeF_2 or CO_2 supercritical drying, solutions which require much more complex and expensive achiness.

4.4. Biological applications of PECVD α-SiC

MEMS materials are suitable for applications in cell culture and tissue engineering [62]. In this direction, a special attention is given to the layers achieved by PECVD deposition especially α-Si_3N_4 [63], [64-66] and α-SiC [37]. The main motivation in using these materials and PECVD as deposition method are related to:

1. Both α-Si_3N_4 and α-SiC can be used for the fabrication of 2-3µm thick membranes using classical micromachining processes.
2. PECVD deposition allows a good control of the residual stress in the layer, a critical aspect in achieving free standing structures.
3. Both α-Si_3N_4 and α-SiC are optically transparent, so the classical inverted microscopes use in biology can be easily used to monitor bio-samples in structures realized with these materials, for example, in assays emplying red/green fluorescence.

4. Using microfabrication, these membranes are attached to a supporting Si ring which makes the structure easy to be handle (comparing with polymeric membranes where no such rim support is present) [66].

5. Porous PECVD α-SiC and α-Si_3N_4 membranes can be fabricated with a good control of the pore size and with a uniform distribution of the pores can be achieved.

6. The commercially available and presently used polymeric membranes are hydrophobic materials, and metabolites (albumin, urea) are absorbed in this membrane with effect on cell viability.

7. The thickness of polymeric porous membrane is usually ranging from tens to hundreds of microns. Consequently, the mass transfer of (bio) chemical substances across/through the membrane is strongly affected by the thickness of the membrane and its nature (hydrophilic/hydrophobic).

8. The presence of functional groups, such as amine (-NH_2), and methyl (-CH_3), on the surface of α-Si_3N_4 and α-SiC membranes can significantly enhance the cell adhesion on these materials.

9. The α-SiC and α-Si_3N_4 membranes can be easily cleaned: complete removal of all organic contaminants can be achieved in piranha solution, followed by subsequent rinsing, drying and if necessary even sterilization can also be performed, in order to ensure complete reusability.

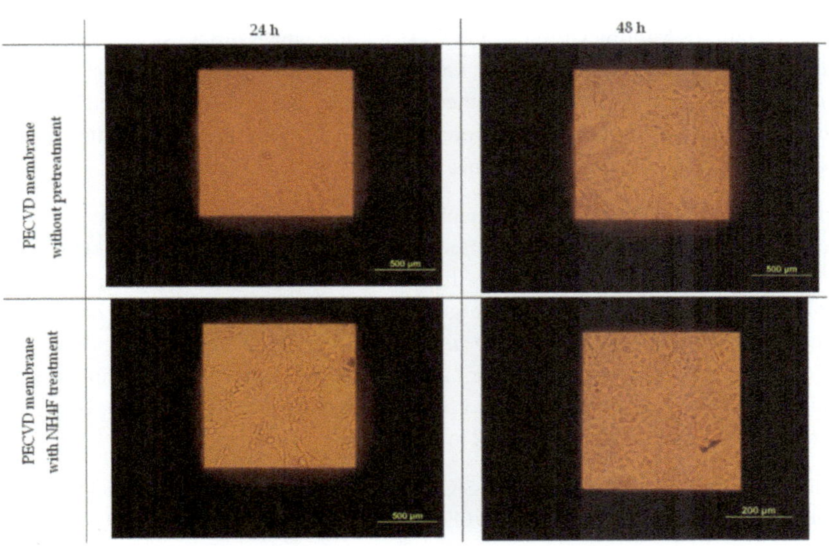

Figure 9. Optical images with the adhesion of fibroblast NIH3T3 cells on PECVD α-SiC membrane treated in NH_4F after 24h and 48h cell culture.

As an illustrative example, chips with 2.5μm-thick PECVD α-SiC membranes were fabricated by bulk micromachining in order to test the biocompatibility of the α-SiC [37]. Fibroblast NIH3T3 cells were used in this study as the model cell line. It was observed that the presence of CH_3 groups on the surface of the membranes improved the cell adhesion. Moreover, dipping for 1 minute in 40% NH_4F, which was performed mainly to reduce the density of native silicon oxide groups on the α-SiC surface, also improved the adhesion of the cells on the membranes' surfaces. Fig. 9 shows cell culture images taken 24 hours and 48 hours after starting the culturing.

5. Concluding remarks

We can conclude that PECVD α-SiC is a very attractive and promising material. Its PECVD deposition enable low temperature processing and thus ensures compatibility with CMOS processing, which is essential when (bio) MEMS/NEMS structures need to be combined together with signal processing and data conditioning circuits, a *sine qua non* condition of realizing "smart sensors". Additionally, PECVD deposition enables the user to modify various deposition parameters (temperature, pressure, precursors gas flow ratios) which can allow the user to optimize –at least in principle- the desired film properties: chemical composition internal average stress and refractive index. However, in practice such multi-dimensional optimization may prove difficult to achieve, and the presence of some extra degrees of freedom is welcome. An example to the point is -in our case- the presence of dual LF and HF modes for the STS Multiplex Pro-CVD PECVD reactor, which provided valuable flexibility in usage and in varying independently various characteristics of the film (e.g. minimizing the stress) without compromising the others (e.g. the refractive index). In some cases, when such a machine is not available, some more maneuvering room is given by adding an anneal process, which always reduces (sometimes significantly) the film's stress.

Application-wise, the long list of extremely desirable properties exhibited by α-SiC (large hardness, high fracture strength, high modulus, excellent wear resistance, superb chemical inertness and excellent stability even at high temperatures and/or under corrosive conditions) ensure it can be used in a very large area of applications, particularly in harsh environments, where Si devices cannot be used. Moreover, the same properties recommend PECVD α-SiC as the material of choice for a large range of MEMS postprocessing fabrication techniques. Thus, it can be used as mold material (or mold-coating protective layer), masking material or active structural material in surface micromachining.

Finally, we should also highlight that α-SiC can has also been employed in bio-related applications. Our own tests have shown that PECVD α-SiC membranes fabricated by bulk micromachining can be used successfully in cell cultures. Here the intrinsic properties of α-SiC ensure very good chemical stability, possibility to easily functionalize the surface for, e.g., increased cell adhesion, and improved mechanical resistance and easy handling.

Author details

Ciprian Iliescu[1*] and Daniel P. Poenar[2]

*Address all correspondence to: ciliescu@ibn.a-star.edu.sg

1 Institute of Bioengineering and Nanotechnology, Singapore

2 Microelectronics Centre, School of Electrical & Electronics Engineering, Nanyang Technological University, Singapore

References

[1] Zhou, W. M., Fang, F., Hou, Z. Y., Yan, L. J., & Zhang, Y. F. (2006). *IEEE Electron Dev. Lett.*, 27(6), 463-465.

[2] Patil, C., Xiao-An, F., Anupongongarch, C., Mehregany, M., & Garverick, S. (2007). *Proc. IEEE Compound Semicond. Int. Circuit Symp.*, 1-4.

[3] Sarazin, N., Morvan, E., di Forte Poisson, M. A., Oualli, M., Gaquiere, C., Jardel, O., Drisse, O., Tordjman, M., Magis, M., & Delage, S. L. (2010). *IEEE Electron Dev. Lett.*, 31(1), 11-13.

[4] Peftitsis, D., Tolstoy, G., Antonopoulos, A., Rabkowski, J., Jang-Kwon, L., Bakowski, M., & Nee, H. (2012). *IEEE Transactions on Power Electr.*, 27(1), 28-36.

[5] Woongje, S., Van Brunt, E., Baliga, B. J., & Huang, A. Q. (2011). *IEEE Electron Dev.Lett.*, 32(7), 880-882.

[6] Nikiforov, V., Tomás García, A. L., Petrushina, I. M., Christensen, E., & Bjerrum, N. J. (2011). *Int. J. Hydrogen Energy*, 36(10), 5797-5805.

[7] Singh, N. B., Wagner, B., Berghmans, A., Knuteson, D. J., Mc Laughlin, S., Kahler, D., Thomson, D., & King, M. (2010). *Crystal Growth & Design*, 10(8), 3508-3514.

[8] Myers, D. R., Cheng, K. B., Jamshidi, B., Azevedo, R. G., Senesky, D. G., Chen, L., Mehregany, M., Wijesundara, M. B. J., & Pisano, A. P. (2009). *J. Micro-Nanolithogr. MEMS MOEMS*, 8(2).

[9] Soo, M. T., Cheong, K. Y., & Noor, A. F. M. (2010). *Sens. Actuator B-Chem.*, 151(1), 39-55.

[10] Borchi, E., Macii, R., Bruzzi, M., & Scaringella, M. (2011). *Nuclear Instruments and Methods in Physics Research Section A: Accelerators, Spectrometers, Detectors and Associated Equipment*, 658(1), 121-124.

[11] Azevedo, G. R., Jones, G. D., Jog, V. A., Jamshidi, B., Myers, R. D., Chen, L., X.-a., Fu, Mehregany, M., Wijesundara, B. J. M., & Pisano, P. A. (2008). *Sens. Actuator A-Phys.*, 145-146(0), 2-8.

[12] Jiang, L., Cheung, R., Hedley, J., Hassan, M., Harris, A. J., Burdess, J. S., Mehregany, M., & Zorman, C. A. (2006). Sens. Actuator A-Phys.(2), 1128(2), 376-386.

[13] Hyun, J. S., Park, J. H., Moon, J. S., Kim, S. H., Choi, Y. J., Lee, N. E., & Boo, J. H. (2005). *Progress in Solid State Chemistry*, 33(2-4), 309-315.

[14] Zorman, A., & Mehregany, M. (2002). *Proc. of IEEE Sensors*, 2, 1109-1114.

[15] Liu, F., Carraro, C., Pisano, A. P., & Maboudian, R. (2010). *J. Micromech. Microeng.*, 20(3).

[16] Sarro, P. M., Deboer, C. R., Korkmaz, E., & Laros, J. M. W. (1998). *Sens. Actuator A-Phys.*, 67(1-3), 175-180.

[17] El Khakani, M. A., Chaker, M., Jean, A., Boily, S., Kieffer, J. C., O'Hern, M. E., Ravet, M. F., & Rousseaux, F. (1994). *J. Mat. Research*, 9(1), 96-103.

[18] Roper, S., Howe, R. T., & Maboudian, R. (2006). *J. Micromech. Microeng.*, 16(12), 2736-2739.

[19] Iliescu, M., Avram, B., Chen, A., Popescu, V., Dumitrescu, D. P., Poenar, A., Sterian, D., Vrtacnik, S., Amon, P., & Sterian, . (2011). *J. Optoel. Adv. Mat.*, 13(4), 387-394.

[20] Sarro, P. M. (2000). *Sens. Actuator A-Phys.*, 82(1-3), 210-218.

[21] Bagolini, L., Pakula, T. L. M., Scholtes, H. T. M., Pham, P. J., French, P. M., & Sarro, . (2002). *J. Micromech. Microeng.*, 12(4), 385-389.

[22] Iliescu, B., & Chen, T. (2008). *J. Micromech. Microeng.*, 18(1), 15024.

[23] Parro, R. J., Scardelletti, M. C., Varaljay, N. C., Zimmerman, S., & Zorman, C. A. (2008). *Solid-State Electronics*, 52(10), 1647-1651.

[24] Iliescu, C., Chen, B. T., Wei, J. S., & Pang, A. J. (2008). *Thin Solid Films*, 516(16), 5189-5193.

[25] Avram, M., Avram, A., Bragaru, A., Chen, B., Poenar, D. P., & Iliescu, C. (2010). *Proc. of the 33rdIEEE Int. Semicond. Conf.*, 239-242.

[26] Pham, H. T. M., De Boer, C. R., Kwakernaak, K., Sloof, W. G., & Sarro, P. M. (2001). *Proc. of SPIE*, 272-279.

[27] Oliveira, R., & Carreno, M. N. P. (2006). *J. Non-Crystalline Solids*, 352(9-20), 1392-1397.

[28] Pandraud, G., French, P. J., & Sarro, P. M. (2008). *Sens. Actuator A-Phys.*, 142(1), 61-66.

[29] Pandraud, G., French, P. J., & Sarro, P. M. (2005). *Microw. Opt. Technol. Lett.*, 47(3), 219-220.

[30] Zhang, H., Guo, H., Wang, Y., Zhang, G., & Li, Z. (2007). *J. Micromech. Microeng.*, 17(3), 426-431.

[31] Slack, G. A., & Bartram, S. F. (1975). *J. Appl. Phys.*, 46(1), 89-98.

[32] Azevedo, R. G., Jingchun, Z., Jones, D. G., Myers, D. R., Jog, A. V., Jamshidi, B., Wijesundara, M. B. J., Maboudian, R., & Pisano, A. P. (2007). *Proc. of IEEE 20th Int. Conf. on Micro Electro Mechanical Systems, MEMS.*, 643-646.

[33] Ivashchenko, V. I., Dub, S. N., Porada, O. K., Ivashchenko, L. A., Skrynskyy, P. L., & Stegniy, A. I. (2006). *Surf. Coat. Technol.*, 200(22-23), 6533-6537.

[34] Behnel, N., Fuchs, T., & Seidel, H. (2009). *Proc. of Int. Conf. on Solid-State Sens., Act. and Microsyst., TRANSDUCERS*, 2009, 740-742.

[35] Chung, C. K., Tsai, M. Q., Tsai, P. H., & Lee, C. (2005). *J. Micromech. Microeng.*, 15(1), 136-142.

[36] Iliescu, C., Tay, F. E. H., & Wei, J. S. (2006). *J. Micromech. Microeng.*, 16(4), 869-874.

[37] Iliescu, C., Chen, B., Poenar, D. P., & Lee, Y. Y. (2008). *Sens. Actuator B-Chem.*, 129(1), 404-411.

[38] Tanaka, S., Rajanna, K., Abe, T., & Esashi, M. (2001). *J. Vac. Sci. Technol. B*, 19(6), 2173-2176.

[39] Chabert, P. (2001). *J. Vac. Sci. Technol. B*, 19(4), 1339-1345.

[40] Schmid, U., Eickhoff, M., Richter, C., Krotz, G., & Schmitt-Landsiedel, D. (2001). *Sens. Actuator A-Phys.*, 94(1-2), 87-94.

[41] Gao, M. B. J., Wijesundara, C., Carraro, R. T., Howe, , & Maboudian, R. (2004). *IEEE Sens. J.*, 4(4), 441-448.

[42] Senesky, D. G., & Pisano, A. P. (2010). *Proc. of 23rd IEEE Int. Conf. on Micro Electro Mechanical Systems (MEMS)*, 352-355.

[43] Pandraud, H. T. M., Pham, P. J., French, P. M., & Sarro, . (2007). *Optics & Laser Technology*, 39(3), 532-536.

[44] Williams, K. R., Gupta, K., & Wasilik, M. (2003). *J. Microelectromech. Syst.*, 12(6), 761-778.

[45] Iliescu, C., Taylor, H., Avram, M., Miao, J., & Franssila, S. (2012). *Biomicrofluidics*, 6(1), 016505-016516.

[46] Iliescu, C., Chen, B., & Miao, J. (2008). *Sens. Actuator A- Phys.*, 143(1), 154-161.

[47] Iliescu, C., Jing, J., Tay, F. E. H., Miao, J., & Sun, T. (2005). *Surf. Coat. Tech.*, 198(1-3), 314-318.

[48] Berthold, F., Laugere, H., Schellevis, C. R., De Boer, M., Laros, R. M., Guijt, P. M., & Sarro, M. J. (2002). *Vellekoop Electrophoresis*, 23(20), 3511-3519.

[49] Iliescu, F. E. H., Tay, J. M., & Miao, . (2007). *Sens. Actuator A-Phys.*, 133(2), 395-400.

[50] Zhang, X., Guo, H., Chen, Z., Zhang, G. B., & Li, Z. H. (2007). *J. Micromech. Microeng.*, 17(4), 775-780.

[51] Poenar, P., Iliescu, C., Carp, M., Pang, A. J., & Leck, K. J. (2007). *Sens. Actuator A-Phys.*, 139(1-2), 162-171.

[52] Zorman, A., & Barnes, A. C. (2012). *in Silicon Carbide Biotechnology*, Elsevier, Oxford, 351-376.

[53] Sharpe, W. N., Jr., Jadaan, O., Beheim, G. M., Quinn, G. D., & Nemeth, N. N. (2005). J. Microelectromech. Syst. ., 14(5), 903-913.

[54] Ashurst, W. R., Wijesundara, M. B. J., Carraro, C., & Maboudian, R. (2004). *Tribology Letters*, 17(2), 195-198.

[55] Flannery, F., Mourlas, N. J., Storment, C. W., Tsai, S., Tan, S. H., Heck, J., Monk, D., Kim, T., Gogoi, B., & Kovacs, G. T. A. (1998). *Sens. Actuator A-Phys.*, 70(1-2), 48-55.

[56] Summers, J. B., Scardelletti, M., Parro, R., & Zorman, C. A. (2007). *in MEMS/MOEMS Components and Their Applications IV, edited by S. A. Tadigadapa, R. Ghodssi and A. K. Henning, Proc. Spie-Int Soc Optical Engineering*, 6464, H4640.

[57] Taylor, D., Boning, C., & Iliescu, . (2011). *J. Micromech. Microeng.*, 21(6).

[58] Gao, X., Yeo, L. P., Chan-Park, M. B., Miao, J. M., Yan, Y. H., Sun, J. B., Lam, Y. C., & Yue, C. Y. (2006). *J. Microelectromech. Syst.*, 15(1), 84-93.

[59] Pakula, L. S., Yang, H., Pham, H. T. M., French, P. J., & Sarro, P. M. (2004). *J. Micromech. Microeng.*, 14(11), 1478-1483.

[60] Carreño, M. N. P., & Lopes, A. T. (2004). *J. Non-Crystalline Solids*, 338-340(0), 490-495.

[61] Carreno, M. N. P., Alayo, M. I., Pereyra, I., & Lopes, A. T. (2002). *Sens. Actuator A-Phys.* [2-3], *100* [2-3], 295-300.

[62] Ni, M., Tong, W. H., Choudhury, D., Rahim, N. A. A., Iliescu, C., & Yu, H. (2009). *Int. J. Mol. Sci.*, 10(12), 5411-5441.

[63] Zhang, S., Tong, W., Zheng, B., Susanto, T. A. K., Xia, L., Zhang, C., Ananthanarayanan, A., Tuo, X., Sakban, R. B., Jia, R., Iliescu, C., , K., Chai, H., Mc Millian, M., Shen, S., Leo, H., & Yu, H. (2011). *Biomaterials*, 32(4), 1229-1241.

[64] Ciarlo, D. R. (2002). *Biomed. Microdev.*, 4(1), 63-68.

[65] Wei, S., Kwong, J., Gaughwin, P., Avram, M., & Iliescu, C. (2009). *Int. J. Comp. Mat. Science Surface Eng.*, 2(3-4), 268-281.

[66] Iliescu, C., Wei, J., Chen, B., & Ong, P. L. (2008). *Rom. J. Inform. Sci. Technol.*, 11(2), 167-176.

Section 3

MOS Processing and Characterizations

Chapter 7

Fundamental Aspects of Silicon Carbide Oxidation

Heiji Watanabe and Takuji Hosoi

Additional information is available at the end of the chapter

1. Introduction

Silicon carbide (SiC), which exhibits a wider band gap as well as a superior breakdown field and thermal conductivity over conventional Si, has gained considerable attention for future power electronics [1]. Among the various types of power devices, metal-oxide-semiconductor field-effect transistors (MOSFETs), which provide a normally-off characteristic, should become a key component for next-generation green electronics. As stated, with the exception of Si, SiC is the only compound semiconductor that yields SiO_2 insulators with thermal oxidation. This makes the device fabrication process easier compared with those for other wide band gap semiconductors. It is commonly believed that carbon impurities within the oxides diffuse out in the form of carbon oxides during high temperature oxidation, but a small amount of carbon impurities remains within the oxide and at the SiO_2/SiC interface. Consequently, the electrical degradation of SiC-MOS devices causing both deteriorated device performance and reliability is the most crucial obstacle to the implementation of SiC-based power electronics.

Unlike mature Si-MOS technology [2-5], a plausible oxidation model of a SiC surface and physical origin explaining electrical degradation of SiC-MOS devices have yet to be established. High-resolution transmission electron microscopy (TEM) observation reported several nm-thick transition layers with an extremely high excess carbon concentration around 20% beneath the SiO_2/SiC interface [6-8]. Although the non-stoichiometric bulk region seems to account for the mobility degradation of SiC-MOSFETs [8], a recent report based on an ion scattering technique pointed out a near-perfect stoichometric SiC region [9].

In addition, there still remain controversial issues of the energy band structure of SiO_2/SiC interfaces, despite the fact that a small conduction band offset significantly increases the gate leakage current, especially under a high electric field and high operation temperatures [10]. For example, while alternative channels, such as a 4H-SiC(000-1) C-face substrate, have proven to provide higher electron mobility than conventional devices on Si-face substrates [11], a further reduction in conduction band offset has been pointed out for the C-face channels [12]. However, details on the physical origins have not been clarified yet.

Furthermore, another important problem of SiC-MOS devices is poor gate oxide reliability, such as low dielectric breakdown field and threshold voltage instability. It has been reported that the location of dielectric breakdown is not correlated with any dislocations in SiC-MOS capacitors with gate oxides formed on 4º-off-angled 4H-SiC(0001) substrates by successive oxidation in N_2O and NO ambient [13]. Our conductive atomic force microscopy (AFM) study on a thermally grown SiO_2/4H-SiC(0001) structure has clearly demonstrated that dielectric breakdowns are preferentially induced at the step bunching [14]. It is generally accepted that the gate oxide breakdown is triggered when electrical defects generated in the oxide by a stress field are connected between the electrode and substrate (percolation model) [15]. Thus, we speculated that a local electric field concentration occurred around the step bunching, resulting in the preferential breakdown due to the acceleration of the defect generation. This suggests that oxidation behavior of step-bunched SiC surface induced by epitaxial growth and high-temperature activation annealing needs to be clarified from the macroscopic point of view, together with the atomic bonding features at SiO_2/SiC interface mentioned above.

This article provides an overview of our recent studies on the thermal oxidation of 4H-SiC substrates and the energy band structure of SiO_2/4H-SiC fabricated on (0001) Si-face and (000-1) C-face surfaces by means of high-resolution synchrotron x-ray photoelectron spectroscopy (XPS). We investigated the correlation between atomic structure and the electrical properties of corresponding SiC-MOS capacitors and discuss the intrinsic and extrinsic effects of the interface structure and the electrical defects on the band offset modulation. In addition, the surface and interface morphology of a thermally grown SiO_2/4H-SiC(0001) structure were systematically investigated using AFM and TEM to clarify the relation between step bunching and oxidation kinetics.

2. Initial oxidation of 4H-SiC(0001)

Synchrotron XPS analysis was performed using photon energy of 686.5 eV at BL23SU in the SPring-8 [16]. The starting substrate was as-grown 4º-off-angled 4H-SiC(0001) wafer with an n-type epitaxially grown layer. After RCA cleaning and subsequent native oxide removal with a diluted hydrofluoric acid (HF) solution, thermal oxidation was conducted in dry oxygen ambient using a conventional tube furnace at 1100ºC. To remove surface contamination due to air exposure, some of the samples were annealed in situ in an analysis chamber under an ultra-high vacuum condition.

Figure 1(a) represents changes in Si 2p core-level spectra as dry oxidation progresses on the 4H-SiC(0001) surface at 1100ºC [17]. Peak intensity was normalized with the bulk signal. Oxide growth on the SiC surfaces was confirmed with an increase in the chemical shift component in the Si 2p core-level spectra at around 104.5 eV. Capacitance-voltage (C-V) measurement of the corresponding Al/SiO_2/SiC capacitors also revealed that oxidation for 10 and 30 min yielded roughly 3.5 and 5.7-nm-thick oxides, respectively. To investigate atomic bonding feature at SiO_2/SiC interfaces, Si 2p signals were analyzed by taking into account spin-orbit splitting. Figure 1(b) shows typical deconvoluted Si $2p_{3/2}$ and $2p_{1/2}$ peak components obtained with the manner adopted in the previous research on SiO_2/Si interfaces [3, 4]. Then, the Si $2p_{3/2}$ spectra taken from SiO_2/SiC were deconvoluted into five components originating from bulk SiC and SiO_2 portions together with intermediate oxide states (Si^{1+}, Si^{2+}, Si^{3+}). High-resolution XPS analysis allows us to detect small amount of intermediate states from an atomically abrupt oxide/substrate interface, and, in addition, these intermediate components can be a good indicator of structural imperfection at SiO_2/SiC interfaces. As shown in Fig. 1(c), we obtained a reasonable curve fitting with these components and confirmed that the total amount of the intermediate states is sufficiently small compared with that of thin thermal oxides. From these results, it is concluded that the physical thickness of the transition layer is as thin as a few atomic layers, which corresponding to areal density of Si-O bonds in the range of a few times 10^{15} cm^{-2}. This indicates formation of a near-perfect SiO_2/SiC interface and coincides well with a recent report based on high-resolution medium energy ion scattering [9].

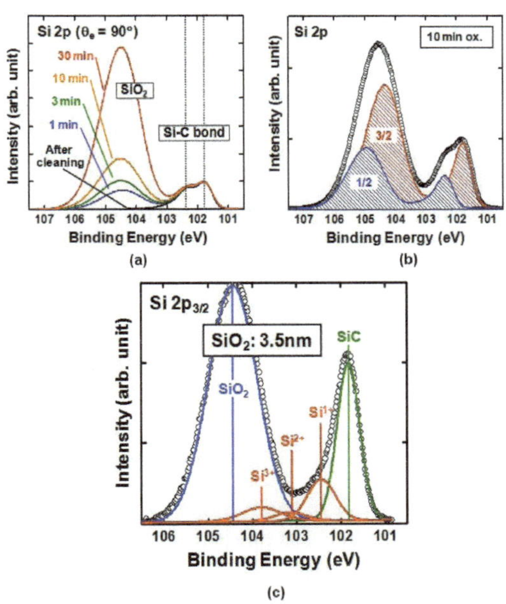

Figure 1. Synchrotron XPS spectra taken from the cleaned and oxidized 4H-SiC(0001) surfaces; (a) change in Si 2p core-level spectra as dry oxidation progresses, (b) peak deconvolution with $2p_{3/2}$ and $2p_{1/2}$ components, (c) result of curve fitting of Si $2p_{3/2}$ core-level with bulk SiC and SiO_2 signals and intermediate oxide states for the SiO_2/SiC sample prepared by 10-min oxidation.

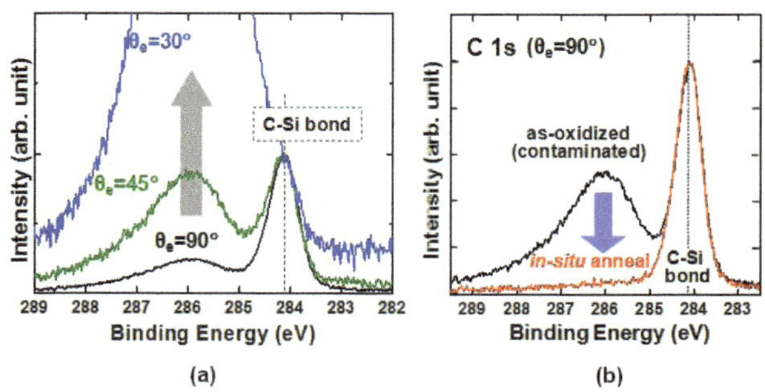

Figure 2. C 1s core-level spectra taken from the oxidized 4H-SiC(0001) surface using synchrotron radiation; (a) angle-resolved XPS analysis, (b) results of in situ vacuum annealing at 500°C.

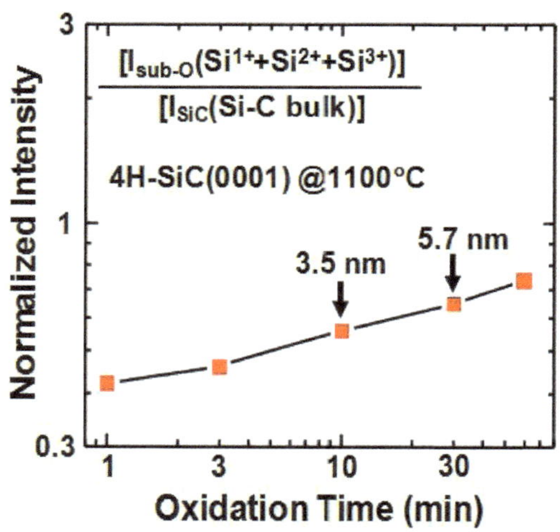

Figure 3. Change in the total amount of intermediate oxide states in Si $2p_{3/2}$ spectra, in which the intensity ratio between the intermediate state and the bulk signal was plotted as a function of oxidation time.

As previously reported, ideal hydrogen passivation of a SiC surface with a diluted HF solution was barely obtained, and the initial sample surface after wet cleaning was partially oxidized and contaminated with adsorbates [18]. This implies that a chemical shift component of C 1s core-level spectra involves unavoidable signals due to surface contamination. Thus, we performed angle-resolved XPS and in situ vacuum annealing prior to XPS analysis in the analysis chamber. Figure 2 represents C 1s core-level spectra taken from the oxidized 4H-SiC(0001) surface [17]. As shown in Fig. 2(a), the chemical shift component originating from carbon-oxides (CO_x) increased with respect to the bulk signal (C-Si bond) under the surface sensitive conditions (at small take-off angle (θ_e)). In addition, it was found that the chemical shift component originating from carbon-oxides was totally removed by vacuum annealing at 500ºC (see Fig. 2(b)). Since stable chemical bonds existing at the SiO_2/SiC interface are hard to decompose under a moderate annealing temperature, we attributed the carbon-oxide signal to surface contamination. This clearly demonstrates that atomic bonding at the thermally grown SiO_2/SiC(0001) interface is dominated by Si-O bonds and that carbon impurity with its oxide form located near the interface is below the detection limit of XPS analysis (about sub-1 atomic percent in general).

Figure 3 shows the change in the total amount of intermediate oxide states in Si $2p_{3/2}$ spectra (Si^{1+}, Si^{2+}, Si^{3+}) [17]. Although a thick transition layer at SiO_2/SiC interface was ruled out, we observed a slight increase in the intermediate oxide states with an increase of the oxide thickness, unlike in the SiO_2/Si interface, which exhibits a perfect interface regardless of oxide thickness [4].

3. Interface structures beneath thick thermal oxides grown on 4H-SiC(0001) Si-face and (000-1) C-face substrates

Figures 4(a) and 4(b) represent typical deconvoluted Si $2p_{3/2}$ spectra obtained from the 40-nm-thick SiO_2/SiC(0001) Si-face and (000-1) C-face substrates, respectively, in which the thick thermal oxides were thinned using a diluted HF solution prior to synchrotron XPS analysis [19]. Similar to the thin thermal oxides (see Fig. 1), the Si $2p_{3/2}$ spectra were fitted well with five components originating from bulk SiC and SiO_2 portions together with intermediate oxide states. It's obvious that, for both cases, the total amount of the intermediate states was sufficiently small compared with that of the remaining oxides (about 3 nm thick). This implies that the physical thickness of the transition layer on the oxide side is as thin as a few atomic layers even for the thick thermal oxides. These experimental results clearly indicate formation of a near-perfect SiO_2/SiC interface with conventional dry oxidation regardless of the substrate orientation and oxide thickness.

Furthermore, for the thick thermal oxide on Si-face substrate, the composition of the bulk SiC region beneath the oxide was estimated from the intensity ratio ([Si 2p]/[C 1s]). We obtained an identical intensity ratio to that of the initial as-grown SiC surface [17]. These experimental results mean that, despite previous literature based on TEM observation [6-8], there exists no thick carbon-rich layer of a high atomic percentage at the SiO_2/SiC interface and that a near-perfect interface dominated by Si-O bonds is formed even for the thick thermal oxidation of the SiC(0001) surface.

Figure 5 compares the change in the total amount of intermediate oxide states in Si $2p_{3/2}$ spectra obtained from SiO_2/SiC interfaces. The intensity ratios between the intermediate states and the bulk signals for thin and thick thermal oxides grown on (0001) Si-face and (000-1) C-face substrates were plotted. Despite that the minimal intermediate oxide states again imply abrupt interface, we observed a slight increase in the intermediate states especially for thick thermal oxide interface on C-face substrates, which suggesting degradation of interface electrical properties of SiC-MOS devices formed on C-face substrates [20].

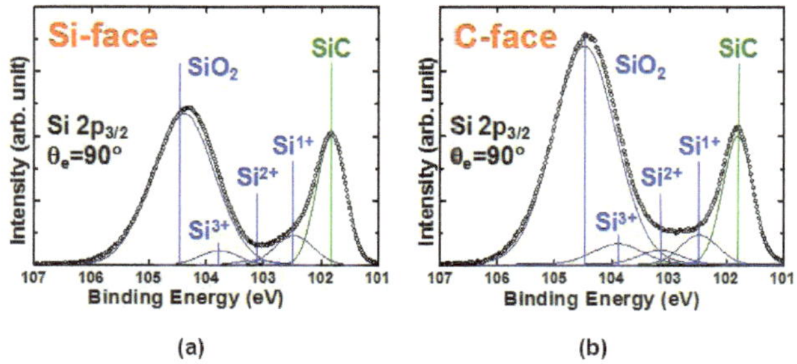

Figure 4. Si $2p_{3/2}$ core-level spectra obtained from the oxidized (a) 4H-SiC(0001) Si-face and (b) (000-1) C-face substrates. Before synchrotron XPS measurement, 40-nm-thick oxide layers were thinned using HF wet etching. The remaining oxide thickness was about 3 nm for both cases.

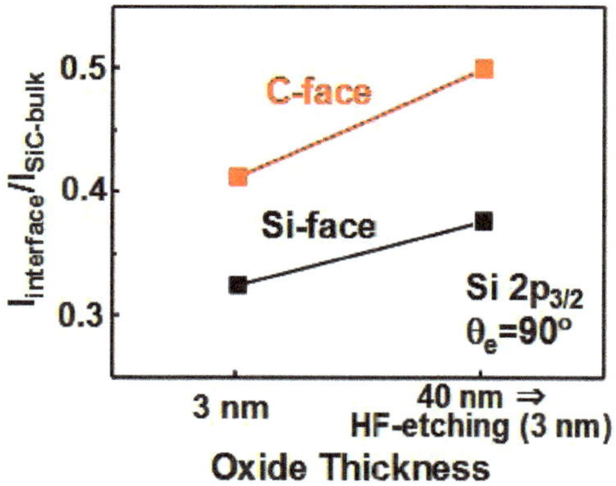

Figure 5. Change in the total amount of intermediate oxide states in Si $2p_{3/2}$ spectra taken from the oxide interfaces grown on SiC(0001) Si-face and C-face substrates.

4. Correlation between atomic structure and electrical properties of SiO_2/SiC interface

Corresponding SiC-MOS capacitors were fabricated with the top aluminum electrode evaporated through a shadow mask after post oxidation annealing at 900ºC in argon ambient. Electrical properties, such as the interface state density (D_{it}) and fixed oxide charge density (Q_{ox}) of the SiO_2/SiC interface, were extracted from the high-frequency C-V characteristics of SiC capacitors. The D_{it} value was extracted with the Terman method, and the Q_{ox} was deduced from the flatband voltage (V_{fb}) shift that depends on the oxide thickness [21]. Figure 6 summarizes the changes in D_{it} and V_{fb} values. Since post-treatment, such as nitrogen and hydrogen incorporation, was not conducted in this experiment, high D_{it} over 10^{12} $cm^{-2}eV^{-1}$ was extracted at an energy level of E_c - E = 0.36 eV. The high D_{it} value indicates degradation of the electrical properties of the SiO_2/SiC interface, especially for thick thermal oxides. In addition, the positive V_{fb} shift in the C-V curves implies the existence of a negative fixed charge within the gate oxides. Assuming that the fixed charge is located at the SiO_2/SiC interface, the Q_{ox} of the SiC-MOS devices estimated from the thickness-V_{fb} slope was 2.3 x 10^{11} cm^{-2} for oxides thinner than 20 nm and 1.2 x 10^{12} cm^{-2} for thick oxides, meaning that the fixed charges also accumulated at the interface probably due to the suppressed out-diffusion of carbon impurities as dry oxidation progressed.

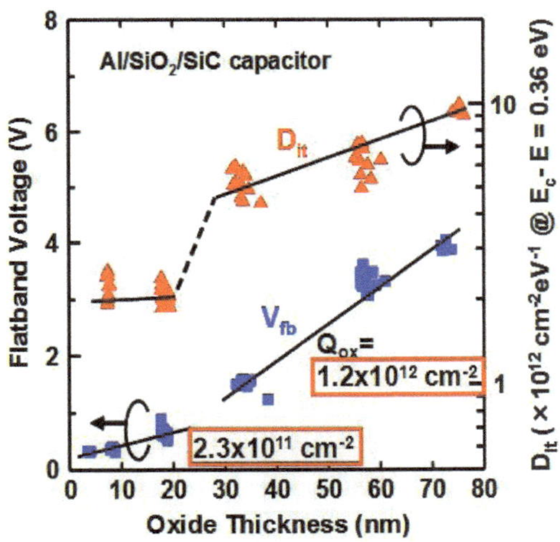

Figure 6. Summary of electrical properties of SiC-MOS capacitors fabricated by dry oxidation. Horizontal axis represents oxide thickness extracted from measured maximum capacitance. The D_{it} and Q_{ox} were estimated using Terman method and V_{fb} shift in C-V curves, respectively.

The correlation between electrical degradation and the atomic bonding feature of the SiO_2/SiC interfaces raises the intrinsic problem of SiC oxidation. This is consistent with the common understanding of SiC-MOS devices, whereas our synchrotron XPS analysis excludes the several-nm-thick transition layer having excess carbon as a physical origin of the electrical degradation. Instead, we think that the electrical defects at the interface, such as D_{it} and Q_{ox}, are partly ascribed to the atomic scale roughness and imperfection identified with the intermediate oxide states in the Si 2p spectra. Moreover, considering the significant mobility reduction in SiC-MOSFETs, we should take into account the various forms of carbon interstitials forming local C-C dimers located on the SiC bulk side as a possible origin of the electrical defects [22]. Therefore, it is concluded that, for improving the performance of SiC-based MOS devices, we should focus our attention on the atomic bonding feature and carbon impurities within the channel region rather than the thick transition layer near the SiO_2/SiC interface.

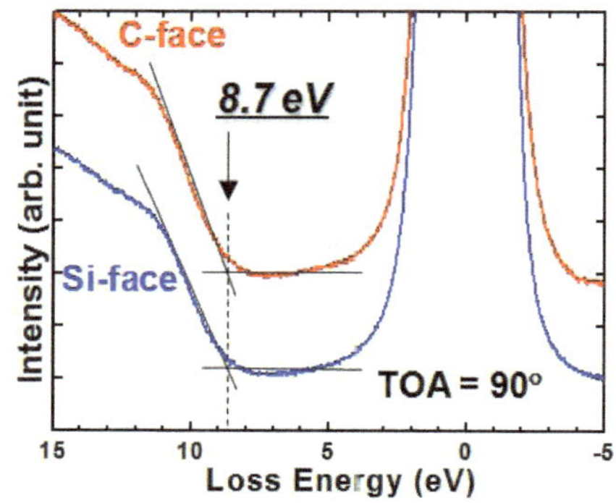

Figure 7. O 1s energy loss spectra for thermal oxides on (0001) Si-face and (000-1) C-face 4H-SiC substrates. The onset of the excitation from the valence to conduction bands (band gap) can be determined from the energy loss.The valence band maximum of SiC substrates and the oxides was determined by the valence spectra taken from SiO_2/SiC structures and a reference SiC surface [24]. Figure 8 represents measured and deconvoluted valence spectra obtained after 3-nm oxidation of the Si-face and C-face substrates, in which the valence band maximum of the thermal oxides was estimated by subtracting the reference SiC spectra ( ) from the measured SiO_2/SiC spectra ( ) both for the Si- and C-face substrates (see

5. Energy band structure of SiO_2/4H-SiC interfaces and its modulation induced by intrinsic and extrinsic interface charge transfer

The energy band structure and interface quality of SiO_2/4H-SiC fabricated on (0001) Si-face and (000-1) C-face substrates was also investigated by means of synchrotron XPS. Thermal oxidation was conducted using a conventional furnace at temperatures ranging from 1000 to 1100ºC. Thin and thick oxide layers of about 3 and 40 nm were prepared by choosing the oxidation temperature and time appropriately for Si-face and C-face substrates. For the thick oxide samples, the oxide layers were thinned to about 3 nm thick by a diluted HF solution. To determine band structures of SiO_2/SiC, we examined the band gap of the thermal oxides and valence band offset at the interface. The energy band gap of these oxides grown on Si-face and C-face substrates was first estimated from O 1s energy loss spectra. Because the photoelectrons generated in oxides suffer energy losses originating from plasmon and electron-hole excitations, the energy band gap can be determined by the threshold energy of an energy loss spectrum for an intense O 1s signal [23]. As shown in Fig. 7, O 1s energy loss spectra for thermal oxides on the Si-face and C-face substrates clearly indicate that the energy band gap of the oxides is identical regardless of substrate orientation (8.7 eV) [24]. Considering the high oxidation temperatures over 1000°C and low concentration of residual carbon impurities within thermally grown oxides on SiC [17], these results seem to be quite reasonable.

spectra -). These results demonstrated that the valence band offset of the SiO_2/SiC for the C-face substrate was about 0.4 eV larger than that for the Si-face. In addition, we conducted similar synchrotron XPS analysis for thick oxide samples to examine the effects of interface defects on energy band modulation as the oxide thickness increased.

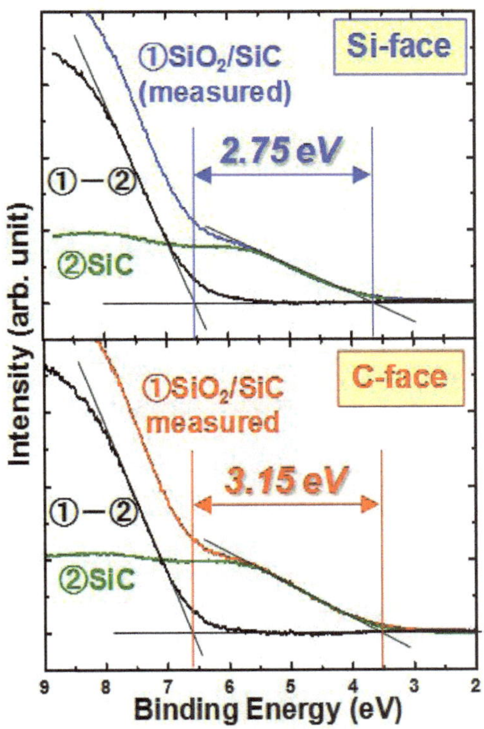

Figure 8. Measured and deconvoluted valence band spectra for SiO_2/SiC structures formed on (0001) Si-face and (000-1) C-face 4H-SiC substrates.

By taking these measured values and the reported energy band gap of 4H-SiC substrate (3.26 eV) into account, we obtained the energy band structures for SiO_2/SiC structures fabricated under various conditions and summarized them in Fig. 9 [24]. Note that the conduction band offset that determines gate leakage current and resultant gate oxide reliability of SiC-MOS devices crucially depends on the substrate orientation and oxide thickness. Since the thin oxide on the C-face substrates exhibits smaller conduction band offsets than those on the Si-face substrates, we conclude that the degraded reliability of SiC-MOS devices fabricated on the C-face surface is an intrinsic problem, which is probably due to the difference in the electronegativity between Si and C atoms bonded with O atoms at the interface.

Furthermore, considering the accumulation of negative fixed charges at the SiO_2/SiC interface, the increase in the conduction band offset for thick oxides both on the Si-face and C-face substrates can be explained by an extrinsic energy band modulation due to the interface defects. This enlarged band offset for the thick MOS devices is preferable from the viewpoint of reducing gate leakage, but electrical defects should negatively impact on the device performance and reliability. Therefore, fundamental tactics, such as applying deposited gate oxides and band engineering by utilizing stacked structures, are indispensable to take advantage of C-face SiC-MOS devices [25-28].

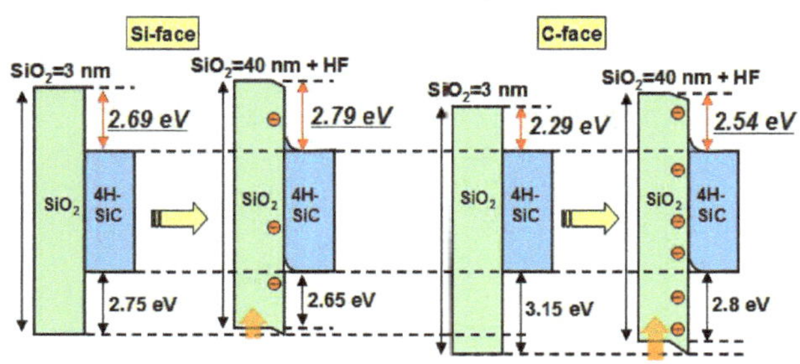

Figure 9. Energy band diagrams of SiO_2/4H-SiC(0001) structures obtained by synchrotron XPS analysis. The measured values of the valence band offsets for SiO_2/SiC interfaces formed under various conditions are indicated.

We also evaluated the modulation of energy band alignment of SiO_2/4H-SiC(0001) structures due to the interface defect passivation [29]. It was found that, although the hydrogen incorporation into the SiO_2/SiC interface is effective in improving the interface property, both XPS analysis and electrical measurements revealed that interface defect passivation induces a reduction of conduction band offset. This indicates that the larger conduction band offset at the as-oxidized SiO_2/SiC interface is attributed to the high density of interfacial carbon related defects.

6. Surface and interface morphology of thermally grown SiO_2 dielectrics on 4H-SiC(0001)

Finally, surface and interface morphology of thermal oxides grown on 4H-SiC(0001) substrates was investigated using AFM and TEM [30]. Thermal oxidation of 4º-off-angled 4H-SiC(0001) Si-face substrate with an n-type epilayer was carried out in dry O_2 ambient at 1100ºC for 12 hours. The root mean square (RMS) roughness of the as-grown surface was estimated to be 0.23 - 0.26 nm (see Fig. 10(a)), suggesting that it is an almost step-bunching-free substrate. However, in some locations, step bunching was observed, and the RMS roughness of the area, including step bunching, was about 2.3 nm. For some samples, high-temperature annealing was performed in an inert ambient at 1700ºC to intentionally emphasize the step bunching prior to the dry oxidation.

As shown in Fig. 10(b), the RMS roughness value of the oxide surface was about 0.36 nm, which is slightly higher than that of the as-grown epilayer surface. In addition, it seems that steps on the oxide surface are more rounded than the initial surface. Figure 11 shows cross-sectional TEM images of as-grown and oxidized samples. Single steps are observed at the SiO_2/4H-SiC interface in contrast to multiple-layer steps for the initial epilayer surface. These findings indicate that the step-terrace structure of an epilayer is enormously transferred to the SiO_2 surface, while the interface roughness decreases by smoothening step bunching. Since the oxidation rate for C-face 4H-SiC is much higher than that for Si-face, it is considered that the step edges will be rounded by enhanced oxidation and that the resulting oxide near the steps will be thicker due to a volume expansion from SiC to SiO_2.

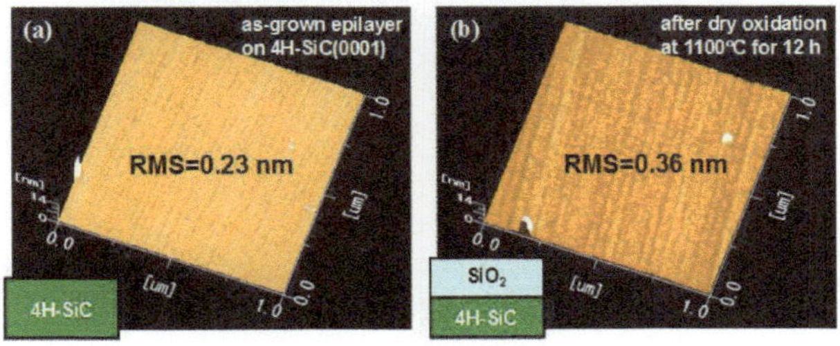

Figure 10. AFM images of (a) as-grown 4H-SiC(0001) epilayer surface and (b) SiO_2 surface formed on sample shown in (a) by dry O_2 oxidation at 1100ºC for 12 h.

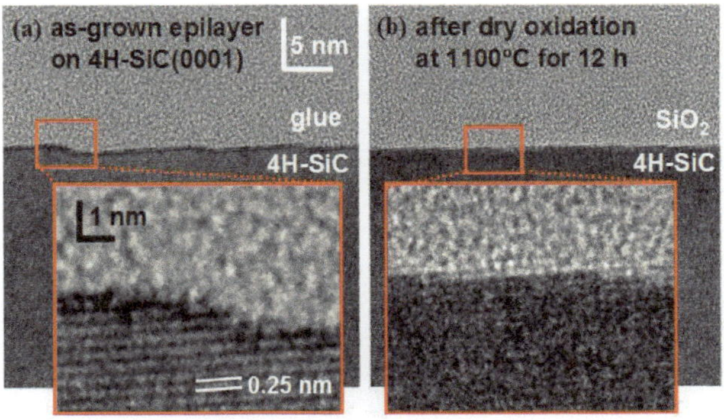

Figure 11. Cross-sectional TEM images of (a) as-grown 4H-SiC(0001) epilayer surface and (b) SiO_2/4H-SiC interface formed on sample shown in (a). The oxide thickness was about 35 nm.

To verify this hypothesis, a large step bunching was intentionally formed by high temperature annealing. As shown in Fig. 12(a), AFM observation of the 4H-SiC(0001) surface annealed at 1700ºC showed that the height of the step bunching ranged from 10 to 20 nm, while the RMS roughness within the terrace was 0.19 nm. After oxidation, a similar step-terrace structure to the one observed in Fig. 12(a) is preserved on the oxide surface, except for the bumps around the step bunching (see Fig. 12(b)). However, the RMS roughness in the flat region that was originally a terrace remains unchanged (<0.19 nm). This result strongly suggests significantly enhanced oxidation at the step face.

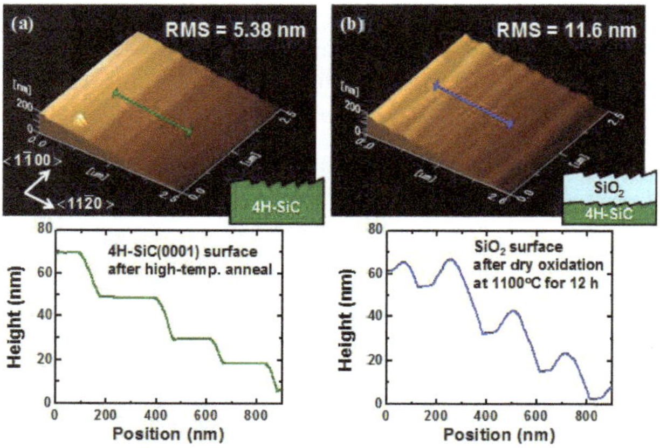

Figure 12. AFM images and cross-sectional profiles of (a) 4H-SiC(0001) surface after annealing at 1700°C and (b) SiO_2 surface formed on the sample shown in (a) by dry oxidation at 1100°C for 12 h.

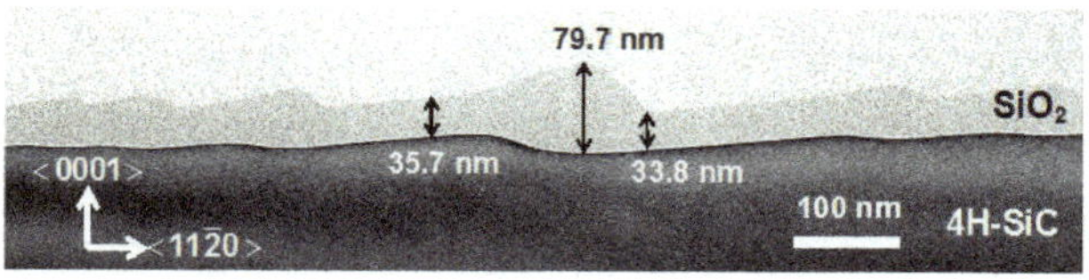

Figure 13. A cross-sectional TEM image of SiO_2/SiC structure shown in Fig. 12(b).

Furthermore, as shown in Fig. 13, a cross-sectional TEM image of the sample shown in Fig. 12(b) also clearly indicates pronounced oxidation at the step bunching. On the other hand, SiO_2/4H-SiC interface morphology is clearly more moderated than that of the SiO_2 surface, implying that the step bunching at the oxide interface became smooth by oxidation. The maximum SiO_2 thickness (~80 nm) was located facing the step, which is more than double that on the terrace. Since the oxidation reaction is controlled by the amount of oxygen molecules diffused through the SiO_2 layer, step edges will become rounded because of the thinner oxide on the terrace. The large oxide thickness fluctuation surely leads to the local electric field concentration around the step bunching during the electrical stressing, thus resulting in a preferential breakdown. Therefore, we can conclude that the surface morphology of the channel region before gate oxide formation is important for improving reliability of SiC-MOS devices.

7. Summary

We have investigated the fundamental aspects of SiC oxidation and SiO_2/SiC interfaces. Despite the literature based on TEM observation, we found that a near-perfect interface dominated by Si-O bonds is formed by dry oxidation of 4H-SiC(0001) substrates. However, atomic scale roughness and imperfection causing electrical degradation of SiC-MOS devices was found to be introduced as oxide thickness increases. We also pointed out the problems regarding oxide reliability originating from the gate leakage. It was found that, although negative fixed charges due to the interface defects enlarge the conduction band offset of SiC-MOS devices, small conduction band offset leading to increased gate leakage is an intrinsic feature, especially for the SiC(000-1) C-face substrates. We have also examined surface and interface morphology of thermally grown oxides to clarify the relation between step bunching and oxide breakdown. Multiple-layer steps as well as step bunching on the wafer surface lead to oxide thickness fluctuation due to the difference in oxidation rate between the terrace and the step face. The bump-like structure of the SiO_2 layer near the step bunching and the relatively thinner oxide on the terrace will cause a local electric field concentration, which enhances the generation of electrical defects in the oxide, indicating that an atomically-flat surface needs to be formed before gate oxide formation.

Acknowledgements

We are grateful to Dr. Takashi Nakamura, Mr. Yuki Nakano, and Mr. Shuhei Mitani for their valuable comments and discussion. Synchrotron XPS experiments were performed at BL23SU in the SPring-8 with the approval of JAEA as Nanotechnology Support Project of the Ministry of Education, Culture, Sports, Science and Technology (MEXT). We also thank Dr. Yuden Teraoka and Dr. Akira Yoshigoe for their assistance for synchrotron XPS measurements.

Author details

Heiji Watanabe* and Takuji Hosoi

*Address all correspondence to: watanabe@mls.eng.osaka-u.ac.jp

Department of Material and Life Science, Graduate School of Engineering, Osaka University, Suita, Osaka 565-0871, Japan

References

[1] Agarwal, A., Ryu, S. H., Palmour, J., Choyke, W. J., Matsunami, H., & Pensl, G. (2004). Silicon CarbideRecent Major Advances. Springer, Berlin.

[2] Gibson, J. M., & Lanzerotti, M. Y. (1989). Observation of interfacial atomic steps during silicon oxidation. *Nature*, 340, 128-131.

[3] Himpsel, F. J., Mc Feely, F. R., Taleb-Ibrahimi, A., & Yarmoff, J. A. (1988). Microscopic structure of the SiO_2/Si interface. *Phys. Rev. B*, 38, 6084-6096.

[4] Ohishi, K., & Hattori, T. (1994). Periodic changes in SiO_2/Si(111) interface structures with progress of thermal oxidation. *Jpn. J. Appl. Phys.*, 33, L 675-L678.

[5] Watanabe, H., Kato, K., Uda, T., Fujita, K., Ichikawa, M., Kawamura, T., & Terakura, K. (1998). Kinetics of initial layer-by-layer oxidation of Si(001) surfaces. *Phys. Rev. Lett.*, 80, 345-348.

[6] Chang, K.C., Nuhfer, N.T., Porter, L.M., & Wahab, Q. (2000). High-carbon concentrations at the silicon dioxide-silicon carbide interface identified by electron energy loss spectroscopy. *Appl. Phys. Lett.*, 77, 2186-2188.

[7] Zheleva, T., Lelis, A., Duscher, G., Liu, F., Levin, I., & Das, M. (2008). Transition layers at the SiO_2/SiC interface. *Appl. Phys. Lett.*, 93, 022108.

[8] Biggerstaff, T. L., Reynolds Jr, C. L., Zheleva, T., Lelis, A., Habersat, D., Haney, S., Ryu, S. H., Agarwal, A., & Duscher, G. (2009). Relationship between 4H-SiC/SiO_2 transition layer thickness and mobility. *Appl. Phys. Lett.*, 95, 032108.

[9] Zhu, X., Lee, H. D., Feng, T., Ahyi, A. C., Mastrogiovanni, D., Wan, A., Garfunkel, E., Williams, J. R., Gustafsson, T., & Feldman, L. C. (2010). Structure and stoichiometry of (0001) 4H-SiC/oxide interface. *Appl. Phys. Lett.*, 97, 071908.

[10] Agarwal, A. K., Seshadri, S., & Rowland, L. B. (1997). Temperature dependence of Fowler-Nordheim current in 6H- and 4H-SiC MOS capacitors. *IEEE Electron Device Lett.*, 18, 592-594.

[11] Kimoto, T., Kanzaki, Y., Noborio, M., Kawano, H., & Matsunami, H. (2005). Interface properties of metal-oxide-semiconductor structures on 4H-SiC{0001} and (1120) formed by N_2O oxidation. *Jpn. J. Appl. Phys.*, 44, 1213-1218.

[12] Suzuki, T., Senzaki, J., Hatakeyama, T., Fukuda, K., Shinohe, T., & Arai, K. (2009). Reliability of 4H-SiC(000-1) MOS gate oxide using N_2O nitridation. *Mater. Sci. Forum*, 557-560.

[13] Matocha, K, Dunne, G, Soloviev, S, & Beaupre, R. (2008). Time-dependent dielectric breakdown of 4H-SiC MOS capacitors and DMOSFETs. *IEEE Trans. Electron Devices*, 55, 1830-1834.

[14] Kozono, K., Hosoi, T., Kagei, Y., Kirino, T., Mitani, S., Nakano, Y., Nakamura, T., Shimura, T., & Watanabe, H. (2010). Direct observation of dielectric breakdown spot in thermal oxides on 4H-SiC(0001) using conductive atomic force microscopy. *Mater. Sci. Forum*, 645-648, 821-824.

[15] Degraeve, R., Groeseneken, G., Bellens, R., Ogier, J. L., Depas, M., Roussel, P. J., & Maes, H. E. (1998). New insights in the relation between electron trap generation and the statistical properties of oxide breakdown. *IEEE Trans. Electron Devices*, 45, 904-911.

[16] Teraoka, Y., & Yoshigoe, A. (2001). Commissioning of surface chemistry end-station in BL23SU of SPring-8. *Appl. Surf. Sci.*, 169-170, 738 -170.

[17] Watanabe, H., Hosoi, T., Kirini, T., Kagei, Y., Uenishi, Y., Chanthaphan, A., Yoshigoe, A., Teraoka, Y., & Shimura, T. (2011). Synchrotron x-ray photoelectron spectroscopy study on thermally grown SiO_2/4H-SiC(0001) interface and its correlation with electrical properties. *Appl. Phys. Lett.*, 99, 021907.

[18] Watanabe, H., Ohmi, H., Kakiuchi, H., Hosoi, T., Shimura, T., & Yasutake, K. (2011). Surface cleaning and etching of 4H-SiC(0001) using high-density atmospheric pressure hydrogen plasma. *J. Nanosci. Nanotechnol.*, 11, 2802-2808.

[19] Watanabe, H., Hosoi, T., Kirino, T., Uenishi, Y., Chanthaphan, A., Yoshigoe, A., Teraoka, Y., Mitani, S., Nakano, Y., Nakamura, T., & Shimura, T. (2012). Synchrotron radiation photoelectron spectroscopy study of thermally grown oxides on 4H-SiC(0001) Si-face and (000-1) C-face substrates. *Mater. Sci. Forum*, 697-702.

[20] Dhar, S., Feldman, L. C., Wang, S., Isaacs-Smith, T., & Williams, J. R. (2005). Interface trap passivation for SiO_2/(000-1) C-terminated 4H-SiC. *J. Appl. Phys.*, 98, 014902.

[21] Sze, S.M. (1981). Physics of Semiconductor Devices. Wiley, New York.

[22] Devynck, F., & Pasquarello, A. (2007). Semiconductor defects at the 4H-SiC(0001)/ SiO_2 interface. *Physica*, B 401-402, 556-559.

[23] Miyazaki, S., Nishimura, H., Fukuda, M., Ley, L., & Ristein, R. (1997). Structure and electronic states of ultrathin SiO_2 thermally grown on Si(100) and Si(111) surfaces. *Appl. Surf. Sci.*, 113-114, 585-589.

[24] Watanabe, H., Kirino, T., Kagei, Y., Harries, J., Yoshigoe, A., Teraoka, Y., Hosoi, T., & Shimura, T. (2011). Energy band structure of SiO_2/4H-SiC interfaces and its modulation induced by intrinsic and extrinsic interface charge transfer. *Mater. Sci. Forum*, 679-680, 386 -389.

[25] Noborio, M, Suda, J, Beljakowa, S, Krieger, M, & Kimoto, T. (2009). 4H-SiC MISFETs with nitrogen-containing insulators. *Phys. Status Solidi A*, 206, 2374-2390.

[26] Hosoi, T., Harada, M., Kagei, Y., Watanabe, Y., Shimura, T., Mitani, S., Nakano, Y., Nakamura, T., & Watanabe, H. (2009). AlON/SiO_2 stacked gate dielectrics for 4H-SiC MIS devices. *Mater. Sci. Forum*, 615-617, 541-544.

[27] Hosoi, T., Kagei, Y., Kirino, T., Watanabe, Y., Kozono, K., Mitani, S., Nakano, Y., Nakamura, T., & Watanabe, H. (2010). Improved characteristics of 4H-SiC MISFET with AlON/nitrided SiO_2 stacked gate dielectrics . *Mater. Sci. Forum*, 645-648, 991 -994.

[28] Hosoi, T., Kagei, Y., Kirino, , Mitani, S., Nakano, Y., Nakamura, T., Shimura, T., & Watanabe, H. (2011). Reduction of charge trapping sites in Al_2O_3/SiO_2 stacked gate dielectrics by incorporating nitrogen for highly reliable 4H-SiC MIS devices. *Mater. Sci. Forum*, 679-680, 496-499.

[29] Hosoi, T., Kirino, T., Chanthaphan, A., Uenishi, Y., Ikeguchi, D., Yoshigoe, A., Teraoka, Y., Mitani, S., Nakano, Y., Nakamura, T., Shimura, T., & Watanabe, H. (2012). Impact of interface defect passivation on conduction band offset at SiO_2/4H-SiC interface . *Meter. Sci. Forum*, 712-720, 721-724.

[30] Hosoi, T., Kozono, K., Uenishi, Y., Mitani, S., Nakano, Y., Nakamura, T., Shimura, T., & Watanabe, H. (2011). Investigation of surface and interface morphology of thermally grown SiO_2 dielectrics on 4H-SiC(0001) substrates. *Mater. Sci. Forum*, 679-680, 342-345.

Chapter 8

Thermal Oxidation Mechanism of Silicon Carbide

Yasuto Hijikata, Shuhei Yagi, Hiroyuki Yaguchi and Sadafumi Yoshida

Additional information is available at the end of the chapter

1. Introduction

Recently, semiconductor devices that work normally for a long time under harsh environments are demanded such as in nuclear power application or in space development field. Especially, after the disaster of Fukushima Dai-ichi nuclear power plant caused by the East-Japan great earthquake on March 11, 2011, the importance of such a *hard* electronic devices has been growing. However, it is difficult to achieve the performance using conventional semiconductors such as Si semiconductor because of its physical property limit.

Prevention of the global warming is one of the most important and urgent subjects in the world. Technologies to reduce energy consumption should be the key for overcoming this problem, and one of which is the improvement in the efficiency of power devices.

Silicon carbide (SiC) semiconductor is one of the wideband gap semiconductors and the use of it is considered as the solution to achieve these performances because it has superior physical properties such as 3 times wider bandgap, 10 times larger electrical break-down field, and 3 times higher thermal conductivity, compared with Si semiconductor [1]. Taking advantages of these properties, on-resistance for unipolar devices such as metal-oxide-semiconductor field-effect-transistors (MOSFETs) can, for example, be reduced by a factor of a few hundreds when replacing Si with SiC semiconductor. In addition, SiO_2 film, utilized as an insulator in MOSFETs, can be grown on the SiC substrate surface by thermal oxidation, which is well compatible with the Si MOS device technologies [2]. Moreover, the power and frequency ranges of SiC MOSFETs are around 1 kV break-down voltage and around 20 kHz switching frequency, respectively, which covers the wide power device application field.

For these reasons, the developments of SiC power MOSFETs have been very popular for a few decades. However, the on-resistances for MOSFETs fabricated practically are beyond the lower limit for Si, however, higher than the SiC limit by a few orders [3]. As a result, conventional Si insulated gate bipolar transistors (IGBTs) still have most of the share in the application fields of power transistors. In the case of 1 kV break-down voltage device, the channel resistance is dominant to the total on-resistance. Therefore, controlling the channel layer, *i.e.* the SiC-SiO_2 interface structure, should be the key technology to realize a SiC -MOSFET with desired performances. Besides, although the long-term reliability of oxide is very important for the practical uses of MOSFETs, that of SiC MOS device are still lower than that of Si by a factor of 1 order. As the creation of interface layers and the characteristics of oxide layers are closely related to the growth mechanism of the oxide, it is safely said that the observation of SiO_2 growing process is very significant work for overcoming these problems.

In previous work, we have, for the first time, performed real-time observation of SiC thermal oxidation using an *in-situ* ellipsometer [4, 5]. The results show that the oxidation-time dependence of oxide thickness can essentially be represented by the Deal-Grove (D-G) model [6], which has been originally proposed for the explanation of Si oxidation. Song *et al.* [7] have modified the D-G model for applying it to SiC oxidation, taking the process of carbon oxidation into account. They have concluded that a linear-parabolic formula can also be applicable to SiC oxidation, although the parabolic term includes the contribution from the diffusion of CO or CO_2 molecules from the SiC-oxide interface to the surface as well as that of oxygen from the surface to the interface. However, our further studies have found that the oxide growth rates in the thin oxide region are higher than those predicted from the D-G model [4, 8, 9, 10, 11]. By the way, it is well known that also the oxidation behavior of Si cannot be explained using the D-G model, *i.e.*, particularly at the initial oxidation stage. Accordingly, several models that describe Si oxidation have been proposed [12, 13, 14, 15, 16].

At the beginning of this chapter, we review the thermal oxidation models for SiC as well as those for Si that have been previously proposed, to elucidate the oxidation mechanism of SiC and then we verify each of these SiC oxidation models by making comparison with the oxide growth rate data with various oxidation conditions and discuss the structure and nature of the SiC-oxide interface layer based on the oxidation model that we have proposed.

2. Thermal oxidation models for Si and SiC

2.1. Deal-Grove model and its related models

The kinetic model of Si oxidation that is most often taken as a reference is the one so-called Deal-Grove model proposed by Deal and Grove [6, 12]. According to this model, the beginning of oxidation is limited to the interfacial oxidation reaction and, after oxidation proceeds, the rate-imiting process is transferred from the interfacial reaction to the diffusion of oxidants in SiO_2. This process is expressed by the following equation given by Deal and Grove as [6, 12]

$$\frac{dX}{dt} = \frac{B}{A + 2X} \tag{1}$$

where B/A and B are denoted as the linear and parabolic rate constants of oxidation, respectively. It is noted that B/A and B are the rate coefficients for the interfacial reaction and the diffusion of oxidants, respectively, *i.e.*,

$$\frac{B}{A}=\frac{kC_{\mathrm{O}}^{0}}{N_0},\ B=\frac{2D_{\mathrm{O}}C_{\mathrm{O}}^{0}}{N_0} \tag{2}$$

where k is the interfacial oxidation rate constant, C^0 the solubility limit in SiO_2, N_0 the molecular density of SiO_2, D the diffusivity in SiO_2, and the subscript means the value for the corresponding atom. By the way, according to the definition in the D-G model, in which the oxide growth proceeds under steady-state and a SiO_2 film is grown only at the SiO_2-Si interface, the equation (1) can also be written as [6, 12]:

$$\frac{dX}{dt}\equiv\frac{kC_{\mathrm{O}}^{\mathrm{I}}}{N_0}\left(=\frac{k}{N_0}\frac{C_{\mathrm{O}}^{0}}{(k/D_{\mathrm{O}})X+1}\right) \tag{3}$$

where C^{I} is the atomic concentration at the interface. Comparing eq. (3) with eqs. (1, 2), $C_{\mathrm{O}}^{\mathrm{I}}$ is almost constant by C_{O}^{0} while $X\le D_{\mathrm{O}}/k$ and, in turn, it reduces inverse proportional to X when $X>D_{\mathrm{O}}/k$ (see the right-end parentheses in eq. (3)). Therefore, we can see that the length $D_{\mathrm{O}}/k(=A/2)$ corresponds to the thickness at which the interfacial reaction rate-limiting step transits to the diffusion of oxidants one. On the contrary, Watanabe *et al.* reconfirmed the validity of eq. (1), but modified the framework of A parameter, based on the consideration that the growth rate at the full range of oxide thickness is limited by the oxygen diffusion, *i.e.*, there is no interfacial reaction-limited step in the Si oxidation process [16].

2.2. Massoud empirical relation

It is well known that, in the case of dry oxidation, the oxidation rate of Si in the thin oxide thickness range cannot be reproduced by the D-G equation [6, 12] and, hence, several models to describe the growth rate enhancement in the thin oxide regime have been proposed. Among them, Massoud *et al.* [13] have proposed an empirical relation for the oxide thickness dependence of oxidation rate, that is, the addition of an exponential term to the D-G equation,

$$\frac{dX}{dt}=\frac{B}{A+2X}+C\exp\left(-\frac{X}{L}\right) \tag{4}$$

where C and L are the pre-exponential constant and the characteristic length, respectively. Also in the case of SiC oxidation, we have found that it is possible to fit the calculated values to the observed ones using eq. (4) much better than using eq. (1) regardless of oxidation temperature, as shown in Fig. 1. However, this empirical equation can only reproduce the observed growth rates numerically, but does not provide a physical meaning.

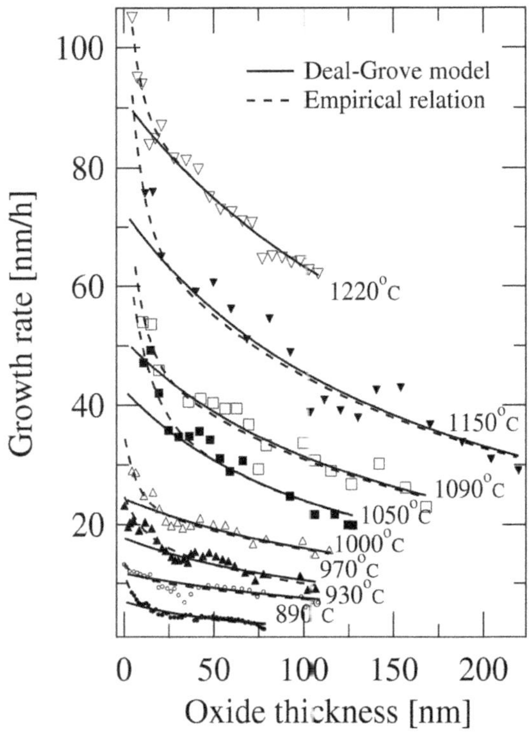

Figure 1. Oxide thickness dependences of oxidation rate at various oxidation temperatures on (000$\bar{1}$) C-face. The solid and dashed lines denote the values derived from the Deal-Grove model (eq. (1) [6]) and those from the empirical relation (eq. (4) [13]), respectively.

2.3. Interfacial Si emission model for Si

Some Si oxidation models that describe the growth rate enhancement in the initial stage of oxidation have been proposed [14, 15, 17]. The common view of these models is that the stress near/at the oxide-Si interface is closely related to the growth enhancement. Among these models, the 'interfacial Si emission model' is known as showing the greatest ability to fit the experimental oxide growth rate curves. According to this model, Si atoms are emitted as interstitials into the oxide layer accompanied by oxidation of Si, which is caused by the strain due to the expansion of Si lattices during oxidation. The oxidation rate at the interface (k in eq. (2)) is initially large and is suppressed by the accumulation of emitted Si atoms near the interface with increasing oxide thickness, *i.e.*, the k is not constant but a function of oxide thickness and the oxidation rate is not enhanced in the thin oxide regime but is quickly suppressed with increasing thickness. To describe this change in the interfacial reaction rate, Kageshima *et al.* introduce the following equation as the interfacial reaction rate, k [14, 17]:

$$k = k_0 \left(1 - \frac{C_{Si}^I}{C_{Si}^0} \right) \tag{5}$$

where k_0 is the initial interfacial oxidation rate.

In the D-G model and the Massoud empirical relation, it has been considered that oxide growth occurs only or mainly at the Si-oxide interface. However, according to the interfacial Si emission model [14], Si atoms are emitted into the oxide layer, some of which encounter the oxidant inside the SiO_2 layer to form SiO_2. In addition to this, when the oxide is very thin, some of the emitted Si atoms can go through the oxide layer and reach the oxide surface, and are instantly oxidized, resulting in the formation of an SiO_2 layer on the oxide surface. Therefore, there are two oxide growth processes other than the interfacial oxide growth, *i.e.*, oxide formation due to oxidation of Si interstitials inside the oxide and on the oxide surface. The total growth rate is given by the sum of these three oxidation processes, as [14],

$$N_0 \frac{dX}{dt} = kC_O^I(1 - \nu_{Si}) + \int_0^X \kappa (C_O)^2 C_{Si} dx + \eta (C_O^S)^2 C_{Si}^S \tag{6}$$

where the ν is the emission ratio, the κ is the oxidation rate of Si interstitials inside SiO_2, $\eta$is the oxidation rate of Si interstitials on the oxide surface, and the superscript 'S' means the position at the oxide surface (x=X). The first, second, and third term in the right-hand side of eq. (6) correspond to the interfacial oxide growth, the oxide growth inside SiO_2, and that on the oxide surface, respectively. The concentrations of Si interstitials and O_2 moleculers in SiO_2 (C_{Si} and C_O, respectively) are derived using a numerical calculation based on the diffusion theory [17]. The diffusion equations will be shown in Sec. 2.4. If the steady-state approximation is assumed for the calculation, the obtained growth rate equation is equivalent to the Massoud empirical relation [14].

2.4. Si and C emission model for SiC

Since the density of Si atoms in 4H-SiC (4.80×10^{22}cm^{-3} [18]) is almost the same as that in Si (5×10^{22}cm^{-3}) and the residual carbon is unlikely to exist at the oxide-SiC interface in the early stage of SiC oxidation, the stress near/at the interface is considered to be almost identical to the case of Si oxidation. Therefore, it is probable that atomic emission due to the interfacial stress also accounts for the growth enhancement in SiC oxidation. In addition, in the case of SiC oxidation, we should take C emission as well as Si emission into account because SiC consists of Si and C atoms.

Recently, we have proposed a SiC oxidation model, termed "Si and C emission model", taking the Si and C emissions into the oxide into account, which lead to a reduction of interfacial reaction rate [19]. Figure 2. schematizes the Si and C emission model. Considering Si and C atoms emitted from the interface during the oxidation as well as the oxidation process of C, the reaction equation for SiC oxidation can be written as,

$$\mathrm{SiC} + \left(2 - \nu_{\mathrm{Si}} - \nu_{\mathrm{C}} - \frac{\alpha}{2}\right)\mathrm{O_2} \rightarrow (1-\nu_{\mathrm{Si}})\mathrm{SiO_2} + \nu_{\mathrm{Si}}\mathrm{Si} + \nu_{\mathrm{C}}\mathrm{C} + \alpha\mathrm{CO} + (1-\nu_{\mathrm{C}}-\alpha)\mathrm{CO_2} \tag{7}$$

where α denotes the production ratio of CO.

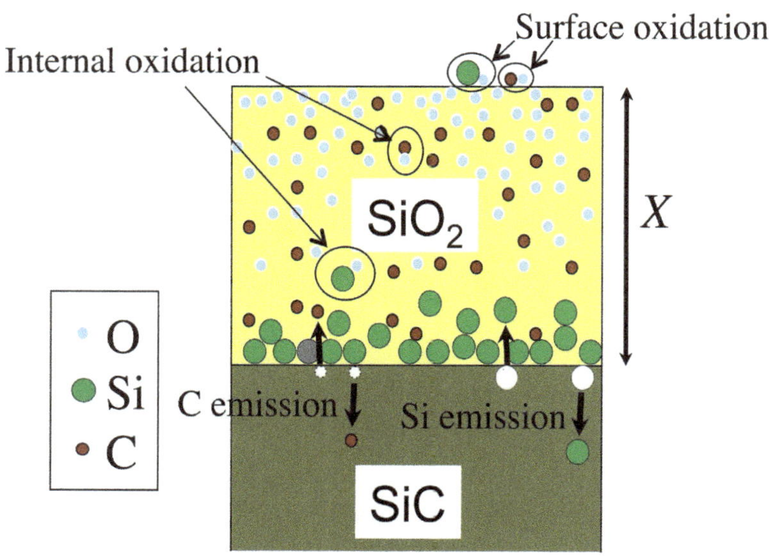

Figure 2. Schematic illustration of the Si and C emission model [19]. It is to be noted that Si and C atoms are practically emitted into not only to the SiO_2-side, but also the SiC-side, as shown in the figure.

In the case of Si oxidation, the interfacial reaction rate (eq. (5)) is introduced by assuming that the value of $C_{\mathrm{Si}}^{\mathrm{I}}$ does not exceed the C_{Si}^{0} though the reaction rate decreases with increase of$C_{\mathrm{Si}}^{\mathrm{I}}$. Based on this idea, the interfacial reaction rate for SiC is thought to be given by multiplying decreasing functions for Si and C [19]:

$$k = k_0\left(1 - \frac{C_{\mathrm{Si}}^{\mathrm{I}}}{C_{\mathrm{Si}}^{0}}\right)\left(1 - \frac{C_{\mathrm{C}}^{\mathrm{I}}}{C_{\mathrm{C}}^{0}}\right) \tag{8}$$

This equation implies that the growth rate in the initial stage of oxidation should reduce by two steps because the accumulation rates for Si and C interstitials should be different from each other, and hence, the oxidation time when the concentration of interstitial saturates should be different between Si and C interstitial. This prediction will be evidenced in the next section.

Diffusion equations for Si and C interstitials, and oxidants can be written by modifying the those given by the interfacial Si emission model [17], that is,

$$
\begin{aligned}
&\frac{\partial C_{\mathrm{Si}}}{\partial t}=\frac{\partial}{\partial x}\left(D_{\mathrm{Si}}\frac{\partial C_{\mathrm{Si}}}{\partial x}\right)-R_1-R_2,\\
&R_1=\eta C_{\mathrm{O}}^{\mathrm{S}}C_{\mathrm{Si}}^{\mathrm{S}},\ R_2=\kappa_1 C_{\mathrm{Si}}C_{\mathrm{O}}+\kappa_2 C_{\mathrm{Si}}(C_{\mathrm{O}})^2,\\
&\frac{\partial C_{\mathrm{C}}}{\partial t}=\frac{\partial}{\partial x}\left(D_{\mathrm{C}}\frac{\partial C_{\mathrm{C}}}{\partial x}\right)-R_1^{'}-R_2^{'},\\
&R_1^{'}=\eta^{'} C_{\mathrm{O}}^{\mathrm{S}}C_{\mathrm{C}}^{\mathrm{S}},\ R_2^{'}=\kappa_1^{'} C_{\mathrm{C}}C_{\mathrm{O}}+\kappa_2^{'} C_{\mathrm{C}}(C_{\mathrm{O}})^2,\\
&\frac{\partial C_{\mathrm{O}}}{\partial t}=\frac{\partial}{\partial x}\left(D_{\mathrm{O}}\frac{\partial C_{\mathrm{O}}}{\partial x}\right)-R_1-R_2-R_1^{'}-R_2^{'}-R_3,\\
&R_3=h\left(C_{\mathrm{O}}^{\mathrm{S}}-C_{\mathrm{O}}^{0}\right)
\end{aligned}
\tag{9}
$$

where the prime means the variation for C atoms. It is noted that the R_2 and $R_2^{'}$ mean absorption of interstitials inside the oxide and they are assumed to be consist of two terms, as suggested by Uematsu *et al.* [20]. From eq. (7), the boundary conditions at the interface are given as,

$$
\begin{aligned}
&D_{\mathrm{Si}}\frac{\partial C_{\mathrm{Si}}}{\partial x}\Big|_{x=0}=-v_{\mathrm{Si}}kC_{\mathrm{O}}^{\mathrm{I}},\ D_{\mathrm{C}}\frac{\partial C_{\mathrm{C}}}{\partial x}\Big|_{x=0}=-v_{\mathrm{C}}kC_{\mathrm{O}}^{\mathrm{I}},\\
&D_{\mathrm{O}}\frac{\partial C_{\mathrm{O}}}{\partial x}\Big|_{x=0}=\left(2-v_{\mathrm{Si}}-v_{\mathrm{C}}-\frac{\alpha}{2}\right)kC_{\mathrm{O}}^{\mathrm{I}}
\end{aligned}
\tag{10}
$$

It has been believed that the oxidation rate in the thick oxide regime is solely limited by the in-diffusion of oxidant and the diffusivity of CO in SiO_2 is much larger than that of O_2. Thus, we assumed that the diffusion process of CO is insensitive to the oxide growth rate. The oxide growth rate is described as eq. (6), but the second term in the right-hand side (*i.e.* the absorption-term of Si interstitials inside the oxide) is modified to be the combination of κ_1 and κ_2 shown in eq. (9) (see eq. (6) in Ref. [19]). We approximated this integral term by the area of a triangle $C_{\mathrm{Si}}^{\mathrm{I}}$ in height and $\partial C_{\mathrm{Si}}/\partial x\,|_{x=0}\,/2$ in gradient of hypotenuse, on the basis of the idea that interstitials are usually distributed according to what resembles a complementary error function or exponential function, and their areal density is given approximately by the area of the triangle [21].

Equations (8-10) were solved numerically using the partial differential equation solver ZOMBIE [22]. The oxide thickness, X, at each time step was obtained from eq. (6). The parameters related to the properties of SiO_2 ($D_{\mathrm{Si}}, D_{\mathrm{O}}$, η, κ_1, $\kappa_2$andC_{Si}^{0}) were set to the same values as those obtained for Si oxidation [17]. The parameters concerning C interstitials (α, v_{C}, D_{C}, $\eta^{'}$, $\kappa_1^{'}$, $\kappa_2^{'}$andC_{C}^{0}) as well as the values of k_0 and v_{Si} were determined by fitting the calculated oxide growth rates to the measured ones.

3. Results and Discussion

3.1. Differences in the oxidation process between C- and Si-face

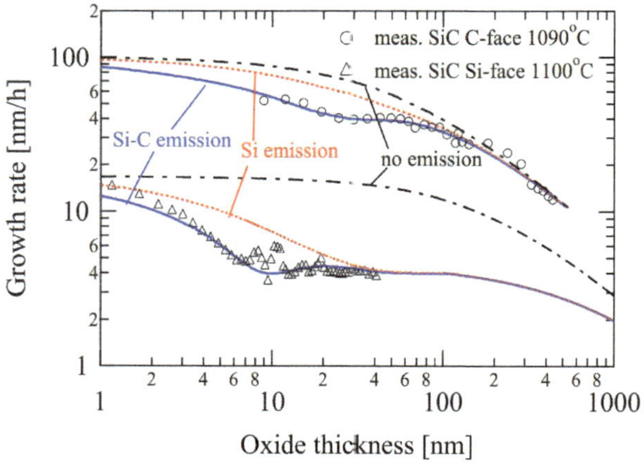

Figure 3. Oxide thickness dependence of growth rates on C- and Si-faces.

Figure 3 shows the oxide growth rates observed for 4H-SiC C face at 1090°C (circles) and Si face at 1100°C (triangles). The oxide growths were executed under dry oxygen ambient at a pressure of 1 atm. The experimental details can be found in the references [8, 9, 10]. Also shown in the figure are the growth rates given by the Si and C emission model (blue solid lines), the Si emission model, and the model that does not take account of both Si and C emission, *i.e.*, the D-G model (red broken line and black double broken line, respectively). We note that the same parameters were used for these three SiC oxidation models.

Figure 3 shows that the Si and C emission model reproduces the experimental values for both the C and Si faces better than the other two models. In particular, the dip in the thickness dependence of the growth rate seen around 20 nm for the C-face and 10 nm for the Si-face, which cannot be reproduced by the Si emission model or the D-G model no matter how well the calculation are tuned, can be well reproduced by the Si and C emission model. These results suggest that the C interstitials play an important role in the reduction of the oxidation rate, similarly to the role of the Si interstitials. Moreover, from the fact that the drop in growth rate in the initial stage of oxidation is larger for the Si and C emission model than in the case of taking only Si emission into account, we found that the accumulation of C interstitials is faster than that of Si interstitials and that the accumulation of C interstitials is more effective in the thin oxide regime.

As mentioned in Sec. 2.1, the growth rate in the thick oxide regime is determined by the parabolic rate constant B as is obvious if we consider the condition that $A \ll 2X$ for eqs. (1, 4). Song *et al.* proposed a modified Deal-Grove model that takes the out-diffusion of CO into account by modifying the parabolic rate constant B by a factor of 1.5 (called 'normalizing factor' [23]), and through this model, they explained the oxidation process of SiC in the parabolic oxidation rate regime [7]. For the Si and C emission model, the normalizing factor corresponds to the coefficient of the oxidant shown in eq. (7), *i.e.* $(2-\nu_{Si}-\nu_C-\alpha/2)$. Since Song's model assumed that there is no interfacial atomic emission (*i.e.* $\nu_{Si}=\nu_C=0$) and carbonaceous products consist of only CO (*i.e.* $\alpha=1$), for this case, the coefficient of the oxidant in eq. (7) equals 1.5, which is the same as the normalizing factor. Actually, our experiments revealed this coefficient to be 1.53 [19], which is almost equal to the measured value from Song *et al.* Therefore, our results support the assumption in the Song's model that $\alpha=1$ and we believe that the obtained B values to be different from that of Si oxidation [5, 7, 23, 24] despite the fact that the native oxide is SiO_2 regardless of Si and SiC were due to the contribution from CO production in SiC oxidation. While in the case of the Si-face, since the influence of oxygen diffusion was significant at thicknesses above several μm as seen in Fig. 3, we could not determine the value ofα. The value of α is thought to depend on the areal density of carbon in the substrate. Thus, we regard the α value of the Si-face as equal to that of C-face.

The parameters to be different between C- and Si-face in the deduced values wereν_{Si}, ν_C, and k_0. In next section, we will discuss the temperature dependence of these three parameters.

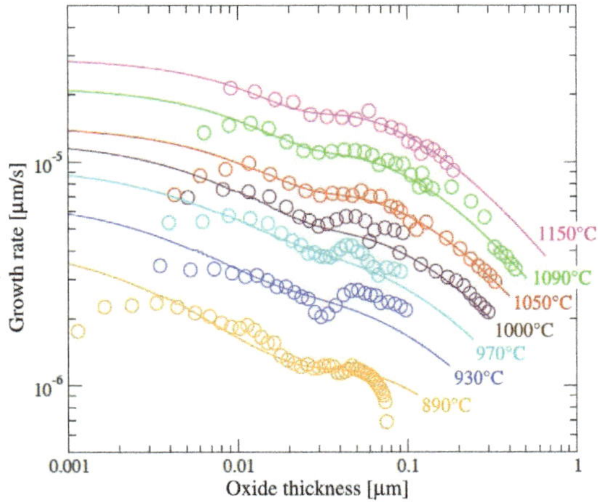

Figure 4. Oxide thickness dependence of oxide growth rates at various oxidation temperatures for (000$\bar{1}$)C-face. The circles and the solid curves denote measured growth rates and calculated ones, respectively.

3.2. Oxidation rate at various oxidation temperatures

Figure 4 shows the oxide growth rates as a function of oxide thickness, observed for the dry oxidation of C-face at various oxidation temperatures (circles) and those given by the Si and C emission model (the solid curves) [25]. The figure indicates that the Si and C emission model reproduces the oxide growth rate curves for all of the temperatures measured. As mentioned above, some articles have suggested that [7, 23, 26] the SiC oxidation can be described by using the D-G model. However, there are several issues in the application of D-G model to SiC oxidation, in which unreasonable parameter values are needed to fit to the measured oxide growth rates. For example, the activation energy of parabolic oxidation-rate constant B for SiC Si-face and $(11\bar{2}0)\alpha$ -face significantly different from that for Si is required to fit the growth rate curve despite the fact that both of oxides on SiC and Si are SiO_2. In accordance with Si and C emission model, on the contrary, the oxide growth rates of SiC for all of the temperatures measured can be derived using just the same diffusivities as those for Si regardless of oxidation temperatures. This shows that the Si and C emission model is more valid than the D-G model to explain SiC oxidation process.

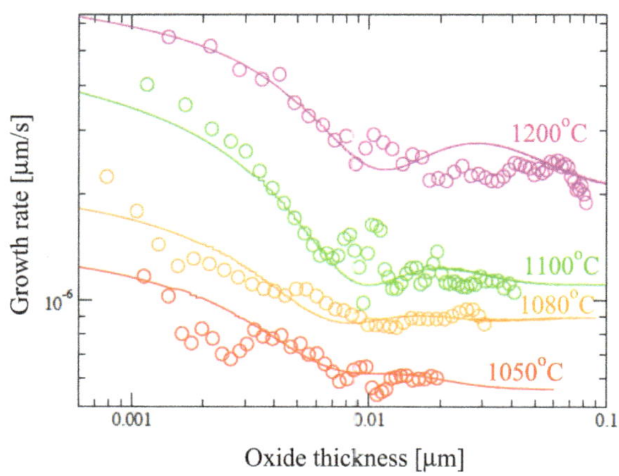

Figure 5. Oxide thickness dependence of oxide growth rates at various oxidation temperatures for (0001)Si-face. The circles and the solid curves denote measured growth rates and calculated ones, respectively.

Figure 5 shows the oxide growth rates for Si-face as a function of the oxide thickness at various oxidation temperatures. The calculated curves also well agree with the measured data. The figure indicates that the values obtained are almost constant in the larger thickness range in this study at any oxidation temperature. The reason for this constant thickness dependence is that, in the case of Si-face, interfacial reaction rate-limiting step continues up to several μm in oxide thickness, as revealed in Sec. 3.1.

Figure 6 shows the temperature dependence of Si and C emission ratios (ν_{Si}andν_C, respectively) and initial interfacial reaction rate (k_0) for C- and Si-face deduced from the curve fits. The figure indicates that the activation energies of $\nu_{Si}$and ν_C are comparable between these polar faces, but the values of them for Si-face are larger than those for C-face, in particular, the ν_{Si} for Si-face is remarkably large. It is noted that the ν_{Si} values for C-face were just the same as those for Si(100)face. On the other hand, the activation energy of k_0 is a little larger for Si-face, while the k_0 values for these faces approach each other as elevating temperature.

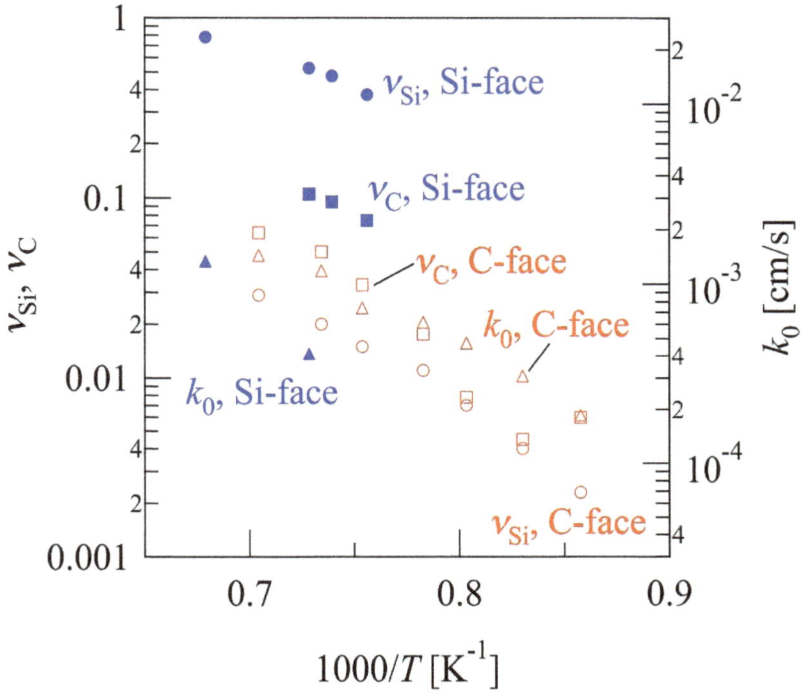

Figure 6. Arrehnius plots ofv_{Si}, v_C, and k_0 (circles, squares, and triangles, respectively) for Si- and C-face.

In the simulation for Si oxidation, Uematsu *et al.* have found that the v_{Si} depends on surface orientation of the substrate [17]. Therefore, it is possible that the differences in v_{Si} and v_C also appear between SiC C- and Si-face. Obviously, a high Si emission ratio leads to a low oxide growth rate, as shown in eqs. (5, 8). Therefore, it is considered that the high Si emission ratio is one of the causes for the low oxidation rate for Si-face [28]. Uematsu *et al.* also found that the k_0 also depends on surface orientation, which is probably due to the difference in surface density of the Si-Si bonds available for the reaction with oxidants [17]. In the case of SiC, the areal density of such reaction-available bonds on Si-face is lower than that on C-face by a factor of three, which is roughly coincident with the difference in observed data (Fig. 6). According to the report from Ref. [27], in the cases of extremely high oxidation temperature (~1500°C), the oxide growth rate is comparable between C- and Si-face, which is different from the fact that the oxide growth rates for the C-face are about eight times larger than those for the Si-face in the several 10 nm thickness region at around 1100°C and atmospheric oxygen pressure. We consider that such a high temperature oxidation proceeds with another mechanism such as 'active oxidation' (*i.e.* oxidizing with evaporation), so that the substrate orientation does not affect the oxidation process any more.

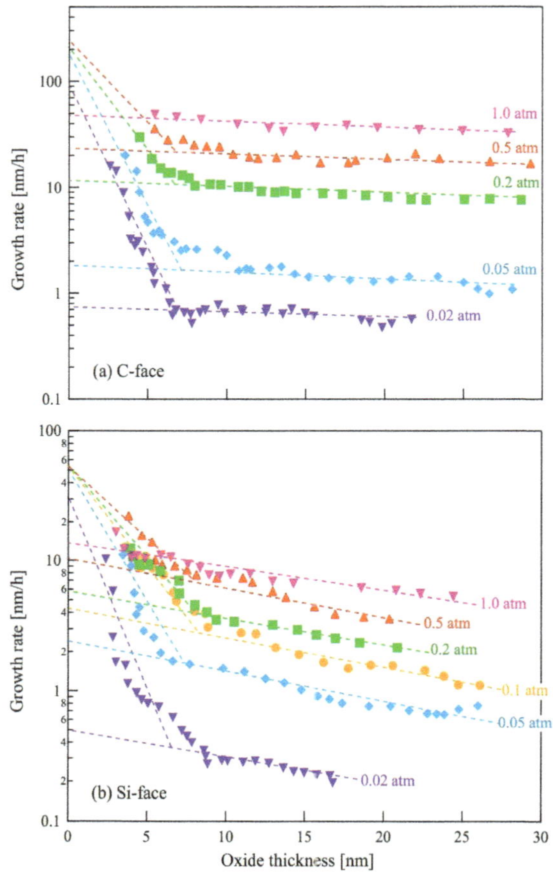

Figure 7. Oxide growth rates as a function of oxide thickness at various oxygen partial pressures on (000$\bar{1}$) C-face (a) and (0001) Si-face (b). The dashed lines are fitted to the experimental data using exponential functions.

Oxide thickness dependence of the oxide growth rates on 4H-SiC C-face (a) and Si-face (b) are shown in Fig. 7. As seen in the figure, we could obtain growth rate data in the extreme-thin oxide region down to a few nm more precisely by reducing the oxygen partial pressure less than 0.1 atm, compared with the case of measurements above 0.1 atm. For both faces, the oxide thickness dependence of growth rate less than 0.1 atm are basically similar to those of above 0.1 atm even if the partial pressure is lowered to 0.02 atm. Namely, just after the oxidation starts, the oxide growth rate rapidly decreases and at around 7 nm in thickness, the deceleration rate changes to gentle one (hereafter each oxidation stage is denoted as rapid and gentle deceleration stage, respectively). Because the growth rates at each of the deceleration stage well ride on a straight line in a semi-log plot (shown by dashed lines in tive to the variation of η and η' when these values are large enough to sink all the interstitials that reach the oxide surface.

3.3. Oxidation rate at various oxygen partial pressures

To determine the oxygen pressure dependence of the SiC oxidation process, *ex-situ* measurements from 10^{-3} to 4 atm have been carried out [23, 27]. However, both of these studies did not examine the initial oxidation process in detail, partly because *in-situ* real-time observations were not possible. We performed *in-situ* real-time measurements at reduced partial pressures between 0.1 and 1.0 atm, and found the presence of the initial growth rate enhancement to be similar to the case at atmospheric pressure [10]. Recently, we observed the SiC oxidation process under low oxygen partial pressures down to 0.02 atm to examine the initial stage of oxidation in more detail [11, 30].

	for SiC
$\eta'[cm^3 \cdot s^{-1}]$	$*5\times10^{-8}$
$D_C[cm^2 \cdot s^{-1}]$	$2.0\times10^{-9}\exp(-0.57eV/k_BT)$
$\kappa_1'[cm^3 \cdot s^{-1}]$	$1.04\times10^{-15}\exp(-1.55eV/k_BT)$
$\kappa_2'[cm^6 \cdot s^{-1}]$	$1.04\times10^{-32}\exp(-1.55eV/k_BT)$
$C_C^0[cm^{-3}]$	$3.6\times10^{23}\exp(-1.07eV/k_BT)$
	common values
$\eta[cm^3 \cdot s^{-1}]$	5×10^{-8}
$D_{Si}[cm^2 \cdot s^{-1}]$	$8.1\times10^{-2}\exp(-3.43eV/k_BT)$
$\kappa_1[cm^3 \cdot s^{-1}]$	$1.46\times10^{-14}\exp(-1.55eV/k_BT)$
$\kappa_2[cm^6 \cdot s^{-1}]$	$1.46\times10^{-31}\exp(-1.55eV/k_BT)$
$D_O[cm^2 \cdot s^{-1}]$	$1.3\times10^{-2}\exp(-1.64eV/k_BT)$
$C_{Si}^0[cm^{-3}]$	$3.6\times10^{24}\exp(-1.07eV/k_BT)$
$C_O^0[cm^{-3}]$	5.5×10^{16}
*Assumed value	

Table 1. Arrhenius equations used for the calculations of SiC oxide growth rate. The common values for SiC and Si are extracted from the reference [17].

In the oxide growth calculations, we gave the same parameters to C- and Si-face in the case that the parameters are related to SiO_2 (*i.e.* η', D_C, κ_1', κ_2', andC_C^0). Through the curve fits for various oxidation temperatures, we obtained their Arrhenius equations. Table 1 addresses the Arrhenius equations for these common parameters such as D_C as well as those related to the properties of SiO_2 from Ref. [17]. It is noticed that the self-diffusivities of Si and O (D_{Si}^{SD} andD_O^{SD}, respectively, in Ref. [17]) have been transformed to ordinary diffusivities. Looking at the table, the activation energy of D_C (0.57 eV) seems to be small. Indeed, *ab-initio* studies have reported the energy between 2-3 eV [29]. We suppose for this discrepancy that our obtained value originates from not a segregation but a migration, that is, it is the diffusivity for C atoms that diffuse as interstitials. Note that the calculated growth rates are quite insensi-

Fig. 7) in respective stage, the oxide thickness dependence of oxide growth rate can be approximated by a sum of the two exponential functions [11], as,

$$\frac{dX}{dt} = R_r \exp(-X / L_r) + R_g \exp(-X / L_g) \tag{11}$$

where, R_r and R_g ($R_r \gg R_g$) are pre-exponential constants, and L_r and L_g ($L_r < L_g$) are characteristic lengths for the deceleration of oxide growth rate in each oxidation stage. The first and the second terms represent the rapid and gentle deceleration stage, respectively. Equation (11) means that, in the thin oxide regime, the oxide growth occurs by two ways and these proceed not in series but in parallel because the growth rate is given by the sum of two terms and is determined mainly by the faster one in each stage. Obviously, the L_r and L_g values correspond to the gradients of the fitted line in the rapid and gentle deceleration stage, respectively. As shown in Fig. 7, the L_r value decreases with decreasing partial pressure, which corresponds to the more remarkable rapid deceleration. In contrast, the L_g value is almost constant regardless of the partial pressure. This suggests that the oxidation process is different between the rapid and gentle deceleration stage. It is noted that the thickness at which the deceleration rate changes from rapid one to gentle one (termed 'X_c') is almost constant around 7 nm regardless of oxygen partial pressure and surface polarity. In the case of Si oxidation, a rapid deceleration stage has also been observed just after oxidation starts, and the thickness corresponding to X_c is also almost independent of the oxygen partial pressure, though the growth rates at X_c depend on the oxygen partial pressure [13, 31]. Therefore, it can be stated that X_c is determined only by the thickness of the oxide layer for both the Si and SiC oxidation cases.

As mentioned above, the existence of a rapid deceleration stage in the oxide growth rate just after oxidation starts (X<10 nm) has been observed also for Si oxidation [13, 31]. However, in the investigations on Si oxidation mechanisms, the cause for the rapid deceleration has not been clarified yet, so far. For SiC oxidation, Yamamoto *et al.* have tried to reproduce the observed data using the Massoud's empirical equation [13]. Here, we will discuss the reasons why two deceleration stages exist in the thickness dependence of oxide growth rate, based on the Si and C emission model [19].

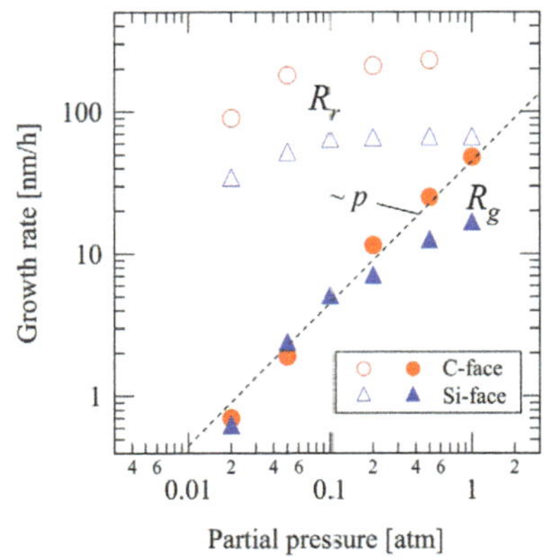

Figure 8. Oxygen partial pressure dependence of the initial growth rate, R_r(unfilled symbols), and the gentle deceleration growth rate, R_g(filled symbols), on C-face (circles) and Si-face (triangles).

We fitted the experimental data at each partial pressure with two straight lines, as shown by the dashed lines in Fig. 7, and derived the initial growth rate of the two deceleration stages, $R_r$andR_g, by extrapolating the straight line to X=0 in the rapid and the gentle deceleration stage, respectively. It is noticed that the meanings of R_r and R_g are the same as those in eq. (11). Figure 8 shows the oxygen partial pressure dependence of R_r and R_g, denoted by unfilled and filled symbols, respectively, on the C- (circles) and Si-faces (triangles). Since the oxide growth in thin thickness region was too fast to follow spectroscopic ellipsometry measurements in the case of 1 atm on C-face, we could not obtain the values of oxide growth rate in the rapid deceleration stage accurately, and thus the value of R_r for C face at 1 atm was not shown in this figure. The broken line in Fig. 8 shows the line proportional to oxygen partial pressure and fitted to the R_g data for C-face. For the both polar faces, the data points of R_g ride almost on the line, suggesting that R_g is proportional to partial pressure, though, for Si-face, $R_g$becomes slightly smaller as seen from the linear relation approaching 1 atm. It should be noted that the rates are almost equal to each other between C- and Si-faces at low pressure region, which is strangely different from the fact that the oxide growth rates for C-face are about eight times larger than those for Si-face at atmospheric oxygen pressure. If the oxide grows chiefly at the interface, the oxide growth rate dX/dt is rewritten by using eq. (6) as follows:

$$N_0 \frac{dX}{dt} \approx k C_{\mathrm{O}}^{\mathrm{I}} \tag{12}$$

The value of k is unlikely to depend on oxygen partial pressure because it corresponds to the rate that one SiC molecule is changed to one SiO_2 molecule, which should not depend on p. In the thin oxide regime discussed here, the interface oxygen concentration $C_{\mathrm{O}}^{\mathrm{I}}$ can be expressed as $C_O^I \propto pC_O^0$ by Henry's low. Therefore, R, the growth rate when the oxide thickness X approaches 0, should be proportional to p, which is in good agreement with the experimental results in the gentle deceleration stage, *i.e.* R_g. Based on the Si emission model for Si oxidation (the Si and C emission model for SiC), the interface reaction rate k decreases with proceeding of oxidation due to the accumulation(s) of Si interstitials (Si and C interstitials). As the number of the accumulated atoms should increase with the proceeding of oxidation and thus is proportional to the quantity of oxidized molecules, *i.e.*, thickness of the oxide X, the variation in k may be approximately given as an exponential function of X in the form of $C\exp(-X / L)$, where C and L are the pre-exponential term and characteristic length, respectively, related to the accumulation(s) of Si (Si and C) interstitials at the interface. From these considerations as well as the fact that R_g is proportional to p, the gentle deceleration of oxide growth rate can be attributed to the accumulation(s) of Si (Si and C) interstitials near the interface, and given approximately as $dX / dt \propto R_g \exp(-X / L_g)$, which is coincident with the second term in eq. (11).

If the initial growth rate R_r in the rapid deceleration stage is also followed by eq. (12), it can be expressed that $R_r = k_0' C_O^I / N_0$, where k_0' is the interfacial reaction rate when oxidation starts. As the value of k_0' is also unlikely to depend on oxygen partial pressure, R_r should be also proportional to oxygen pressure. While, as seen in Fig. 8, R_r is not proportional to p, though it decreases with decreasing p in the low p region. Also in the case of Si oxidation, the experimental data show almost no dependency of R_r with respect to p [13]. Here, we consider the reason why R_r is not proportional to but almost independent of oxygen partial pressure both for Si and SiC oxidation.

It has been considered that oxide growth occurs only or mainly at the Si-oxide (SiC-oxide) interface, so far. However, according to the interfacial Si emission model [14, 17] for Si oxidation and the Si and C emission model [19] for SiC oxidation, Si atoms (Si and C atoms) emit into the oxide layer, and some of which meet with oxidant inside the oxide to form SiO_2. When the thickness of the oxide is very thin, a part of the emitted Si atoms can go through the oxide layer and reach the oxide surface, and then are instantly oxidized, resulting in the formation of a SiO_2 layer on the oxide surface. Therefore, there are two oxide growth processes other than the interfacial oxide growth, *i.e.*, the oxide formation due to the oxidations of Si interstitials on the oxide surface and inside the oxide, and the total growth rate is given by the sum of these three oxidation processes, as revealed in eq. (6). In the case of oxidation inside the oxide, the possibility that an emitted Si interstitial encounter oxygen inside the oxide should be proportional to the oxygen concentration in the oxide. Therefore, this oxidation process should be proportional to p like R_g, and thus, can be excluded as a candidate of the origin of R_r. In contrast, in the case of oxidation on the oxide surface, the amount of oxygen is thought to be enough to oxidize all the Si atoms emitted and appearing on the surface, because the number of oxygen molecules impinging onto the surface from the gaseous atmosphere is several orders larger than that of emitted Si atoms transmitted through oxide even if the oxygen pressure is as low as 0.02 atm[1]. Therefore, the oxide growth rate for the oxidation on the oxide surface should be independent of the oxygen partial pressure, which is in good agreement with the behavior of R_r. Besides, the possibility that Si interstitials go through the oxide and reach the oxide surface is considered to decrease rapidly with increasing oxide thickness, and can be given as a form of $\exp(-X / L_r)$, where L_r ($< L_g$) is the escape depth of Si atoms from the oxide layer. From these considerations, the rapid deceleration stage of oxide growth rate observed just after oxidation starts is thought to be due to oxidation of Si interstitials on the oxide surface. The value of X_c obtained from the experiments, around 7 nm, indicates that the escape depth of Si from the oxide is estimated to be several nm at 1100°C.

As has been mentioned above, in the rapid deceleration stage, the growth rate is determined with the oxidation rate of the emitted Si interstitials on the oxide surface (termed 'surface oxide growth'), while in the gentle deceleration stage, it is determined by the oxidation rate at the SiC-oxide interface and that of the emitted Si interstitials inside the SiO_2 layer (termed 'interfacial oxide growth' and 'internal oxide growth', respectively). The surface oxide growth rate depends little on partial pressure and, in contrast, the interfacial and internal oxide growth rates are proportional to partial pressure. It is therefore predicted from these pressure dependence that the rapid deceleration becomes more remarkable (*i.e.* the smaller L_r) at lower partial pressures, which is confirmed in Fig. 7. In addition, the reason why the growth rates in the rapid deceleration stage at low pressures are not so much different between C- and Si-faces is that the surface oxide growth is dominant to the oxide growth in this stage, so that the oxidation on the oxide surface may proceed independently of the surface polarity.

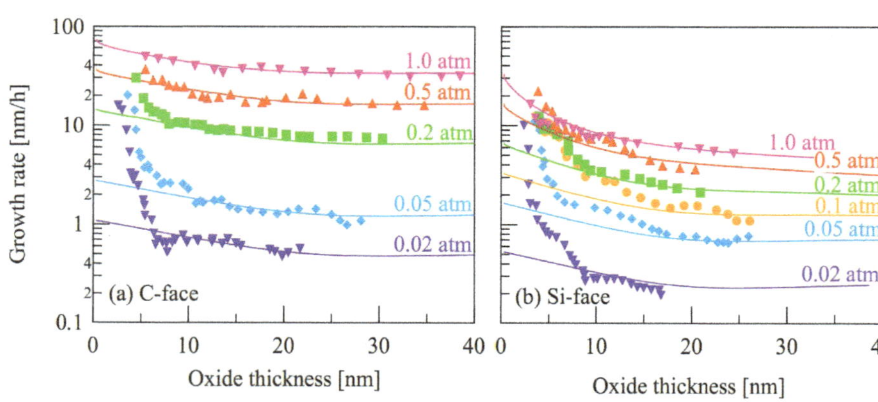

Figure 9. Oxide growth rates as a function of oxide thickness at various oxygen partial pressures on (000$\bar{1}$) C-face (a) and (0001) Si-face (b). The solid lines denote growth rates given by the Si and C emission model [19].

---

1 The oxygen flux impinging from a gaseous atmosphere of pressure p to the solid surface is $3\mathrm{x}10^{22}p[\mathrm{m}^{-2}\mathrm{s}^{-1}]$. Since the areal density of Si atoms on SiO_2 is about $8\mathrm{x}10^{18}[\mathrm{m}^{-2}]$, the flux in the case that p=0.02 atm corresponds to 75 [monolayer/s], which is the oxygen flux necessary for the oxide growth rate of 105 nm/h.

Figure 9 compares the observed oxide growth rates at various oxygen partial pressures on SiC (000$\bar{1}$) C-face (a) and (0001) Si-face (b) with the growth rates given by the Si and C emission model (solid lines). It is noted that all the parameters used in the calculations other than the solubility limit of oxygen in SiO_2 [2] are the same regardless of partial pressures. The figures indicate that the calculated curves successfully reproduce the observed growth rates though restricted in the gentle deceleration region and that the calculated curves show the rapid reduction more remarkable as the higher partial pressure, which is opposite tendency to that of the observed data. Also in the case of Si oxidation, the interfacial Si emission model [14, 17] cannot reproduce the growth rate in the thin oxide region at sub-atmospheric pressure, as pointed out by Farjas and Roura [31]. We consider that the accurate description for the surface oxide growth may dissolve the disagreement between the calculated and observed oxide growth rates in the cases of low partial pressures. To verify this idea, we separately calculated the oxide growth rate on the surface, inside the SiO_2 layer, and at the SiC-SiO_2 interface (denoted asF_{sur}, F_{ox}, andF_{int}, respectively) using the Si and C emission model.

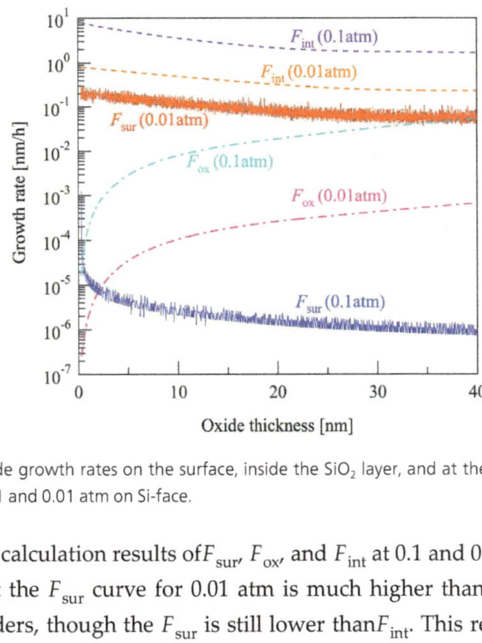

Figure 10. Simulated oxide growth rates on the surface, inside the SiO_2 layer, and at the SiC-SiO_2 interface (F_{sur}, F_{ox}, andF_{int}, respectively) at 0.1 and 0.01 atm on Si-face.

Figure 10 shows the calculation results ofF_{sur}, F_{ox}, and F_{int} at 0.1 and 0.01 atm on Si-face. The figure indicates that the F_{sur} curve for 0.01 atm is much higher than that for 0.1 atm by a factor of about 5 orders, though the F_{sur} is still lower thanF_{int}. This result confirms that the low pressure oxidation enhances the oxide growth on the surface, as discussed above, which is probably due to the reduction in the oxygen concentration at the interface and inside the oxide. Conversely, the F_{ox} and F_{int} curves for 0.01 atm are lower than those for 0.1 atm by factors of 2 orders and 1 order, respectively. Since the F_{int} is higher than F_{ox} in general, it is

2 The corresponding solubility limit of oxygen is derived by multiplying partial pressure by that for 1 atm, *i.e.*, $C_O^0 = p \times C_O^0$ (1 atm).

found that the growth rate in the gentle deceleration stage to be proportional to pressure is due to the proportional increase inF_{int}. Moreover, a careful look confirms the deceleration of F_{int} in the very-thin oxide region (*ca.*< 5 nm) to be enhanced as elevating the pressure, which has been seen in the calculated curves in Fig. 9. Therefore, it is necessary for the Si and C emission model, as well as for the interfacial Si emission model that an appropriate description for the oxide growth on the surface is introduced to the model so as to enhance the surface growth rate in the initial oxidation region up to over the interfacial growth rate.

3.4. Discussion on Si and C emission phenomenon and its relation to the interface layer

So far, we specify the oxidation mechanism of SiC from the viewpoint of the Si and C emission phenomenon. In this section, we will discuss the structure and the formation mechanism of the interface layer between SiC-SiO_2, which has been considered as the crucial issue for the practical use of SiC-MOSFET, in the light of interfacial Si and C emission.

In previous work, we have found that [32, 33] the photon energy dependence of the optical constants *n* and *k* of the interface layer derived from the complex dielectric constants between 2 and 6 eV, covering the direct interband transition energy E_0 of 4H-SiC of 5.65 eV, is similar to that of bulk 4H-SiC, though the absolute values of *n* are about 1 larger than those of SiC [34]. We also found from the real time observation of SiC oxidation process using an *in-situ* spectroscopic ellipsometry that [30] the interface layer thickness increases with increasing oxide thickness and saturates around 1.5 nm at the oxide thickness of around 7 nm, and the refractive indices of the layer are also saturated at around 7 nm in oxide thickness to the similar values obtained from *ex-situ* measurements. The similarity in the energy dispersion of the optical constants of the interface layers suggests that the interface layer is not a transition layer between SiC and SiO_2, such as SiO_x or SiC_xO_y. Rather, it is a layer having a modified structure and/or composition compared to SiC, such as a stressed or interstitials-incorporated SiC layer, locating not on the SiO_2 side but the SiC side of the SiC-oxide interface.

According to the results from the real time observation [30], the thickness at which the interface layer thickness and the refractive index become constant (*i.e.*, 7 nm) is determined not from the surface polarity or oxygen partial pressure but from the oxide thickness. The Si and C emission model describes this behavior by considering that Si and C atoms are emitted into both directions of the SiC-oxide interface accompanying oxidation at the interface, *i.e.*, into not only the oxide layer but also the SiC layer (see Fig. 2), and accumulation of interstitial Si and/or C atoms emitted into the SiC substrate may form a layer having similar optical properties as SiC but larger refractive indices compared to SiC. Since accumulation of interstitials is linked to the growth of the oxide, it is considered that growth of the interface layer is saturated at some intrinsic oxide thickness even if the oxygen pressure is changed. Takaku *et al.* have found from the real time observation that the use of relatively low oxidation temperature or oxygen pressure leads to no formation of the interface layer [35], which can be explained by considering that the accumulations of Si and/or C interstitials are prevented by lowering the temperature or pressure.

Hashimoto *et al.* have found that there is a strong correspondence between the refractive index of the interface layer and interface state density obtained from the MOS capacitors fabricated on the same sample for ellipsometry measurements [32]. Hence, we believed that the accumulations of Si and C interstitials near/at the interface are closely related to the formation of interface states. According to the report from Afanasev *et al.* [36], a graphite-like carbon layer forms near/at the interface, which causes interface states over the whole range of forbidden energy band and, in addition, an intrinsic SiO_2 defect (NIT: near interface trap), presumably originating from oxygen deficiency, exists regardless of Si or SiC-polytypes and gives rise to the interface states in the vicinity of the 4H-SiC conduction band edge, which cause a channel degradation in MOSFETs. Since the oxygen deficiency can be regarded as a Si interstitial in SiO_2, Si (Si and C) emission into SiO_2 layer may be the origin for this SiO_2 defect. According to the *ab-initio* studies performed by Knaup *et al.* [37], the origin of NITs is Si interstitials or C dimers originating from C interstitials inside SiO_2. They also reported that C dimers in the SiC-side give rise to interface states at the energy range between mid gap and valence band edge. On the other hand, according to the *ab-initio* studies from Devynck *et al.* [38], C-C pair in SiC originating from C interstitials and a Si-C-O structure give rise to a broad peak of interface state density, and a Si_2-C-O structure give rise to a sharp peak near the conduction band edge, which is compatible with NIT. Cochrane *et al.* have performed an Electrically detected magnetic resonance (EDMR) techniques and found Si vacancies (or C dangling bonds) in the SiC-side near the interface [40], which is presumably attributed to the Si emission during oxidation. Very recently, Shen and Pantelides have reported that they identified the origin of interface states that degrade the SiC MOS channel mobility as 'C di-interstitials', which are formed by a combination of the two C interstitials injected into the SiC [41]. Accordingly, we tried to simulate the formation of the interface layer based on the diffusion theory and, as a result, the calculations reproduced the interface layer with the thickness of around 1 nm [39], which agrees with the values often reported [18, 34, 39]. Anyway, we would like to emphasize here that further understanding of the interfacial Si and C emission phenomenon during oxidation of SiC might be the key to realize the intrinsic performances of SiC MOSFETs.

Finally, we would like to introduce a recent innovation using oxidation about elimination of point defects. Hiyoshi and Kimoto invented an epoch-making idea, in which as-grown deep-level states such as a $Z_{1/2}$ center can be eliminated by oxidation of SiC substrates [43]. Taking into account the report from Storasta *et al.* [44] in which the origin of $Z_{1/2}$ is perhaps C vacancy, Hiyoshi and Kimoto considered the mechanism of deep-level state elimination by oxidation as follows: The excess silicon and carbon interstitials may be generated at the oxidizing interface, and a part of the emitted silicon and carbon interstitials may diffuse into the epilayer, and consequently, recombination of the interstitials with vacancies present in the epilayers may take place [45]. Therefore, the Si and C emission surely takes place during oxidation and the survey of Si and C interstitial movement is very important to utilize thermal oxidation as a device fabrication process, *e.g.* Ref. [46], even in the developments in bipolar devices of over 10 kV class. Very recently, we found that stacking-faults are formed or extended by thermal oxidation, which is perhaps induced by the interfacial strain due to thermal oxidation [47].

4. Conclusion

We review the oxidation mechanism of SiC and the formation of the interface layer between SiC and SiO_2. Though most of articles have reported that the Deal-Grove model also describes the SiC oxidation process, we pointed out the discrepancy in the application of Deal-Grove model, in which the oxide growth rate in the thin oxide region cannot be reproduced with this model. Aimed at the elucidation of SiC oxidation process in this thin oxide region, we proposed a kinetic model that accounts for SiC oxidation, termed the 'Si and C emission model', and showed that the model well reproduces the oxide growth rate over the entire thickness range including the thin oxide region for both the C and Si faces. The results indicated that the oxidation and emission of C and the emission of Si all need to be taken into account to describe the oxide growth process in SiC. A comparison of the parameters obtained for the C and Si faces from the curve fits revealed that the differences in initial interfacial reaction rate and Si emission ratio are contributing to the large difference in oxide growth rate between these polar faces.

We tried to apply the Si and C emission model to the oxide growth rate data at various oxidation temperatures and found that the model reproduces the oxide growth rate curves for all of the temperatures measured for both of the Si- and C-faces. Comparing with the parameters deduced from the curve fits, we discussed the differences in oxidation process between Si- and C-face. We also showed the parameters obtained in this study, some of which such as a diffusivity and solubility limit of C in SiO_2 were firstly reported.

We have studied the oxygen partial pressure dependence of the SiC oxidation process on the Si- and C-faces. The oxide thickness dependence of the growth rate at sub-atmospheric oxygen partial pressures down to 0.02 atm show that, just after oxidation starts, the oxide growth rate rapidly decreases and the deceleration-rate changes to a gentle mode at around 7 nm in oxide thickness, which are probably the oxide surface growth mode and oxide internal/interfacial growth mode, respectively. We tried to reproduce the pressure dependence of oxide growth rates, however, it cannot be achieved, which is perhaps due to the inaccurate description for the oxide growth on the oxide surface.

Finally, we discussed the structure and formation mechanism of the SiC–oxide interface layer in terms of the Si and C emission phenomenon. We also introduced a recently invented technology on defect using oxidation, *i.e.* point defect elimination by oxidation, and suggested that understanding of oxidation mechanism is also very important in the developments of SiC bipolar devices.

Acknowledgements

The authors would like to thank Professor Uematsu of Keio University and Dr. Kageshima of NTT for their helpful advice and technical support for the simulation. This work was partially supported by a Grant-in-Aid for Scientific Research (24560365) from the Japan Society for the Promotion of Science.

Author details

Yasuto Hijikata[1*], Shuhei Yagi[1], Hiroyuki Yaguchi[1] and Sadafumi Yoshida[2]

*Address all correspondence to: yasuto@opt.ees.saitama-u.ac.jp

1 Division of Mathematics Electronics and Informatics, Graduate School of Science and Engineering, Saitama University, Japan

2 Advanced Power Electronics Research Center, National Institute of Advanced Industrial Science and Technology, Japan

References

[1] Yoshida, S. (2000). *Electric Refractory Materials*, ed. Y. Kumashiro (Dekker, New York), 437.

[2] Matsunami, H. (2004). Technological Breakthroughs in Growth Control of Silicon Carbide for High Power Electronic Devices. *Jpn. J. Appl. Phys.*, 43(1), 6835-6847.

[3] Afanas'ev, V. V., & Stesmans, A. (1997). H-complexed oxygen vacancy in SiO_2: Energy level of a negatively charged state. *Appl. Phys. Lett.*, 71, 3844-3846.

[4] Hijikata, Y., Yaguchi, H., & Yoshida, S. (2011). *Properties and Applications of Silicon Carbide*, ed. R. Gerhardt (INTECH open access publisher, 2011) chapter 4.

[5] Kakubari, K., Kuboki, R., Hijikata, Y., Yaguchi, H., & Yoshida, S. (2006). Real Time Observation of SiC Oxidation using In-Situ Ellipsometer. *Mater. Sci. Forum*, 527-529, 1031-1034.

[6] Deal, B. E., & Grove, A. S. (1965). General Relation Ship for the Thermal Oxidation of Silicon. *J. Appl. Phys.*, 36, 3770-3778.

[7] Song, Y., Dhar, S., Feldman, L. C., Chung, G., & Williams, J. R. (2004). Modified Deal Grove model for the thermal oxidation of silicon carbide. *J. Appl. Phys.*, 95, 4953-4957.

[8] Yamamoto, T., Hijikata, Y., Yaguchi, H., & Yoshida, S. (2007). Growth Rate Enhancement of (0001^-)-Face Silicon-carbide Oxidation in Thin Oxide Regime. *Jpn. J. Appl. Phys*, 46, L770-L772.

[9] Yamamoto, T., Hijikata, Y., Yaguchi, H., & Yoshida, S. (2008). Oxide Growth Rate Enhancement of Silicon Carbide (0001) Si-Faces in Thin Oxide Regime. *Jpn. J. Appl. Phys.*, 47, 7803-7806.

[10] Yamamoto, T., Hijikata, Y., Yaguchi, H., & Yoshida, S. (2008). Oxygen-Partial-Pressure Dependence of SiC Oxidation Rate Studied by In-situ Spectroscopic Ellipsometry. *Mater. Sci. Forum*, 600-603, 667-670.

[11] Kouda, K., Hijikata, Y., Yaguchi, H., & Yoshida, S. (2010). In-situ Spectroscopic Ellipsometry Study of SiC Oxidation at Low Oxygen-Partial-Pressures. *Mater. Sci. Forum*, 645-648, 813-816.

[12] Grove, A. S. (1967). *Physics and Technology of Semiconductor Devices*, John Wiley & Sons, New York, 31.

[13] Massoud, H. Z., Plummer, J. D., & Irene, E. A. (1985). Thermal Oxidation of Silicon in Dry Oxygen: Growth-Rate Enhancement in the Thin Regime: I. Experimental Results. *J. Electrochem. Soc.*, 132, 2685-2693, Thermal Oxidation of Silicon in Dry Oxygen: Growth-Rate Enhancement in the Thin Regime: II. Physical Mechanism, ibid 132, 2693-2700.

[14] Kageshima, H., Shiraishi, K., & Uematsu, M. (1999). Universal Theory of Si Oxidation Rate and Importance of Interfacial Si Emission. *Jpn. J. Appl. Phys.*, 38(2), L971-L974.

[15] Ogawa, S., & Takakuwa, Y. (2006). Rate-Limiting Reactions of Growth and Decomposition Kinetics of Very Thin Oxides on Si(001) Surfaces Studied by Reflection High-Energy Electron Diffraction Combined with Auger Electron Spectroscopy. *Jpn. J. Appl. Phys.*, 45, 7063-7079.

[16] Watanabe, T., Tatsumura, K., & Ohdomari, I. (2006). New Linear-Parabolic Rate Equation for Thermal Oxidation of Silicon. *Phys. Rev. Lett.*, 96, 196102-1-196102-4.

[17] Uematsu, M., Kageshima, H., & Shiraishi, K. (2001). Simulation of wet oxidation of silicon based on the interfacial silicon emission model and comparison with dry oxidation. *J. Appl. Phys.*, 89, 1948-1953.

[18] Hijikata, Y., Yaguchi, H., Yoshida, S., Takata, Y., Kobayashi, K., Nohira, H., & Hattori, T. (2006). Characterization of Oxide Films on 4H-SiC Epitaxial ($000\bar{1}$) Faces by High-Energy-Resolution Photoemission Spectroscopy: Comparison between Wet and Dry Oxidation. *J. Appl. Phys.*, 100, 053710-1-053710-6.

[19] Hijikata, Y., Yaguchi, H., & Yoshida, S. (2009). A Kinetic Model of Silicon Carbide Oxidation Based on the Interfacial Silicon and Carbon Emission Phenomenon. *Appl. Phys. Express*, 2, 021203-1-021203-3.

[20] Uematsu, M., Kageshima, H., & Shiraishi, K. (2000). Simulation of High-Pressure Oxidation of Silicon Based on the Interfacial Silicon Emission Model. *Jpn. J. Appl. Phys.*, 39, L952-L954.

[21] Sze, S. M. (2002). *Semiconductor Devices Physics and Technology*, John Wiley & Sons, Inc., New Jersey, 2nd ed., Chap. 13, 457.

[22] Jügling, W., Pichler, P., Selberherr, S., Guerrero, E., & Pötzel, H. W. (1985). Simulation of Critical IC Fabrication Processes Using Advanced Physical and Numerical Methods. *IEEE Trans. Electron. Devices*, 32, 156-167.

[23] Ray, E. A., Rozen, J., Dhar, S., Feldman, L. C., & Williams, J. R. (2008). Pressure dependence of SiO_2 growth kinetics and electrical properties on SiC. *J. Appl. Phys.*, 103, 023522-1-023522-7.

[24] Hijikata, Y., Yamamoto, T., Yaguchi, H., & Yoshida, S. (2009). Model Calculation of SiC Oxidation Rates in the Thin Oxide Regime. *Mater. Sci. Forum*, 600-603, 663-666.

[25] Hijikata, Y., Yaguchi, H., & Yoshida, S. (2010). Model calculations of SiC oxide growth rate at various oxidation temperatures based on the silicon and carbon emission model. *Mater. Sci. Forum*, 645-648, 809-812.

[26] Szilágyi, E., Petrik, P., Lohner, T., Koós, A. A., Fried, M., & Battistig, G. (2008). Oxidation of SiC investigated by ellipsometry and Rutherford backsccatering spectrometry. *J. Appl. Phys.*, 104, 014903-1-014903-7.

[27] Hijikata, Y., Yaguchi, H., & Yoshida, S. (2009). Model Calculation of SiC Oxide Growth Rate based on the Silicon and Carbon Emission Model. *Mater. Sci. Forum*, 615-617, 489-492.

[28] Zheng, Z., Tressler, R. E., & Spear, K. E. (1990). Oxidation of Single-Crystal Silicon Carbide. *J. Electrochem. Soc.*, 137, 854-858.

[29] Krafcsik, O. H., Vida, G., Pócsik, I., Josepovits, K. V., & Deák, P. (2001). Carbon Diffusion through SiO_2 from a Hydrogenated Amorphous Carbon Layer and Accumulation at the SiO_2 / Si Interface. *Jpn. J. Appl. Phys.*, 40, 2197-2200.

[30] Kouda, K., Hijikata, Y., Yagi, S., Yaguchi, H., & Yoshida, S. Oxygen partial pressure dependence of the SiC oxidation process studied by in-situ spectroscopic ellipsometry. *J. Appl. Phys*, accepted for publication.

[31] Farjas, J., & Roura, P. (2007). Oxidation of silicon: Further tests for the interfacial silicon emission model. *J. Appl. Phys.*, 102, 054902-1-054902-8.

[32] Hashimoto, H., Hijikata, Y., Yaguchi, H., & Yoshida, S. (2009). Optical and Electrical Characterizations of 4H-SiC-Oxide interfaces by Spectroscopic Ellipsometry and Capacitance-Voltage measurements. *Appl. Surf. Sci.*, 255, 8648-8653.

[33] Seki, H., Hijikata, Y., Yaguchi, H., & Yoshida, S. (2009). Characterization of 4H-SiC-SiO_2 Interfaces by a Deep Ultraviolet Spectroscopic Ellipsometer. *Mater. Sci. Forum*, 615-617, 505-508.

[34] Iida, T., Tomioka, Y., Midorikawa, M., Tsukada, H., Orihara, M., Hijikata, Y., Yaguchi, H., Yoshikawa, M., Itoh, H., Ishida, Y., & Yoshida, S. (2002). Measurement of the depth profile of the refractive indices in the oxide films on SiC by spectroscopic ellipsometry. *Jpn. J. Appl. Phys.*, 41(1), 800-804.

[35] Takaku, T., Hijikata, Y., Yaguchi, H., & Yoshida, S. (2009). Observation of SiC Oxidation in Ultra-Thin Oxide Regime by In-situ Spectroscopic Ellipsometry. *Mater. Sci. Forum*, 615-617, 509-512.

[36] Afanasev, V. V., Bassler, M., Pensl, G., & Shulz, M. (1997). Intrinsic SiC/SiO_2 Interface States. *Phys. stat. sol. (a)*, 162, 321-337.

[37] Knaup, J. M., Deák, P., Frauenheim, Th., Gari, A., Hajnal, Z., & Choyke, W. J. (2005). Theoretical study of the mechanism of dry oxidation of 4H-SiC. *Phys. Rev. B*, 71, pp. 235321-1-9, Defects in SiO_2 as the possible origin of near interface traps in the SiC-SiO_2 system: A systematic theoretical study, ibid, 72, 115323-1-9.

[38] Devynck, F., Alkauskas, A., Broqvist, P., & Pasquarello, A. (2011). Defect levels of carbon-related defects at the SiC/SiO_2 interface from hybrid functionals. *Phys. Rev. B*, 83, pp. 195319-1-11, Charge transition levels of carbon-, oxygen-, and hydrogen-related defects at the SiC/SiO_2 interface through hybrid functionals, ibid, 84, 235320-1-18.

[39] Cochrane, C. J., Lenahan, P. M., & Lelis, A. J. (2011). An electrically detected magnetic resonance study of performance limiting defects in SiC metal oxide semiconductor field effect transistors. *J. Appl. Phys.*, 109, 014506-1-014506-12.

[40] Shen, X., & Pantelides, S. T. (2011). Identification of a major cause of endemically poor mobilities SiC/SiO_2 structures. *Appl. Phys. Lett.*, 98, 0535071-1-0535071-3.

[41] Hijikata, Y., Yaguchi, H., & Yoshida, S. (2011). Theoretical studies for Si and C emission into SiC layer during oxidation. *Mater. Sci. Forum*, 679-680, 429-432.

[42] Hijikata, Y., Yaguchi, H., Yoshikawa, M., & Yoshida, S. (2001). Composition analysis of SiO_2/SiC interfaces by electron spectroscopic measurements using slope-shaped oxide films. *Appl. Surf. Sci.*, 184, 161-166.

[43] Hiyoshi, T., & Kimoto, T. (2009). Reduction of Deep Levels and Improvement of Carrier Lifetime in n-Type 4H-SiC by Thermal Oxidation. *Appl. Phys. Express*, 2, pp. 041101-1-3.

[44] Storasta, L., Tsuchida, H., & Miyazawa, T. (2008). Enhanced annealing of the $Z_{1/2}$ defect in 4H-SiC epilayers. *J. Appl. Phys.*, 103, 013705-1-013705-7.

[45] Hiyoshi, T., & Kimoto, T. (2009). Elimination of the Major Deep Levels in n- and p-Type 4H-SiC by Two-Step Thermal Treatment. *Appl. Phys. Express*, 2, 091101-1-091101-3.

[46] Kawahara, K., Suda, J., & Kimoto, T. (2012). Analytical model for reduction of deep levels in SiC by thermal oxidation. *J. Appl. Phys.*, 111, 053710-1-053710-9.

[47] Yamagata, H., Yagi, S., Hijikata, Y., & Yaguchi, H. (2012). Micro-Photoluminescence study on the influence of oxidation on stacking faults in 4H-SiC epilayers. *Appl. Phys. Express.*, 5, 051302-1-051302-3.

Chapter 9

Materials and Processing for Gate Dielectrics on Silicon Carbide (SiC) Surface

Sanjeev Kumar Gupta, Jitendra Singh and Jamil Akhtar

Additional information is available at the end of the chapter

1. Introduction

Dielectrics are the materials that do not conduct current in the presence of an electric field. The applications of this material in semiconductor industries are very broad in various capacities. Nowadays, extensive research and development (R&D) are in progress to grow high quality high-k gate dielectrics on semiconductors surface. Applications include extending the limit of transistor gate capacitance beyond that of ultra thin silicon dioxide (SiO_2) and to improve the gate dielectric reliability in wide band gap semiconductor devices. SiO_2 is one of the best gate dielectric, which is continuously investigated rigorously since long lime back for silicon based metal-oxide-semiconductor (MOS) device technology. The thermally grown SiO_2 offers several key advantages in microelectronics device processing including thermodynamically and electrically stable high quality interface state density as well as better electric insulation properties. New materials are endlessly researched to fulfill the limitation of silicon across a wide spectrum of industrial applications.

Silicon Carbide (SiC) has been proven to be most suitable material, offering significant potential advantages both in high temperature as well as high power device technology. Moreover, SiC is the only material that can be thermally oxidized to grow high quality SiO_2, which enables to fabricate the MOS structures. A large drawback of SiO_2 is its low dielectric constant, which is about 2.5 times lower than that of SiC and also poorer interface properties at SiO_2/SiC. This causes a proportionally large electric field enhancement in the dielectric compared to that in the semiconductor, which is a reason why new dielectrics with dielectric constant at least similar to that of SiC and lower interface states densities are desired for device applications. There have been few successful high-k dielectrics including silicon nitride (Si_3N_4), Oxynitride, aluminum nitride (AlN), hafnium dioxide (HfO_2), tin oxide (SnO_2), cesium oxide (CeO_2), titanium oxide (TiO_2), tungsten oxide (WO_3), aluminum oxide (Al_2O_3), lanthanum oxide (La_2O_3), gadolinium oxide and others have been attempted in SiC technology. This chapter covers the selection of gate dielectrics, their processing, interface properties, their electronic structure, flat band voltage shifts and electronic defects.

2. Selection of gate dielectrics

Most of the gate dielectrics material fallows the trend of decrement in their bandgap energy, when the value of their dielectric constant goes towards higher end (high-k). Therefore, the band offset alignment at the interface of gate dielectric and SiC interfaces is an important issue while integrating a high-k gate dielectric in SiC-based devices because SiC belongs to the family of wide gap semiconductor (>3 eV). Figure 1 shows the tendency of bandgap energy as a function of dielectric constant. The bandgap energy of dielectric material has a direct correlation to the leakage current through the band edge offset. Wider bandgap energy means a better chance for a larger conduction or valence band offsets at the interface of semiconductor and the gate dielectric. Furthermore, the band gap energy of high-k oxides (5-6 eV) is modest as an insulator, which may result in large leakage current because of insufficient barrier height at interface. The value of low band offsets at the high-k/SiC interface may be overcome by introducing an ultrathin SiO2 interfacial layer in between dielectric and SiC layer (Mahapatra R., 2008). Figure 2 shows the energy band diagram of metal-gate dielectric-SiC structures as well as stack layers of HfO_2/SiO_2/4H-SiC. In figure 2, HfO_2 was considered as gate dielectric while 4H-SiC was the base substrate. In a situation when the theoretical calculations of energy band diagrams are not available, nor there are measured differences between insulator and SiC energy bands, the size of the bandgap allows for a rough evaluation of the probable usability of the gate dielectric. In ideal case, symmetrical offsets for electrons and holes of the order of 2 eV, the bandgap energy of the gate dielectric should be at least 7 eV. The mechanical and structural properties of ideal gate dielectrics are as important as its electrical performance. The most recommended form of the materials is mono-crystalline structures but those are often grown at very high temperatures and precise pressures. In case of passivation, when the dielectric material is considered to be deposited as the last processing step at low temperatures, the material should have an amorphous composition. This would prevent a possible current conduction through the grain boundaries of the polycrystalline material. Moreover, a lot of research is oriented towards nano-crystalline structures and application of those could provide materials with e.g. larger bandgap, modified by small grains dimension. Other mechanical feature like surface roughness, purity, or right stoichiometry of insulators implies that a good control over the deposition process and correct film uniformity is achieved. To eliminate the mechanical stress caused by device operation at high temperature, the thermal expansion coefficient and thermal conductivity of the dielectric and SiC should be similar. The insulator should also be hard, resistant to cracks and should not be influenced by the surrounding atmosphere.

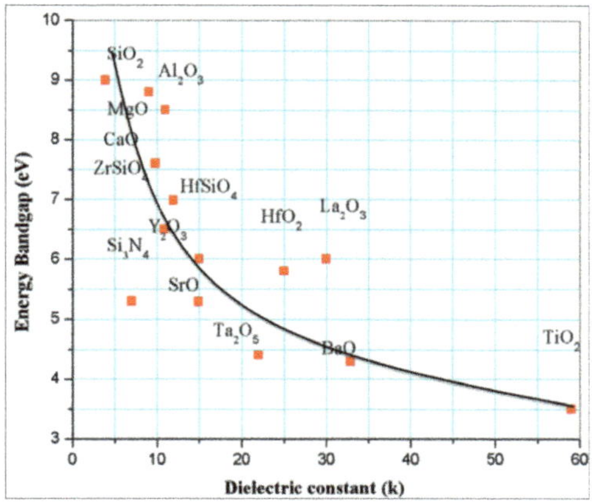

Figure 1. Variation of bandgap energy of different dielectric materials as function of its dielectric constant

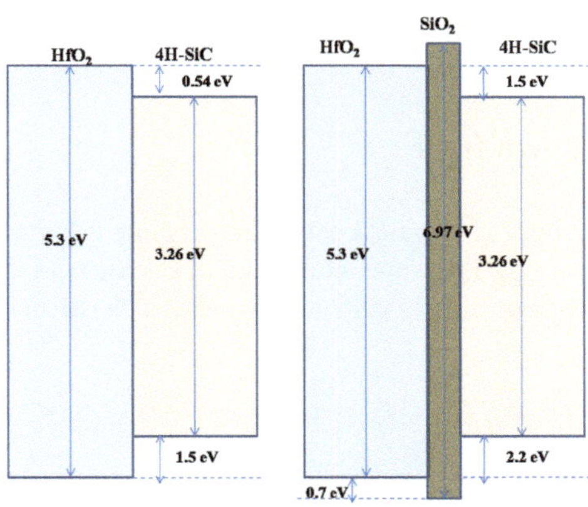

Figure 2. Energy band diagram of HfO_2/4H-SiC and HfO_2/SiO_2/4H-SiC system

However, a significant higher interface state density and inferior electrical properties were found at the SiC/oxide interface because of the interface imperfections. It has been demonstrated by many researchers that a proper annealing of gate dielectrics can reduce the density of traps and passivate the defects level. The process parameters of annealing (temperature, ambient, time etc) can also be a key factor on interface properties. The incorporation of atomic nitrogen shows favorable effects on the structural stability of gate dielectric layers (Chen Q., 2008). At high temperature operation the quality of gate dielectric degrade as the result interface properties show the poor performance. Therefore, the understanding of proper band alignment and the thermal stability at the interface is critical for the application of high-k/SiC-base stacks for high temperature applications (Weng M-H, 2006). Another issue of interest is surface preparation techniques prior to dielectric deposition. There are standard process in Si technology is wet chemical etching in hydrofluridic acid (HF) to remove the native oxide and initial impurities. Unfortunately, this procedure does not provide much satisfactory outcome in the case of SiC as a base material. Many methods are incorporated to clean the SiC surface. The most common method is UV light cleaning, which has a sufficient energy to break carbon clusters that are present on the SiC surface.

Some basic features of gate dielectric, which can be implemented on SiC surface, are as follows

1. The value of dielectric constant (k) must have enough high that may be used for long time of year of scaling.
2. The interface of dielectric layer with SiC surface should thermally stable.
3. It must have a sufficient barrier height and band offset with SiC surface so carrier chare injection into its band can be minimize.
4. It should be compatible with processing technology.
5. It must for good and stable electrical interface with SiC surface.

3. Dielectric breakdown field

A capability to sustain a high electric field without any failure is known a dielectric strength. The dielectric breakdown (E_{BR}) is the maximum limit of electric field that dielectric can tolerate under the influence of high supply voltage. In general dielectric field can be defined as

$$E_{BR} = \frac{V_{BR}}{d} \tag{1}$$

Where V_{BR} is the breakdown voltage and d is the dielectric thickness. Dielectric strength is the inherent phenomenon of dielectric material and it mainly depends on structural properties. In the development of dielectric material, growth condition (material thickness, voltage ramp speed, critical vacuum, growth/deposition rate) and environmental condition (Temperature, humidity) are very important.

4. Materials chemistry of high K oxides

4.1. Silicon dioxide (SiO_2)

A high quality thin SiO_2 is most popular gate dielectric from the SiC based microelectronics industries to make the fabrication process cost effective. Various oxidation processes has been implemented such as dry oxidation, wet oxidation, chemical vapour deposition (CVD), and pyrogenic oxidation in order to achieve the most suitable process to realize the SiC-based MOS structures (Gupta S.K, 2011[a]). This condition produced a lot of effort into the implementation of SiO_2/SiC interface, in the fabrication of MOS transistor. The intricacy of SiO_2/SiC interface, in comparison to the Si based structure, causes severe problems even though the mobility is reduced by 5% of the theoretical value. The best oxide quality is obtained by the means of dry oxidation process performed at temperatures more than 1100 °C. The growth mechanism of oxide on Si substrate is limited by the diffusion of oxygen at SiO_2/Si interface. However, in case of SiC system this diffusion process countenances difficulties because of the presence of C atoms, which are present along with Si atoms. The actual growth mechanism of SiO_2 on SiC surface is not well understood yet. Hypotheses propose migration of free C atoms in almost every direction. The most probable is out-diffusion of CO_2 or CO through the grown silicon oxide but also formation of carbon clusters at the SiO_2/SiC interface and even diffusion of C into bulk SiC are possible. (Song Y., 2004) and his team have proposed a model of the thermal oxide growth on hexagonal SiC in the frame of deal and grove model. The work assumes two competitive processes influencing SiO_2 formation, one is the in-diffusion of oxygen towards the interface and the other one is the out-diffusion of CO. However, by experimental data one cannot prove that some of the carbon atoms do not stay at the interface and form very stable carbon cluster (Kobayashi H. 2003 and Wang S., 2001). Further investigation was also carried out by of thermal oxidation and re-oxidation with different by using different oxygen isotopes, which seems to confirm that unknown carbon structures exist at the interface (Cantin, J.L.2004). Atomic layer deposition (ALD) has proved a potential method for materials deposition (Leskela M., 2002). Using this technique very well controlled growth is possible, almost atomic layer by atomic layer, of the desired species from gaseous precursors. Unluckily, very few publications report efficient SiO_2 deposition using ALD technique on 4H-SiC substrate (Perez I., 2000). (Amy F., 1999) and his co worker has deposited thin Si layer on SiC surface and later thermal oxidation of the Si layer was performed. X-ray photoelectron spectroscopy (XPS) was further employed to study of such an attempt on 6H-SiC and the material formed by overplayed oxidation shows less Si and C related species in comparison with thermally oxidized samples. This method shows a less complex oxidation mechanism by comparing the case. (Avanas'ev V.V., 1997) and his team has performed verity of experiment to characterize the interface properties of SiO_2/SiC. Finally, this research group was investigated the basic mechanism of interface states distribution for SiC system. In such a system the interface traps density may arise from three main sources i.e. graphite-like carbon, carbon clusters and oxide traps. However, a similar type of paper was again presented by the same author in 2005 (Avanas'ev V.V., 1997). He concluded that that during 8 years of intensive studies this complex problem of oxidation and interface properties is still unsolved. At present time also Si and SiO_2 are very useful system, but electric field strength in SiC can reach the values 10 times higher than those observed in case of Si. In case of SiC as base material the potential barrier height between SiO_2 and SiC is even smaller, indicates toward a serious problem. Moreover, SiC based structures can operate at much more higher temperature than that of Si based structures. Additionally, due to poorer interface properties and low value of dielectric constant of SiO_2 is not seems to be implement in future MOS structures on SiC substrate.

4.2. Hafnium dioxide (HfO_2)

HfO_2 is second most promising dielectric on SiC surface after SiO_2 due to its high dielectric constant and very high breakdown voltage. A good quality and desired thickness can be easily achieved in laboratory that is why HfO_2 based MOS device are seems to me future devices. Pure form of HfO_2, and its silicate are the potential candidate SiO_2 as a gate material in a scaled down MOS technology. A continuous research and development (R&D) on this material is considerably seems to be more advanced compared to other high-k dielectrics (Avanas'ev V.V., 1997; Tanner C. M., 2007). Moreover, K Y Cheong et al has observed a significant improvement in the performance of HfO_2/SiO_2 stack gate dielectric on 4H-SiC surface (Cheong K.Y., 2007). Atomic layer deposition (ALD) is the most advisable and recommended process to deposit HfO_2 (Cho M., 2002). However, large variations in growth rate, dielectric constant, and fixed charge are reported for HfO_2 deposited on silicon substrate. The interface stability is one of the most important issues in the deposition process. When HfO_2 is considered to be deposited on SiC surface, the accurate knowledge of thermal stability at elevated temperatures is a must.

4.3. Titanium dioxide (TiO_2)

Titanium dioxide (TiO_2) is another gate dielectric, which is explained in this chapter. The electronic bandgap energy of this material is relatively small (3.5 eV), but dielectric constant can be varied from 40 to 110. TiO_2 exist in two important phases, Anatase and Rutile, which depends on growth process. Rutile phase of TiO_2 is the thermally stable phase that presents the higher dielectric constant around 80. Other form i.e. Anatase is a thermally unstable phase, which shows a lower dielectric constant. The Anatase form can be transforming to Rutile phase by annealing the deposited material at temperatures more than 600° C. A high leakage current values and higher interface density are the most drawback of this material, which is unacceptable in the fabrication of transistor structure. In order to minimize these problems it is interesting to employ a stack layer of thin SiO_2 and TiO_2 on SiC substrate. In this way, the interface quality can be improved and the other problems may be minimized, turning this material viable and very attractive to substitute the current dielectric material on SiC surface. Variable-energy positron annihilation spectroscopy (VEPAS) was employed to investigate the atomic scale structure of TiO_2/SiO_2T gate dielectric stack on 4H-SiC surface (Coleman P.G., 2007). In this study a vacancy type defects was observed. Thin film of TiO_2 film can be deposited with many techniques likes chemical vapour deposition (CVD), RF sputtering, e-beam evaporation, metal-organic chemical vapour deposition (MOCVD) and so on. The dielectric constant of TiO_2 was reported to be 31, which is stable in the frequency range from 100 Hz to 1 MH. The critical breakdown field is 3 MV/cm. TiO_2 is seems to of be very promising material in the development of gas sensors particularly Hydrogen sensors (Weng M-H, 2006; Shafiei M., 2008)

4.4. Aluminium oxide (Al_2O_3)

Aluminium oxide (Al_2O_3) is another gate dielectric, which has proven the demanded gate material SiC MOS structures. This material has a broad scope in semiconductor industry and the single crystal wafer of Al_2O_3 is commercially available. Crystalline form of Al_2O_3 is known as called sapphire, or α-Al_2O_3, which has the rhombohedral symmetry. The application of sapphire as passivation material for SiC is very hard due to crystalline mismatch and polycrystalline Al_2O_3 may cause large leakage trough grain boundaries of material. This material belongs to the family of wide bandgap (8.8 eV) and having the potential barrier of 2.8 eV with Si conduction band. The calculated conduction band offset for 4H-SiC system is about 1 eV, which is smaller than that of the measured on Si system. But this value is high enough to effectively prevent carrier injection at interface. However, amorphous form of Al_2O_3 seems to be an attractive candidate as a gate dielectric for SiC based structures. This material may be deposited by many different techniques such as sputtering (Jin p., 2002), plasma deposition (Werbowy A., 2000), Atomic layer deposition LD (Gao, K.Y. 2005) and so on with the suitable gaseous inlet of the precursors. ALD is seems to have the largest interest for fabrication of devices. K.Y. Gao et al has demonstrated a very good result and explained very nicely (Gao, K.Y. 2005). Post deposition annealing of Al_2O_3 in presence of H_2 environment at 500°C demonstrates a effective reduction in interface states density in the mid bandgap of the 6H-SiC. World-wide numbers of researchers (Avice M. 2007; She J., 2000) are intensively working to explore the Al2O3/SiC interface properties. As the result the first 4H-SiC MESFET with Al2O3 as a gate dielectric was successfully demonstrated (Hino S., 2007).

4.5. Aluminium nitride (AlN)

Aluminium nitride (AlN) is also one of the very promising gate dielectric materials, which can be associated with SiC system. Its lower bandgap of 6 eV in comparison with Al_2O_3 or SiO_2 might be disappointing, but a lattice mismatch to SiC of only 1%, almost the same thermal expansion up to 1000 °C and a high dielectric constant are more encouraging. Generally, AlN is used as a buffer layer prior to grow GaN structures on SiC substrates. This is the basic cause for largest number of research associated with the epitaxial growth at very high temperatures. Low temperature deposition is also possible over verity of substrate like other techniques that are of interest for low temperature deposition of passivation layers like atomic layer deposition, RF- sputtering pulsed laser deposition. There are not so many studies were focused on electrical characterization of AlN layers on SiC surface. Some results, however, shows satisfactory insulating properties for mono-crystalline AlN with acceptable leakage currents of the order of 10-9 A/cm^2 and a breakdown field of around 4 MV/cm (Onojima N., 2002). AlN/SiC interface do not shows the promising characteristics because of charge trapping at interface. However ozone cleaning and HCl pre-treatment of SiC surface shows a tremendous improvement of the properties of dielectric layer and provides interface quality sufficient for the fabrication of MOS structures. Introduction of thin SiO_2 as a buffer layer between SiC and AlN is an additional barrier to prevent electron injection from semiconductor to dielectric, which may further decrease leakage current. This type of stack layer of $AlN/SiO_2/6H\text{-}SiC$ was presented in (Biserica O., 2000) and reveals a low charging effect when 100 Å SiO_2 layer was used.

5. Depositions method

SiO_2 can be thermally grown by Thermal oxidation and this growth processes have the great advantage. In similar way, high *K* dielectric must be grown/deposited. It is well known that deposited oxides are never as good as compared to grown one.

5.1. Physical Vapour Deposition (PVD)

In the process of PVD based dielectric material deposition, e-Beam evaporation and Sputtering have been intensively used. The basis different between these two methods is the step coverage: e-Beam shows negligible step coverage while Sputtering produce a film with good step coverage as shown in Figure 3. Normally pure metal like Ti, Pt, Au, Ni, Al are deposited by e-beam evaporation method followed by an oxidation at suitable temperature.

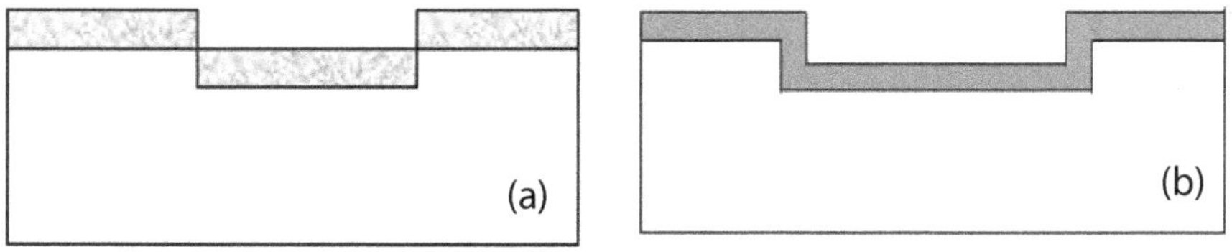

Figure 3. Schematic showing step coverage (a) poor step coverage using e-Beam method (b) good step coverage using Sputtering

In the e-beam evaporation method, a focused electron beam is used to heat a metal target to evaporate. In the high vacuum chamber, the evaporated metal radiates out from the metal target of which some portion is deposited on the mounted substrate. Generally, the target is placed at bottom and substrate is placed at the top of the vacuum chamber as shown in Figure 4. A method for producing highly pure, thin oxides is to evaporate metal by electron beam (e-beam) which is highly controllable to small thickness, and to oxidize the deposited metal by ozone or UV assisted oxidation. The advantage of this process is that it produces less damage than oxide sputtering and should produce the purest oxide. But it is not an exact method for commercial production.

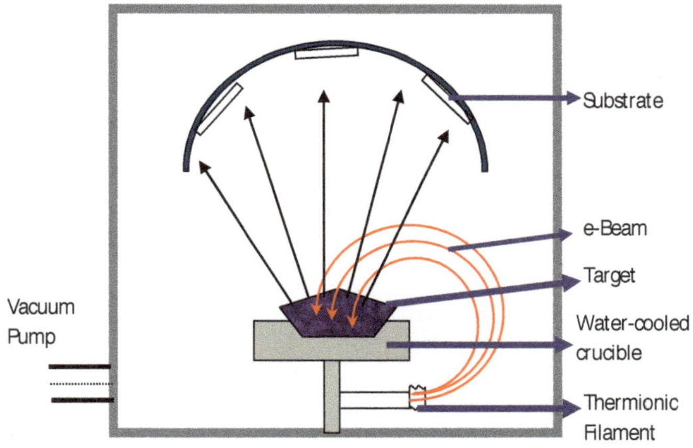

Figure 4. Schematic diagram of e-Beam evaporation system

Sputtering is generally used to deposit refractory materials, compound and alloys, which are difficult to evaporate by e-Beam method. Sputtering exists in the category of the Physical Vapor Deposition (PVD) process, in which metals are removed from the solid cathode. The whole process is carried out by bombarding the cathode with positive ions emitted from rare gas discharge. When ions with high kinetic energy are incident on the cathode, the subsequent collision knocks loose or sputters atoms from materials. The schematic of Sputtering system is given in Figure 5. Its advantage is that it is broadly available and can produce pure oxides. Its disadvantages are that oxides are insulators so sputtered oxides tend to have plasma-induced damage. Also, PVD methods deposit in line of sight, so they do not give good coverage.

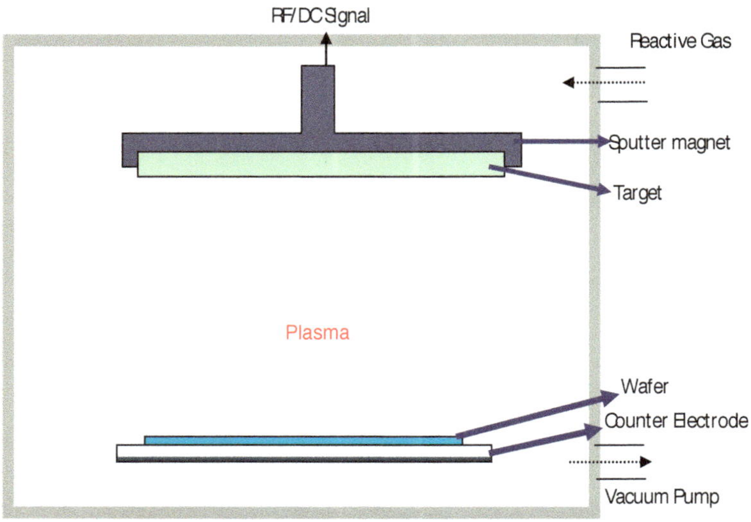

Figure 5. Schematic diagram of sputtering system

5.2. Chemical Vapour Deposition (CVD)

Chemical vapour deposition (CVD) and Atomic Layer Deposition (ALD) are preferred industrial method to deposit gate dielectrics. CVD involves the formation of a thin solid gate dielectric on a desired substrate by a chemical reaction of vapour-phase precursors. In CVD process generally a volatile metal compound as a precursor is introduced into the process chamber/tube and oxidized during deposition onto the desired substrate. CVD is widely used in the electronics industry for most of insulator deposition. It gives a conformal coverage even though a three dimension shapes because it is not just line of sight. The other major advantage is that the deposition rate is controllable over a wide range from very slow to high. It can thus be distinguished from physical vapour deposition (PVD) processes, such as evaporation and reactive sputtering, which involve the adsorption of atomic or molecular species on the substrate. The chemical reactions of precursor species occur both in the gas phase and on the substrate. Reactions can be promoted or initiated by heat (thermal CVD), higher frequency radiation such as UV (photo-assisted CVD) or plasma (plasma-enhanced CVD). Figure 6 shows the schematic diagram of horizontal CVD reactor.

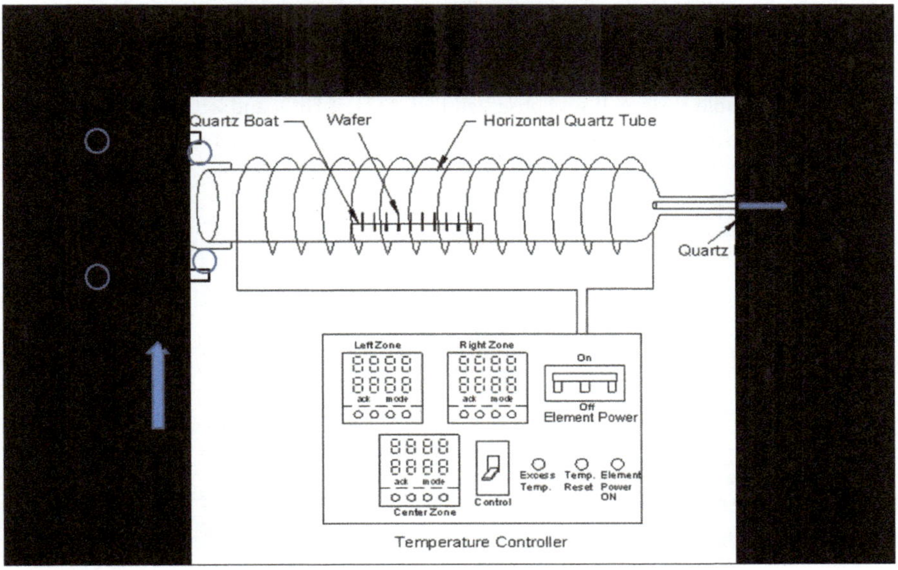

Figure 6. Schematic diagram of CVD system

5.3. Atomic Layer Deposition (ALD)

ALD was developed to recover the shortcoming, which arise due to CVD process. ALD produce a highly conformal, pinhole-free highly insulating film. It has many advantages over CVD method like able to grow the thinnest films even though all deposition methods, and the most conformal films even into deep trenches. Atomic layer deposition is a method of cyclic deposition and oxidation. In this process, the desired surface is exposed to the suitable precursor, which is further absorbed as a saturating monolayer. The rest of the precursor is then purged from the tube/chamber by passing Ar/N_2 gas. A pulse of oxidant such as $H2O_2$, ozone or H_2O, is then introduced in the chamber/tube, which must then fully oxidize the adsorbed layer to the oxide and a volatile by-product. The excess oxidant is then purged by a pulse of Ar, and the cycle is repeated. Figure 7 represent a cyclic process of ADL press. ZrO2 and HfO2 was shown as example in figure7. Slow growth rate is a major disadvantage of this process but some time it is very useful to control the thickness of films. It has been seemed that some impurities like Cl, C and H also introduce in the film during deposition process, depending on used precursor. A compatible annealing methodology is needed to remove such type of impurity and densify the deposited oxides films. ALD is an excellent method for producing many high *K* oxides. An addition of an oxide layer, which is usually much less than an atomic layer thick in each and every cycle of ALD process adds an oxide layer, justify its nomenclature. ALD is usually carried out on a native oxide (SiO_2) surface followed by ozone cleaning of SiC surface. This limits the crucial lowest EOT that ALD can presently attain. H-terminated surface, which arises by the HF-cleaning treatment procedure, is not favorable surface. It was observed that ALD of HfO_2 and ZrO_2 from chloride or other organic precursors do not easily nucleate on HF treated SiC surface.

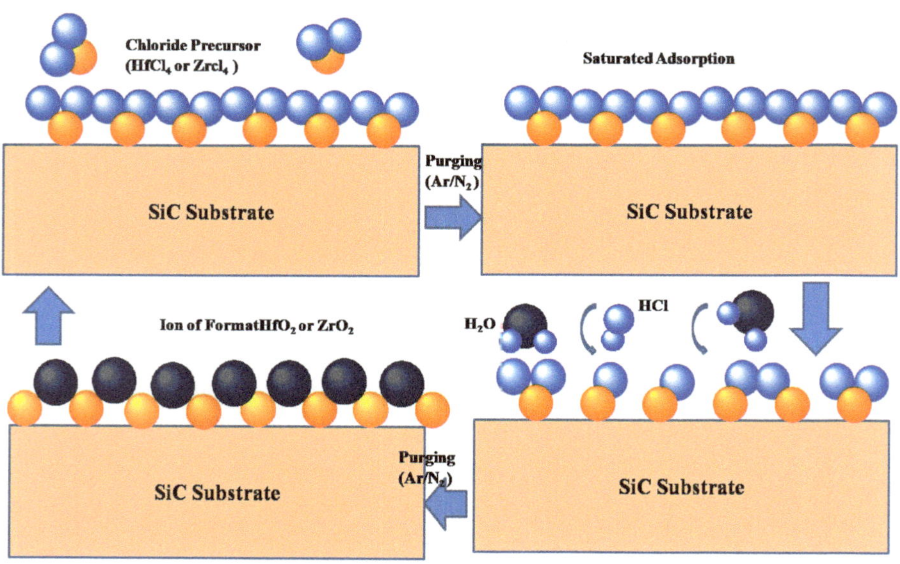

Figure 7. Schematic of the cyclic process of Atomic layer deposition process

6. Electrical behavior of dielectric material

The quality of dielectric material can be electrically characterized by current-voltage (I-V) and capacitance-voltage (C-V) technique. In order to employ the above techniques, the grown/deposited layers on SiC should be sandwiched between two metal electrodes. A different type of current conduction mechanisms were observed and presented in this section, based on dielectric thickness and applied electric field across electrode. In this section SiO_2 was considered as a gate dielectric on SiC surface. Gupta, S.K. and his research group are continuously working to investigate the current conduction mechanism and charge management of gate dielectric material and SiC system (Gupta, S.K. 2010[a]; Gupta, S.K. 2010[b]; Gupta, S.K. 2011[b] ; Gupta, S.K. 2012).

6.1. Direct tunneling

Schrödinger equation describes that there is a finite probability that a particle can tunnel through a non-infinite potential barrier. As the width of potential barrier decreases, the probability of particles (electrons and holes) penetrating through the barrier by quantum-mechanical tunneling, rises exponentially. In sufficiently thin oxides (below 5 nm), direct quantum mechanical tunneling through the potential barrier can occur. This quantum-mechanical phenomenon can easily be understood by recognizing that the electron or hole wave function cannot immediately stop at the barrier (SiO_2/4H-SiC interface), but rather it decreases exponentially into the barrier with a slope determined by the barrier height. If the potential barrier is very thin, there is non-zero amplitude of the wave function remaining at the end of the barrier means a non-zero probability for the electron or hole to penetrate the barrier. It is well known that the barrier heights of hole tunneling in the SiO_2 layer from the metal gate and from the Si substrate are higher than the corresponding values for electrons, moreover, the hole mobility in SiO_2 is lower than the electron mobility, therefore the main contribution to conduction in SiO_2 is due to electrons. Since in n-type 4H-SiC mobility of electrons is much higher than that of hole, therefore, the described conduction mechanism in case of Si can be fully applied to 4H-SiC. The metal/SiO_2 and SiO_2/4H-SiC interfaces are at the position $X = 0$ and $X = t_{oxt}$ respectively, in our notation. $V_{oxt} = V(0) - V(t_{oxt})$ is the voltage drop in the oxide layer, where V(x) is the potential in the oxide at position X.

At low gate voltages (figure 8 (a)), electrons can move from the gate metal through SiO_2 to the 4H-SiC substrate only by tunneling directly the entire oxide thickness i.e. by tunneling the trapezoidal potential barrier between gate and 4H-SiC substrate. The quantum-mechanical phenomenon of a trapezoidal barrier tunneling is termed as direct tunneling effect. It contributes significantly to the conduction through the SiO_2 only in ultra thin oxide layers ($t_{oxt} < 5$ nm). At higher gate voltage (figure 8 (b)), the band bending causes the potential barrier shape to become triangular. Electron tunnel from the gate to the SiO_2 conduction band, through the triangular potential barrier and finally, moves in the SiO_2 conduction band to the 4H-SiC substrate. The conduction mechanism through a triangular potential is called Fowler-Nordheim (F-N) tunneling, which is described in next section.

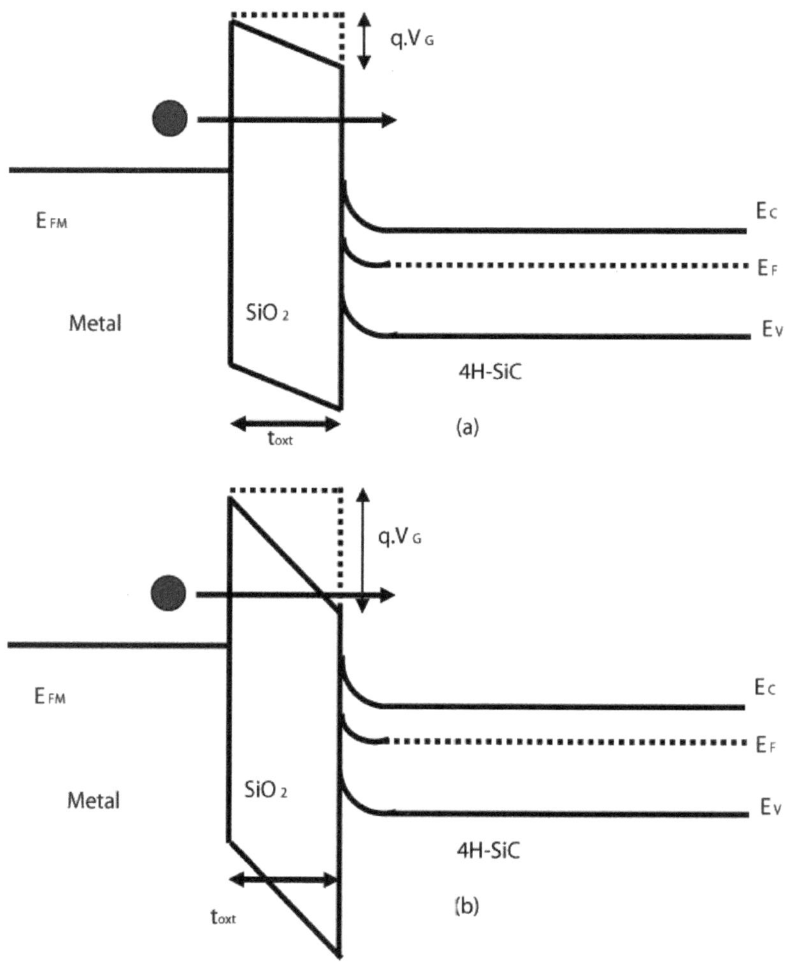

Figure 8. a) The band diagram of negative gate bias the closec circle represents one electron injected from gate to the SiC conduction band through trapezoidal energy barrier (b) tunneling across triangular energy barrier

6.2. Fowler-Nordheim tunneling

Fowler-Nordheim tunneling in metal oxide semiconductor (MOS) can be observed when the oxide thickness will be less than 50 nm, the oxide potential barrier is usually assumed to be a triangular one, free of charge gate insulator. As a consequence, when we apply a uniform electric field across the MOS structure and thickness of the potential barrier at the semiconductor Fermi level and the potential $\phi(x)$, at the distance x from semiconductor/oxide interface vary linearly with the applied voltage.

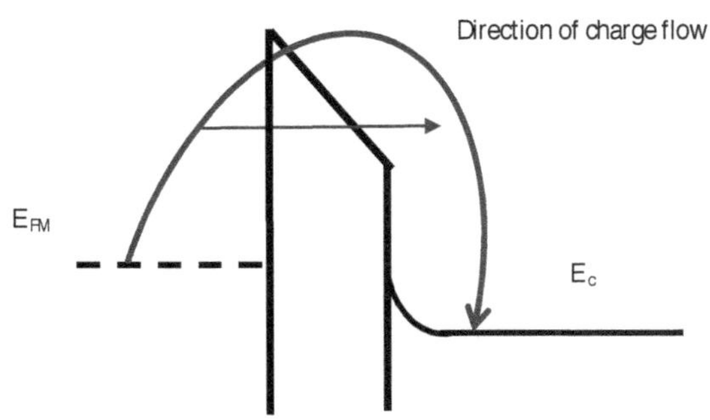

Figure 9. Energy band diagram for F-N tunneling

The insulating region is separated by an energy barrier with barrier height $q\phi_B$, measured from the Fermi energy of metal to the conduction band edge of the insulating layer. The distribution functions at both sides of the barrier are indicated as in the figure 9. In the derivation of current density (J) as a function of applied voltage we have to consider some assumptions like effective-mass approximation, parabolic bands and conservation of parallel momentum (Chung, G. Y., 2001). The net tunneling current density from metal to semiconductor can be written as the net difference between current flowing from the metal region to the semiconductor region and *vice versa.* This expression for current density is usually written as an integral over the product of two independent parts, which only depend on the energy perpendicular to the interface: the transmission coefficient T (*E*) and the supply function N (*E*).

$$J = \frac{4\pi m_{eff} q}{h^3} \int_{E_{min}}^{E_{max}} T(E) N(E) dE \qquad (2)$$

This expression is known as Tsu-Esaki formula. This model has been proposed by Duke and was used by Tsu and Esaki for the modeling of tunneling current in resonant tunneling devices. The calculation of current density requires not only the knowledge of the energy dependent transfer coefficient, but also the energy dependent electron probability (supply function). Using theTsu-Esaki formula for current density the Fowler-Nordheim formula can be derived as:

$$J\alpha AE^{2}_{diel}\exp\left(-\frac{B}{E_{diel}}\right) \tag{3}$$

Where,E_{diel} denotes the electric field in dielectric, A and B are constants dependent on barrier configuration for oxide layer separated by metal and semiconductor, the constants *A* and *B* are given by:

$$A=\frac{q^{3}m_{\text{eff}}}{8\pi m_{\text{diel}}hq\Phi_{\text{B}}}\quad B=\frac{4\sqrt{2m_{\text{diel}}(q\Phi_{\text{B}})^{3}}}{3\hbar q} \tag{4}$$

Where, ϕ_B is the height of the potential barrier measured from the Fermi level of metal to the conduction band in the dielectric, m_{iff} is the effective electron mass in electrode material and m_{diel} is the effective electron mass in the dielectric material. This physical model has been directly applied to in order to establish the validation of Fowler-Nordheim tunneling with the oxide thickness limit.

6.3. Schottky emission

The Schottky emission is an electrode limited process occurring across the interface between a semiconductor (or metal) and an insulating film as a result of barrier lowering due to the applied electric field and the image force as shown in figure 10. Normally, the S-E current conduction process is an electrode-limited conductivity that depends strongly on the barrier between the metal and insulator and has the proclivity to occur for insulators with fewer defects.

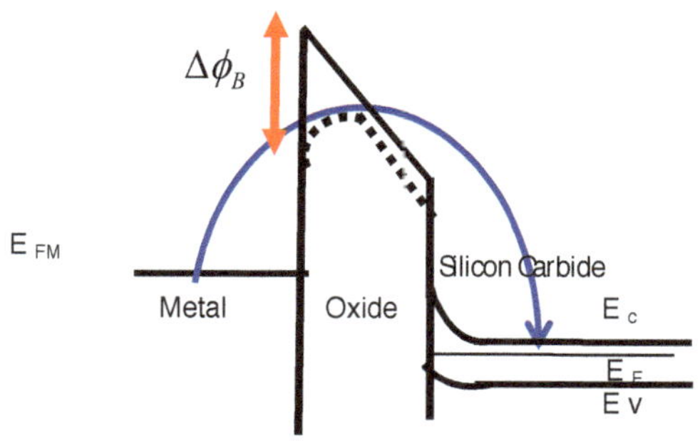

Figure 10. Energy band diagram for Schottky emission in metal oxide silicon carbide (MOSiC) structure

Schottky emission from the metal cathode or from the oxide states is assumed to be the limiting mechanism for filling or emptying the oxide traps. For the emission from a semiconductor the Schottky emission current conduction is given by (Chang S.T., 1984)

$$J \alpha A^{*} T^{2} E \exp\left[\frac{-q\left(\varphi_{B}-\sqrt{qE/4\pi\varepsilon_{i}}\right)}{kT}\right] \tag{5}$$

Where, J is the current density; A* iseffective Richardson constant; T, the absolute temperature; q, the electronic charge; ϕ_B, the potential barrier at the metal and insulator interface; E, electric field in insulator; ε_i, dielectric constant; and k, the Boltzmann constant.

The potential barrier lowering in the MOSiC structures caused by image forces as shown in figure 10 is often neglected in the calculation of the tunneling current, based on an argument that for large barriers in the case of semiconductor and insulator the image-force lowering of the barrier is very small, and this was supported by experimental evidence at the time. In case of very thin oxides, however, this might not be the case, and the barrier lowering can have an impact on the calculation of the tunneling current.

The potential barrier lowered with respect to ideal structure has been termed to an effective trapezoidal barrier in order to account for image force effect. The image-barrier height lowering can be described by:

$$\Delta\varphi_{B}=\sqrt{\frac{qE_{m}}{4\pi\varepsilon_{0}\varepsilon_{r}}} \tag{6}$$

Where, $\Delta\Phi B$is the image-barrier height lowering, and E $_m$is the applied electric field at the metal-semiconductor interface, ε_0 is permittivity of vacuum and ε_r relative dielectric constant of insulating layer.

6.4. Poole-Frenkel conduction

6.4.1. Classical theory of Poole-Frenkel conduction

In ideal metal oxide semiconductor diode, it is assumed that current conduction through the insulator is zero. Real insulator, however, show the current conduction mechanism which may the function of thickness of the insulator or applied electric field or both. In the classical Poole-Frenkel conduction model effective mechanism can be analyzed by the analogue of Schottky emission (Wright P. J., 1989) where as transport of charge carriers is governed by trapping and de-trapping in the forbidden band gap of an insulator, which reduces the barrier on one side of the trap. At zero electric field amount of free charge carriers can be determined by the trapped ionization energy (qϕ), which is the amount of energy required for the trapped electron to escape the influence of the positive nucleus of the trapping center when no field is applied. When electric field is applied, the ionization energy of trapping center decreases in the direction of applied electric field by the amount of $\Delta\phi_B=\beta E^{1/2}$ as shown in figure 11. As the electric field increases, the potential barrier decreases on the right side of the trap, making it easier for the electron to vacate the trap by thermal emission and enter the quasi-conduction band of the crowd material. Quasi-conduction band edge is the energy at which the electron is just free from the influence of the positive nucleus. The term quasi-conduction is generally used in amorphous solid, which have no real structure. In MOSiC structure, of course, electron would escape from gate metal to the conduction band of semiconductor through the insulator. Since we will deal here with the P-F mechanism in amorphous dielectrics, we will refer to its quasi-conduction band. For the Poole-Frenkel conduction mechanism to occur the trap must be neutral when filled with an electron, and positively charged when the electron is emitted, the interaction between positively charged trap and electron giving rise to the Coulombic barrier. On the other hand, a neutral trap that is, a trap which is neutral when empty and charged when filled will not show the Poole-Frenkel effect.

According to the Poole-Frenkel model, the magnitude of the reduction of trap barrier height due to the applied electric field as shown in figure 11 is given by

$$\Delta q\varphi = \beta\sqrt{E} \tag{7}$$

Where β is Poole-Frenkel constant, is given by

$$\beta = \sqrt{\frac{q^3}{\pi\varepsilon_0\varepsilon_r}} \tag{8}$$

Finally, quasi conduction band edge is lowered or in other words, the trap barrier height is reduced due to the applied electric field. From Equation 8, it is clear that β is a material parameter, depending on the dielectric constant. Therefore, materials with larger dielectric constants will be less sensitive to the field-induced trap barrier lowering effect in the P-F conduction.

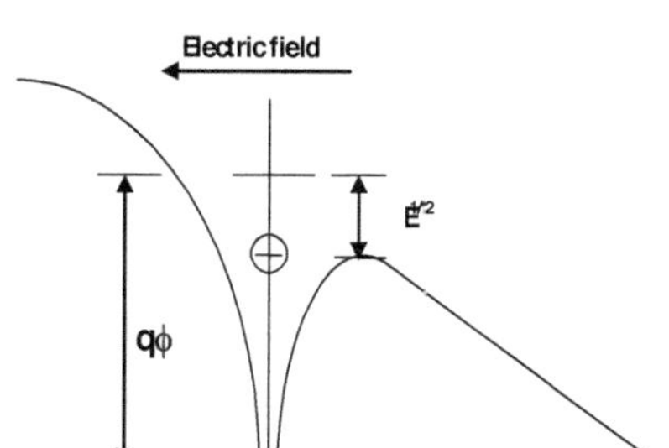

Figure 11. Figure 11.Coulombic potential distribution in the presence of applied electric field showing the Poole-Frenkel effect

6.4.2. Poole-Frenkelconduction in MOSiC structure

P-F conduction mechanism is most often observed in amorphous materials, particularly dielectrics, because of the relatively large number of defect centers present in the energy gap. In fact, the particular host material, where the defects reside, can basically be viewed as acting only as a medium for localized defect states. Transformation of charge is therefore, mainly between localized electronic states (Lenzlinger, M., 1969). Thus it is reasonable to expect the P-F effect to occur, at least to some extent, in any dielectric. The main physical properties effecting the current conduction in different dielectrics are the relative dielectric constant, ε_r, and the ionization potential. The P-F conduction effect has been observed in many dielectric materials, which are used in microelectronic device fabrication. For example, in Si_3N_4 films, the dominating current transport mechanism is the P-F conduction. The thin films with high dielectric constants, such as Ta_2O_5 and $BaSrTiO_3$, which hold great potential for use as the gate oxide in DRAMs, have shown that current conduction in these materials is bulk-limited which is governed by the P-F conduction. Currently, one of the most important dielectric materials used in microelectronics is SiO_2, which can be easily thermally grown on SiC substrate.

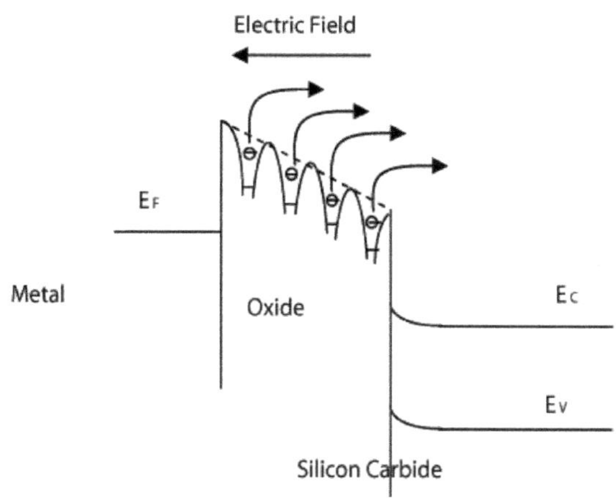

Figure 12. Energy band diagram for Poole-Frenkel conduction in MOSiC structure having multiple Coulombic traps

Figure 12 shows the P-F conduction in MOSiC structure that is basically a parallel plate capacitor. The trapezoidal band diagram of MOSiC structure drawn for the silicon dioxide layer is replaced in figure 11 by a random distribution of Coulombic traps in the vicinity of the quasi-conduction band edge as shown in figure 12. The dashed line indicates the quasi-conduction band of the oxide in the absence of any traps. When an electric field is applied as shown, the trapped electrons can enter the oxide's quasi-conduction band by the Poole-Frenkel mechanism and flow from the oxide across the SiC/SiO_2 interface into the silicon carbide conduction band edge. The Poole-Frenkel effect can be observed at the high electric field. The standard quantitative equation for P-F conduction is

$$J \alpha E \exp\left[\frac{-q\left(\varphi_B - \sqrt{qE / \pi\varepsilon_i}\right)}{kT}\right] \tag{9}$$

Where, J is the current density; T, the absolute temperature; q, the electronic charge; ϕ_B, the potential barrier at metal and the insulator interface; E, electric field in insulator; ε_i, dielectric constant; and k, the Boltzmann constant.

7. Charge management in SiO_2 on 4H-SiC

7.1. Origin and basic theory of oxide charges in SiO_2

In the prospects of technological issues on Silicon carbide based MOS system, the almost similar consideration has been adopted to investigate the charge management as silicon based MOS system. In this section, the oxide charges associated with Ni/SiO_2/4H-SiC systems have been examined with varying oxide thickness. There are general four types of charges associated with the SiO_2-Si system as shown in figure 13. They are fixed oxide charge, mobile oxide charge, oxide trappedcharge and interface trapped charge (Afanas'ev, V, V., 1996 and Schroder, D. K., 2006). The basic origin of all oxide charges and experimental methods in order to calculate these charge are presented here one by one.

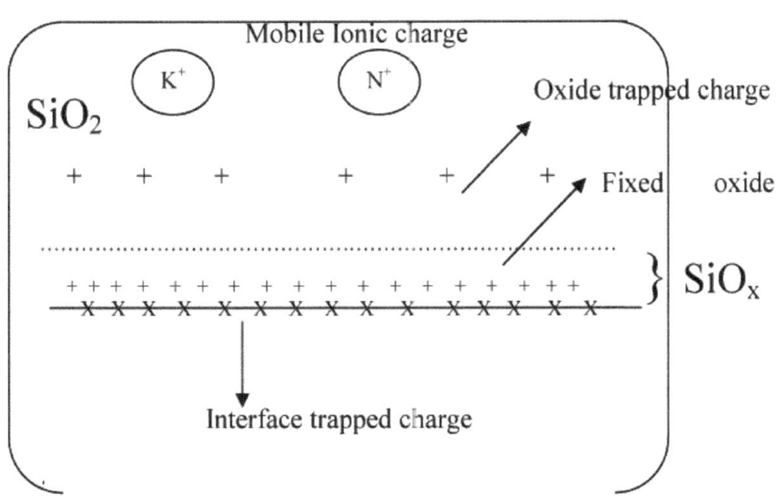

Figure 13. Allocation of different oxide charges associated with SiC MOS system

7.2. Fixed oxide charges (Q_{fix})

These are positive or negative charges located near the SiO_2/4H-SiC (less than 25 Å from the interface) interface due to primarily structural defects in the oxide layer. The origin of fixed charge density is related to the oxidation process, oxidation ambient and temperature, cooling condition of the furnace and also on polytypes of Silicon carbide. Q_{fix} highly depends on the final oxidation temperature. For higher oxidation temperature, a lower Q_{fix}will be observed. However, if it is not permissible to oxidize the wafer at high temperatures, it is also possible to reduce the $Q_{fix}$by annealing the oxidized wafer in a nitrogen or argon ambient after oxidation.

The fixed charge can be determined by comparing the flatband voltage shift of an experimental C–V curve with a theoretical curve, provided by the oxide thickness and work function differences of metal and 4H-SiC. Fixed charge related to the flatband voltage is given by:

$$Q_{fix} = (\varphi_{ms} - V_{FB})C_{oxt} \quad (10)$$

Where, ϕ_{MS} is the difference of work function between metal and semiconductor, which must be known in order to determine the value of $Q_{fix.}$

7.3. Oxide trapped charge (Q_{oxt})

This oxide charge may be positive or negative due to the hole and electron trapped in the bulk of the oxide. These trapping may results from ionizing radiation, avalanche injection, Fowler-Nordheim tunneling or other similar processes. Unlike the fixed charge, this chare can also be reduced by annealing treatment. Oxide charges can be trapped in the oxide during device operation, even if not introduced during device fabrication. During the device operation electrons and/or holes can be injected from the substrate or from the gate material. Energetic radiation also produces electron-hole pairs in the oxide and some of these electrons and/or holes are subsequently trapped in the oxide. The oxide trapped charge is usually not located at the oxide/4H-SiC, but is distributed through the oxide. The distribution of Q_{oxt}must be known for proper interpretation of C–V curves. Oxide trapped charge can be determined by:

$$Q_{oxt} = -V_{FB}.C_{oxt} \quad (11)$$

Where, the symbols have their usual meaning.

7.4. Mobile oxide charge (Q_{mob})

The origin of this oxide charge is due to the presence of ionic impurities such as Na^+, Li^+, K^+ and possible H^+ in the oxide films. These ionic impurities may be resulted from the ambient, which, 7
was used for thermal oxidation. Negative ions and heavy metals ions may also contribute to this

charge. Sodium ion is the dominant contaminant. The other ionic impurities like potassium may be introduced during chemical-mechanical polishing. For mobile charge calculation the measurement temperature must be sufficient high so the charge to be mobile. Typically, the devices are heated to 200^{0}C to 300^{0}C. A gate bias, to produce an oxide field of around 10^6 V/cm is applied for a sufficiently long time in order to drift charge from interface. The mobile charge can be determined from the flatband voltage shift, according to the equation:

$$Q_{mob} = -\Delta V_{FB} . C_{oxt} \tag{12}$$

7.5. Interface trap level density (D_{it})

These are positive or negative charges, due to structural defects, oxidation-induced defects, metal impurities, or other defects caused by radiation or similar bond breaking processes (*e.g.*, hot electrons). The interface trapped charge is located at the SiC–SiO_2 interface. Unlike the fixed charge or trapped charge, interface trapped charge is in electrical communication with the underlying SiC. Interface traps can be charged or discharged, depending on the surface potential.This charge type has been also called surface states, fast statesand interface states and so on.

There are three main approaches to investigate the problem of interface state.

1. By the comparison of measured high frequency capacitance with a theoretical capacitance with no interface traps.
2. By the comparison of measured low frequency capacitance with a theoretical capacitance with no interface traps.
3. By the comparison of measured high frequency capacitance with measured low frequency capacitance.

Interface trap change their charges sate depending on whether they are filled or empty. Acceptor interface traps are negative when filled, and neutral when empty, whereas donor interface traps are neutral when filled and positive when empty. Both types of interface traps may exist, perhaps simultaneously in the same device.

$$D_{it} = \frac{1}{q}\left[\left\{\frac{1}{C_{LF}} - \frac{1}{C_{oxt}}\right\}^{-1} - \left\{\frac{1}{C_{HF}} - \frac{1}{C_{oxt}}\right\}^{-1}\right] \tag{13}$$

8. The Metal Oxide Silicon Carbide (MOSiC) band diagram

8.1. An ideal band diagram of MOSiC capacitor

In accord to the ideal case, the value of interface traps (Q_{it}) should be zero. The relationship of surface potential to the gate voltage having interface traps zero is known as an ideal MOSiC capacitor as shown in figure 14.

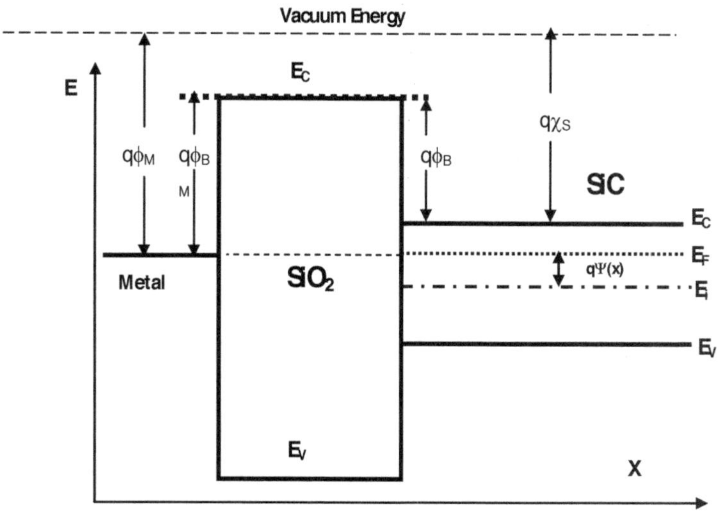

Figure 14. Flatband energy band diagram of an ideal MOSiC structure

An ideal MOS diode is defined as follows:

1. The energy difference between the metal work-functionand the semiconductor work-function is zero. Under this condition, the Fermi levels of the metal and semiconductor are aligned at equilibrium. This is equivalent with no charge flowing when they are put in contact.

2. The only charges that can exist in the structure under any biasing conditions are those in the semiconductor and those with the equal but with opposite sign on the metal surface adjacent to the insulator.

3. The semiconductor Fermi level is constant from the SiC-bulk toward the interface. It is determine by the shallow doping (usually nitrogen N).

4. There are no traps at the metal/SiO_2 or SiO_2/4H-SiC interface. The SiO_2 is free of defects (structural defects, impurities, vacancies, etc.). The only allowed charge in the structure exists in the semiconductor and, with opposite sign in the metal.

5. The resistivity of the insulator (oxide) is infinity so that there are no carrier transports under Bias conduction.

8.2. The real MOSiC capacitor

In real, the oxides of any MOS capacitor features a number of charges for example fixed oxide charge, mobile charge, oxide trap charge and interface trap level density. There is also a non-zero difference between the gate metal and semiconductor work function. The electric fields produced are compensated by a corresponding charge of the semiconductor. Since the ideal dielectric does not conduct any current, the semiconductor Fermi level remains flats. However, the bands are bending in compliance with the applied and created fields. To compensate this bending and to reach the flatband situation (Ψ=0), a gate bias has to be applied. This bias will shift the C-V characteristics of the MOS capacitor. The flatband (Ψ=0) situation is reached, when the flatband bias V_{FB} is applied:

$$V_{FB} = \varphi_{ms} - \frac{Q_{oxt} + Q_{fit} + Q_{it}(\psi_S)\big|_{\psi_S=0}}{C_{oxt}} \tag{14}$$

The effective charge density at the interface N_{eff} is obtained from a C-V measurement and has the form:

$$N_{eff} = V_{eff} \times C_{oxt} = -(V_{FB} - \varphi_{ms}).C_{oxt} = Q_{oxt} + Q_{fit} + Q_{it}(\psi_S = 0)$$

When the interface trap density is high (>10^{11} cm-2), the flatband biases for the opposite sweep directions are different. This difference id called hysteresis:

$$\Delta VH = V_{FB}(accum \rightarrow depl) - V_{FB}(depl \rightarrow accum) \tag{15}$$

The applied gate bias is the sum of the potential drop over the oxide Voxt, the flatband voltage V_{FB} and the potential at the SiC surface Ψ_S.

$$V_G = V_{oxt} + V_{FB} + \psi_S \tag{16}$$

The oxide capacitance corresponds to the accumulated charge at the gate divided by the potential drop over the oxide

$$C_{oxt} = \frac{Q_G}{V_{oxt}} \tag{17}$$

In the ideal case, this charge equals to the space charge of the semiconductor with negative sign Q_G=-Q_{SC} (Ψ_S), whereas the space charge of the semiconductor is a function of the surface potential. All these considerations lead to the following relationship between the applied gate voltage and the surface potential. Figure 15 shows the work functions of various metals used as gate dielectric together with the energy position SiC valence and conduction band edge.

$$V_G - V_{FB} = -\frac{Q_{SC}(\psi_S)}{C_{oxt}} + \psi_S \tag{18}$$

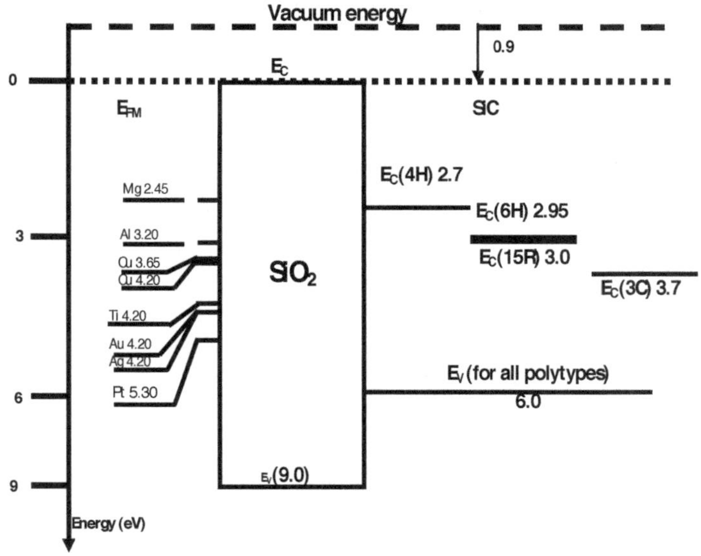

Figure 15. Work functions of various metals used as gate together with the energy position SiC valence and conduction band edge

Author details

Sanjeev Kumar Gupta, Jitendra Singh and Jamil Akhtar

CSIR-Central Electronics Engineering Research Institute (CEERI), Pilani, India

References

[1] Amy, F., Soukiassian, P., Hwu, Y., Brylinski, K., & , C. (1999). SiO2/6H-SiC (0001) 3x3 initial interface formation by Si overlayer oxidation,Applied Physics Letters, , 75(21), 3360.

[2] Afanasev, V. V., Bassler, M., Pensl, G., Schulz, M., & (1997, . (1997). Intrinsic SiC/SiO2 Interface States. Physica Status Solidia), , 162, 321.

[3] Afanas'ev, V. V. ., Ciobanu, F., Dimitrijev, S., Pensl, G., & and, Stesmans. A. (2005). SiC/SiO2 Interface States: Properties and Models, Materials Science Forum, 483-485, 563.

[4] Afanas'ev, V. V., Stesmans, A., Chen, F., Campbell, S. A., & Smith, R. (2003). HfO2-based insulting stacks on 4H-SiC (0001),Applied physics Letter, , 82, 922.

[5] Afanas'ev, V. V., Bassler, M., Pensl, G., Schultz, M. J., & Kamienski, E. S. (1996). J. Appl. Phys. , 79, 3108.

[6] Avice, M., Grossner, U., Monakhov, E. V., Grillenberger, J., Nielsen, O., Fjellvåg, H., & Svensson, B. G. (2005). Electrical Properties of Aluminum Oxide Films Grown by AtomicLayer Deposition on n-type 4H-SIC, Materials Science Forum, , 483-485, 705.

[7] Avice, M., Grossner, U., Pintillie, L., Svensson, B. G., Servidori, M., & (2007, . (2007). Electrical properties of Al2O3/4H-SiC structures grown by atomic layer chemical vapor deposition. *Journal of Applied Physics*, 102, 054513.

[8] Biserica, O., Godignon, P., Jorda, X., Montserrat, J., Mestres, N., & Hidalgo, S. (2000). Study of AlN/SiO2 as dielectric layer for SiC MOS structures, CAS 2000 Proceedings,International , 1, 205.

[9] Cantin, J. L., Bardeleben, H. J., Shishkin, Y., Ke, Y., Devaty, R. P., & Choyke, W. J. (2004). Identification of the Carbon Dangling Center at the 4H-SiC/SiO2 Interface by an EPR Study in Oxidized Porous SiC, Phys. Rev. Lett., , 92, 015502.

[10] Chang, S. T., Johnson, N. M., Lyon, S. A., & (1084, . Capture and tunnel emission of electrons by deep levels in ultra thinnitrided oxides on silicon, Appl. Phys. Lett., , 44, 316.

[11] Chen, Q., Feng, Y. P., Chail, J. W., Zhang, Z., Pan, J. S., & Wang, S. J. (2008). Energy-band alignment of HfO2 /SiO2 /SiC gate dielectric stack, Applied physics Letter, , 93, 052104.

[12] Cheong, K. Y., Moon, J. H., Park, T. J., Kim, J. H., Hwang, C. S., Kim, H. J., Bhang, W., & Kim-K, N. (2007). Improved Electronic Performance of HfO2/SiO2 stacking gate dielectric on 4H-SiC, IEEE Transaction on Electronic Devices, , 54, 3409-3413.

[13] Cho, M., Park, J., park, H. B., Hwang, C. S., Jeong, J., & Hyun, K. S. (2002). Chemical interaction between atomic-layer-deposited HfO_2 thin films and the si substrate,. Applied physics Letter, , 81, 334.

[14] Chung, G. Y. (2001). Improved inversion channel mobility for 4H-SiC MOSFETs following high temperature anneals in nitric oxide. IEEE Electron Device Lett., , 22(4), 176-278.

[15] Deal, B. E., Grove, A. S., & (1965, . (1965). General relationship for the thermal oxidation of Silicon, J.Appl.Phy. , 36, 3770.

[16] Gao, K. Y., Seyller, Th., Emtsev, K. ., Ley, L., Ciobanu, F., & Pensl, G. (2005). ALD Deposited Al2O3 Films on 6H-SiC(0001) after Annealing in Hydrogen Atmosphere, Materials Science Forum, , 483-485, 559.

[17] Gupta, S. K., Azam, A., Akhtar, J., & (2010a, . (2010a). Experimental Analysis of I-V and C-V characteristics of Ni/SiO2/4H-SiC system with varying oxide thickness.Microelectronic International, , 27, 106-112.

[18] Gupta, S. K., Azam, A., & Akhtar, J. (2010b). Experimental Analysis current conduction through thermally grown SiO_2 on thick epitaxial 4H-SiC employing Poole-Frenkel mechanism, . Pranama-Jour nal of Physics, , 74, 325-330.

[19] Gupta, S. K., & Akhtar, J. (2011a). Thermal oxidation of Silicon Carbide (SiC); experimentally observed facts, Book chapter in "Silicon Carbide-Materials, processing and applications in electronic devices, Intech open access publisher, , 207-230.

[20] Gupta, S. K., Azam, A., & Akhtar, J. (2011b). Variation of interface trap level charge density within the bandgap of 4H-SiC with varying oxide thickness. Pranama-Journal of Physics, , 76, 165-172.

[21] Gupta, S. K., Azam, A., Akhtar, J., & (2012, . (2012). Effect of post oxidation annealing on Electrical characteristics of Ni/SiO2/4H-SiC capacitor with varying oxide thickness, . Semicondu ctor, , 46, 545-551.

[22] Hino, S., Hatayama, T., Miura, N., Oomori, T., Tokumitsu, E., & (2007, . (2007). Fabrication and characterization of 4H-SiC MOSFET with MOCVD grown Al2O3 gate insulator, Material Science forum, Vols. 556-557, , 787-790.

[23] Jin, P., Xu, G., Tazawa, M., Yoshimura, K., Music, D., Alami, J., & Helmersson, U. (2002). Low temperature deposition of a-Al2O3 thin films by sputtering using a Cr_2O_3 template, Journal of Vacuum Science and Technology A, , 20(6), 2134.

[24] Jakschik, S., Schroeder, U., Hecht, Th., Krueger, D., Dollinger, G., Bergmaier, A., Luhmann, C., & and, Bartha. J. W. (2003). Physical characterization of thin ALD-$Al2O_3$ films. *Applied Surface Science*, 211, 352.

[25] Kobayashi, H., Sakurai, T., Takahashi, M., & Nishioka, Y. (2003). Interface states at SiO_2/6H-SiC (0001) interfaces observed by x-ray photoelectron spectroscopy measurements under bias: Comparison between dry and wet oxidation, Phys. Rev. B, , 67, 115305.

[26] Leskela, M., Ritala, M., & (2002, . (2002). Atomic layer deposition (ALD): from precursors to thin film structures. *Thin Solid Films*, 409, 138.

[27] Mahapatra, R., Chakraborty, Amit. K., Horsfall, A. B., Wright, N. G., Beamson, G., & Coleman, Karl. S. (2008). Energy-band alignment of HfO2 /SiO2 /SiC gate dielectric stack, . Applied physics Letter, , 92, 042904.

[28] Onojima, N., Suda, J.and., & Matsunami, H. (2002). Heteroepitaxial growth of insulating AlN on 6H-SiC by MBE. *Materials Science Forum*, 389-393, 1457.

[29] Perez, I.; Elam, J.; George, S.; Groner, M.; Torvik, J. T. and Van Zeghbroeck, B. (2000).Fabrication and characterization of 4H-SiC MOS capacitors with atomic layer

deposited(ALD) SiO2, Proceedings 2000 IEEE/ Cornell Conference on High Performance Devices, p. 144-7.

[30] Schroder, D. K., "Semiconductor material and device characterization", John Wiley & Sons, Inc., Hoboken, New Jersey 3rd edition 319-70, 2006.

[31] She, J., Inoue, T., Ueno, K., & (2000, . (2000). Multilayer Al2O3/SiC ceramics with improved mechanical behavior. Journal of the European Ceramic Socity, , 20, 1771-1775.

[32] Shafiei, M., Sadek, A. Z., Yu, J., Arsat, R., Kalantar-Zadeh, K., Yu, X. F., Partridge, J. G., & Wlodarski, W. (2008). Pt/anodized TiO_2/SiC-based MOS device for hydrocarbon sensing, Proc SPIE. K., 7268 EOF.

[33] Song, Y., Dhar, S. L., Feldman, C., Chung, G., & Williams, J. R. (2004). Modified deal and grove model for the thermal oxidation of silicon car. bide, J. Appl. Phys., , 95, 4953-4957.

[34] Tanner, C. M., Choi, J., & Chang, J. P. (2007). Electronic structure and band alignment at the HfO2/4H-SiC interface. *Journal of applied physics*, 101, 034108.

[35] Wang, S. ., Di Ventra, M., Kim, S. G., & Pantelides, S. T. (2001). Atomic scale dynamics of the formation and dissolution of carbon clusters in SiO_2, Phys. Rev. Lett., , 86, 5946-5949.

[36] Weng, ., -H, M., Horsfall, A. B., Mahapatra, R., & Wright, N. G. (2006). First observation of Hydrogen sensing by trap assisted conduction current in Pd/TiO2/SiC capacitor at high temperature, presented at 5th IEEE conference on Sensors on October , 22-25.

[37] Weng-H, M., Horsfall, A., mahapatra, R., & Wright, N. (2006). First observation of Hydrogen sensing by trap Assisted conduction current in Pd/TiO2/SiC capacitor at high temperature, IEEE Sensors, Exco, Daegu, Korea/October , 22-25.

[38] Werbowy, A., Zdunek, K., Dusinski, E., Szmidt, J., & Elert, M. (2003). Impulse plasma deposition of aluminum oxide layers for Al2O3/Si, SiC, GaN systems. Surface and-Coatings Technology, , 174-175, 170.

[39] Wright, P. J., & Saraswat, K. C. (1989). The effect of fluorine in silicon dioxide gate dielectric,. IEEE Trans. Electron Devices, , 36, 879.

Chapter 10

Tailoring Oxide/Silicon Carbide Interfaces: NO Annealing and Beyond

John Rozen

Additional information is available at the end of the chapter

1. Introduction

We live in an energy-hungry world in which industrialization and globalization have accelerated the demand for resources that now doubles approximately every 40 years. Today, we consume about 18 TW (18×10^{12} Watts) which is equivalent to 97 billion barrels of crude oil yearly. While renewable energy sources offer an environmentally conscious alternative to fossil fuels, they account for only about 10% of this total [64]. In parallel to the advent of clean energy, an effort has to be made to curb consumption, which can in part be achieved by improving system efficiency. In this Section, we will discuss in such terms why high-volume sectors such as transportation, electricity generation, and distribution, can benefit from SiC-based electronics.

First, it should be recognized that the adoption of a new technology will be driven mainly by component cost and end-user benefits. Silicon carbide electronics is no exception and only makes sense if it can deliver on these fronts. A good example is the recent introduction of the pricier fluorescent light sources which make financial sense in the long term since they consume a fraction of the energy of incandescent bulbs and last some 20 times longer, proving that efficiency and reliability can justify the investment. So what are the key parameters that influence SiC device cost and efficiency?

Cost - Substrate size and availability have benefited from the boom in LED demand as III-nitride blue diodes can be fabricated on SiC. Indeed, the diameter of commercially available substrates has steadily increased from the release of two inch (50 mm) wafers in September 1997 to the recent unveiling of six inch (150 mm) wafers in August 2012 by Cree, inc., a very fast pace compared to Si evolution [100]. Also, tremendous quality improvements have been achieved together with increased process rate and uniformity. One of the many challenges facing SiC production has been the reduction of extended defects such as micropipes [29, 49]. Today, substrates are virtually free of such defects, optimizing device

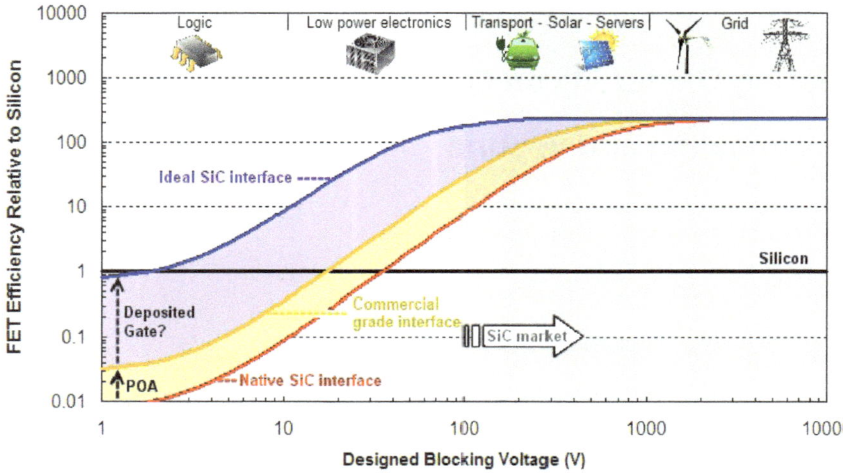

Figure 1. DC efficiency of SiC-based FETs relative to Si devices at given designed blocking voltages. While commercially available switches using NO-annealed thermal gate oxides have improved efficiency, one suggested route is the use of deposited oxides to achieve optimum properties [109].

yield. Demand and production costs have thus progressively driven down the price of the material, which has translated into cheaper and higher quality components for optoelectronic and high power applications.

Efficiency - While investment costs have diminished, SiC-based devices are still more expensive than their Si counterparts. Their efficiency is what can make them attractive in the long run. As illustrated in Fig. 1, the energy consumption of metal-oxide field-effect transistors (MOSFETs) can be orders of magnitude lower when using silicon carbide as a substrate to control high blocking voltages. Industries that would benefit from the widespread of such components include transportation, electricity distribution, grid coupling, high-performance computing, etc. Indeed, automakers have invested heavily in SiC research, targeting the implementation of SiC-based inverters in hybrid vehicles. To get an idea of how single device consumption will translate into system efficiency, let's take the example of photovoltaic (PV) power converters. PV inverters are used to convert the DC current from solar sources to feed it to the AC grid. They are made of power diodes and switches. A typical residential system has a nominal power of 5 kW at 400 V AC. Such Si-based converters can operate at over 95% efficiency but replacing Si components by commercially available SiC Schottky diodes and power MOSFETs can cut the loss by about 50%, yielding a saving of the order of $100 a year per household [22, 23]. Moreover, they can operate at higher temperature, so that limited cooling and volume requirements go in favor of system prices which can indeed prove beneficial over the years for the consumer choosing to adopt the new technology.

Further improvements in SiC device efficiency will make the case even stronger. Among the key building blocks at the device level is the oxide/semiconductor interface. Figure 1 highlights how it affects consumption, especially at low biases. In this Chapter, we will derive important parameters defining SiC devices from physical properties, and discuss the role and formation of the oxide/semiconductor interface, covering thermal oxides, post-oxidation annealing, and deposited dielectrics.

2. Breakdown field and device efficiency

Let us compare vertical double-implanted MOSFETs (DMOSFETs) designed to control the same bias, one Si-based, the other SiC-based, as shown in Fig. 2(b), using the constants of Fig. 2(a). The key differences between the two materials can be traced back to the Si-Si bond and the Si-C bond, respectively. The stronger interaction between silicon and carbon atoms is evidenced by the shorter bond length of 1.89 Å when compared to 2.35 Å for Si-Si. The proximity of atoms in SiC yield a more pronounced splitting of bonding and antibonding levels, which translates into a wider band gap in the periodic crystalline structures. The diatomic base of silicon carbide also explains the better thermal conductivity of the material because its vibration modes, i.e. phonons, are more energetic on average, as reflected in the Debye temperature. Ultimately, it is the phonon distribution that explains the higher critical field of silicon carbide, ξ_c, that can be used to derive a key parameter impacting device efficiency in high power electronics, the drift component of the specific ON resistance.

The breakdown field of a material is indeed not directly related to its band gap E_g. While, to first order, the free carriers need to reach a kinetic energy of at least $3/2\ E_g$ to induce the cascading impact ionization phenomenon, called avalanche, that multiplies the number of carriers and therefore the conductivity, the limiting factor in the bulk is phonon coupling [84, 86]. If the net velocity of carriers $\bar{v}$, proportional to the current, is smaller than or equal to the thermal velocity $v_{th} = \sqrt{3k_bT/m^*}$, the electron-phonon system is in equilibrium because of the ability of phonons to thermalize the carriers. In that regime, phonon scattering damps the energy gain of free carriers whose distribution in the bands can be visualized as a Fermi sphere slightly shifted in the direction of the electric field. However, if the field increases and reaches ξ_c, the rate at which carriers gain energy becomes too high to allow equilibrium with the lattice vibrations. Hot carriers then achieve phonon runaway. Their motion is no longer damped and they can accelerate freely from v_{th} to the critical speed $v_c \approx \sqrt{3E_g/m^*}$ allowing the avalanche process to start. It is worth noting here that in thin films, an additional constraint comes from the time the carriers take to accelerate to v_c, so that ξ_c can become larger than the bulk value, as illustrated in Fig. 3(a).

Constant	4H-SiC	Silicon
Bandgap E_g (eV)	3.26	1.12
e^- *mobility* μ_e $(cm^2/V.s)$	1000	1400
e^- *effective mass* (m^*/m_0)	0.36	0.26
Critical field ξ_c (MV/cm)	2.2	0.3
Dielectric constant (ϵ/ϵ_0)	9.7	11.7
Debye temperature T_d (K)	1300	640
Th. conductivity θ $(W/cm.K)$	3.7	1.5

(a)

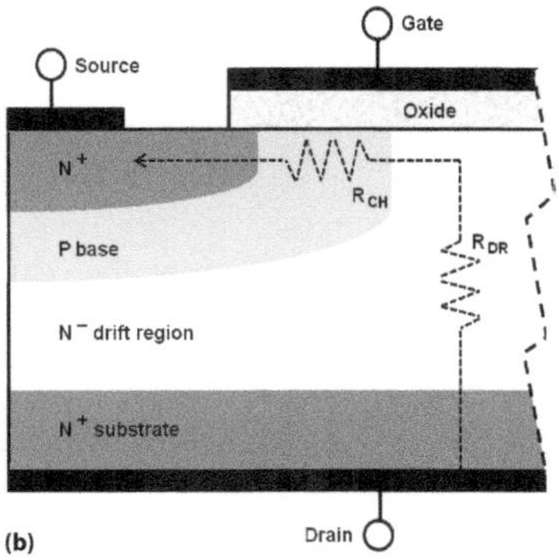

(b)

Figure 2. (a) Properties of 4H-SiC and Si with 10^{15}-10^{16} cm^{-2} doping at 300 K [30, 65, 66, 84, 121]. (b) Vertical power DMOSFET.

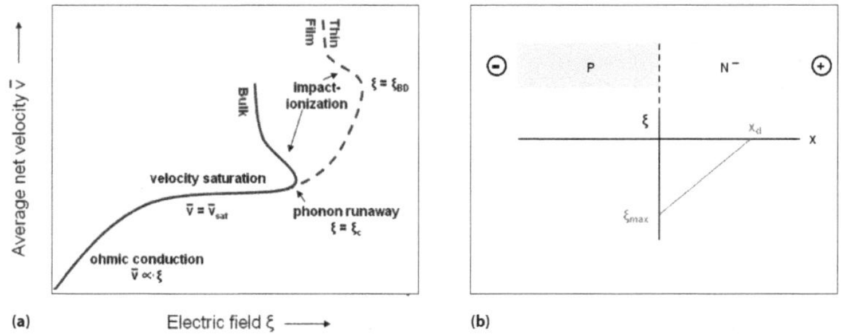

Figure 3. (a) Hot carrier velocity saturation and dielectric breakdown [105]. (b) Field distribution in a one-sided reverse-biased PN junction.

The critical field ξ_c of a semiconductor, can be used to design the most efficient device for a given blocking voltage V_d. When a DMOSFET is in the OFF state, the positive bias applied to the drain is entirely dropped in the N^- drift region [84, 124]. Indeed, together with the P base, it forms a reverse-biased one-sided PN junction, represented in Fig. 3(b). For a large blocking voltage, $V_d \gg E_g/q$, the extent of the depletion region on the lowly-doped side in the step-junction approximation is

$$x_d = \sqrt{\frac{2\epsilon_s V_d}{qN_d}} \tag{1}$$

where N_d is the density of donor atoms in the drift region, ϵ_s is the permittivity of the semiconductor, and q is the elementary charge. The peak electric field, at the boundary, being

$$\xi_{max} = 2\frac{V_d}{x_d} = \sqrt{\frac{2qN_dV_d}{\epsilon_s}} \tag{2}$$

The highest doping concentration that can sustain V_d is therefore

$$N_d^* = \xi_c^2 \frac{\epsilon_s}{2qV_d} \tag{3}$$

obtained by substituting for the critical field in Eq. (2). Then, from Eqs. (1) & (3), the minimum thickness of the drift region is given by

$$x_d^* = 2\frac{V_d}{\xi_c} \tag{4}$$

Accordingly, the drift region of a 4H-SiC DMOSFET can be substantially thinner and more highly doped than a Si-based device designed to control the same bias. Neglecting the current spread [9], the ratio of the optimal ON resistance components from the drift region can thus be estimated by

$$\frac{R_{dr}^{*}\,|_{SiC}}{R_{dr}^{*}\,|_{Si}} = \left.\frac{N_d^{*}\,\mu_{dr}}{x_d^{*}}\right|_{Si} \left.\frac{x_d^{*}}{N_d^{*}\,\mu_{dr}}\right|_{SiC} = \frac{\mu_{dr}\,\epsilon_s\,\xi_c^{3}\,|_{Si}}{\mu_{dr}\,\epsilon_s\,\xi_c^{3}\,|_{SiC}} \tag{5}$$

where μ_{dr} is the drift mobility in the bulk of the semiconductor. Plugging the respective constants of the bulk semiconductors into Eq. (5) implies that in ideal devices the energy dissipated using 4H-SiC would be several hundred times lower compared to the Si equivalent for a given ON current [15].

In a real device, however, there are other components to the resistance such as the contact resistance and the channel resistance. Here we will discuss only the channel specific resistance which contains the contribution from the oxide/semiconductor interface of interest in this Chapter. It can be calculated using the long channel approximation as

$$R_{ch} = \frac{V_d}{I_{sat}}\,P^2 = \frac{LP}{nq\mu_{ch}} \tag{6}$$

where L is the channel length, P is the channel width, or the square cell pitch, μ_{ch} is the inversion mobility, and n is the minority carrier density [9]. The total specific ON resistance is then

$$R_{tot} = R_{dr}^{*} + R_{ch} = \frac{x_d^{*}}{N_d^{*}\,q\mu_{dr}} + \frac{LP}{nq\mu_{ch}} \tag{7}$$

which has the following dependence on designed blocking voltage according to Eqs. (3) & (4)

$$R_{tot} = \frac{4V_d^2}{\mu_{dr}\epsilon_s\xi_c^3} + \frac{LP}{nq\mu_{ch}} \tag{8}$$

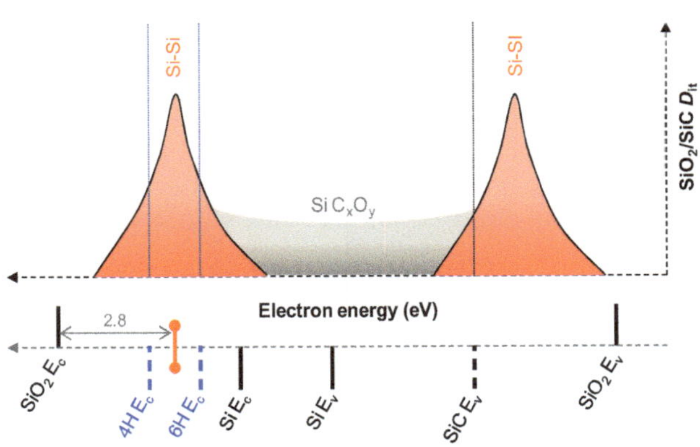

Figure 4. Schematic of the density of interface traps at the SiO_2/SiC interface.

From Eqs. (7) & (8), it can be seen that the smaller the designed blocking voltage, the smaller the width of the necessary drift region, and the larger the contribution from the channel resistance. In Si, that does not have a major impact in power devices because μ_{ch} can be as high as 50% of the bulk value μ_{dr}. However, SiC channels suffer from a mobility that can be less than 1% of μ_{dr} at the native SiO_2/4H-SiC interface. Therefore, interface quality can affect performance even in the kV range and the full potential of the SiC material cannot be reached. This is highlighted in Fig. 1, where the ratio of ON resistances was calculated using Fig. 2(a) constants, $L = 1\mu m$, $P = 10\mu m$, $n = 10^{15}$ cm^{-2}, and the following SiC μ_{ch}: 5, 50, and 500 cm^2/V.s. The significance of those mobility values are discussed in the next Sections.

3. Channel mobility and interface state density

Mobility is a measure of the ease a carrier can be moved in a solid under the application of an electric field ξ. It can be related to the speed of the carrier which is limited by scattering events occurring at average time intervals τ. Several types of scattering processes affect transport but the ones yielding the most frequent disturbances define the mobility value which can be written explicitly as

$$\frac{1}{\mu} = \frac{\xi}{\bar{v}} = \frac{m^*}{q} \sum_i \frac{1}{\tau_i} \tag{9}$$

where m^* is the effective mass, $\bar{v}$ is the net drift velocity, and τ_i corresponds to mean scattering times associated with various processes.

Because of the sudden termination of the semiconductor periodic lattice at the oxide interface, the channel mobility, μ_{ch}, is expected to be lower than the bulk mobility. Indeed, electrically active levels can appear in the band gap and act as recombination centers or Coulomb scattering centers. Moreover, the free carriers can interact remotely with charged border states in the oxide, further reducing τ_{Cb}. Other major damping mechanisms include surface-phonon and surface-roughness scattering represented by τ_{ph} and τ_{sr}, respectively. We note that the coupling of carriers to scattering effects depends strongly on their velocity such that Coulomb scattering dominates at low fields, while surface roughness scattering becomes dominant at higher fields. Because of these interface phenomena, the best SiO_2/Si devices display a channel mobility of about 700 cm^2/V.s, equivalent to 50% of the bulk mobility [78]. In the case of 4H-SiC however, the native oxide interface yields mobilities of about 5 cm^2/V.s, or less than 1% of the bulk value. So why is SiC so affected by the formation of a thermally grown interface? Let us discuss it from the point of view of Coulomb scattering, so that the question becomes: why is the density of interface traps (D_{it}) so much more prominent in SiC?

First, a down side of having a wider band gap is that it is sensitive to a wider range of defects. To first order, only the corresponding levels falling inside the band gap can be charged and yield Coulomb scattering. Since there is no evidence that SiO_2 formed on SiC is any different from thermally oxidized silicon [99], it can contain the same type of defects, some having energy levels affecting only SiC carriers. Let's take the example of the oxygen vacancy, detected by electron spin resonance (ESR) in both systems, which yields Si-Si bonds [75, 106]. The energy split of that dimer is inversely proportional to the proximity between atoms. Long Si-Si bonds, i.e. bigger than 2.35 Å, can have energy levels inside both the Si and SiC gaps. But the splitting between short Si-Si levels can be outside the silicon gap while being electrically active at a SiO_2/SiC interface [90, 130], as illustrated in Fig. 4. The same analysis is true when comparing defect levels in the 6H- and 4H-SiC polytypes which have energy gaps of 3 and 3.3 eV, respectively. Because their valence bands are lined up, the conduction band of 4H-SiC is 0.3 eV higher. This is why the major defect affecting 4H-SiC, located 0.1 eV below its conduction band, has a limited impact on 6H-SiC devices [114]. It is interesting to note that this defect corresponds to an energy of 0.4 eV above the Si conduction band and that a similar trap level has been observed in SiO_2 on Si even though it does not limit channel mobility [3, 4, 8].

The second reason explaining the large defect density at SiC interfaces is the oxidation process

$$2\,SiC + 3\,O_2 \rightarrow 2\,SiO_2 + 2\,CO \tag{10}$$

Like silicon oxidation, it follows the Deal-Grove reaction-diffusion model, so that the thermal oxide thickness as a function of time can be written as

$$x[t] = -\frac{1}{2}\frac{B}{B/A} + \frac{B}{2}\sqrt{\frac{1}{(B/A)^2} + \frac{4(t+t_0)}{B}} \tag{11}$$

where B/A is the linear rate constant, B is the parabolic rate constant, and t_0 is an offset time constant [45]. For details on SiC oxidation kinetics and parameters corresponding to various orientations, pressures, and temperatures, see Refs. [43, 68, 98, 113, 122, 127, 135, 136]. But unlike Si oxidation, Eq. (10) implies the release of carbon. Because of this complex multi-step process, a variety of atomic defects involving C can result from oxide formation if CO molecules do not all find their way to the gas phase. [48, 71, 90]

Now that we have reviewed the impact and origin of defects at the SiO_2/SiC interface, a comprehensive picture of associated trap levels limiting inversion mobility can be put forward, as shown in Fig. 4. Thermal oxidation of silicon carbide results in a SiC_xO_y inter-layer that includes threefold and fourfold coordinated Si and C atoms [8, 44, 90, 130]. Some generate dangling bonds whose energy spreads across the band gap because it is determined by the environment surrounding the defects. Another stable configuration yielding a trap level in the gap is a split C interstitial. When substituting for a Si site in the semiconductor, it can be viewed as a small C-aggregate or "C-cluster" comprising 6 atoms. A carbon-rich interface has indeed been observed by techniques such as Rutherford backscattering (RBS) [46, 57, 89, 97], x-ray photoelectron spectroscopy (XPS) [54, 62, 67, 77, 128], electron energy-loss spectroscopy (EELS) [33–35], Raman spectroscopy [80], and in-situ spectroscopic ellipsometry [63]. The dominant defects however, are likely oxide-related. Indeed, Si-Si bonds of various lengths can extend into SiO_2 yielding interface and border traps [6, 17, 50, 72, 92, 112]. As mentioned before, the majority of corresponding bonding and antibonding states are located outside of the silicon gap but the former induce a distribution centered between the 4H and 6H-SiC conduction bands. The slightly smaller density of levels at lower energies is probably due to the majority of antibonding levels residing within the valence band of SiC whose edge is common in different polytypes [1]. The combination of trap levels associated with the SiC_xO_y inter-layer and Si-Si bonds, yields the U-shaped D_{it} distribution. It rises sharply towards the SiC band edges because of the Si-related defects not dominant at silicon interfaces. Therefore, the efficiency of passivation techniques are expected to be very different at interfaces formed on the two semiconductors.

In the following Sections, we will discuss how to reduce D_{it} and its relationship with mobility. Although there is extensive literature dedicated to various orientations and polytypes, this overview is dedicated to devices fabricated on the (0001) Si-face of 4H-SiC.

4. Argon anneal

Oxidation conditions and post-oxidation annealing (POA) can affect the trap density at the SiO_2/4H-SiC interface. Both Ar anneal performed at growth temperature and re-oxidation at lower temperatures (e.g. 900 °C) have indeed proven to slightly reduce the amount of deep states [40, 129]. This can be explained by the removal of excess carbon without additional oxide formation, as corresponding atomic configurations yield defects populating interface states toward the middle of the gap. Since it does not reduce the density of levels close to the conduction band edge of 4H-SiC, Ar POA alone is not sufficient to enable efficient SiC devices. Nevertheless, it is typically used after thermal oxidation and before other annealing schemes.

5. Hydrogen passivation

Wet oxidation of 4H-SiC also yields a small reduction of interface states with energies away from the semiconductor band edges when compared to SiO_2 formation in dry oxygen [2, 26, 55, 90, 130]. It correlates well with the effects of H_2 POA. While at silicon interfaces hydrogen annealing yields a D_{it} from about 10^{11} to 10^{10} $cm^{-2}eV^{-1}$ in the middle of the gap and a mobility close to half the one of the bulk [20, 21], its impact at SiC interfaces is much less efficient, highlighting the differences between the two semiconductors [27, 58, 96]. Molecular hydrogen can indeed passivate Si- or C- dangling bonds, and insert long Si-Si bonds [28]. But it does not significantly affect split carbon interstitials and slow near interface states which populate the majority of the D_{it} at the 4H-SiC band edges. Like Ar annealing, wet oxidation and/or H_2 POA can be used together with other annealing techniques to optimize the trap density throughout the semiconductor band gap.

6. Nitridation

6.1. NO annealing

In 1997, the group of Prof. Dimitrijev, at Griffith University in Australia, demonstrated that high temperature (1100 °C) nitric oxide (NO) annealing reduces the D_{it} at SiO_2/6H-SiC interfaces [76]. In 2000, Chung *et al.* published results on the effects of NO at the SiO_2/4H-SiC interface revealing that, in addition to removing deep states, it is also very efficient at reducing the density of slow states (by a factor of ≈ 10), and yields an order of magnitude increase in the channel mobility from about 5 to 50 cm^2/V.s along the Si-face [40, 41]. This breakthrough discovery, which originated from the joint effort between Auburn University and Vanderbilt University, led to the adoption of the NO process by the scientific and industrial communities as it enables the fabrication of high-quality oxide-based SiC power devices, facilitating their commercial release (Fig. 1).

The benefits of NO annealing have been directly correlated with the incorporation of nitrogen, which is confined to the SiO_2/SiC interface, as detected by various techniques such as secondary ion mass spectroscopy (SIMS) [82, 106], nuclear reaction analysis (NRA) [81], electron energy loss spectroscopy (EELS) [33], and medium energy ion scattering (MEIS) [47, 137]. To study the impact of nitrogen, the amount incorporated can be tailored by the NO annealing time as illustrated in Fig. 5(a). The N density is then extracted by integrating SIMS interface peaks resulting from 1175 °C NO exposure of a dry thermal oxide for up to 2 hours. The nitrogen content is found to saturate around 6×10^{14} cm^{-2} or about a half monolayer coverage of the SiC surface. The nitridation kinetics result from a balance between N incorporation and removal. Indeed, at this temperature, NO decomposes partially into N_2 and O_2. While 1175 °C is required to enable NO diffusion to the interface and subsequent nitridation, the presence of oxygen limits its effect as interfacial nitrogen is unstable against the slow re-oxidation occurring in parallel [37]. Moreover, additional defect formation can also result from the presence of the excess oxygen.

Progressive reduction of the D_{it} across the 4H-SiC band gap corresponding to the tailored introduction of nitrogen has been measured in metal-oxide-semiconductor capacitors (MOSCAPs), as shown in Fig. 5(b). The density of states shows a strong correlation to the nitrogen content and is reduced by up to an order of magnitude close to the conduction band edge [82, 106, 108]. The sensitivity of the inversion mobility of electrons to the D_{it} reduction was studied in lateral field-effect transistors containing different amounts of nitrogen at the SiO_2/4H-SiC interface. From the results depicted in Figs. 6(a) & 10, the peak field-effect mobility is found to be inversely proportional to the density of charged states, which reveals a Coulomb-scattering-limited transport. It is important to note that this is true even in devices with the lowest D_{it} so that further defect passivation is projected to increase the mobility from 50 to more than 100 cm^2/V.s, which cannot be achieved by NO POA alone as nitrogen density becomes saturated. These conclusions are in agreement with separate mobility studies using Hall effect measurements on nitrided samples [13, 114, 125]. Such experiments also reveal that at higher fields, mobility becomes limited by surface-roughness scattering. Although NO POA has been shown to yield smoother interfaces [51], it is not clear what else can be done to further reduce that particular component.

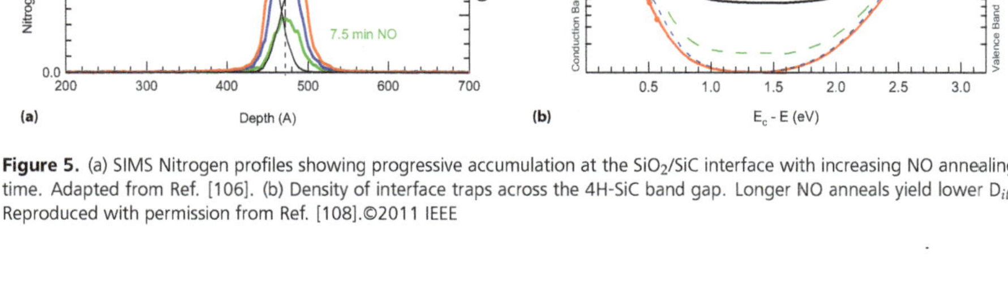

Figure 5. (a) SIMS Nitrogen profiles showing progressive accumulation at the SiO_2/SiC interface with increasing NO annealing time. Adapted from Ref. [106]. (b) Density of interface traps across the 4H-SiC band gap. Longer NO anneals yield lower D_{it}. Reproduced with permission from Ref. [108].©2011 IEEE

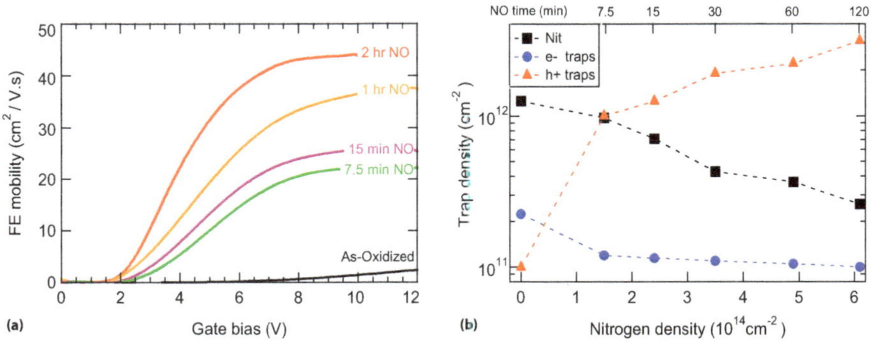

Figure 6. (a) Field-effect mobility extracted from lateral n-channel MOSFETs on 4H-SiC fabricated using different N doses. Reproduced with permission from Ref. [107]. ©2010 TTP (b) Dependence of integrated interface trap density (N_{it}), electron-generated levels, and positive oxide traps, on N content. Reproduced with permission from Ref. [107]. ©2010 TTP

The benefits of nitrogen incorporation also extend to the reliability of SiC devices when it comes to electron injection in the gate oxide, which is inherent to n-channel transistor operation, as the dielectric is exposed to leakage currents and charge tunneling from the inversion layer towards the biased gate contact [31, 32, 52, 73, 132]. While the resulting degradation can take a long time to develop under normal operation, accelerated techniques can be used to study device response to excess carriers penetrating the gate structure. To best simulate actual bias conditions, electrons were injected at low oxide fields (< 2 MV/cm) using a mercury lamp promoting carriers from the negatively-biased gate metal to the conduction band of 4H-SiC in MOSCAPs fabricated using dry oxidation followed by various NO annealing times [103, 106]. As shown in Fig. 7(a), the density of trapped negative charge, extracted from the flatband voltage shift of capacitance-voltage (CV) curves, can be plotted as a function of the integrated gate current, i.e. the injected electron density. If no nitrogen is present at the oxide/semiconductor interface, device characteristics continuously drift towards positive voltages. From the observation of hysteresis behavior of CV curves, this has been correlated to electron-induced acceptor state generation at the interface [5]. However, the presence of even the smallest amount of nitrogen can suppress the degradation, exposing the secondary component of the negative charge, the bulk electron traps in SiO_2 [103], Fig. 6(b).

When it comes to positive charge trapping however, the presence of nitrogen proves to be detrimental to device stability [73, 74]. This can be of concern even for n-channel transistors to which a negative gate bias can be applied to ensure that it is OFF in its idle state. Several methods have been used to accelerate hole exposure of the gate oxide such as x-ray irradiation [52], Fowler-Nordheim tunneling, and internal photoemission [104]. Figs. 6(b) & 7(b) show the positive trapped charge as a function of the injected hole carrier density using such a technique with samples containing different amount of nitrogen. There is a clear correlation between nitrogen content and oxide trap density. Because the nitrogen is contained at the interface and the charge is stable against bias reversal, it is attributed to near-interface traps in the SiON layer. ESR experiments have ruled out oxygen vacancies as the main positive trap in the oxide, another indication that atomic configurations involving nitrogen must be at play [106]. In fact, similar conclusions have been drawn from N-induced negative bias temperature instability (NBTI) on silicon which has been attributed to Si-N bonds [24, 25], which also suggests that the defects must originate from the SiO_2 conversion into a nitride.

From the impact of nitrogen incorporation on D_{it}, electron traps, and hole traps, summarized in Fig. 6(b), it is possible to paint a picture of what happens at the atomic level. First, it is important to note that although the term "passivation" is often used to describe the effects of nitrogen, the large density introduced by the NO process more likely yields a complete reconfiguration of the interface from SiO/SiC to SiON/SiC. But for simplicity, let us look at the effect of nitrogen on single defects. As mentioned previously, the majority of traps at the thermally grown interface are considered to be single and split carbon interstitials, as well as Si-Si bonds. Unlike hydrogen, nitrogen can reduce the energy of most of these atomic configurations by substituting for threefold-coordinated carbons or by inserting short suboxide bonds [90, 130]. While it can suppress acceptor levels close to the 4H-SiC conduction band edge, the donor nature of nitrogen and its 5 valence electrons can then yield states close to, or even within, the semiconductor valence band [104]. If the resulting atomic configurations are located on the oxide side of the interface, they can therefore act as stable hole traps. Let us take the example of a short Si-Si suboxide bond expected to have an unoccupied state close to the 4H-SiC conduction band. When inserted by NO, it yields a Si-NO-Si bridge and moves the trap level close the valence band. Theory suggests that the nitrogen lone electron pair leads to a partially occupied state that is a favorable hole trap, since giving away an electron makes the atomic configuration reduce its energy by a few eV. As mentioned earlier, nitrogen might not discriminate between defects and stoichiometric oxide sites as it indeed converts SiO to SiON. Therefore, the same configuration could result from N insertion in bridging Si-O-Si, which can create trap levels where there was none before.

In summary, the presence of nitrogen at the oxide/semiconductor interface is beneficial as it reduces D_{it}, increases mobility, and suppresses electron-induced interface state generation. But on the other hand, re-oxidation during NO POA limits the amount of N that can be inserted, and nitrogen generates a quantity of new traps by bonding in the near-interface

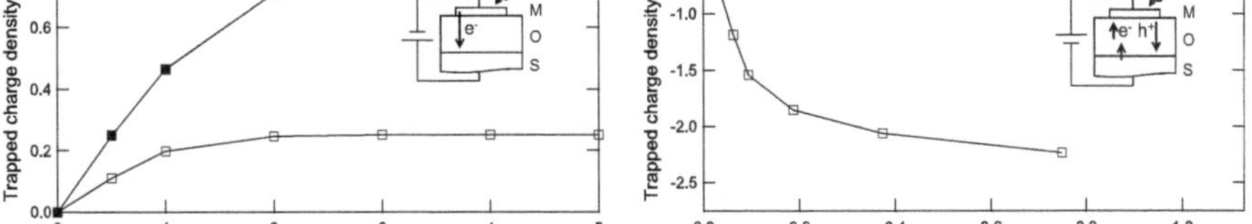

Figure 7. Effective trapped charge as a function of (a) injected electron density and (b) emitted hole density by internal photoemission. NO POA suppresses electron-induced interface state generation but increases the amount of hole traps in the oxide. Adapted from Refs. [103, 104]

region of the oxide. Since we have shown that more nitrogen would increase the mobility even further, other nitridation methods could maximize its density while confining it to the interface boundary.

6.2. N_2O and NH_3

Ammonia (NH_3) POA shows benefits as well in terms of D_{it} reduction close to the conduction band edge of 4H-SiC [39, 53, 133]. However, it yields unnecessary incorporation of nitrogen throughout the oxide, totalling a density $\sim$ 100 times larger than NO POA. This compromises the integrity and reliability of the gate dielectric, as evidenced by the lowering of the breakdown voltage.

Nitrous oxide (N_2O) oxidation or POA also improves the properties of the oxide/semiconductor interface, but to a lesser extent than NO [12, 37, 70, 76, 81]. This is because it decomposes at high temperature into NO, O_2, and N_2. While the resulting NO incorporates nitrogen at the 4H-SiC/SiO_2 interface, the larger fraction of background oxygen mitigates this greatly. Indeed, since incorporated N is unstable against the slow re-oxidation occurring in parallel, N_2O POA yields about an order of magnitude less nitrogen than pure NO POA at similar process temperature. It reduces D_{it} by about a factor of 2 close to E_c and leads to a peak field-effect mobility of up to $\approx$ 25 $cm^2/V.s$ in 4H-SiC transistors, as indicated in Fig. 10. However, nitrous oxide is sometimes preferred over NO for safety reasons as it comes with less demanding handling requirements.

6.3. N implants and radicals

Following the improvements induced by NO POA, other methods were developed to introduce N at the SiO_2/SiC interface. Although they may be more involved, they can yield NO-like properties for oxide-based devices formed on 4H-SiC and bring their own contribution to understanding the role of nitrogen.

One such nitridation technique is implantation. N^+ ions can be inserted in the top semiconductor layer that will subsequently be consumed by thermal oxidation, yielding the presence of nitrogen at the interface. The amount of N atoms can be tuned by implantation dose and energy. Studies have revealed that similarly to NO POA, the higher the nitrogen density at the thermally-formed interface, the lower the D_{it}, and the higher the field-effect mobility [42, 91, 93]. In fact, Poggi *et al.* have reported about an order of magnitude reduction of electrically active defects close to the conduction band edge of 4H-SiC and a room temperature field-effect mobility of up to 42 $cm^2/V.s$ in lateral nFETs fabricated on the (0001) surface [94, 95], Fig. 10. While the progressive increase of μ_{FE} with N dose is consistent with the reduction of Coulomb scattering, Hall mobility measurements reveal that in devices with the higher nitrogen content, μ_{Hall} decreases with temperature. This implies that, unlike for the NO process, another dominant scattering mechanism appears following high implant doses. This has been attributed to induced damages in SiC and residual N interstitials left within the semiconductor [16]. Also, note that the process temperature can be kept at or below 1100 °C following implantation, to avoid activation that would convert N atoms into donors in SiC. But activation of a minority of dopants in the tail end of the implant can never be ruled out.

Another elegant way to introduce nitrogen is the exposure of thermal oxides to nitrogen radicals [116, 134, 137]. It can be achieved using a remote plasma generating highly reactive N^+ ions. SIMS measurements have shown that, like NO-POA, this results in nitrogen accumulation strictly at the interface between the oxide and the semiconductor. One advantage being that it can potentially occur without re-oxidation, allowing for N maximization. Studies on devices fabricated on the Si-face of 4H-SiC again show that the D_{it} is reduced proportionally to the amount of incorporated nitrogen, in line with results from NO POA. In fact, similar and even better performance in terms of peak field-effect mobility has been demonstrated using that technique. However, prolonged plasma exposure can also reduce the integrity of the gate oxide, implying that this promising nitridation method still requires optimization.

7. Phosphorus

In 2009, about a decade after the introduction of NO annealing, Okamoto *et al.*, from the Nara Institute of Science and Technology in Japan, proposed another post-oxidation annealing technique that significantly reduces D_{it} at SiO_2/4H-SiC interfaces. As mentioned in the previous section, implantation of nitrogen in SiC prior to oxidation has proved to be a beneficial nitridation technique. Hence, Prof. Yano and his group cleverly extended this logic to a screening method for various potential passivating species [87]. This is how phosphorus caught their attention as oxidation of P-implanted SiC also showed a lower density of electrically active defects than as-oxidized un-implanted interfaces. Following this discovery, they implemented a more gentle way to introduce P at the interface in order to avoid ion-induced damages and undesirable doping of the substrate, by flowing gas through a $POCl_3$ bubbler during a high temperature post-oxidation anneal.

When performed at 1000 °C on SiO_2, grown on the Si-face of 4H-SiC, $POCl_3$ POA leads to a D_{it} below 10^{11} $cm^{-2}eV^{-1}$ close to E_c, or several times lower than following NO POA [88]. This is reflected in the efficiency of lateral nFETs as the peak value of the field-effect mobility almost doubles compared to NO POA to about 90 cm^2/V.s. This has been correlated with the presence of phosphorus at the interface. Another proposed method to reach similar mobility values is exposing a thermal oxide to P_2O_5 extracted from a solid phosphosilicate glass (PSG) diffusion source [115]. Device properties following $POCl_3$ or PSG POA are reported in Figs. 8(a), 8(b) & 10.

Note that from SIMS analysis, it is found that both $POCl_3$ and PSG POA convert the dielectric into a phosphosilicate by yielding phosphorus throughout the gate. This compromises the reliability of the devices. Recently, forming a thin P-containing interfacial oxide, using $POCl_3$ and O_2, followed by dielectric deposition, was shown to reduce trapping by narrowing the phosphorus profile [11].

The benefits of phosphorus at SiO_2/SiC interfaces represent a milestone for silicon carbide research; not only because of mobility improvements, but also because it shines light on the nature of passivation at the atomic level. Indeed, both N and P are among the group V elements of the periodic table, possessing similar chemistry due to their 5 valence electrons. For example, it has fueled the discussion of the role of sub-surface SiC doping in improving device characteristics [36]. But while the physics of N and P binding at interfaces is still being debated, we are one step closer to a more comprehensive understanding of post-oxidation annealing mechanisms.

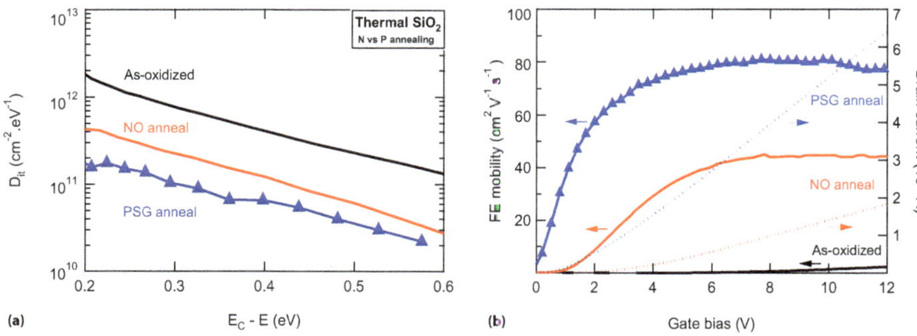

Figure 8. (a) D_{it} reduction compared to as-oxidized films using NO POA or a phosphorus source. Adapted from Ref.[115]. (b) Transfer characteristics and extracted field-effect mobilities showing the efficiency of P. Adapted from Ref.[115].

8. Deposited oxides

Considering all the efforts necessary to improve the interface, boasting that silicon carbide should be preferred over other wide band gap semiconductors because it can grow a stable thermal oxide may no longer be a valid argument. First, the dominating trap level close to the conduction band edge of 4H-SiC has been associated with slow near-interface defects most likely residing inside SiO_2, possibly oxygen vacancies. Second, evidence is mounting that the oxidation process itself, which creates C-related defects at the oxide/semiconductor interface, also yields a transition region that extends within SiC, potentially affecting transport properties inside the inversion layer of transistors [10, 17, 19, 117].

In particular, it has been proposed by the group of Prof. Kimoto, from Kyoto University in Japan, that thermal oxidation results in the emission of C atoms from the interface towards the bulk of the semiconductor [61]. Indeed, deep level transient spectroscopy (DLTS) measurements reveal a strong similarity between the effects of thermal oxidation and the ones of C implantation in irradiated samples, as both processes yield the suppression of a level labeled $Z_{1/2}$ thought to be associated with C vacancies in silicon carbide [69, 123]. A refined oxidation model including the kinetics of emitted atoms was then put forward by Prof. Hijikata's research group at Saitama University in Japan; it successfully models their in-situ ellipsometry observations of thermal oxide growth rate on silicon carbide in both the thin and thick film regimes [59].

Because of these drawbacks of thermal oxidation, it should not come as a surprise that using deposited oxides has yielded encouraging results. For example, chemical vapor deposition of a thin SiN buffer layer on the Si-face of 4H-SiC followed by SiO_2 deposition and N_2O annealing, has been reported to lower the D_{it} and increase the field-effect mobility to above 30 $cm^2/V.s$ in lateral n-channel devices [85]. This is an improvement in efficiency compared to gates formed thermally in N_2O, highlighting the benefits of deposited dielectrics. In that particular case, thinning the SiN layer has also proven to be key as thick oxynitride films resulting from oxygen insertion can possess a high density of positive charges which lower mobility and move the threshold voltage to negative values.

8.1. Alternate dielectrics

The versatility added in device fabrication from using deposited oxides allows to explore a variety of deposition techniques, temperatures, and most importantly the ability to use gate materials other than SiO_2. It is interesting to note that the reason for considering alternate dielectrics is to move away from thick thermal oxides to reach higher mobility, which is fundamentally different from the evolution of gate fabrication in silicon technology where the motivation comes from scaling and the need for physically thicker oxides at a given capacitance to maintain gate control while minimizing leakage, often at the expense of mobility [56, 131]. This has led to the development and integration of materials which possess a higher dielectric constant, or higher k, that are used in the new generations of Si electronics. Regardless of the dichotomy, it comes at a very opportune time for SiC device research which looks to take advantage of the acquired expertise in deposited gate dielectrics.

Another factor to consider when using high-k oxides on a wide band gap material like SiC however, is the reduced band offsets between the semiconductor and the dielectric. The conduction band offset relative to 4H-SiC for example goes from ≈ 3 eV with a SiO_2 gate to ≈ 1 eV with a HfO_2 gate [1, 102]. This has implications in terms of leakage current and reliability as such small barriers promote tunneling of carriers into the insulator, even more so that SiC devices are expected to perform at high temperature, which exacerbates the issue [105]. So if HfO_2 is to be considered as a gate material, a thin SiO interlayer is necessary to achieve reasonable leakage. Indeed, Afanas'ev *et al.* have demonstrated that such structures have good interface properties, without the need for nitridation, but that it comes at the expense of a maximum surface field of about 3 MV/cm in the semiconductor to ensure gate integrity, dangerously close to transistor minimum requirements [7]. Ultimately, a balancing act between SiO and HfO_2 thicknesses yields a compromise between performance and reliability [38].

But unlike Si technology, we are not aiming for the highest possible dielectric constant, so we can somewhat move away from this compromise as long as a quality dielectric can be obtained on SiC. A promising candidate in this respect is Al_2O_3; it has a dielectric constant close to 9 and a band gap only a few tens of eV narrower than SiO_2, yielding a conduction band offset relative to 4H-SiC that is still above 2 eV. Most importantly, it does not possess the same dominating trap level as SiO_2, so that high electron mobility can be achieved in n-channel devices. Indeed, peak field-effect mobility values measured at room temperature on the Si-face can exceed 100 $cm^2/V.s$ [60, 79]. A key observation is again that a thin thermal SiO or SiON layer is still required, not so much to reduce gate leakage but to increase efficiency by providing a progressive transition between the semiconductor and the deposited oxide. Since it acts as a passivating layer, it is no surprise that while it is needed, the thinner it is, the better, so that a good interface can be formed while reducing the impact of remote thermal oxide traps on channel transport properties. In practice, the stability of a high-k material on SiO has to be considered carefully to avoid intermixing of the atomic species during subsequent device fabrication steps. To mitigate this, a SiN barrier layer can be used between the thermal oxide and Al_2O_3, blocking Al diffusion to the interface [83], or the learning from Si technology can be extended to the use of low-temperature gate-last processing schemes [14].

8.2. Surface conditioning

In the previous Sections, we have shown that while moving away from a thermally-formed gate oxide improves transport properties in 4H-SiC devices, a thin SiO transition layer still brings a much-needed passivated interface. While this is also true for deposited dielectrics on Si, the remarkable difference is that the thinner the thermal oxide is, the better. The next logical step is therefore to avoid thermally-formed SiO entirely, circumventing the negative impact of oxidation-induced defects and near-interface traps. But how can this be done so that there is a clean transition between the SiC surface and the deposited oxide?

It all comes down to surface preparation of the semiconductor, which cannot simply end with a wet clean since atomic ordering and bonding are paramount. A common surface preparation technique that has been used prior to epitaxial deposition or graphene formation on SiC might provide part of the answer. Hydrogen exposure of the (0001) face of hexagonal SiC to high temperatures (e.g. 1350 °C) etches the surface, smoothing it and yielding atomic reconstruction [18, 101, 120, 126]. However, if a gate dielectric is deposited directly on it, there appears to be no reduction of the D_{it} when compared to an HF-last surface, which, as expected, is even worse than a thermally formed interface, at least when looking at SiO_2 deposition on 4H-SiC. But Shirasawa *et al.* observed on 6H-SiC that when high temperature H_2 exposure is directly followed by N_2 exposure at the same temperature, it results in an ordered surface terminated by a monolayer of SiN topped by a monolayer of SiO [118, 119]. These thin films have well defined band gaps, a sign of their electrical integrity, and are estimated to be stable without the existence of dangling bonds at the SiC surface.

Our group at the Central Research Institute of Electronic Industry, located in Yokosuka, Japan, then speculated that such a technique could be applied to 4H-SiC in order to form an ideal seed layer for subsequent deposition [110, 111]. Indeed, this "nitrogen conditioning" process is expected to yield an interface saturated by N atoms which are highly localized within a monolayer, a very promising scenario since we know that maximizing nitrogen content at the interface while limiting its presence in the oxide is key. Moreover, the single oxide layer formed by Si-O-Si bridges bonded to the nitrogen provides the bare minimum silicon oxide, while still offering a clean transition to a deposited layer.

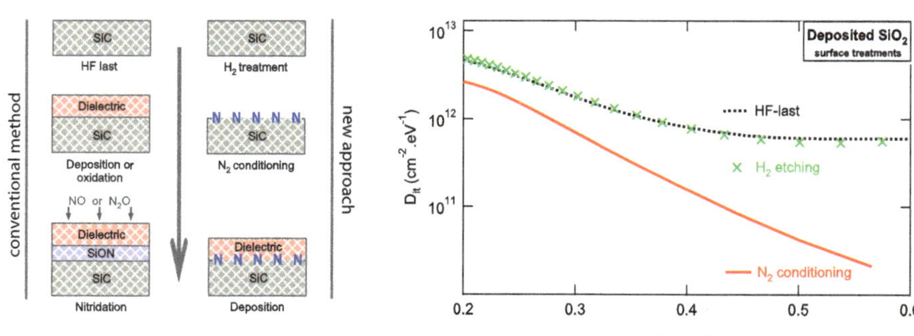

Figure 9. Schematic of the nitrogen conditioning technique for deposited oxides vs. conventional nitridation methods. On the right is the corresponding D_{it} from deposited SiO_2 on surfaces which were subjected to various treatments. Adapted from Refs. [110, 111].

XPS and SIMS measurements confirm that a SiON layer can be formed using nitrogen conditioning on the Si-face of 4H-SiC. More importantly, we find that the nitrogen remains localized at the interface following 900 °C SiO_2 deposition by CVD. N is exclusively bonded to silicon, and no C-O, C-N, or N-O bonds are detected, which is consistent with the structure mentioned above, and atomically cleaner than interfaces resulting from either thermal oxidation or NO POA [77]. Electrical characterization shows that the density of interface states of a deposited oxide is lower than the one of a thermal oxide only when nitrogen conditioning is used, Fig. 9. While this is a notable improvement over as-deposited oxides on a HF-last surface, even better results are expected with alternate dielectrics and low temperature deposition methods such as atomic layer deposition (ALD). Indeed, if the seed layer can be kept intact and a quality oxide can be deposited, this smooth and clean interface should translate into higher inversion mobility because of reduced Coulomb and surface-roughness scattering.

9. Summary and conclusions

In order to realize the full potential of SiC-based devices across the voltage range depicted in Fig. 1, the quality of the oxide/semiconductor interface must be improved. Figure 10 summarizes both mobility and D_{it} from thermal oxide gates following the various annealing processes described in this Chapter. Despite increasingly efficient POA processes from N_2O, to NO, to phosphorus exposure, the wide band gap and complex oxidation process yield a large density of states that still limits the mobility in even the best oxide-based SiC transistors. While rigourous experiments have led to that conclusion, the trend depicted in Fig. 10 shows a clear dependence between carrier transport and interface states. It suggests that the significant increase in mobility compared to NO POA induced by PSG or $POCl_3$ annealing is

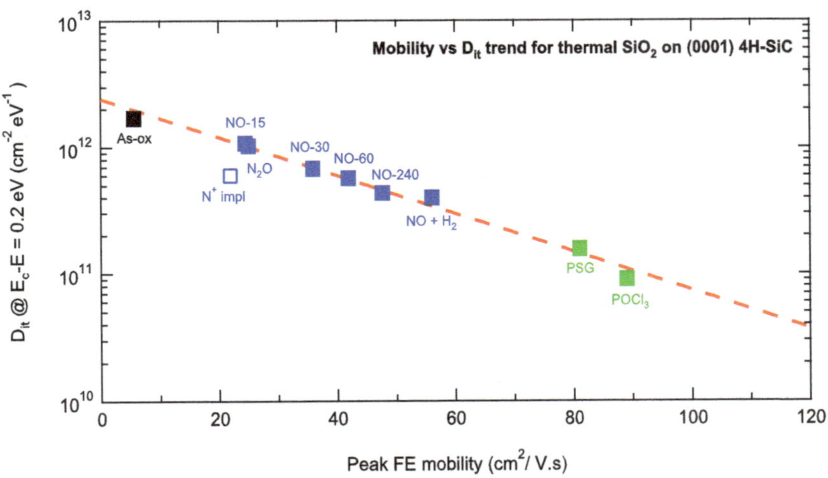

Figure 10. Peak field-effect mobilities extracted from lateral MOSFETs fabricated on (0001) 4H-SiC using thermal oxides with various POAs and the corresponding D_{it} @ 0.2 eV from the conduction band. The plot suggests that Coulomb scattering is the dominant mechanism affecting carrier transport. Further reduction of the interface trap density will keep improving device efficiency. Data gathered from Refs. [12, 88, 94, 106, 108, 115, 130].

still hampered by trap-induced Coulomb scattering. Moreover, it is another indication that if D_{it} can be decreased further, the mobility could potentially reach values well above 100 $cm^2/V.s$.

One can then speculate whether the best properties will be obtained from passivating a defective thermally-formed SiO_2/SiC interface or using alternate dielectrics like promising Al_2O_3 gate stacks that have demonstrated such high mobilities. Both of these top-down and bottom-up approaches come with their respective challenges going from process integration to device stability and reliability. Thermal oxide POA must optimize the density of passivating species while limiting their presence inside the oxide, indeed D_{it} is inversely proportional to the amount of nitrogen but the quantity of hole traps increases with it. As for alternate dielectrics, benefits could arise by moving away from pre- or post-deposition oxidation, using instead efficient surface preparation techniques.

If the steady progress made during the past couple decades is any indication, silicon carbide research, development, and technology, have a bright future ahead.

Acknowledgements

The author would like to thank his friends and colleagues at Vanderbilt University, Auburn University, the Central Research Institute of Electric Power Industry, and IBM, for discussions, challenges, and opportunities that have contributed to the work summarized here.

Author details

John Rozen*

* Address all correspondence to: jrozen@us.ibm.com

IBM T.J. Watson Research Center, Yorktown Heights, NY, USA

References

[1] Afanas'ev, V. V., Bassler, M., Pensl, G., Schulz, M. J. & von Kamienski, E. S. [1996]. Band Offsets and Electronic Structure of SiC/SiO_2 Interfaces, *J. Appl. Phys.* 79: 3108–3114.

[2] Afanas'ev, V. V., Bassler, M., Pensl, G. & Schulz, M. [1997]. Intrinsic SiC/SiO_2 Interface States, *Phys. Status Solidi A* 162(1): 321–337.

[3] Afanas'ev, V. V. & Stesmans, A. [1997a]. Interfacial Defects in SiO_2 Revealed by Photon Stimulated Tunneling of Electrons, *Phys. Rev. Lett.* 78(12): 2437–2440.

[4] Afanas'ev, V. V. & Stesmans, A. [1997b]. Photon-Stimulated Tunnelling of Electrons in SiO_2: Evidence for a Defect-Assisted Process, *J. Phys.: Condens. Matter* 9(6): L55–L60.

[5] Afanas'ev, V. V., Stesmans, A., Bassler, M., Pensl, G., Schulz, M. J. & Harris, C. I. [1999]. SiC/SiO_2 Interface-State Generation by Electron Injection, *J. Appl. Phys.* 85(12): 8292–8298.

[6] Afanas'ev, V. V., Stesmans, A., Bassler, M., Pensl, G. & Schulz, M. J. [2000]. Shallow Electron Traps at the 4H-SiC/SiO_2 Interface, *Appl. Phys. Lett.* 76(3): 336–338.

[7] Afanas'ev, V. V., Stesmans, A., Chen, F., Campbell, S. A. & Smith, R. [2003]. HfO_2-Based Insulating Stacks on 4H-SiC (0001), *Appl. Phys. Lett.* 82(6): 922–924.

[8] Afanas'ev, V. V., Ciobanu, F., Pensl, G. & Stesmans, A. [2004]. *Silicon Carbide: Recent Major Advances*, Springer, chapter Contributions to the Density of Interface States in SiC MOS Structures, pp. 343–372.

[9] Agarwal, A., Ryu, S.-H. & Palmour, J. [2004]. *Silicon Carbide: Recent Major Advances*, Springer, chapter Power MOSFETs in 4H-SiC: Device Design and Technology, pp. 785–811.

[10] Agarwal, A. & Haney, S. [2008]. Some Critical Materials and Processing Issues in SiC Power Devices, *J. Electron. Mater.* 37(5): 646–654.

[11] Akagi, T., Yano, H., Hatayama, T. & Fuyuki, T. [2012]. Effect of Interfacial Localization of Phosphorus on Electrical Properties and Reliability of 4H-SiC MOS Devices, *Proceedings of the 9th European Conference on Silicon Carbide and Related Materials*, TuP–61.

[12] Allerstam, F., Gudjónsson, G., Ólafsson, H. Ö., Sveinbjörnsson, E. Ö., Rödle, T. & Jos, R. [2007]. Comparison Between Oxidation Processes Used to Obtain the High Inversion Channel Mobility in 4H-SiC MOSFETs, *Semicond. Sci. Technol.* 22(4): 307–311.

[13] Arnold, E. & Alok, D. [2001]. Effect of Interface States on Electron Transport in 4H-SiC Inversion Layers, *IEEE Trans. Electron Dev.* 48(9): 1870–1877.

[14] Auth, C., Buehler, M., Cappellani, A., Choi, C., Ding, G., Han, W., Joshi, S., McIntyre, B., Prince, M., Ranade, P., Sandford, J. & Thomas, C. [2008]. 45nm High-k+Metal Gate Strain-Enhanced Transistors, *Intel Tech. J.* 12(2): 77–86.

[15] Baliga, B. J. [2006]. *Silicon Carbide Power Devices*, World Scientific Publishing Co.

[16] Basile, A. F., Dhar, S. & Mooney, P. M. [2011]. Electron Trapping in 4H-SiC MOS Capacitors Fabricated by Pre-Oxidation Nitrogen Implantation, *J. Appl. Phys.* 109: 114505.

[17] Basile, A., Rozen, J., Williams, J. R., Feldman, L. C. & Mooney, P. M. [2011]. Capacitance-Voltage and Deep-Level-Transient Spectroscopy Characterization of Defects near SiO_2/SiC Interfaces, *J. Appl. Phys.* 109(6): 064514–064514.

[18] Bernhardt, J., Schardt, J., Starke, U. & Heinz, K. [1999]. Epitaxially Ideal Oxide-Semiconductor Interfaces: Silicate Adlayers on Hexagonal (0001) and (000-1) SiC surfaces, *Appl. Phys. Lett.* 74(8): 1084–1086.

[19] Biggerstaff, T. L., Reynolds Jr, C. L., Zheleva, T., Lelis, A., Habersat, D., Haney, S., Ryu, S. H., Agarwal, A. & Duscher, G. [2009]. Relationship Between 4H-SiC/SiO_2 Transition Layer Thickness and Mobility, *Appl. Phys. Lett.* 95: 032108.

[20] Brower, K. L. [1988]. Kinetics of H_2 Passivation of Pb Centers at the (111) Si/SiO_2 Interface, *Phys. Rev. B* 38(14): 9657–9666.

[21] Brower, K. L. [1990]. Dissociation Kinetics of Hydrogen-Passivated (111) Si/SiO_2 Interface Defects, *Phys. Rev. B* 42(6): 3444–3453.

[22] Burger, B., Kranzer, D. & Stalter, O. [2008]. Cost Reduction of PV-Inverters with SiC-DMOSFETs, *5th International Conference on Integrated Power Systems*, VDE.

[23] Burger, B. & Kranzer, D. [2009]. Extreme High Efficiency PV-Power Converters, *13th European Conference on Power Electronics and Applications*, IEEE.

[24] Campbell, J. P., Lenahan, P. M., Cochrane, C. J., Krishnan, A. T. & Krishnan, S. [2007]. Atomic-Scale Defects Involved in the Negative Bias Temperature Instability, *IEEE Trans. Dev. Mat. Rel.* 7(4): 540–557.

[25] Campbell, J. P., Lenahan, P. M., Krishnan, A. T. & Krishnan, S. [2007]. Identification of Atomic-Scale Defect Structure Involved in the Negative Bias Temperature Instability in Plasma-Nitrided Devices, *Appl. Phys. Lett.* 91: 133507.

[26] Campi, J., Shi, Y., Luo, Y., Yan, F. & Zhao, J. H. [1999]. Study of Interface State Density and Effective Oxide Charge in Post-Metallization Annealed SiO_2/SiC Structures, *IEEE Trans. Electron Dev.* 46(3): 511–519.

[27] Cantin, J. L., von Bardeleben, H. J., Shishkin, Y., Ke, Y., Devaty, R. P. & Choyke, W. J. [2004]. Identification of the Carbon Dangling Bond Center at the 4H-SiC/SiO_2 Interface by an EPR Study in Oxidized Porous SiC, *Phys. Rev. Lett.* 92(1): 15502.

[28] Cantin, J. L. & von Bardeleben, H. J. [2006]. Forming Gas Annealing of the Carbon P_{bC} Center in Oxidized Porous 3C- and 4H-SiC: an EPR Study, *Mater. Sci. Forum* 527: 1015–1018.

[29] Carter Jr, C. H., Tsvetkov, V. F., Glass, R. C., Henshall, D., Brady, M., Müller, S. G., Kordina, O., Irvine, K., Edmond, J. A., Kong, H. S., Singh, R., Allen, S. T. & Palmour, J. W. [1999]. Progress in SiC: from Material Growth to Commercial Device Development, *Mat. Sci. Eng. B-Solid* 61: 1–8.

[30] Casady, J. B. & Johnson, R. W. [1996]. Status of Silicon Carbide (SiC) as a Wide-Bandgap Semiconductor for High-Temperature Applications: A Review, *Solid State Electron.* 39(10): 1409–1422.

[31] Chakraborty, S., Lai, P. T. & Kwok, P. C. K. [2002]. MOS Characteristics of NO-grown Oxynitrides on n-type 6H-SiC, *Microelectron. Reliab.* 42: 455–458.

[32] Chanana, R. K., McDonald, K., Di Ventra, M., Pantelides, S. T., Feldman, L. C., Chung, G. Y., Tin, C. C., Williams, J. R. & Weller, R. A. [2000]. Fowler-Nordheim Hole Tunneling in p-SiC/SiO_2 Structures, *Appl. Phys. Lett.* 77(16): 2560–2562.

[33] Chang, K. C., Cao, Y., Porter, L. M., Bentley, J., Dhar, S., Feldman, L. C. & Williams, J. R. [2005]. High-Resolution Elemental Profiles of the Silicon Dioxide/ 4H-Silicon Carbide Interface, *J. Appl. Phys.* 97: 104920.

[34] Chang, K. C., Nuhfer, N. T., Porter, L. M. & Wahab, Q. [2000]. High-Carbon Concentrations at the Silicon Dioxide-Silicon Carbide Interface Identified by Electron Energy Loss Spectroscopy, *Appl. Phys. Lett.* 77: 2186.

[35] Chang, K. C., Porter, L. M., Bentley, J., Lu, C. Y. & Cooper, J., J. [2004]. Electrical, Structural, and Chemical Analysis of Silicon Carbide-Based Metal-Oxide-Semiconductor Field-Effect-Transistors, *J. Appl. Phys.* 95: 8252.

[36] Chatterjee, A., Matocha, K., Tilak, V., Fronheiser, J. & Piao, H. [2010]. Multiscale Modeling and Analysis of the Nitridation Effect of SiC/SiO_2 Interface, *Mater. Sci. Forum* 645: 479–482.

[37] Chatty, K., Khemka, V., Chow, T. P. & Gutmann, R. J. [1999]. Re-Oxidation Characteristics of Oxynitrides on 3C-and 4H-SiC, *J. Electron. Mater.* 28(3): 161–166.

[38] Cheong, K. Y., Moon, J. H., Park, T. J., Kim, J. H., Hwang, C. S., Kim, H. J., Bahng, W. & Kim, N. K. [2007]. Improved Electronic Performance of HfO_2/SiO_2 Stacking Gate Dielectric on 4H-SiC, *IEEE Trans. Electron Dev.* 54(12): 3409–3413.

[39] Chung, G., Tin, C. C., Williams, J. R., McDonald, K., Di Ventra, M., Chanana, R. K., Pantelides, S. T., Feldman, L. C. & Weller, R. A. [2000]. Effects of Anneals in Ammonia on the Interface Trap Density near the Band Edges in 4H-Silicon Carbide Metal-Oxide-Semiconductor Capacitors, *Appl. Phys. Lett.* 77(22): 3601–3603.

[40] Chung, G. Y., Tin, C. C., Williams, J. R., McDonald, K., Ventra, M. D., Pantelides, S. T., Feldman, L. C. & Weller, R. A. [2000]. Effect of Nitric Oxide Annealing on the Interface Trap Densities near the Band Edges in the 4H Polytype of Silicon Carbide, *Appl. Phys. Lett.* 76(13): 1713–1715.

[41] Chung, G. Y., Tin, C. C., Williams, J. R., McDonald, K., Chanana, R. K., Weller, R. A., Pantelides, S. T. & Feldman, L. C. [2001]. Improved Inversion Channel Mobility for 4H-SiC MOSFETs Following High Temperature Anneals in Nitric Oxide, *IEEE Electron Dev. Lett.* 22(4): 176–178.

[42] Ciobanu, F., Pensl, G., Afanas' ev, V. & Schöner, A. [2005]. Low Density of Interface States in n-type 4H-SiC MOS Capacitors Achieved by Nitrogen Implantation, *Mater. Sci. Forum* 483: 693–696.

[43] Costello, J. A. & Tressler, R. E. [1986]. Oxidation Kinetics of Silicon Carbide Crystals and Ceramics: I, In Dry Oxygen, *J. Am. Ceram. Soc.* 69(9): 674–681.

[44] Deak, P., Knaup, J. M., Hornos, T., Thill, C., Gali, A. & Frauenheim, T. [2007]. The Mechanism of Defect Creation and Passivation at the SiC/SiO_2 Interface, *J. Phys. D. Appl. Phys.* 40: 6242–6253.

[45] Deal, B. E. & Grove, A. S. [1965]. General Relationship for the Thermal Oxidation of Silicon, *J. Appl. Phys.* 36: 3770.

[46] Dhar, S., Song, Y. W., Feldman, L. C., Isaacs-Smith, T., Tin, C. C., Williams, J. R., Chung, G., Nishimura, T., Starodub, D., Gustafsson, T. & Garfunkel, E. [2004]. Effect of Nitric Oxide Annealing on the Interface Trap Density near the Conduction Band Edge of 4H-SiC at the Oxide/(11-20) 4H-SiC Interface, *Appl. Phys. Lett.* 84(9): 1498–1500.

[47] Dhar, S. [2005]. *Nitrogen and Hydrogen Trap Passivation at the SiO_2/4H-SiC Interface*, PhD thesis, Vanderbilt University, Nashville, TN.

[48] Di Ventra, M. & Pantelides, S. T. [1999]. Atomic-Scale Mechanisms of Oxygen Precipitation and Thin-Film Oxidation of SiC, *Phys. Rev. Lett.* 83(8): 1624–1627.

[49] Dimitriev, V. A. & Spencer, M. G. [1998]. *SiC Materials and Devices*, Academic Press, chapter SiC Fabrication Technology: Growth and Doping, pp. 21–76.

[50] Dimitrijev, S., Tanner, P. & Harrison, H. B. [1999]. Slow-Trap Profiling of NO and N_2O Nitrided Oxides Grown on Si and SiC Substrates, *Microelectron. Reliab.* 39: 441–449.

[51] Dimitrijev, S., Harrison, H. B., Tanner, P., Cheong, K. Y. & Han, J. [2004]. *Silicon Carbide: Recent Major Advances*, Springer, chapter Properties of Nitrided Oxides on SiC, pp. 373–386.

[52] Dixit, S. K., Dhar, S., Rozen, J., Wang, S., Schrimpf, R. D., Fleetwood, D. M., Pantelides, S. T., Williams, J. R. & Feldman, L. C. [2006]. Total Dose Radiation Response of Nitrided and Non-nitrided SiO_2/4H-SiC MOS Capacitors, *IEEE Trans. Nucl. Sci.* 53(6): 3687–36892.

[53] Dunn, G. J. & Wyatt, P. W. [1989]. Reoxidized Nitrided Oxide for Radiation-Hardened MOS devices, *IEEE Trans. Nucl. Sci.* 36(6): 2161–2168.

[54] Ekoue, A., Renault, O., Billon, T., Di Cioccio, L. & Guillot, G. [2003]. Study of the Wet Re-Oxidation Annealing of SiO_2/4H-SiC (0001) Interface Properties by AR-XPS Measurements, *Materials Sci. Forum* 433: 555–558.

[55] Friedrichs, P., Burte, E. P. & Schörner, R. [1996]. Interface Properties of Metal-Oxide-Semiconductor Structures on n-type 6H and 4H-SiC, *J. Appl. Phys.* 79: 7814.

[56] Gusev, E. P., Narayanan, V. & Frank, M. M. [2006]. Advanced High-κ Dielectric Stacks with poly-Si and Metal Gates: Recent Progress and Current Challenges, *IBM J. Res. Dev.* 50(4-5): 387–410.

[57] Hayton, D. J., Jenkins, T. E., Bailey, P. & Noakes, T. C. Q. [2002]. Optical and Ion-Scattering Study of SiO_2 Layers Thermally Grown on 4H-SiC, *Semicond. Sci. Technol.* 17(7): L29–L32.

[58] Helms, C. R. & Poindexter, E. H. [1994]. The Silicon-Silicon Dioxide System: Its Microstructure and Imperfections, *Rep. Prog. Phys.* 57(8): 791–852.

[59] Hijikata, Y., Yaguchi, H. & Yoshida, S. [2010]. Model Calculations of SiC Oxide Growth Rate at Various Oxidation Temperatures Based on the Silicon and Carbon Emission Model, *Mater. Sci. Forum* 645: 809–812.

[60] Hino, S., Hatayama, T., Kato, J., Tokumitsu, E., Miura, N. & Oomori, T. [2008]. High Channel Mobility 4H-SiC Metal-Oxide-Semiconductor Field-Effect Transistor with Low Temperature Metal-Organic Chemical-Vapor Deposition Grown Al_2O_3 Gate Insulator, *Appl. Phys. Lett.* 92(18): 183503.

[61] Hiyoshi, T. & Kimoto, T. [2009]. Elimination of the Major Deep Levels in n- and p-type 4H-SiC by Two-Step Thermal Treatment, *Appl. Phys. Express* 2: 091101.

[62] Hornetz, B., Michel, H.-J. & Halbritter, J. [1994]. ARXPS Studies of SiO_2-SiC Interfaces and Oxidation of 6H-SiC Single Crystal Si-(001) and C-(00-1) Surfaces, *J. Mater. Res.* 9(12): 3088–3094.

[63] Iida, T., Tomioka, Y., Yoshimoto, K., Midorikawa, M., Tukada, H. & Orihara, M. [2002]. Measurements of the Depth Profile of the Refractive Indices in Oxide Films on SiC by Sprectroscopic Ellipsometry, *Jpn. J. Appl. Phys.* 41(2): 800–804.

[64] *International Energy Outlook* [2011]. *Technical report*, Energy Information Administration.

[65] *Ioffe Physico-Technical Institute. Electronic Archive of New Semiconductor Materials; Characteristics and Properties.* URL: *http://www.ioffe.ru/SVA*

[66] Jarrendahl, K. & Davis, R. F. [1998]. *SiC Materials and Devices*, Academic Press, chapter Materials Properties and Characterization of SiC, pp. 1–20.

[67] Jernigan, G. G., Stahlbush, R. E. & Saks, N. S. [2000]. Effect of Oxidation and Reoxidation on the Oxide-Substrate Interface of 4H-and 6H-SiC, *Appl. Phys. Lett.* 77: 1437.

[68] Kakubari, K., Kuboki, R., Hijikita, Y., Yaguchi, H. & Yoshida, S. [2006]. Real Time Observation of SiC Oxidation Using an in-situ Ellipsometer, *Mater. Sci. Forum* 527: 1031–1034.

[69] Kawahara, K., Suda, J. & Kimoto, T. [2012]. Analytical Model for Reduction of Deep Levels in SiC by Thermal Oxidation, *J. Appl. Phys.* 111: 053710.

[70] Kimoto, T., Kanzaki, Y., Noborio, M., Kawano, H. & Matsunami, H. [2005]. Interface Properties of Metal-Oxide-Semiconductor Structures on 4H-SiC(0001) and (11-20) Formed by N_2O Oxidation, *Jpn. J. Appl. Phys.* 44: 1213–1218.

[71] Knaup, J. M., Deák, P., Frauenheim, T., Gali, A., Hajnal, Z. & Choyke, W. J. [2005a]. Theoretical Study of the Mechanism of Dry Oxidation of 4H-SiC, *Phys. Rev. B* 71(23): 235321.

[72] Knaup, J. M., Deák, P., Frauenheim, T., Gali, A., Hajnal, Z. & Choyke, W. J. [2005b]. Defects in SiO_2 as the Possible Origin of Near Interface Traps in the SiC/SiO_2 System: A Systematic Theoretical Study, *Phys. Rev. B* 72(11): 115323.

[73] Krishnaswami, S., Das, M. K., Agarwal, A. K. & Palmour, J. W. [2004]. Reliability of Nitrided Oxides in n-and p-type 4H-SiC MOS Structures, *Spring Meeting Proceedings*, Vol. 815, MRS, p. J8.4.

[74] Krishnaswami, S., Ryu, S.-H., Heath, B., Agarwal, A., Palmour, J., Geil, B., Lelis, A. & Scozzie, C. [2006]. A Study on the Reliability and Stability of High Voltage 4H-SiC MOSFET Devices, *Mater. Sci. Forum* 527-529: 1313–1316.

[75] Lenahan, P. M. & Dressendorfer, P. V. [1984]. Hole Traps and Trivalent Silicon Centers in Metal/Oxide/Silicon Devices, *J. Appl. Phys.* 55: 3495–3499.

[76] Li, H., Dimitrijev, S., Harrison, H. B. & Sweatman, D. [1997]. Interfacial Characteristics of N_2O and NO Nitrided SiO_2 Grown on SiC by Rapid Thermal Processing, *Appl. Phys. Lett.* 70(15): 2028–2030.

[77] Li, H., Dimitrijev, S., Sweatman, D., Harrison, H. B., Tanner, P. & Feil, B. [1999]. Investigation of Nitric Oxide and Ar Annealed Interfaces by X-Ray Photoelectron Spectroscopy, *J. Appl. Phys.* 86(8): 4316–4321.

[78] Liang, M., Choi, J., Ko, P. & Hu, C. [1986]. Inversion-Layer Capacitance and Mobility of Very Thin Gate-Oxide MOSFETs, *IEEE Trans. Electron Dev.* 33(3): 409–413.

[79] Lichtenwalner, D., Misra, V., Dhar, S., Ryu, S. & Agarwal, A. [2009]. High-Mobility Enhancement-Mode 4H-SiC Lateral Field-Effect Transistors Utilizing Atomic Layer Deposited Al_2O_3 Gate Dielectric, *Appl. Phys. Lett.* 95: 152113.

[80] Lu, W., Feldman, L. C., Song, Y., Dhar, S., Collins, W. E., Mitchel, W. C. & Williams, J. R. [2004]. Graphitic Features on SiC Surface Following Oxidation and Etching Using Surface Enhanced Raman Spectroscopy, *Appl. Phys. Lett.* 85: 3495.

[81] McDonald, K., Huang, M. B., Weller, R. A., Feldman, L. C., Williams, J. R., Stedile, F. C., Baumvol, I. J. R. & Radtke, C. [2000]. Comparison of Nitrogen Incorporation in SiO_2/SiC and SiO_2/Si Structures, *Appl. Phys. Lett.* 76(5): 568.

[82] McDonald, K., Weller, R. A., Pantelides, S. T., Feldman, L. C., Chung, G. Y., Tin, C. C. & Williams, J. R. [2003]. Characterization and Modeling of the Nitrogen Passivation of Interface Traps in SiO_2/4H-SiC, *J. Appl. Phys* 93(5): 2719–2722.

[83] Moon, J., Yim, J., Seo, H., Lee, D. H., Kim, C., Kim, H., Cheong, K., Bahng, W. & Kim, N. [2010]. Effects of Rapid Thermal Annealing on Al_2O_3/SiN Reaction Barrier Layer/Thermal-Nitrided SiO_2 Stacking Gate Dielectrics on n-type 4H-SiC, *Appl. Phys. Lett.* 96: 122108.

[84] Muller, R. & Kamins, T. [2003]. *Device Electronics for Integrated Circuits*, Wiley, New York.

[85] Noborio, M., Suda, J., Beljakowa, S., Krieger, M. & Kimoto, T. [2009]. 4H-SiC MISFETs with Nitrogen-Containing Insulators, *Phys. Status Solidi (A)* 206(10): 2374–2390.

[86] O'Dwyer, J. J. [1973]. *The Theory of Electrical Conduction and Breakdown in Solid Dielectrics*, Oxford University Press.

[87] Okamoto, D., Yano, H., Hatayama, T. & Fuyuki, T. [2010]. Systematic Investigation of Interface Properties in 4H-SiC MOS Structures Prepared by over-Oxidation of Ion-Implanted Substrates, *Mater. Sci. Forum* 645: 495–498.

[88] Okamoto, D., Yano, H., Hirata, K., Hatayama, T. & Fuyuki, T. [2010]. Improved Inversion Channel Mobility in 4H-SiC MOSFETs on Si Face Utilizing Phosphorus-Doped Gate Oxide, *IEEE Electron Dev. Lett.* 31(7): 710–712.

[89] Okawa, T., Fukuyama, R., Hoshino, Y., Nishimura, T. & Kido, Y. [2007]. Kinetics of Oxynitridation of 6H-SiC (11-20) and the Interface Structure Analyzed by Ion Scattering and Photoelectron Spectroscopy, *Surf. Sci.* 601(3): 706–713.

[90] Pantelides, S. T., Wang, S., Franceschetti, A., Buczko, R., Di Ventra, M., Raashkeev, S. N., Tsetseris, L., Evans, M. H., Batyrev, I. G., Feldman, L. C., Dhar, S., McDonald, K., Weller, R. A., Schrimpf, R. D., Fleetwood, D. M., Zhou, X. J., Williams, J. R., Tin, C. C., Chung, G. Y., Isaacs-Smith, T., Wang, S. R., Pennycook, S. J., Duscher, G., Van Benthem, K. & Porter, L. M. [2006]. Si/SiO_2 and SiC/SiO_2 Interfaces for MOSFETs: Challenges and Advances, *Mater. Sci. Forum* 527: 935–948.

[91] Pensl, G., Beljakowa, S., Frank, T., Gao, K., Speck, F., Seyller, T., Ley, L., Ciobanu, F., Afanas' ev, V., Stesmans, A., Kimoto, T. & Schoner, A. [2008]. Alternative Techniques to Reduce Interface Traps in n-type 4H-SiC MOS Capacitors, *Phys. Status Solidi (b)* 245(7): 1378–1389.

[92] Pippel, E., Woltersdorf, J., Ólafsson, H. Ö. & Sveinbjörnsson, E. Ö. [2005]. Interfaces Between 4H-SiC and SiO_2: Microstructure, Nanochemistry, and Near-Interface Traps, *J. Appl. Phys.* 97(3): 034302.

[93] Poggi, A., Moscatelli, F., Hijikata, Y., Solmi, S. & Nipoti, R. [2007]. MOS Capacitors Obtained by Wet Oxidation of n-type 4H-SiC pre-Implanted with Nitrogen, *Microelectron. Eng.* 84(12): 2804–2809.

[94] Poggi, A., Moscatelli, F., Solmi, S. & Nipoti, R. [2008]. Investigation on the Use of Nitrogen Implantation to Improve the Performance of n-Channel Enhancement 4H-SiC MOSFETs, *IEEE Trans. Electron Dev.* 55(8): 2021–2028.

[95] Poggi, A., Moscatelli, F., Solmi, S., Armigliato, A., Belsito, L. & Nipoti, R. [2010]. Effect of Nitrogen Implantation at the SiO_2/SiC Interface on the Electron Mobility and Free Carrier Density in 4H-SiC Metal-Oxide-Semiconductor Field-Effect-Transistor Channel, *J. Appl. Phys.* 107: 044506.

[96] Poindexter, E. H. [1989]. MOS Interface States: Overview and Physicochemical Perspective, *Semicond. Sci. Technol.* 4(12): 961–969.

[97] Radtke, C., Baumvol, I. J. R. & Stedile, F. C. [2002]. Effects of Ion Irradiation in the Thermal Oxidation of SiC, *Phys. Rev. B* 66(15): 155437.

[98] Ray, E. A., Rozen, J., Dhar, S., Williams, J. R. & Feldman, L. C. [2008]. Pressure Dependence of SiO_2 Growth Kinetics and Electrical Properties on SiC, *J. Appl. Phys.* 103: 023522.

[99] Raynaud, C. [2001]. Silica Films on Silicon Carbide: a Review of Electrical Properties and Device Applications, *J. Non-Cryst. Solids* 280(1-3): 1–31.

[100] Richmond, J., Hodge, S. & Palmour, J. [2004]. Silicon Carbide Power Applications and Device Roadmap, *Power Electronics Europe* 7: 17.

[101] Riedl, C., Coletti, C., Iwasaki, T., Zakharov, A. A. & Starke, U. [2009]. Quasi-Free-Standing Epitaxial Graphene on SiC Obtained by Hydrogen Intercalation, *Phys. Rev. Lett.* 103(24): 246804.

[102] Robertson, J. [2000]. Band Offsets of Wide-Band-Gap Oxides and Implications for Future Electronic Devices, *J. Vac. Sci. Technol. B* 18(3): 1785–1791.

[103] Rozen, J., Dhar, S., Wang, S., Pantelides, S. T., Afanas'ev, V. V., Williams, J. R. & Feldman, L. C. [2007]. Suppression of Interface State Generation upon Electron Injection in Nitrided Oxides Grown on 4H-SiC, *Appl. Phys. Lett.* 91: 153503.

[104] Rozen, J., Dhar, S., Dixit, S. K., Roberts, F. O., Dang, H. L., Wang, S., Pantelides, S. T., Afanas'ev, V. V., Williams, J. R. & Feldman, L. C. [2008]. Increase in Oxide Hole Trap Density Associated with Nitrogen Incorporation at the SiO_2/SiC Interface, *J. Appl. Phys.* 103: 124513.

[105] Rozen, J. [2008]. *Electronic Properties and Reliability of the SiO_2/SiC Interface*, PhD thesis, Vanderbilt University, Nashville, TN.

[106] Rozen, J., Dhar, S., Zvanut, M. E., Williams, J. R. & Feldman, L. C. [2009]. Density of Interface States, Electron Traps, and Hole Traps as a Function of the Nitrogen Density is SiO_2 on SiC, *J. Appl. Phys.* 105(12): 124506.

[107] Rozen, J., Zhu, X., Ahyi, A. C., Williams, J. R. & Feldman, L. C. [2010]. The Limits of Post Oxidation Annealing in NO, *Mater. Sci. Forum* 645-648: 693–696.

[108] Rozen, J., Ayayi, A. C., Zhu, X., Williams, J. R. & Feldman, L. C. [2011]. Scaling Between Channel Mobility and Interface State Density in SiC MOSFETs, *IEEE Trans. Electron Dev.* 58(11): 3808–3811.

[109] Rozen, J. [2012]. Energy Efficiency in High Power Electronics: Role and Challenges of Silicon Carbide Interfaces, *Human Photonics International Forum, Saitama University*.

[110] Rozen, J., Nagano, M. & Tsuchida, H. [2012a]. Enhancing Interface Quality by Gate Dielectric Deposition on a Nitrogen-Conditioned 4H-SiC Surface, *J. Mater. Res.* [Online early access]: DOI: 10.1557/jmr.2012.269.

[111] Rozen, J., Nagano, M. & Tsuchida, H. [2012b]. Improved Deposited Oxide Interfaces from N_2 Conditioning of Bare SiC Surfaces, *Mater. Sci. Forum* 717: 729–732.

[112] Rudenko, T. E., Osiyuk, I. N., Tyagulski, I. P., Ólafsson, H. Ö. & Sveinbjörnsson, E. Ö. [2005]. Interface Trap Properties of Thermally Oxidized n-type 4H-SiC and 6H-SiC, *Solid-State Electron.* 49(4): 545–553.

[113] Rys, A., Singh, N. & Cameron, M. [1995]. Modeling and Characterization of Thermally Oxidized 6H Silicon Carbide, *J. Electrochem. Soc.* 142: 1318.

[114] Saks, N. S. [2004]. *Silicon Carbide: Recent Major Advances*, Springer, chapter Hall Effect Studies of Electron Mobility and Trapping at the SiC/SiO_2 Interface, pp. 387–410.

[115] Sharma, Y. K., Ahyi, A. C., Issacs-Smith, T., Shen, X., Pantelides, S. T., Zhu, X., Feldman, L. C., Rozen, J. & Williams, J. R. [2012]. Phosphorous Passivation of the SiO_2/4H-SiC Interface, *Solid-State Electron.* 68: 103–107.

[116] Sharma, Y. K. [2012]. *Advanced SiO_2/SiC Interface Passivation*, PhD thesis, Auburn University, Auburn, AL.

[117] Shen, X. & Pantelides, S. T. [2011]. Identification of a Major Cause of Endemically Poor Mobilities in SiC/SiO_2 Structures, *Appl. Phys. Lett.* 98(5): 053507.

[118] Shirasawa, T., Hayashi, K., Mizuno, S., Tanaka, S., Nakatsuji, K., Komori, F. & Tochihara, H. [2007]. Epitaxial Silicon Oxynitride Layer on a 6H-SiC (0001) Surface, *Phys. Rev. Lett.* 98(13): 136105.

[119] Shirasawa, T., Hayashi, K., Yoshida, H., Mizuno, S., Tanaka, S., Muro, T., Tamenori, Y., Harada, Y., Tokushima, T., Horikawa, Y., Kobatashi, E., Kinoshita, T., Shin, S., Takahashi, T., Ando, Y., Akagi, K., Tsuneyuki, S. & Tochihara, H. [2009]. Atomic-Layer-Resolved Bandgap Structure of an Ultrathin Oxynitride-Silicon Film Epitaxially Grown on 6H-SiC (0001), *Phys. Rev. B* 79(24): 241301.

[120] Sieber, N., Seyller, T., Ley, L., James, D., Riley, J., Leckey, R. & Polcik, M. [2003]. Synchrotron X-Ray Photoelectron Spectroscopy Study of Hydrogen-Terminated 6H-SiC (0001) Surfaces, *Phys. Rev. B* 67(20): 205304.

[121] Son, N. T., Chen, W. M., Kordina, O., Konstantinov, A. O., Monemar, B., Janzen, E., Hofman, D. M., Volm, D., Drechsler, M. & Meyer, B. K. [1995]. Electron Effective Masses in 4H-SiC, *Appl. Phys. Lett.* 66(9): 1074–1076.

[122] Song, Y., Dhar, S., Feldman, L. C., Chung, G. & Williams, J. R. [2004]. Modified Deal Grove Model for the Thermal Oxidation of Silicon Carbide, *J. Appl. Phys.* 95: 4953.

[123] Storasta, L., Tsuchida, H., Miyazawa, T. & Ohshima, T. [2008]. Enhanced Annealing of the $Z_{1/2}$ Defect in 4H-SiC Epilayers, *J. Appl. Phys.* 103(1): 013705.

[124] Sze, S. M. [1981]. *Semiconductor Devices, Physics and Technology*, Wiley, New York.

[125] Tilak, V., Matocha, K. & Dunne, G. [2007]. Electron-Scattering Mechanisms in Heavily Doped Silicon Carbide MOSFET Inversion Layers, *IEEE Trans. Electron Dev.* 54(11): 2823–2829.

[126] Tsuchida, H., Kamata, I. & Izumi, K. [1997]. Infrared Spectroscopy of Hydrides on the 6H-SiC Surface, *Appl. Phys. Lett.* 70(23): 3072–3074.

[127] Vickridge, I. C., Tromson, D., Trimaille, I., Ganem, J. J., Szilágyi, E. & Battistig, G. [2002]. Oxygen Isotopic Exchange Occurring During Dry Thermal Oxidation of 6H-SiC, *Nucl. Instrum. Meth. B* 190(1-4): 574–578.

[128] Virojanadra, C. & Johansson, L. I. [2004]. Studies of Oxidized Hexagonal SiC Surfaces and the SiC/SiO_2 Interface Using Photoemission and Synchrotron Radiation, *J. Phys.: Condens. Matter* 16(17): 1783–1814.

[129] von Kamienski, E. G. S., Portheine, F., Stein, J., Golz, A. & Kurz, H. [1996]. Charge Trapping in Dry and Wet Oxides on n-type 6H-SiC Studied by Fowler-Nordheim Charge Injection, *J. Appl. Phys.* 79(5): 2529–2534.

[130] Wang, S., Dhar, S., Wang, S.-r., Ahyi, A. C., Franceschetti, A., Williams, J. R., Feldman, L. C. & Pantelides, S. T. [2007]. Bonding at the $SiC\text{-}SiO_2$ Interface and the Effects of Nitrogen and Hydrogen, *Phys. Rev. Lett.* 98(2): 026101.

[131] Wilk, G., Wallace, R. & Anthony, J. [2001]. High-κ Gate Dielectrics: Current Status and Materials Properties Considerations, *J. Appl. Phys.* 89(10): 5243–5275.

[132] Xu, J. P., Lai, P. T., Chan, C. L., Li, B. & Cheng, Y. C. [2000]. Improved Performance and Reliability of N_2O-grown Oxynitride on 6H-SiC, *IEEE Electron Dev. Lett.* 21(6): 298–300.

[133] Yankova, A., Do Thank, L. & Balk, P. [1987]. Effects of Thermal Nitridation on the Trapping Characteristics of SiO_2 Films, *Solid-State Electron.* 30(9): 939–946.

[134] Yano, H., Furumoto, Y., Niwa, T., Hatayama, T., Uraoka, Y. & Fuyuki, T. [2004]. Radical Nitridation of Ultra-Thin SiO_2/SiC Structure, *Mater. Sci. Forum* 457: 1333–1336.

[135] Zheng, Z., Tressler, R. E. & Spear, K. E. [1990a]. Oxidation of Single-Crystal Silicon Carbide. Part I. Experimental Studies, *J. Electrochem. Soc.* 137(3): 854–858.

[136] Zheng, Z., Tressler, R. E. & Spear, K. E. [1990b]. Oxidation of Single-Crystal Silicon Carbide. Part II. Kinetic Model, *J. Electrochem. Soc.* 137(9): 2812–2816.

[137] Zhu, X., Ahyi, A., Li, M., Chen, Z., Rozen, J., Feldman, L. & Williams, J. [2011]. The Effect of Nitrogen Plasma Anneals on Interface Trap Density and Channel Mobility for 4H-SiC MOS Devices, *Solid-State Electron.* 57(1): 76–79.

Section 4

SiC Devices

Chapter 11

Silicon Carbide Power MESFET

Yintang Yang, Baoxing Duan and Xianjun Zhang

Additional information is available at the end of the chapter

1. Introduction

The wide band gap materials, such as silicon carbide (SiC) [1-3] and gallium nitride (GaN) [4-6], are the third generation semiconductor materials, which had been developed after the Silicon (Si) and gallium arsenide (GaAs) materials. Especially, the SiC material is very well-suited for the high voltage, high power and high temperature applications due to its superior material properties. Silicon carbide has been known investigated since 1907 after Captain H. J. Round demonstrated yellow and blue emission by application bias between a metal needle and SiC crystal. The potential of using SiC in semiconductor electronics was already recognized about a half of century ago. The most remarkable SiC properties include the wide band gap, very large avalanche breakdown field, high thermal conductivity, high maximum operating temperature and chemical inertness and radiation hardness.

Therefore the microwave power devices based on 4H-SiC have received increasing attention. The MESFETs (Metal Semiconductor Field Effect Transistor) is a hot research topic [7-10].

For the SiC power MESFETs, the breakdown voltage is a very important parameter that allows the power devices to achieve a specific power density and power conversion. Prior research has proposed many techniques to improve the breakdown voltage [11-15]. The conventional termination techniques include of the Field Plate structure in the source or drain electrode [16], RESURF (Reduced Surface field) technology [17], floating metal rings [18-19], p-epi guard rings [20], etc [21-23]. In order to optimize the surface electric field and improve the breakdown voltage, the new technologies had been proposed, which includes of the REBULF (REduced BULk Field) [24] and complete 3D RESURF [15]. For the new power devices based on the silicon materials, the trade-off relationship had been broken between the breakdown voltage and specific on resistance by the complete 3D RESURF [15]. The high breakdown voltage had been obtained on the ultra thin epitaxial layer with the REBULF technology [24]. It can be sure that these new technologies can be transplanted directly to the SiC power MESFETs. So, several new SiC power MESFETs had been designed to optimize the characteristic of the breakdown voltage, specific on resistance, frequency and transconductance.

2. Operation principle for the 4H-SiC Power MESFETs

For the *n*-channel MESFETs, which including normally-on (depletion mode) and normally-off (enhancement mode) devices, as are shown in Fig.1, an *n*-type channel connects the drain and source regions with the n^+ type doping. The depletion layer under the metal-semiconductor contact determines the current flow across the channel between the source and drain electrodes. The thickness of the channel conductivity is modulated by the gate bias-dependent depletion region.

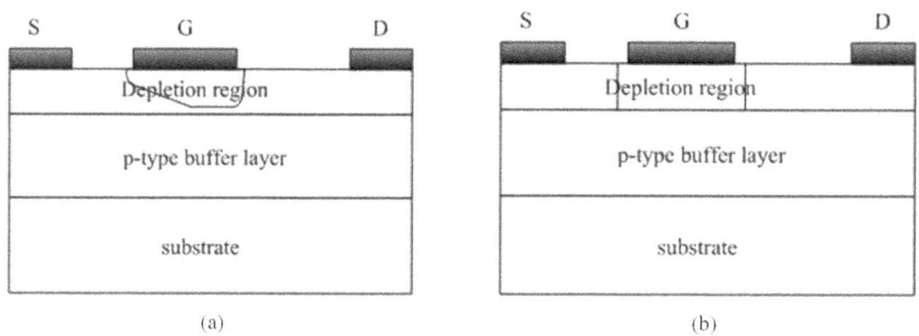

Figure 1. Normally-on (a) and normally-off (b) MESFETs at zero gate bias.

The channel of normally-off MESFETs is totally depleted by the gate build-in potential even at zero gate bias, and its threshold voltage is positive. In contrast, the normally-on MESFETs have a finite cross-section of conducting channel at zero gate bias and the negative threshold voltage.

The basic operation principle will be discussed in this section for the 4H-SiC power MESFETs.

For the normally-on MESFETs, usually, the source is grounded, and the gate and drain are biased negatively and positively, respectively. A schematic diagram of the depletion region under the gate of MESFETs for a finite drain-to-source voltage is shown in Fig. 2. On this condition, the electrons will flow from the source to the drain and a current flow (I_{ds}) occurs in the channel. When the negative gate bias is changed, or an altering-current (AC) voltage signal is superposed on the direct- current (DC) gate bias, the thickness of the depletion layer or the width of the conducting channel, which determining the resistance of the channel, will be modulated, and thus the current flow in the channel is regulated. So, the MESFETs is actually a voltage-controlled electric device by the means that the gate bias modulates the conducting channel resistance, and thus controls the current flow in the channel.

For the *p*-channel normally-on MESFETs, the gate is biased positively, so as to ensure that the gate is reverse biased. Note that it is the electrons for the *n*-channel MESFETs, but the holes for the *p*-channel MESFETs that transport as the carriers in the channel. However, whether in the *n*-channel or *p*-channel MESFETs, the unique majority carriers undertake transporting the current. So, the MESFETs is an unipolar device.

3. New 4H-SiC Power MESFETs

In this section several new structures for the 4H-SiC power MESFETs are provided in which the surface electric field and breakdown characteristics are optimized.

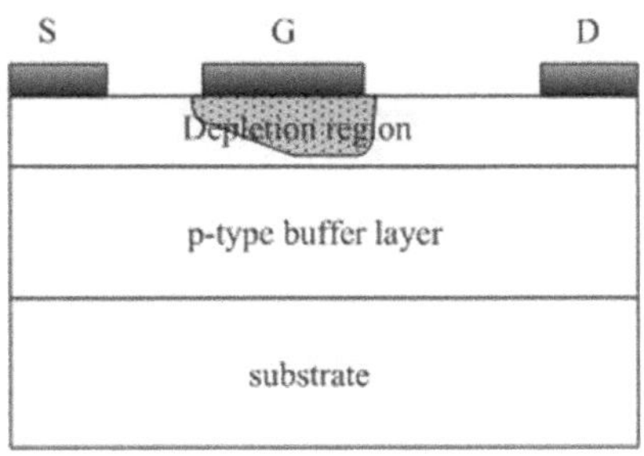

Figure 2. Depletion region in MESFETs with positive drain bias.

3.1. Field-Plated 4H-SiC MESFETs structure

Fig. 3 is the schematic diagram of the 4H-SiC MESFETs with the field-plate [25], which is the same as the channel-recessed device except the Si_3N_4 layer on top of the surface. The gate length (L_g) is 0.5 μm, and the space of gate-source (L_{gs}) and gate-drain (L_{gd}) are 0.5 μm and 1.0 μm, respectively. The thickness of Si_3N_4 is 100 nm.

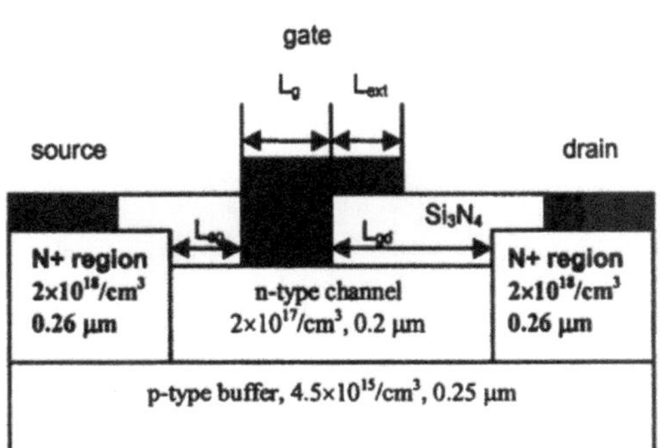

Figure 3. Field-plate SiC MESFETS structure.

Fig.4 shows that the breakdown voltage is dependent on the extension length of the field-plate toward the drain side L_{ext} and show a peak at L_{ext}=0.35 μm. The breakdown voltage of 240-250V is obtained at L_{ext}=0.35 μm, which is 100 V higher than that of the conventional structure without the field-plate. This is because that the gate leakage current is decreased due to the surface field peak significantly lowered at the gate corner while raised at the drain corner, as is shown in Fig.5.

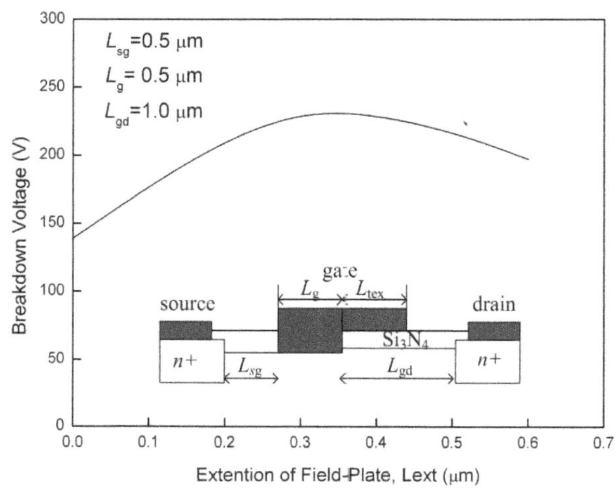

Figure 4. Breakdown voltage versus the extension length of the field-plate toward the drain side L_{ext}.

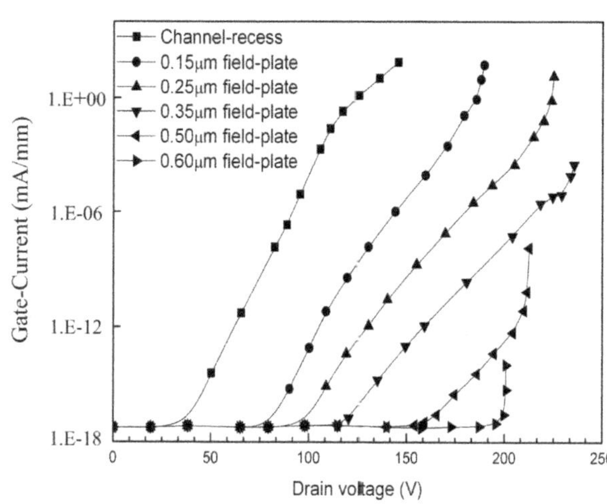

Figure 5. Behavior of the gate leakage current versus the drain voltage as a function of L_{ext}.

3.2. Double-recessed 4H-SiC MESFETs structure with recessed source/drain drift region

Fig. 6 shows the schematic cross-section of the 4H-SiC MESFETs [26]. Fig. 6a and 6b show the improved DR and conventional DR structures, respectively. Compared with the conventional DR structure, additional source/drain drift region recess will be formed for the improved DR structure.

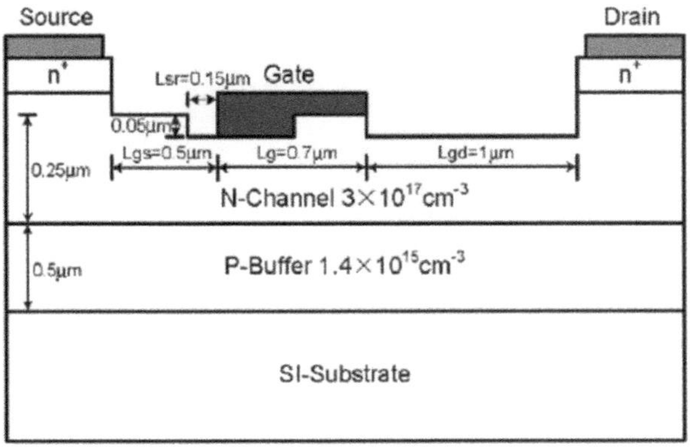

(a)

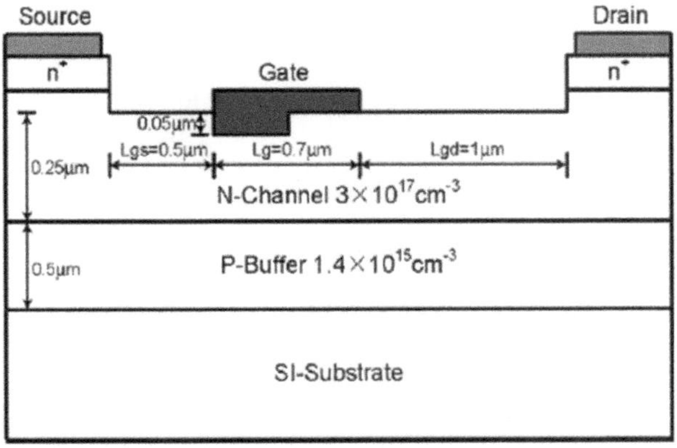

(b)

Figure 6. Schematic cross-section of the 4H-SiC MESFETs structures (a) Improved DR structure (b) double-recessed (DR) structure.

It can be seen from Fig.7 that the breakdown voltage (V_b) of the improved DR structure is increased compared to that of the conventional DR structure. A further investigation shows that the breakdown happened at gate corner near drain electrode due to the electric field crowding at that corner for these two structures. Actually, the maximum electrical field at gate corner near drain of the improved DR structure is reduced, compared with that of the conventional DR structure. Therefore, it clearly shows that the increase of breakdown voltage is attributed to reduced electric field crowding at gate corner near the drain due to thin channel thickness between the gate and drain electrodes with recessed drain drift region.

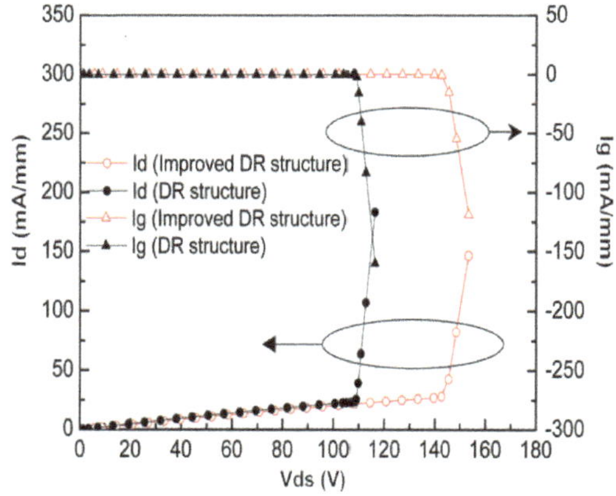

Figure 7. Simulated three-terminal breakdown characteristics.

3.3. Buffer-Gate 4H-SiC MESFETs structure

Fig. 8 is the schematic diagram of the structure of the Buffer-Gate SiC MESFETs structure [27]. Compared with the conventional 4H-SiC MESFETs, a low doped gate-buffer layer is introduced between the gate and channel layer.

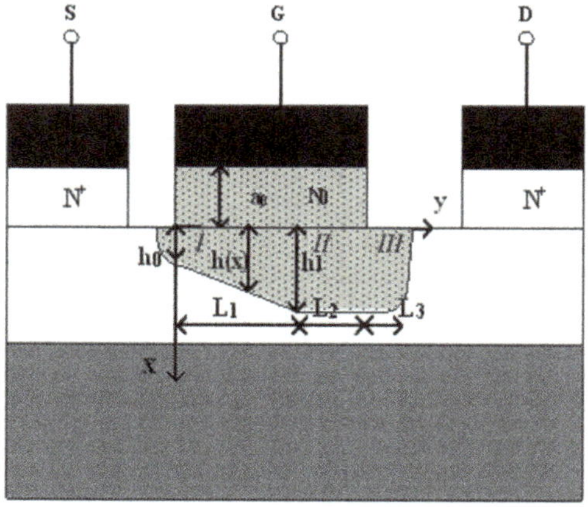

Figure 8. The schematic diagram of the buffer-gate structure.

In Buffer-Gate 4H-SiC MESFETs, the gate length and width are 0.7 μm and 332 μm. Meanwhile, the thickness and doping concentration are 0.26 μm and 1.7×10^{17} cm^{-3} for the channel layer, and 0.2 μm and 1×10^{14} cm^{-3} for the gate-buffer layer between the gate and channel.

Fig. 9 shows the dependence of the current in the channel on the gate-buffer layer. It reveals that the thicker the gate-buffer layer is, the larger the drain current is. This is because the channel width is increased owing to the decrease of the thickness of the depletion layer in the channel when the gate-buffer layer doping concentration and thickness are increased. It is also shown in Fig. 9 that when the gate-buffer layer is increased sufficiently, the drain current increases slowly because the thickness of the depleted layer in the channel layer decreases, and thus the width of the conduction channel increases more and more slowly with increasing the thickness of the gate-buffer layer.

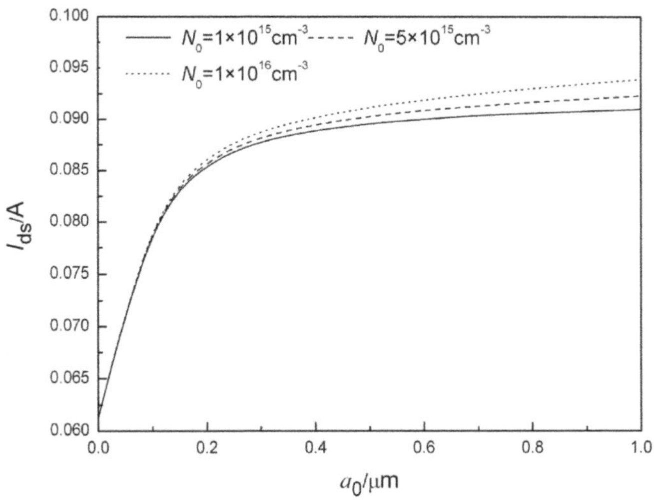

Figure 9. The dependence of the drain current on the gate-buffer layer for $N_0=1\times10^{15}cm^{-3}$(solid line), $N_0=5\times10^{15}cm^{-3}$(dash line) and $N_0=1\times10^{16}cm^{-3}$(dot line). V_{gs}=0 V, V_{ds}=5 V.

Fig.10(a). is the breakdown characteristics for the two structures. It can be seen that the breakdown voltage (V_b) is significantly increased, compared with that of the conventional structure. In fact, the breakdown happens at the gate corner near to the drain side due to the electric field crowding here for both two structures. However, the electric field peak is significantly lowered at the gate corner, i.e., the surface electric field is more uniform, owing to the inserted lower doped gate-buffer layer when compared with that of the conventional 4H-SiC MESFETs. Fig.10 (b) plots the corresponding electric field distribution. The reason for the suppression of the electric field at the gate corner is similar to that for the weakened surface electric field caused by lightly doped drain (LDD) in the MOSFETs [28].

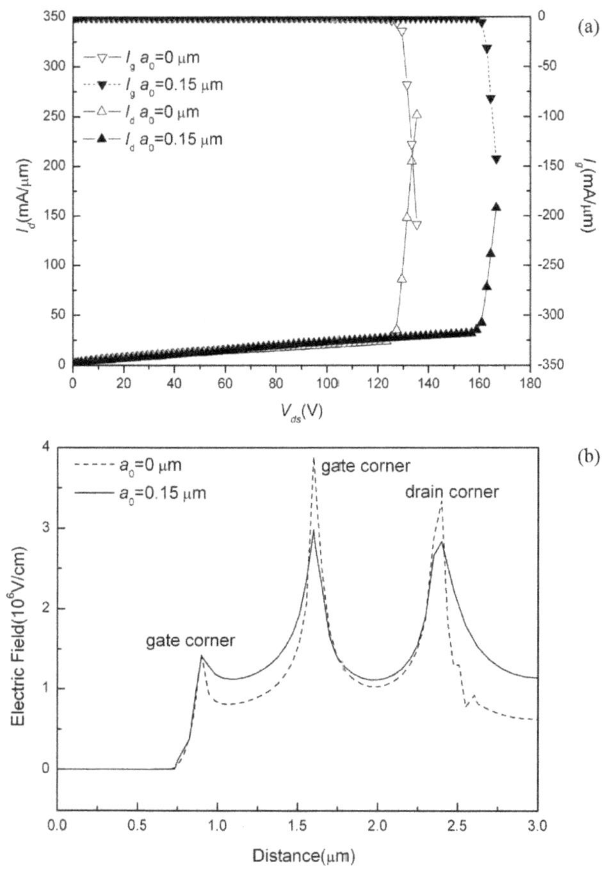

Figure 10. (a) The simulated breakdown characteristics for a_0=0 μm (open) and a_0=0.15 μm (filled), (b) The distribution of the surface electric field for a_0=0 μm (dash) and a_0=0.15 μm (solid).

3.4. Gate-drain surface epitaxial layer MESFETs structure (GDSE)

The schematic diagram of the GDSE structure is shown in Fig.11 [29]. Compared with the conventional 4H-SiC MESFETs, a low doped p-type surface epitaxial layer is introduced between the gate and drain when its doping concentration is lower than that of the channel layer by two orders. Firstly, there appears a build-in potential in the p-n junction between the p-type epitaxial layer and n-type channel layer or n^+ cap layer, which reduces the electric field peak at the gate corner. Thus it improves the electric field distribution. Secondly, the most part of the depletion layer in the p-n junction lies in the p-type epitaxial layer due to its doping concentration much lower than that of the n-type channel layer. Therefore, the gate-drain p-type epitaxial layer has little bad effect on the current density.

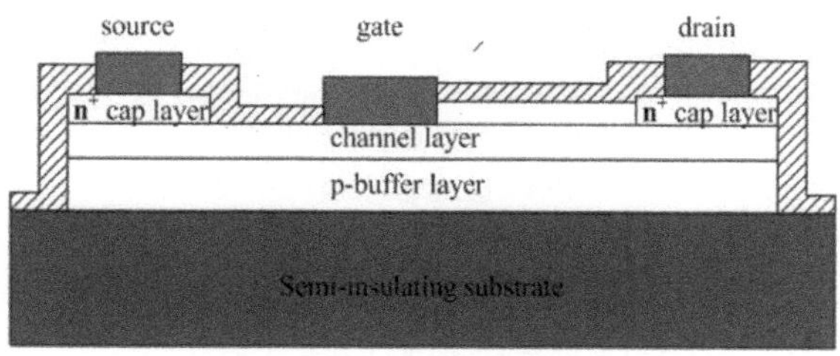

Figure 11. The schematic diagram of the MESFETs with gate-drain surface epi-layer.

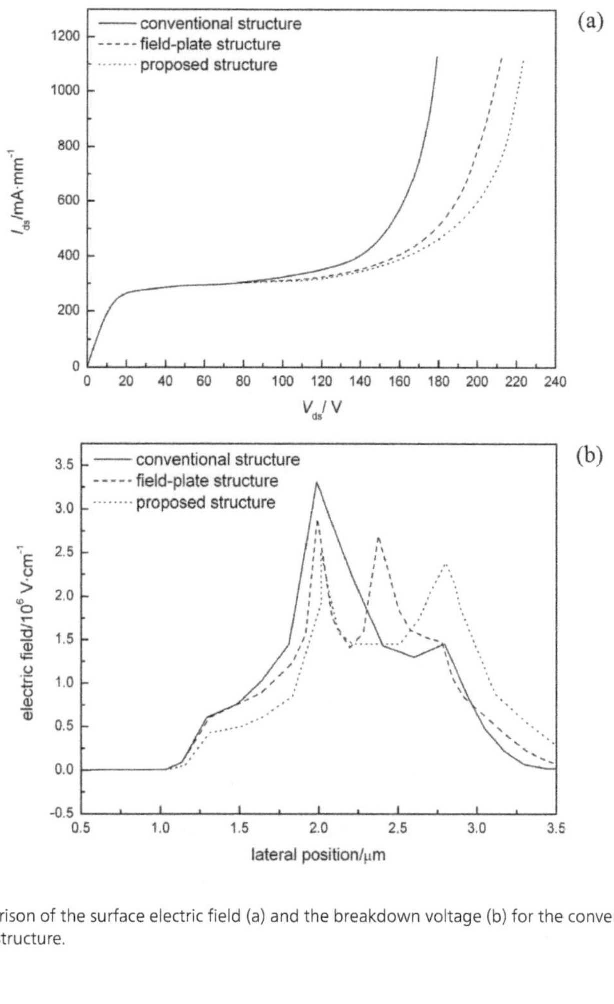

Figure 12. Comparison of the surface electric field (a) and the breakdown voltage (b) for the conventional, Field-plated, and the GDSE structure.

The breakdown characteristics and surface electric field distribution are shown in Fig.11. Fig.12 (a) shows that the breakdown voltage of the GDSE structure is the largest one for the three structures. The reason for the breakdown increased is that the electric field peak is significantly lowered at the gate corner, i.e., the surface electric field is more uniform, owing to the inserted lower doped p-type epitaxial layer when compared with that of the other two SiC MESFETs, as is shown in Fig.12 (b).

4. New 4H-SiC Power MESFETs Modeling

In most cases, the constant mobility is adopted for simplicity to develop the analytical models of the MESFETs which describes device operation. However, the constant mobility approximation is not adequate to describe the electron transport in the low field region and leads to an inaccurate estimation of the drain current and device characteristics of the 4H-SiC MESFETs. In this section, the improved analytical model for the conventional 4H-SiC MESFETs is provided which adopts the field-dependent mobility of the electrons, and takes into account the ungated high field region between the gate and drain which is usually omitted [30].

For the 4H-SiC material, the dependence of the mobility of the electron on the electric field can be described by Caughey–Thomas model [31].

$$\mu(E)=\frac{\mu_0}{\left[1+\left(\frac{\mu_0 E}{v_s}\right)^{\beta}\right]^{1/\beta}} \tag{1}$$

where μ_0 is the low field mobility, v_s the saturation velocity, E the electric field strength and β a curvature parameter of the order of 1 describing the saturation tendency. For SiC material, the measured value of β is 1.2 in the experiment [32]. For simplicity, β=1 is a reasonably good approximation [33].

Fig. 13 is the schematic diagram of the 4H-SiC MESFETs. The depletion layer thickness $h(x)$ under the gate, x away from the source is obtained by solving the one dimensional (1-D) Poisson's equation.

$$h(x)=\left(\frac{2\varepsilon[V(x)+V_G+V_{bi}]}{qN_D}\right)^{\frac{1}{2}} \tag{2}$$

where N_D is the uniformly doping concentration of the channel layer, and ε the dielectric constant. $V(x)$ is the potential difference between the source and the point x away from the source, V_G the gate bias, and V_{bi} the build-in voltage.

When the drain voltage is low, the electric field in the channel is less than the saturation field E_s, and only region ◎ exists. Based on integrating over the channel length L, the channel current is obtained

$$I_C(V_G, V_D) = I_P \frac{[3(u_d^2 - u_0^2) - 2(u_d^3 - u_0^3)]}{1 + Z(u_d^2 - u_0^2)} \tag{3}$$

where u_0 (u_d) is the depletion layer width h_0 (h_d) at the source (drain) end of the channel normalized to the epi-layer channel layer thickness a. I_P, Z and V_P are constants expressed by

$$I_p = \frac{q^2 N_D^2 a^3 \mu_0 W}{6\varepsilon L},\ V_p = \frac{q N_D a^2}{2\varepsilon},\ Z = \frac{q N_D a^2 \mu_0}{2\varepsilon L\ v_s} \tag{4}$$

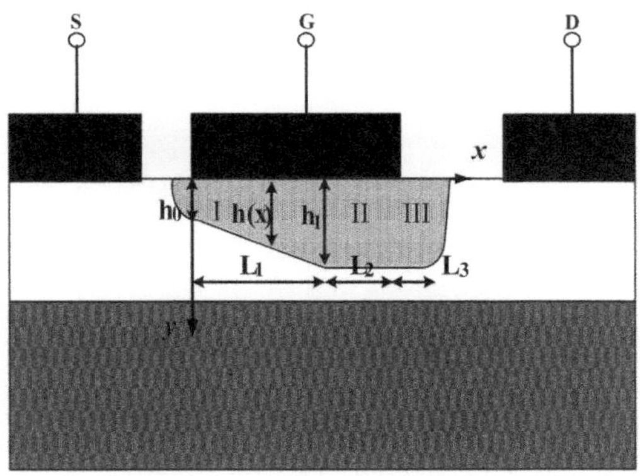

Figure 13. The schematic diagram of the 4H-SiC MESFETs.

With the drain bias increasing gradually, the lateral electric field increases and the electron velocity rises toward its saturation value. At enough high drain bias, the channel is in saturation regime, and can be divided into three regions which are shown in Fig. 13. In region with its length of L_1 and nearer to the source under the gate, the electric field is low and the electron velocity is less than the saturation velocity (v_s). The saturation region below the gate and the saturation region between the gate and the drain are labeled as regions and , respectively. The saturation channel current is

$$I_{csat} = q N_D W a \gamma v_s (1 - u_1) \tag{5}$$

where u_1 is the normalized depletion layer thickness at the point where the electron reaches the saturation velocity, γ=0.99.

$$u_1(V_G, V_D) = \frac{h_1}{a} = \frac{1}{a}\sqrt{\frac{\varepsilon}{q N_D}[V(L_1) + V_G + V_{bi}]} = \sqrt{\frac{[V(L_1) + V_G + V_{bi}]}{V_P}} \tag{6}$$

where h_1 is the depletion layer thickness at the point where the electron reaches the saturation velocity, $V(L_1)$ is the potential difference between the source and the point at $x=L_1$, where the saturation velocity $v(L_1)=\gamma v_s$.

To obtain the saturation current in the channel, there requires other equations involving u_1 or L_1, which can be obtained by solving 2-D Poisson's equation.

The potential drop across regions and achieved by solving the 2-D Poisson's equation is

$$V(L+L_3)-V(L_1)=\left(\frac{2au_1}{\pi}+\frac{L_3}{3}\right)E_s\sinh\left(\frac{\pi L_2}{2au_1}\right)\exp\left(\frac{-\pi L_3}{2au_1}\right) + \frac{E_s L_3}{3}\left(2\exp\left(\frac{\pi L_2}{2au_1}\right)+1\right) \quad (7)$$

The equation involving L_3 is

$$L_3^2\left[\frac{qN_D au_1}{\varepsilon}-E_s\exp\left(\frac{\pi L_2}{2au_1}\right)+E_s\sinh\left(\frac{\pi L_2}{2au_1}\right)\right] = (au1)^2 E_s\left[\exp\left(\frac{\pi L_2}{2au_1}\right)-1-\sinh\left(\frac{\pi L_2}{2au_1}\right)\exp\left(\frac{-\pi L_3}{2au_1}\right)\right] \quad (8)$$

where
$V(L+L_3)=V_D-I_C R_D$.

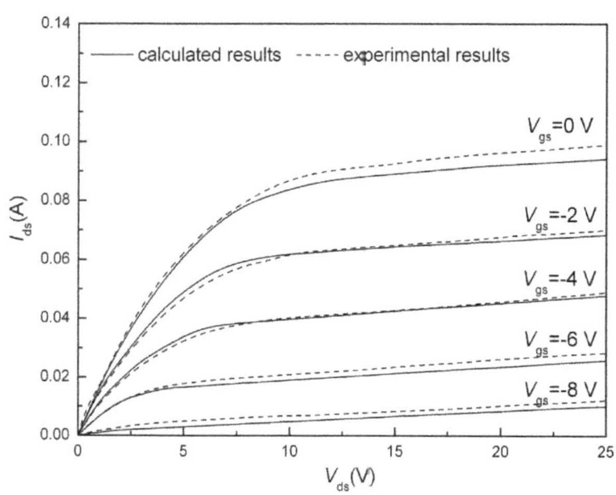

Figure 14. Calculated and experimental *I-V* characteristics, at V_{gs}=0,-2,-4,-6,-8 V.

From the analysis above, the drain current can be achieved when the structure parameters (L, W, a, a_0, N_D, N_0) and bias voltage (V_D, V_G) are given.

Using the obtained model, the *I-V* characteristics are calculated, as is shown in Fig.14. It can be seen that the *I–V* characteristics calculated using the obtained model are in agreement with the experimental data, which verifies the validity of the model.

Adopting the same approach, Yang et al developed analytical models for the new Buffer-Gate MESFETs [27]. The analytical models describing the direct-current (DC) and alternating current (AC) characteristics are as follows:

Under low drain voltage, the channel current is

$$I_C(V_G, V_D)=I_P\frac{[3(u_d^2-u_0^2)-2(u_d^3-u_0^3)]-{3a_0}/{a}\cdot[(u_d^2-u_0^2)-2(u_d-u_0)]}{1+Z(u_d^2-u_0^2)+Z\cdot{2a_0}/{a}\cdot(u_d-u_0)} \tag{9}$$

At high drain bias, the channel is in saturation mode, the saturation channel current is I_{csat}.

$$I_{csat}=qN_DWa\gamma v_s(1-u_1) \tag{10}$$

where u_1 is the depletion layer thickness normalized to the epilayer channel layer thickness a at the point where the electron reaches the saturation velocity and given by equation (11)

$$u_1(V_G, V_D)=\frac{h_1}{a}=\sqrt{(1-\frac{N_0}{N_D})\frac{a_0^2}{a^2}+\frac{(V(L_1)+V_G+V_{bi})}{V_p}}-\frac{a_0}{a} \tag{11}$$

where h_1 is the depletion layer thickness at the point where the electron reaches the saturation velocity, $V(L_1)$ is the potential difference between the source and the point at $x=L_1$, where the saturation velocity $v(L_1)=\gamma v_s$.

To evaluate the high frequency performance conveniently, it is important to describe the small signal parameters analytically.

From Eq.(9), the device's drain conductance for the linear region can be derived which is given in equation (12)

$$g_{dl}=\frac{3I_P\cdot(u_d+{a_0}/{a})}{m_dV_P}\cdot\frac{\left[(1-u_d)-Z(u_d-{a_0}/{a})(u_d^2-u_0^2)+{2}/{3}\cdot Z(u_d^3-u_0^3)-2Z\cdot{a_0}/{a}\cdot u_d\cdot(u_d-u_0)\right]}{\left[1+Z(u_d^2-u_0^2)+2Z\cdot{a_0}/{a}\cdot(u_d-u_0)\right]^2} \tag{12}$$

where

$m_d=u_d+{a_0}/{a}$.

Similarly, the transconductance for the linear region is equation (13)

$$g_{ml} = \frac{3I_P \cdot (u_d + {}^{a_0}\!/_a)}{m_d V_P} \cdot \frac{\left[(1-u_d) - Z(u_d - {}^{a_0}\!/_a)(u_d^2 - u_0^2) + {}^2\!/_3 \cdot Z(u_d^3 - u_0^3) - 2Z \cdot {}^{a_0}\!/_a \cdot u_d \cdot (u_d - u_0)\right]}{\left[1 + Z(u_d^2 - u_0^2) + 2Z \cdot {}^{a_0}\!/_a \cdot (u_d - u_0)\right]^2}$$
$$- \frac{3I_P \cdot (u_0 + {}^{a_0}\!/_a)}{m_0 V_P} \cdot \frac{\left[(1-u_0) - Z(u_0 - {}^{a_0}\!/_a)(u_d^2 - u_0^2) + {}^2\!/_3 \cdot Z(u_d^3 - u_0^3) - 2Z \cdot {}^{a_0}\!/_a \cdot u_0 \cdot (u_d - u_0)\right]}{\left[1 + Z(u_d^2 - u_0^2) + 2Z \cdot {}^{a_0}\!/_a \cdot (u_d - u_0)\right]^2} \quad (13)$$

where

$m_0 = u_0 + {}^{a_0}\!/_a.$

From Eq.(10), the drain conductance for the saturation region is obtained expressed by Eq.(14)

$$g_{ds} = -\frac{3\gamma I_P}{2Z m_{L_1} V_P} f \quad (14)$$

where

$$f^{-1} = 1 + \frac{1}{2m_{L_1} V_P}\left[\frac{2E_S a}{\pi}\sinh\frac{\pi(L - L_1)}{2(au_1 + a_0)} - \frac{E_S a(L - L_1)}{(au_1 + a_0)}\cosh\frac{\pi(L - L_1)}{2(au_1 + a_0)}\right] -$$
$$\frac{E_S LZ}{2m_{L_1} V_P}\cosh\frac{\pi(L - L_1)}{2(au_1 + a_0)} \times \quad m_{L_1} = u_1 + {}^{a_0}\!/_a.$$
$$\left\{\left[\frac{2u_1(1-\gamma)}{\gamma} + \frac{(u_1^2 - u_0^2) - {}^2\!/_3(u_1^3 - u_0^3)}{\gamma(1-u_1)^2}\right] - \frac{a_0}{\gamma a}\left[\frac{(u_1^2 - u_0^2) - (u_1 - u_0)}{(1-u_1)^2} - 2(1-\gamma)\right]\right\}$$

The expression of transconductance for the saturation region is equation (15)

$$g_{ms} = -\frac{3\gamma I_P}{2Z m_{L_1} V_P}(k + 1) \quad (15)$$

where

$$h=-\frac{1}{2m_{L_1}V_P}\left[\frac{2E_S a}{\pi}\sinh\frac{\pi(L-L_1)}{2(au_1+a_0)}-\frac{E_S a(L-L_1)}{(au_1+a_0)}\cosh\frac{\pi(L-L_1)}{2(au_1+a_0)}\right]+$$

$$\frac{E_S LZ}{2m_{L_1}V_P}\cosh\frac{\pi(L-L_1)}{2(au_1+a_0)}\times$$

$$\left\{\left[\frac{2u_1(1-\gamma)}{\lambda}+\frac{(u_1^2-u_0^2)-{}^2/_3(u_1^3-u_0^3)}{\gamma(1-u_1)^2}\right]-\frac{a_0}{\gamma a}\left[\frac{(u_1^2-u_0^2)-(u_1-u_0)}{(1-u_1)^2}-2(1-\gamma)\right]\right\}.$$

$$-\frac{u_0 E_S LZ}{m_0 V_P}\cosh\frac{\pi(L-L_1)}{2(au_1+a_0)}\left[\frac{1-u_0}{\gamma(1-u_1)}-1\right]\left(1+\frac{a_0}{au_0}\right)$$

To find the gate-source capacitance, the magnitude of the charge in the depletion layer under the gate is needed. It can be derived from the potential distribution in the depletion layer which is obtained by solving the 2-D Poisson's equation.

The expression of the gate-source capacitance is equation (16)

$$C_{gs}=C_{gs1}+C_{gs2}+C_{gs3} \tag{16}$$

where

$$C_{gs1}=\frac{4\varepsilon WL\,V_P I_P}{aID}\left\{\begin{array}{l}\frac{-g_m}{I_D}\left[(1-\frac{a_0}{a})(u_1^3-u_0^3)-\frac{3}{4}(u_1^4-u_0^4)+\frac{3a_0}{a}(u_1^2-u_0^2)\right]\\+\frac{3u_1}{2m_{L_1}V_P}(u_1+\frac{a_0}{a})(1-u_1-\frac{ZI_D}{3I_P})(k+1)-\frac{3u_0}{2m_0V_P}(u_0+\frac{a_0}{a})(1-u_0-\frac{ZI_D}{3I_P})\end{array}\right\}$$

$$C_{gs2}=\frac{2\varepsilon WLZu_1}{a}\left\{\frac{g}{2Zm_{L_1}u_1}(k+1)+\frac{u_0}{m_0}\left[\frac{u_1-u_0}{\gamma(1-u_1)}+\frac{1-\gamma}{\gamma}\right]\left(1+\frac{a_0}{au_0}\right)\right\}$$

$$C_{gs3}=-\varepsilon E_S Wa\left\{\begin{array}{l}\frac{k+1}{2m_{L_1}V_P}\left\{\begin{array}{l}1-\cosh\frac{\pi(L-L_1)}{2(au_1+a_0)}-\frac{\pi(1-u_1)}{2(au_1+a_0)}\sinh\\\frac{\pi(L-L_1)}{2(au_1+a_0)}\left[\frac{a(L-L_1)}{au_1+a_0}-\frac{L\cdot g-L+L_1}{u_1}\right]\end{array}\right\}\\+\frac{\pi u_0 LZ(1-u_1)}{2m_0V_P(au_1+a_0)}\sinh\frac{\pi(L-L_1)}{2(au_1+a_0)}\left(\frac{1-u_0}{\gamma(1-u_1)}-1\right)\left(1+\frac{a_0}{au_0}\right)\end{array}\right\}$$

$$g=1-\frac{L_1}{L}-Zu_1\left\{\begin{array}{l}\left[\frac{2u_1(1-\gamma)}{\gamma}+\frac{(u_1^2-u_0^2)-{}^2/_3(u_1^3-u_0^3)}{\gamma(1-u_1)^2}\right]-\\\frac{a_0}{\gamma a}\left[\frac{(u_1^2-u_0^2)-(u_1-u_0)}{(1-u_1)^2}-2(1-\gamma)\right]\end{array}\right\}.$$

Using their model, Yang et al calculate the *I-V* characteristics and DC small signal characteristics of the proposed 4H-SiC MESFETs with a gate-buffer layer of 0.2 μm. The results are shown in Fig.15 to Fig.18.

Fig.15 shows the *I-V* characteristics of the proposed 4H-SiC MESFETs. For a specified gate bias, the drain currents achieved in their proposed device are larger than that in the conventional one. For instance, the maximum saturation drain current is 0.18 A, which is larger than 0.1 A in the conventional structure. Since the total depletion layer is composed of the completely depleted gate-buffer layer and the depletion layer in the channel layer, the part in the channel layer shrinks when there exists a gate-buffer layer, compared to the conventional structure, as shown in Fig. 8. As a result, the effective channel width is increased and a larger current occurs in the Buffer-Gate MESFETs than in the conventional structure.

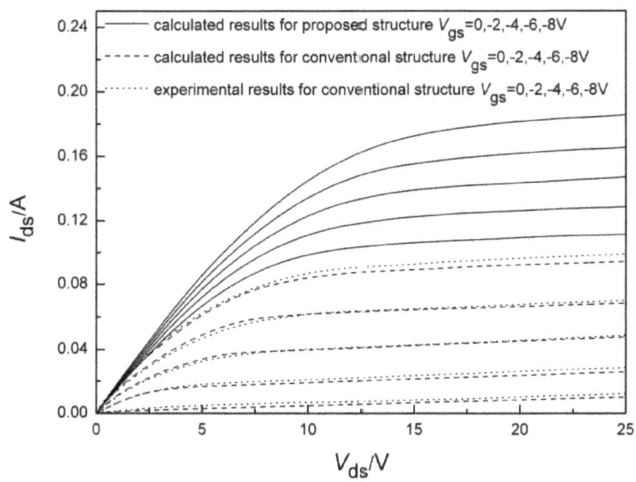

Figure 15. Calculated and experimental *I-V* characteristics, for the conventional and buffer-gate structures, V_{gs}=0,-2,-4,-6,-8 V.

Fig. 16, 17 and 18 show the effect of the gate-buffer layer thickness on the small signal parameters. The transconductance, gate-source capacitance and channel resistance are decreased, while the drain conductance is increased with the inserted *lower doped* gate-buffer layer thickness (completely depleted) because the total depletion layer thickness under the gate is increased but the part in the channel is decreased, as are depicted in Fig.16 and Fig. 17. The cutoff frequency (f_T) is increased with the gate-buffer layer thickness slowly because the decrease of the transconductance is overcome by the decrease of the gate-source capacitance. Since the decrease of the channel resistance is more dominant than the increase of the drain conductance with the gate buffer layer thickness, the maximum oscillation frequency (f_{max}) is increased more rapidly than the cutoff frequency. The results show that f_T and f_{max} are significantly improved, compared with those of the conventional structure. Therefore, the buffer-gate structure has better frequency performance than that of the similar device based on the conventional structure.

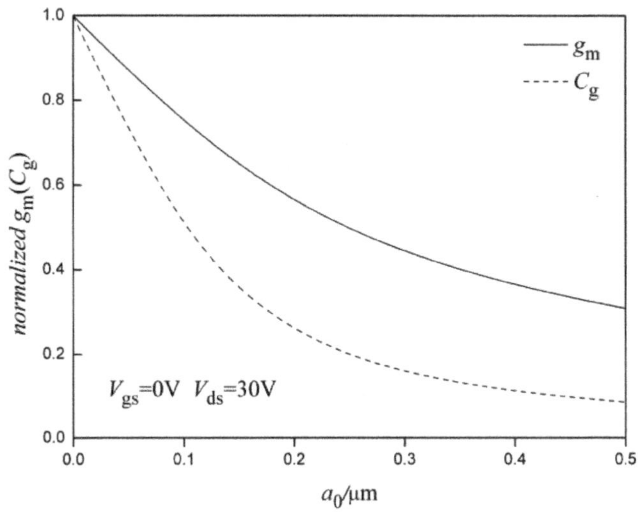

Figure 16. Dependence of transconductance g_m and gate-source capacitance C_{gs} on the gate buffer layer thickness, V_{gs}=0 V, V_{ds}=30 V.

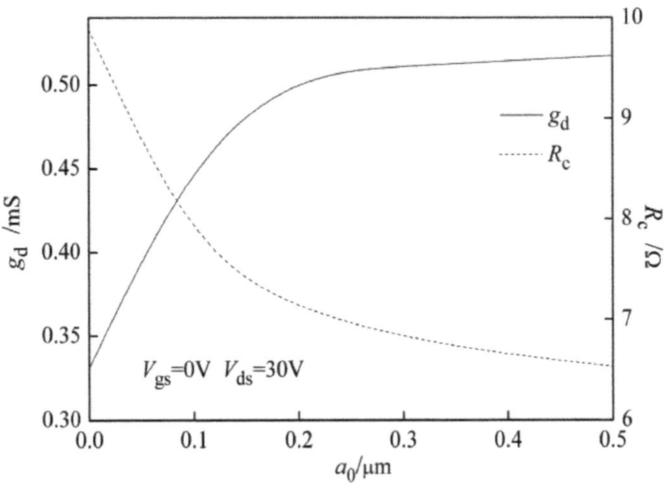

Figure 17. Dependence of drain conductance g_d and channel resistance R_c on the gate buffer layer thickness, V_{gs}=0 V, V_{ds}=30 V.

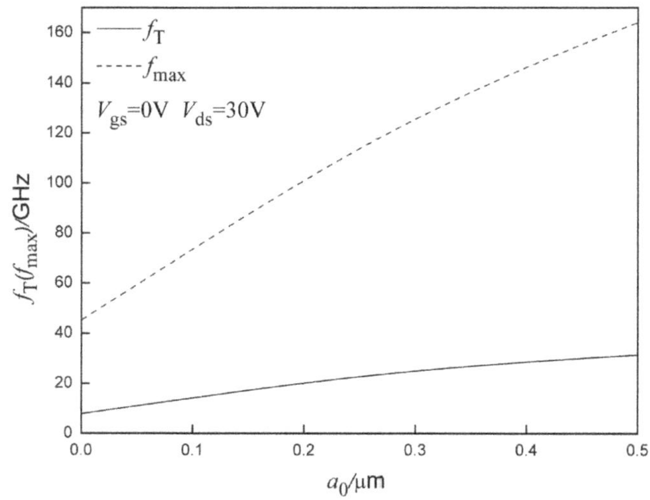

Figure 18. Dependence of cutoff frequency f_T and maximum oscillation frequency f_{max} on the gate buffer layer thickness, V_{gs}=0 V, V_{ds}=30 V.

Notes

In this chapter, the significant results, which are obtained recently, are reviewed for the 4H-SiC MESFETs (Metal Semiconductor Field Effect Transistor). First, the basic operation principle for the 4H-SiC power MESFETs is briefly reported. Then, several new 4H-SiC power MESFETs structures are discussed, focusing on the surface electric field, breakdown voltage and frequency characteristics. Finally, the models with the electric field-dependent mobility are also reported for the conventional and buffer gate SiC MESFETs.

Author details

Yintang Yang[1*], Baoxing Duan[1] and Xianjun Zhang[2]

*Address all correspondence to: ytyang@xidian.edu.cn

1 School of Microelectronics, Xidian University, China

2 College of Automation & Electronic Engineering, Qingdao University of Science & Technology, China

References

[1] Levinshtein, M. E., Rumyantsev, S. L., & Shur, M. S. (2001). *Properties of Advanced Semiconductor Materials: GaN, AlN, InN, BN, SiC, SiGe,* Wiley & Sons, Inc., New York.

[2] Wu, J., Fursin, L., Li, Y., Alexandrov, P., & Zhao, J. H. (2004). 4,308V, 20.9mΩcm^2 4H-SiC MPS diodes based on a 30μm drift layer. *Mat. Sci. Forum,* 1109-1112.

[3] Weitze, C. E., Palmour, J. W., Carter, C. H., Moore, K., Nordquist, K. J., Allen, S., & Thero, C. (1996). Silicon carbide high-power devices. *IEEE Trans. Electron Devices,* 43, 1732-1741.

[4] Yong, Cai, Yugang, Zhou, Chen, Kevin J., & Kei, May Lau. (2005). High-Performance Enhancement-Mode AlGaN/ GaN HEMTs Using Fluoride-Based Plasma Trentment. *IEEE Electron Device Lett.,* 26, 435-437.

[5] Di , Song, Jie, Liu, Zhiqun, Cheng, Tang, Wilson C. W., Kei, May Lau, & Chen, Kevin J. (2007). Normally Off AlGaN/GaN Low-Density Drain HEMT (LDD-HEMT) With Enhanced Breakdown Voltage and Reduced Current Collapse. *IEEE Electron Device Lett.,* 28, 189-191.

[6] Baoxing, Duan, & Yintang, Yang. (2012). New $Al_{0.25}Ga_{0.75}N$/GaN HEMT structure with the partial silicon doping. *Micro &Nano Letter,* 7, 9-11.

[7] Wright, N. G., Knights, A. P., O'Neill, A. G., & Johnson, C. M. (1999). Evaluation of 4H-SiC varactor diodes for microwave applications. *Proceedings of EDMO-99, IEEE Symposium on High Performance Electron Devices for Microwave and Optoelectronic Applications,* 301-306.

[8] Na, H. J., Moon, J. H., Yim, J. H., Lee, J. B., & Kim, H. J. (2006). Fabrication and characterization of 4H-SiC planar MESFETs. *Microelectronic Engineering,* 83, 160-164.

[9] Sriram, S., Hagleitner, H., Namishia, D., Alcorn, T., Smith, T., & Pulz, B. (2009). High-Gain SiC MESFETs Using Source-Connected Field Plates. *IEEE Electron Device Lett.,* 30(9), 952-953.

[10] Xian, Zhang-Jun, Yin, Yang-Tang, Bao, Duan-Xing, Chang, Chai-Chun, Song, Kun, & Chen, Bin. (2012). Modeling of the drain induced barrier lowering effect and optimization for dual-channel 4H silicon carbide metal semiconductor field effect transistor. *Chinese Physics B.*

[11] Chen, X. B., & Johnny, K. O. S. (2001). Optimization of the Specific On-Resistance of the COOLMOS^{TM} . IEEE Trans., *IEEE Transactions on Electron Devices,* 48, 344-348.

[12] Sameh, G., Nassif, K., & Salama, C. A. T. (2003). Super-junction LDMOST on a silicon-on-sapphire substrate. *IEEE Trans. Electron Devices,* 50, 1385-1391.

[13] Duan, B. X., Zhang, B., & Li, Z. J. (2006). New Thin-Film Power MOSFET'S with a Buried Oxide Double Step Structure. *IEEE Electron Device Lett.,* 27, 377-379.

[14] Duan, B. X., Yang, Y. T., Zhang, B., & Hong, X. F. (2009). Folded Accumulation LDMOST (FALDMOST): New Power MOS Transistor with Very Low Specific On-resistance. *IEEE Electron Device Lett.*, 12, 1329-1331.

[15] Duan, B. X., & Yang, Y. T. (2011). Low Specific On-Resistance Power MOS Transistor with Multi-Layer Carrier Accumulation breaks the limit line of silicon. *IEEE Transactions on Electron Devices*, 58, 2057-2060.

[16] Saxena, V., Jian, Nong Su, & Steckl, A. J. (1999). High-voltage Ni- and Pt-SiC Schottky diodes utilizing metal field plate termination. *IEEE Trans. Elect. Dev.*, 46, 456-464.

[17] Kinoshita, K., Hatakeyama, T., Takikawa, O., Yahata, A., & Shinohe, T. (2002). Guard ring assisted RESURF: a new termination structure providing stable and high breakdown voltage for SiC power devices. *ISPSD*, 253-256.

[18] Ueno, K., Urushidani, T., Hashimoto, K., & Seki, Y. (1995). The guard-ring termination for the high-voltage SiC Schottky barrier diodes. *IEEE Electr. Dev. Lett.*, 16, 331-332.

[19] Sheridan, D. C., Niu, G., Merrett, J. N., Cressler, J. D., Tin, C., Ellis, C. C., & Siergiej, R. R. (2000). Simulation and fabrication of high-voltage 4H-SiC diodes with multiple floating guard ring termination. *Materials Science Forum*, 1339-1342.

[20] Ueno, Katsunori, Urushidani, Tatsuo, Hashimoto, Kouichi, & Seki, Yasukazu. (1995). Al/Ti Schottky barrier diodes with the guard-ring termination for 6H-SiC. *ISPSD'95*, 107-111.

[21] Brezeanu, G., Badila, M., Millan, J., Godignon, Ph, Locatelli, M. L., Chante, J. P., Lebedev, A., Dilimot, G., Enache, I., Bica, G., & Banu, V. (1999). A nearly ideal SiC Schottky barrier device edge termination. *International Semiconductor Conference*, 183-186.

[22] Alok, D., Baliga, B. J., & Mc Larty, P. K. (1994). A simple edge termination for silicon carbide devices with nearly ideal breakdown voltage. *IEEE Electr. Dev. Lett.*, 15, 394-395.

[23] Kimoto, I. T., & Matsunami, H. (1996). Excellent Reverse Blocking Characteristics of High-Voltage 4H-SiC Schottky Rectifiers with Boron-Implanted Edge Termination. *IEEE Eleclr. Dev. Lett.*, 17, 139-141.

[24] Baoxing, Duan, Yintang, Yang, & Bo, Zhang. (2010). High Voltage REBULF LDMOS with N+-Buried Layer. *Solid-State Electronics*, 54, 685-688.

[25] Cha, H. Y., Choi, Y. C., Eastman, L. F., & Spencer, M. G. (2004). Simulation study on breakdown behavior of field-plate SiC MESFETs. *International Journal of High Speed Electronics and Systems*, 14(3), 884-9.

[26] Zhang, J. P., Luo, X. R., Li, Z. J., & Zhang, B. (2007). *Microelectron Eng.*, 84(12), 2888-2891.

[27] Zhang, X. J., Yang, Y. T., Duan, B. X., Chen, B., Chai, C. C., & Song, K. (2012). New 4H-SiC MESFET with the Buffer Layer between the Gate and Channel. *Chinese Physics B*, 21(1), 017201-1-7.

[28] Sze, S. M. (1981). *Physics of Semiconductor Devices*, New York, Wiley.

[29] Song, K., Chai, C. C., Yang, Y. T., Zhang, X. J., & Chen, B. (2012). Improvement in breakdown characteristics of 4H-SiC MESFET with a gate-drain surface epi-layer and optimization of the structure. *Acta. Phys. Sin.*, 61(2), 027202-1-6.

[30] Zhu, C. L., Rusli, C. C., Tin, S. F., & Yoon, J. Ahn. (2006). A three-region analytical model for short-channel SiC MESFETs. *Microelectron Eng.*, 83(1), 96-9.

[31] Caughey, D. M., & Thomas, R. E. (1967). Carrier mobilities in silicon empirically related to doping and field. *Pro IEEE*, 52, 2192-2193.

[32] Khan, I. A., & Cooper Jr, J. A. (2000). Measurement of high-field electron transport in silicon carbide. *IEEE Tran Electron Dev.*, 47(2), 269-273.

[33] Murray, S. P., & Roenker, K. P. (2002). An analytical model for SiC MESFETs. *Solid-State Electron.*, 46(10), 1495-1505.

Chapter 12

Conductance Deep-Level Transient Spectroscopic Study of 4H-SiC MESFET and Traps

Malek Gassoumi and Hassen Maaref

Additional information is available at the end of the chapter

1. Introduction

Silicon carbide (SiC) is an important material for fabricating high-power, high-temperature and high frequency devices [1]. The semi-insulating (SI) form of 4H-SiC is useful for making microwave devices [2] because it helps in lowering the stray device capacitances, thereby increasing the maximum operating frequency of the device. Selective area ion implantation is regarded as an attractive doping method for fabricating MESFETs in bulk SI 4H-SiC [3–5] due to the ease of inter-device isolation without the loss of planarity.

The silicon carbide (SiC) MESFET becomes very promising candidate for high power microwave applications in commercial and military communications.

However, SiC MESFETs are not without trapping problems associated with both the surface and with the layers underlying the active channel, which influence power performance through the formation of quasistatic charge distributions. This parasitic charge acts to restrict the drain current and voltage excursions, thereby limiting the high-frequency power output [6, 7].

Over the past few years, the vanadium-doped semi-insulating SiC substrate has attracted much attention in explaining the deterioration of the SiC MESFET microwave performance. Recently, the concern has shifted more towards surface traps due to the introduction of high-purity semi-insulating substrates, which have eliminated the larger part of the problems associated with the substrate [8,9]. The presence of surface states in the ungated channel regions between drain and source terminals has modulated the depletion of the channel under the device surface, and has resulted in the frequency dispersion of the transconductance (gm) and gate lag transient [10, 12]. These anomalies make the device characteristics much more complicated, and make some troubles in circuit design.

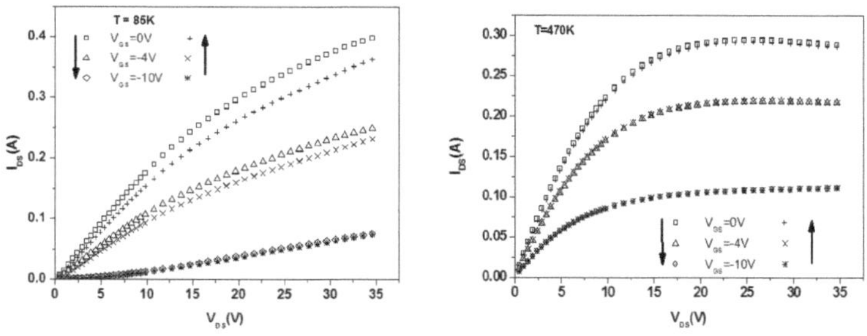

Figure 1. a: Static output characteristics (IDS-VDS-VGS) at T=85K. The gate voltage is first increased from 0 to -10 V (open symbols) and next decreased from -10V to 0V (cross symbols), b: Same as figure 1.a for T=475K. The drain current decrease is no present for the second set of curves

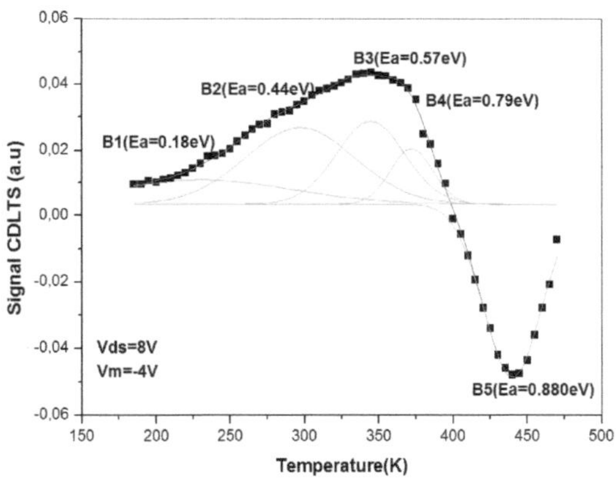

Figure 2. Conductance DLTS spectra for a 4H-SiC MESFET.

Then trapping effects due to the substrate deep levels can occur. On the other hand, it is well known that SiC / insulator interface presents high density of surface states. As an example, when using a SiO_2 passivation, a high interface state density (in the 10^{11} to 10^{12} cm^{-2} range) between SiO_2 and SiC is still present. Consequently, the second hypothesis is consistent with the well known remaining passivation problem for SiC devices. Understanding the nature of the surface states in MESFETs could be a key to solve the problems. However, only a few works were reported on this topic. Conductance Deep Level Transient Spectroscopy (CDLTS), as capacitance DLTS, is an efficient tool to obtain information about traps in semiconductors, such as the activation energy, the capture cross section and the density of traps (in the case of Capacitance DLTS). The advantage of CDLTS over DLTS is the possibility to perform the measurements on a device with a small gate area and in operating conditions. However, to the best of our knowledge, no work has been conducted using this technique for 4H-SiC MESFETs. Subsequently, in this study we have investigated the origin of parasitic effect in DC characteristics of 4H-SiC MESFETs with a gate length of 1µm by CDLTS.

In this study, we have used an old but not yet reported technique to investigate the origin of non stationnarity in 4H-SiC MESFETs. This technique is called Current (or Conductance) Deep Level Transient Spectroscopy (CDLTS). Indeed this technique is more suitable for the study of transistors than the classical capacitance DLTS for two main reasons.

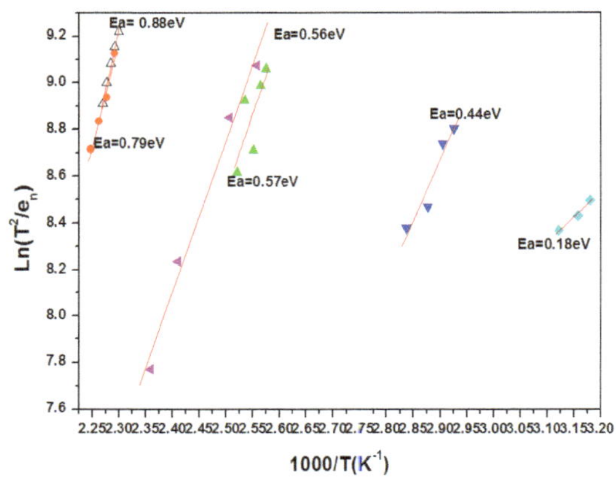

Figure 3. Arrhenius plot for the deep levels observed in the 4H-SiC MESFET.

2. Experimental procedure

The 4H-SiC MESFETs studied in this work were realized at THALES Research and Technology (TRT) in Orsay (France) using the same fabrication process for all the transistors. The epitaxial layer structures were prepared by CVD on semi-insulating substrates supplied by CREE. The layer stack consists of three layers: a P-type buffer layer with a thickness of 0.3 µm and a doping level of 1.10^{16} cm^{-3}, an N-type active layer with 0.3-0.4 µm thickness and doping level N_d = 1-2.10^{17} cm^{-3}, and an N^+ contact layer with a thickness and doping of 0.2 µm and $N_d = 10^{19}$ cm^{-3}. The lift-off process for the fabrication of the device has already been described elsewhere [5]. The first step consists in reactive ion etching for the channel recess. An evaporation of Ti/Pt/Au stack is realized for the gate contact. The surface is passivated by a deposited oxide layer. The measurements presented in this work have been realized on short test transistors with a gate of 1µm and a gate width of 100 µm. For all the transistors the source gate distance is 0.5 µm and the gate drain distance is 2 µm.

The current (or conductance) deep level transient spectroscopy (CDLTS) measurements were performed by applying both a drain to source voltage of 8 V, in the linear regime, and a drain to source voltage of 18V in the saturation region. The steady-state reverse bias and filling pulse voltages applied to the gate were -10V and 0V. The drain source current transient were recorded using a numerical multimeter (HP 34401 A). The transient were treated numerically using the Lang method as in classical capacitance DLTS. The measurements were carried out between 80K and 600K in a nitrogen cooled cryostat.

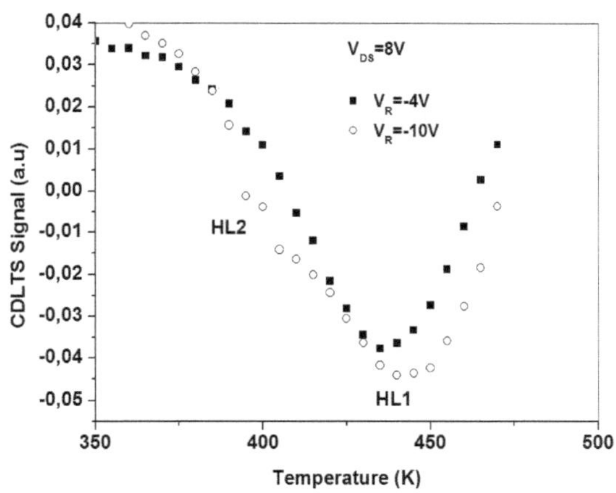

Figure 4. Conductance DLTS under gate pulse spectrum with V_{gs} = -4V and -10V.

3. Results and discussion

3.1. Output Characteristics

Drain-source current voltage (I_{DS}-V_{DS}) measurements as a function of gate voltage and temperature have been performed. Output characteristics registered at different temperatures show several parasitic effects. The first anomaly observed on I_{DS}-V_{DS} characteristics consist in decreasing of the current when we perform the V_{GS} sweep consecutively increasing the negative gate voltage (i.e pinching the channel) and next decreasing V_{GS} to zero (i.e opening the channel). This effect observed at 85K becomes more and more important for low gate voltages (Fig. 1a), i.e. when the current flows in the physical channel near the surface. This degradation in current vanishes progressively when the temperature is raised and totally disappears above 470K (Fig. 2b). Obviously, this behavior is due to a thermally activated effect. In a previous work [17] we observed also an effect of current collapse, but more pronounced for high gate voltage, when the current flows near the substrate. We attributed this to the presence of deep center located in the SI substrate. Basically, the explanation here can be the same, but the trap location is different. Indeed, the degradation in current can be due to the presence of deep centers located in the vicinity of the channel surface. Trapping/ detrapping phenomena on these centers change the charge density near the surface. When we apply V_{DS} for the first measurement we force these traps to be charged. In the case of electron traps, for instance, this charge is negative and a parasitic depletion at the channel surface reduces the drain current. At low temperature the emission kinetic of the traps is very slow. Consequently, when the second measurement is performed, the parasitic depletion region is still present and the current is reduced. When the temperature is increased, the traps are thermally activated and the phenomenon disappears. The same effect of gate lag has been observed recently and was undoubtedly attributed to surface trapping phenomena by an other group. Indeed, in their work, Cha et al [18] noticed that the phenomenon depends strongly on the channel surface: current lag is lower for recessed gate and buried gate geometry than for unrecessed and channel recessed ones. Another parasitic effect observed on some of our transistors consists in an important leakage current increasing with the temperature as can be seen for V_{GS} = -10 V in Fig. 1 a and 1b. We will focus in the following on the transistors showing these parasitic effects in output characteristics.

I-V-T measurements have been performed on the Schottky gate of defective MESFETs in the temperature range 100K-460K. On these diodes we have not observed any anomalous behavior like multi barrier height in direct characteristics. The ideality factor for all transistors ranges from 2 at 300K to 1.3 at 400K. These values show that the dominant mechanism of transport in our case is the generation recombination (G-R) mechanism. This result confirms the presence of (G-R) centers in the structure.

3.2. Part A

The current DLTS measurements were performed on the MESFET by applying a small drain to source voltage, V_{ds} of 8V while keeping the source grounded. This low V_{ds} was used to ensure that the MESFET operates within the linear region of the current-voltage characteristics.

With a gate bias of 4V in the neighbourhood of the threshold voltage, the drain current DLTS is sensitive to traps in the channel.

Figure 2 gives current DLTS spectrum under a gate pulse, showing five traps called (B1, B2, B3, B4 and B5). The apparent activation energies and capture cross-sections are deduced from the Arrhenius plot of Ln (T^2/e_n) versus 1000/T (Figure 3) of all observed traps. The nature and localization of traps B_1 to B_4 has been discussed in a previous publication [6]. On the other hand, with a gate bias of -10V the device is biased closer to the threshold voltage, the drain-current DLTS is sensitive to traps in the channel and at the buffer-channel interface. A current DLTS spectrum under a gate pulse (Vgs switching from 0-4V and 0-10V) is depicted in Figure 4. The peak amplitude of the B1 (0.18eV), B2 (0.44eV) B3 (0.57eV) and B4 (0.79) traps is invariant with the change of the gate bias which improves their localisation at the channel surface [13].

The anomalous hole trap like peaks observed in current DLTS spectra of GaAs MESFET have been previously studied and have been associated with surface states present in the ungated regions between the gate and the source-drain electrode, acting as trapping sites for electrons emitted from the gate edge during the reverse bias. Such an explanation is supported by the fact that the temperature dependence of the thermal emission current from the gate edge to the ungated region observed in the DLTS measurements induces the change of the DLTS peak height of the hole-like trap when the value of t_2 was $4t_1$ at the constant value of V_p and V_m (Figure 5).

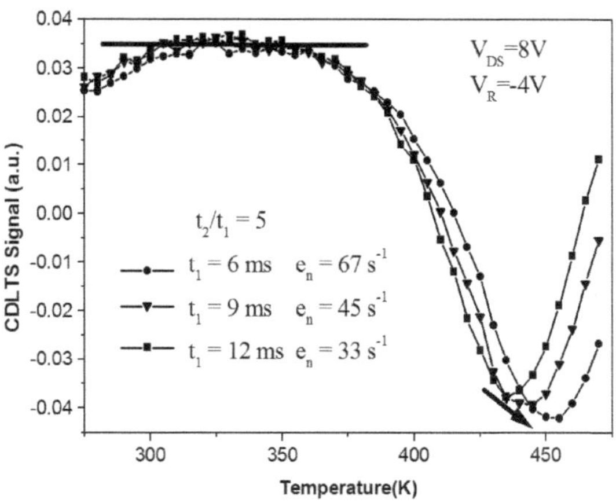

Figure 5. Conductance DLTS spectra in the linear regime (VDS = 8 V) at V_R = -4 V and for deferent emission rates

Thermal emission from the bulk electron traps, if present, in the active channel region below the gate would reduce the depletion width under the gate, this would in turn cause an increase in the measured source to drain current I_{ds}. However, the emission of electrons from the interface electron traps into the surface conduction channel would increase I_s. As the surface channel forms an additional conduction path between the gate and the drain (source), the interface trap emission would thus decrease I_{ds}. This is because the change in the surface current I_s and the change in the bulk channel flow in opposite directions.

The appearance of the hole-like trap signal B4 (Ea=0.78eV) only for V_m lower than -10V near the pinch-off voltage leads us to conclude that the hole-like trap is not related to the surface states at the ungated regions, but is related to channel-buffer interface.

The independence of the CDLTS signal associated to all the traps with tp lead us to conclude the absence of all types of micropipes which can be responsible for the high power device deficiencies [13].

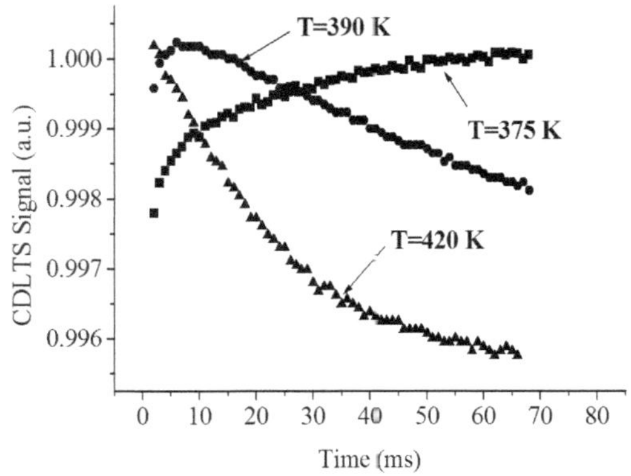

Figure 6. Ids transient showing the progressive change from emission (375 K) to capture (420 K).

3.3. Part B: Surface Traps in 4H-SiC MESFET

In this pat we can study the surface traps in MESFET 4H-SiC we can used CDLTS. Therefore, in our samples, we expect to observe emission from the channel (or buffer/channel interface) electron traps. In this case, the depletion width under the gate (or at the buffer/channel interface) is reduced and then an increase in Ids is observed. The corresponding C-DLTS signal is positive. Nevertheless, a progressive change from emission to capture was observed in the Ids transient measurement between 375 K and 420 K (Fig. 6). The corresponding C-DLTS is shown on Fig. 6. A broad positive peak due to electron emission by at four different levels is observed. The signatures of these traps have been reported in a previous work [15]. The interesting point comes from the negative peak which is clearly observed. We will focus here, on this level.

Different mechanism can explain the presence of a negative peak:

- In capacitance DLTS they can be due to a measurement artifact when a high frequency modulation is used for the differential capacitance measurement ($R^2C^2w^2$>>1). This obviously cannot be the case in C-DLTS measurement.
- Even if this kind of behavior is frequently called "hole like", a hole emission process is not possible in our samples. Where do these holes could come from?
- A mid-gap amphoteric level can exchange with both bands. The activation energy of 0.9 eV for level B5 rules out this explanation.
- The last explanation is the presence of a conductive layer at the channel/passivating layer interface in the ungated regions. This conductive layer constitutes a tank for electrons which can be captured by a trap located close to the interface.

This phenomenon has been previously observed and is well-known in the case of GaAlAs/Si_3N_4 interface for GaAs MESFETs [16]. The presence of a large amount of interfacial traps (in the 10^{12} cm^{-2} range) at the SiC/SiO_2 interface is consistent with this explanation. These defects acting as a conductive layer at the channel surface can also explain the leakage current observed on DC characteristics (surface current flowing from the gate to the drain). To strengthen the hypothesis of a capture process on surface states, an additional C-DLTS measurement in the saturation regime (Vd=18V) has been performed. From The result displayed on Fig. 7 we clearly observe the absence of the negative peak. Indeed, in these polarization conditions the channel between the gate and the drain is almost fully deserted even for Vg = 0 V and then, it is no more modulated by the gate pulses. As the gate-drain distance is 4 times higher than the source gate one (respectively 2 μm and 0.5 μm), the response in C-DLTS measurement is almost insensitive to the channel/SiO_2 interface. This is why peak B_5 is no more observed.

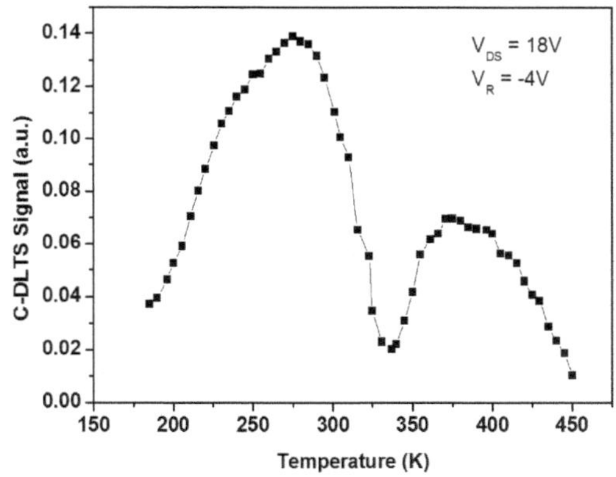

Figure 7. Current DLTS spectra in the saturation regime (V_{DS} = 15V). Only positive peaks, corresponding to electron emission are observed in this case.

4. Conclusion

In summary, deep levels in 4H-SiC MESFET were directly measured by means of the drain-current DLTS technique. Three kinds of electron traps called B1, B2, and B3 with the activation energies of 0.18eV, 0.44eV and 0.57eV respectively. These traps are located in the channel surface.

In part B we have been used Conductance DLTS for 4H-SiC MESFETs characterization and revealed capture phenomenon of electrons present at the channel/passivating layer (SiC/SiO_2) interface or at the channel/buffer or buffer/substrate interface. One kind of hole-like trap signal with activation energy of 0.90 eV is observed in conductance DLTS measurements on 4H-SiC MESFETs when the device is biased in the linear regime with a gate reverse pulse in the neighbourhood of the threshold voltage. This result shows the interest of this technique for the analysis of trapping phenomena due to SiC/SiO_2 interfacial defects. The conductance DLTS using a gate pulse closer to the pinch-off voltage shows one additional hole trap HL2 with activation energies of 0.56eV located at the channel/buffer or buffer/SI substrate interface.

The understanding of trapping phenomena due to surface and interface states and the surface passivation breakout is of main interest for the future industrial development of high power RF transistors on wide band gap materials (SiC MESFETs and AlGaN/GaN HEMTs).

Acknowledgements

We thank Jean-Marie BLUET, INL-Lyon-France for his collaboration

Author details

Malek Gassoumi* and Hassen Maaref

*Address all correspondence to: gassoumimalek@yahoo.fr

Laboratoire de Micro-Optoélectroniques et Nanostructures, Université de Monastir, Monastir 5000, Tunisia

References

[1] Chow, T. P., & Ghezzo, M. (1996). *Mater. Res. Soc. Symp. Proc.*

[2] Allen, S. T., Palmour, J.W., Tsvetkov, V.F., Macko, S.J., Carter Jr, C.H., Weitzel, C.E., Moore, K.E., Nordquist, K.J., & Pond, L.L. (1995). *3rd Annual Dev Res Conf Digest 102.*

[3] Tucker, J.B., Papanicolaou, N., Rao, M.V., & Holland, O.W. (2002). *Diamond Relat Mater.*

[4] Kimoto, T., Itoh, A., Inoue, N., Takemura, O., Yamamoto, T., Nakajima, T., & Matsunami, H. (1998). *Int Conf Silicon Carbide, III Nitrides and Related Materials 97, Mater Sci Forum.*

[5] Tucker, J.B., Mitra, S., Papanicolaou, N., Siripuram, A., Rao, M.V., & Holland, O.W. (2002). *Diamond Relat Mater.*

[6] Sghier, N., Bluet, J.M., Souifi, A., Guillot, G., Morvan, E., & Brylinski, C. (2003). *IEEE Trans. Electron Dev.*, 50-297.

[7] Steven, C., Klein, P.B., & Kazior, E.T. (2002). *Proc. of the IEEE*, 90-1048.

[8] Fang, Z.Q., Claflin, B., Look, B. C., Polenta, L., & Mitchel, W. C. (2005). *J. Electron. Mater.*, 34-336.

[9] Sriram, S., Ward, A., Janke, C., Alcorn, T., et al. (2003). *Mater. Sci. Forum*, 457-460, 1205-1208.

[10] Ho-Young, C., Thomas, C.I., Koley, G., Lester, F., Eastman, M., & Spencer, G. (2003). *IEEE Trans. Electron Dev.*, 50-1569.

[11] Andersson, K., Sudow, M., Nilsson, P.A., & Sveinbjornsson, E. (2006). *IEEE Trans. Electron Dev. Lett.*, 27-573.

[12] Sankha, S.M., Syed, S.I., & Robert, J.B. (2004). *Solid-State Electron*, 48-1709.

[13] Gassoumi, M., Bluet, J. M., Dermoul, I., Maaref, H., Guillot, G., Morvan, E., Dua, C., & Brylinski, C. (2006). *Solid-State Electronics*, 50.

[14] Gassoumi, M., et al. (2005). *Materials Science Forum Vols.*, 483-485, 865-868.

[15] Gassoumi, M., Dermoul, I., Chekir, F., Sghaier, N., Maaref, H., Bluet, J. M., Guillot, G., Morvan, E., Noblanc, O., Dua, C., & Brylinski, C. (2004). *Proc. 24th MIEL Conference (Nis, Serbia and Montenegro, 16-19 May*, 2, 417.

[16] Balakrishnan, V.R., Kumar, V., & Gosh, S. (1998). *Semicond. Sci. Technol.*, 13, 1094.

[17] Sghaier, N., Bluet, J.M., Souifi, A., Guillot, G., Morvan, E., & Brylinski, C. (2003). Study of trapping phenomenon in 4H-SiC MESFETs : Dependence on substrate purity. *IEEE Trans.Electron Devices*, 50(2), 297-302.

[18] Cha, H., Thomas, C., Koley, G., Eastman, L.F., & Spencer, M.G. (2003). Reduced trapping effect and improved electrical performance in buried-gate 4H-SiC MESFETs. *IEEE Trans.Electron Devices*, 50(7), 1569-74.

Chapter 13

High-Power Hexagonal SiC Device: A Large-Signal High-Frequency Analysis

Moumita Mukherjee

Additional information is available at the end of the chapter

1. Introduction

Among the microwave and MM-wave solid-state devices, IMPact Avalanche Transit Time (IMPATT) device has emerged as the most efficient and powerful source of MM-wave power. Most of the current research activities for MM-wave systems are focused on the design and development of IMPATT devices at MM-wave window frequencies, i.e., 35, 94, 140, 220 GHz, where atmospheric attenuation is relatively low. These devices are finding important applications in tracking radars, missile guidance, battle field communication, collision avoidance system and radiometers. IMPATT diodes based on Si, GaAs, InP have been experimentally realized to provide sufficient power at MM-wave frequencies. For realizing higher RF power (P_{RF}) from an IMPATT device, one should choose a semiconductor material that has higher value of critical electric field (E_c), saturated drift velocity (v_s) & thermal conductivity (K), since P_{RF} from an IMPATT device is proportional to $E_c{}^2.v_s{}^2$. The excellent material properties of WBG semiconductors suggest that WBG semiconductor based IMPATT devices are the future MM-wave sources. With the advent of new technologies for growth of SiC crystals, researchers are showing renewed interest in exploring the possibilities of extracting more power from SiC-based IMPATT devices. The experimental research on the development of SiC-IMPATTS is underway. On the other hand, IMPATT device technology based on Si is well established over a wide frequency range. The authors have therefore chosen Si/SiCbased hetero-structure IMPATT diodes, to simulate the large signal properties of the device at W-band (75-110GHz). The authors have developed a generalized technique based on self-consistentmodel for large-signal simulation of SiC DDR IMPATT devices.

Several methods for the large-signal analysis of IMPATTs and other negative resistance devices are reported in the literature [1-5]. Earlier reported large-signal modelling are basically analytical modelling of *Read-diodes* with some simplifications and restrictive assumptions, such as, equal carrier velocities, ionization rates of electrons and holes, punched through depletion layer boundaries, non-inclusion of mobile space charge effects. Consequently in their analysis, the generated power increases monotonically with the increasing RF amplitudes i.e. those analyses do not exhibit saturation effects and this constitutes an important limitation in applying those type of models. The authors have assumed a sinusoidal current at the input of the Si/SiC double drift device and obtained the corresponding voltage response to calculate the device impedance. In the present paper the author has formulated a simple and generalized method for large-signal simulation of Si/SiC IMPATTs at 94 GHz based on current excitation at the input of the device. The large-signal impedance as a function of frequency has been obtained by considering the fundamental frequency and higher harmonic terms.

2. Modelling and Simulation Technique

Numerous computer based models, those are appeared in the earlier literatures, may be divided into two main categories:

1. Analytical modeling of Read diodes with some simplifications and restrictive assumptions, such as, equal carrier velocities, ionization rates of electrons and holes, punched through depletion layer boundaries, non inclusion of mobile space charge effects. Consequently in their analysis, the generated power increases monotonically with the increasing RF amplitudes i.e. those analyses do not exhibit saturation effects and this constitutes an important limitation in applying this type of model.
2. The second type of large signal model is defined as 'full-scale-simulation' with again some assumption.

Though the full scale simulation is applicable to wide variety of structures and they can allow for realistic physical processes at large RF amplitudes, they are difficult to program and time consuming to run on a routine basis. In addition to this, numerical instability appears to be a common problem with these simulations, particularly at larger RF amplitudes.

The purpose of this article is to describe an alternative to the generally unrealistic and/or time consuming and complicated approach of earlier analytical and numerical methods. In essence, the present full-scale program employs a generalized, non-linear analysis of a p^+ p n n^+ type SiC based Double Drift IMPATT device, without any drastic assumption. In this self-consistent single frequency analysis of IMPATT diodes, the modified 'field-maximum method'is used to obtain the detailed 'snap-shot' of the electric field, hole and electron current density as function of active region width during one complete cycle of steady-state oscillation. This program takes into account the non-linear model that contains the differential equations for the carrier concentrations, current density equations, un-equal values of field dependent carrier ionization rates, mobile space charge effects, the behavior of charge carriers and their interactions with electric filed as well as most of the physical effects, such as elevated temperature effects, parasitic effects etc., pertinent to IMPATT operation. With this program it is possible to obtain large signal admittance and impedance of the diode, RF power output as well as other important properties of the device at larger amplitudes of RF signal. A drift-diffusion model has been used for the large-signal analysis. A time varying electric field is assumed in the form, $\xi = \xi_p + m_x \xi_p \sin(\omega t) + m_x^2 \xi_p \sin(2\omega t) + \ldots\ldots\ldots$, where ξ_p is the DC peak electric field and m_x is the modulation factor. The value of m_x can be varied to study the effect of field modulation on the large signal properties of IMPATT device. The physical phenomena take place in the semiconductor bulk along the symmetry axis of the mesa structure of IMPATT diodes.

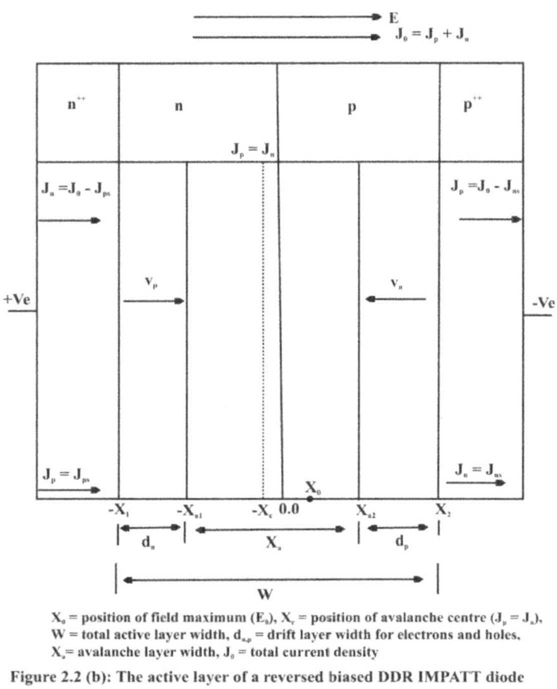

Figure 1. One-dimensional model of DDR IMPATT device.

The fundamental device equations i.e., Poison's equation (equation (1)), continuity equations (equations (2)& (3)) and current density equations (equations (4)& (5)) involving mobile space charge in the depletion layer are simultaneously solved under large-signal condition by using appropriate boundary conditions by using a double iterative field maximum computer method described below. The fundamental device equations are given below.

$$\frac{d\xi(x,t)}{dx} = \frac{q}{\varepsilon}\left(N_D - N_A + p(x,t) - n(x,t)\right) \tag{1}$$

$$\frac{\partial p(x,t)}{\partial t} = -\frac{\partial J_p(x,t)}{\partial x} + q\left(n(x,t)\alpha_n(x,t)v_n(x,t) + p(x,t)\alpha_p(x,t)v_p(x,t)\right) \tag{2}$$

$$\frac{\partial n(x,t)}{\partial t} = \frac{\partial J_n(x,t)}{\partial x} + q\left(n(x,t)\alpha_n(x,t)v_n(x,t) + p(x,t)\alpha_p(x,t)v_p(x,t)\right) \tag{3}$$

$$J_p(x,t) = qp(x,t)v_p(x,t) - qD_P\left(\frac{\partial p(x,t)}{\partial x}\right) \tag{4}$$

$$J_n(x,t) = qn(x,t)v_n(x,t) + qD_n\left(\frac{\partial n(x,t)}{\partial x}\right) \tag{5}$$

Where the symbols α_n, α_p, v_n, v_p, μ_n, μ_p, etc. have their usual significance.

In the above mentioned simulation method, the computation starts from the field maximum near the metallurgical junction. The boundary conditions for the electric field at the depletion layer edges are given by,

$$\begin{aligned} \xi(-x_1,t) &= 0 \\ \xi(+x_2,t) &= 0 \end{aligned} \tag{6}$$

Similarly the boundary conditions for normalized current density $P(x,t) = (J_p(x,t) - J_n(x,t)) / J_0(t)$ (where, $J_0(t) = J_p(x,t) + J_n(x,t)$) at the depletion layer edges ie, at $x = -x_1$ and $x = x_2$ are given by,

$$\begin{aligned} P(-x_1,t) &= \left(\frac{2}{M_p(-x_1,t)} - 1\right) \\ P(x_2,t) &= \left(1 - \frac{2}{M_n(x_2,t)}\right) \end{aligned} \tag{7}$$

$M_n(x_2,t)$ and $M_p(-x_1,t)$ are the electron and hole multiplication factors at the depletion layer edges are given by.

$$M_p(-x_1,t) = \frac{J_0(t)}{J_p(-x_1,t)}$$
$$M_n(x_2,t) = \frac{J_0(t)}{J_n(x_2,t)} \tag{8}$$

The initial values of electron and hole densities are either furnished by a previous run or are given by the quiescent zero bias solution, i.e. solution when

$$J_n(x,t) = J_p(x,t) = 0$$
$$\frac{\partial n(x,t)}{\partial t} = \frac{\partial p(x,t)}{\partial t} = 0 \tag{9}$$

The breakdown voltage at any time instant is calculated by integrating the spatial field profile over the total depletion layer width, i.e.

$$V_B(t) = \int_{-x_1}^{x_2} \xi(x,t)\,dx \tag{10}$$

The magnitude of peak field at the junction (ξ_p), the widths of avalanche and drift zones (x_A and x_D; where $x_D = d_n + d_p$), breakdown voltage (V_B) and the voltage drops across these zones (V_A, $V_D = V_B - V_A$) are obtained from double iterative DC simulation program at $t = 0$. The snap-shots of electric field and current density profiles in the depletion layer of IMPATT diodes can be obtained from the simultaneous numerical solution of the basic device equations (equations (1) – (5)) subject to appropriate boundary conditions (equations (6)& (7)). A software package has been developed for simultaneous numerical solution of the above said equations. Boundary conditions are imposed at the contacts (at the n^+-n& p^+-p contacts) by setting up the appropriate restrictions in equations (1) – (5).

The evaluation of current and voltage in time & space domain of the device and the equivalent circuit has been obtained through the developed simulation program. The total terminal current (i.e. the external current) is given by,

$$I(t) = -C_d \frac{dV(t)}{dt} + I_e(t) \tag{11}$$

Where, C_d is the depletion region capacitance of the diode, $V(t)$ is the AC terminal voltage and $I_e(t)$ is the terminal current induced by the transport of carriers through the diode (Fig. 2). The optimized design parameters are obtained after several computer runs. Once the 1
snap-shots of electric field and current profiles at different time & space over a complete cy-

cle are obtained, the optimized values have been used for simulation of large-signal impedance and admittance characteristics of the device. The diode admittance (including the depletion region capacitance (C_d) at the oscillation frequency ω is given by the fundamental frequency components of current and voltage by,

$$Y_d = j\omega C_d - \frac{I_{e,1}(\omega)}{V_1(\omega)} \tag{12}$$

Finally the expression governing the terminal voltage $V(t)$ is given as [7],

$$V(t) = \frac{A_0 l_d}{\varepsilon\omega A_r}\cos\left\{\omega\left(t+\frac{\tau_d}{2}\right)\right\} - \frac{2A_0 l_d \sin\left(\frac{\omega\tau_d}{2}\right)}{\varepsilon\omega^2\tau_d A_r}\cos(\omega t) - \frac{\tau_a E_c l_d}{2m}\,\frac{A_0\omega\cos\left\{\omega\left(t+\frac{\tau_d}{2}\right)\right\}}{I_{dc}+A_0\sin\left\{\omega\left(t+\frac{\tau_d}{2}\right)\right\}} \tag{13}$$

Where,

$$A_0 = I_{RF}\frac{\left(\frac{\omega\tau_d}{2}\right)}{\sin\left(\frac{\omega\tau_d}{2}\right)} \tag{14}$$

Simultaneously,

$$I_e(t) = I_{dc} + I_{RF}\sin(\omega t) \tag{15}$$

$$I_c(t) = I_{dc} + A_0\sin\left\{\omega\left(t+\frac{\tau_d}{2}\right)\right\} \tag{16}$$

The fundamental frequency component of $V(t)$ is found by a Fourier analysis of (13) and is given as,

$$V_1(t) = \frac{A_0 l_d}{\varepsilon\omega A_r}\cos\left\{\omega(t+\frac{\tau_d}{2})\right\} - \frac{2A_0 l_d\sin(\frac{\omega\tau_d}{2})}{\varepsilon\omega^2\tau_d A_r}\cos(\omega t) - \frac{\tau_1 E_c l_d\omega}{mA_0}(I_{dc}-\sqrt{I_{dc}^2-A_0^2})\cos\left\{\omega(t+\frac{\tau_d}{2})\right\} \tag{17}$$

After $I_e(t)$ and $V_1(t)$ in equations (15) and (17) are represented as phasors and substituted into equation (12), the diode admittance is found to be,

$$Y_d = j\omega C_d + \frac{j\omega C_d A_0 I_{RF} \exp\left(\frac{-j\omega\tau_d}{2}\right)}{A_0^2 - A_0^2 \frac{\sin\left(\frac{\omega\tau_d}{2}\right)}{\left(\frac{\omega\tau_d}{2}\right)} \exp\left(\frac{-j\omega\tau_d}{2}\right) - 2I_{dc}\frac{\omega^2}{\omega_a^2}\left(I_{dc} - \sqrt{I_{dc}^2 - A_0^2}\right)} \qquad (18)$$

A generalized large-signal program has been developed to solve the equations (12), (13), (17) and (18). The symbols have their usual significance. The simulation is carried out by considering a space division of *500* steps and time scale is varied from *100-150* to increase the accuracy of the simulation. Material parameters of SiC is taken from NSM archive [6].

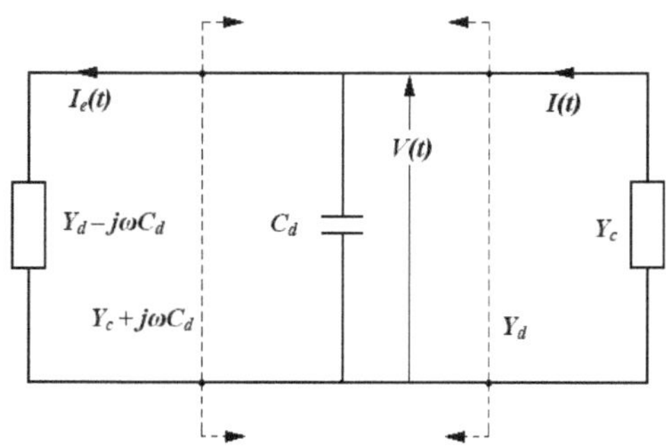

Figure 2. Terminal voltage and current for an IMPATT diode showing the separation of the depletion region capacitance C_d from the diode admittance Y_d. Y_c is the circuit admittance including the diode package.

3. Results and Discussions

The optimized structural parameters of Si/SiC DD hetero-structure IMPATTs for operation at W-band are given in Table 1 for the bias current density of $8x10^8$ A/m^2. Figures 3(a-c) denotes the voltage and current waveform for SiC IMPATT device operating under large-signal condition. The plots depict the 180^0 phase-shift between voltage and terminal current, essential criteria for IMPATT oscillation. This proves the validity of the simulation software. Large-signal simulation provides the snap-shots of electric-field profiles at different phase angles as shown in figures 4(a) to 4(e). The electric field increases from t=0 and attains its peak value in the positive half-cycle at t =T/4 (E_{max} = $5x10^8$ V/m) as shown in figure 4(b). It then decreases and attains the same magnitude of negative peak in the negative half-cycle at t = 3T/4., as shown in figure 4(d). The program is also run for the second and consecutive cycles and it is observed that the above nature of variation of electric field is repeated in each and every cycle.

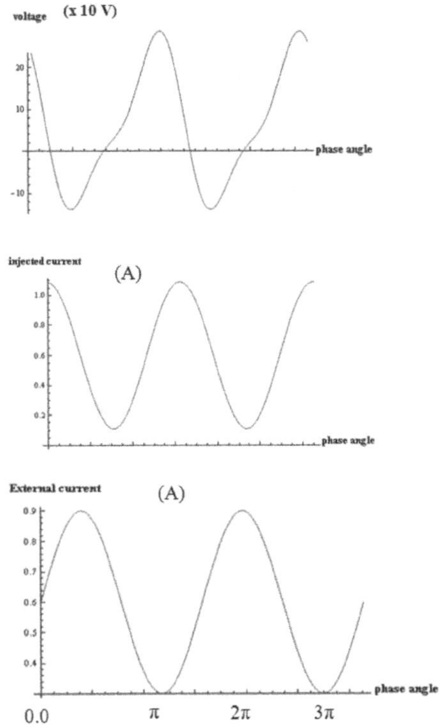

Figure 3. a-c) : Voltage and current waveforms of Si/SiC IMPATTs under large-signal analysis

The effect of voltage modulation on the large-signal negative resistance, reactance, RF power, efficiency, negative conductance and Q-factor of the device has been studied and the results are presented in this paper. Large-signal admittance plots (conductance versus susceptance) for different modulation factors are shown in figure 5. It is observed that the magnitude of peak negative conductance decreases from $37.5\text{x}10^6$ S/m^2 at 94 GHz to 5.0 $\text{x}10^6$ S/m^2 when voltage modulation increases from 10% to 50%, i.e., corresponding RF voltage increases from 20.0V to 104.0 V. In the limiting case of RF-voltage being very low, the large-signal peak negative conductance value should approach the small-signal value. When RF voltage modulation is very low, i.e., 2% and the RF voltage amplitude is 4.0V, the simulated large-signal negative conductance is $43.5\text{x}10^6$ S/m^2. The authors have also carried out small-signal simulation of the Si/SiCdevice based on Gummel-Blue approach and obtained the admittance plot with the same design structural parameters to verify whether the peak negative conductance under small-signal condition approaches that under large-signal condition with negligibly small voltage modulation of 2%. It is observed from Figure 5 that the large-signal admittance plot for lowest voltage modulation almost coincides with the simulated small-signal admittance plot which verifies the validity of the proposed large-signal modeling of the device.

Structure	Diode conductance (-G x10^6) (Sm^{-2})	Diode Susceptance (B x 10^6) (Sm^{-2})	-ZR (for 2% modulation (Ω)	R_S (Ω)	Expected load conductance (G_L x 10^6) (Sm^{-2})	Rs, total (including p^+& n^+ohmic contact)
Si/SiCDD hetero-structure	35.0	80.0	6.0	0.5	32.0	2.3

Table 1. Optimised design parameters of Si/SiC DD hetero-structure IMPATTs for large-signal analysis.

Fig. 6 shows the variation of RF output power with RF voltage. It is interesting to observe that under large-signal condition RF power initially increases with the increasing voltage modulation, reaches a peak value at 50% voltage modulation and then decreases with further increase of voltage modulation. Figure 6 shows the variation of efficiency with RF voltage. It is observed that the efficiency increases with increase in RF voltage, attains a peak value corresponding to 50% voltage modulation and then starts decreasing. It is observed that, with the increasing amplitude of RF voltage from 21.0V to 104.0 V, the magnitude of negative resistance of the device decreases from 10.0 Ω to 6.0 Ω. The analysis shows the variation of negative reactance of the device for different voltage modulation. It is observed that the magnitude of negative reactance increases from 13.0 ohm to 14.5 ohm when RF voltage increases from 10% to 50%. The variation of negative reactance with RF voltage is sharper than that of the negative resistance with RF voltage. At 94 GHz window, the increase in Q-factor from 1.0 to 6.0 with change in RF voltage from 21.0V to 104V, as expected. Large-signal Q-factor for a particular RF voltage indicates the overall RF performance of the device.

Diode	n and p epilayers doping conc. ($N_{D,A}$) (10^{23} m^{-3})	Width of n and p epilayers ($W_{n,p}$) (μ)	Depletion layer capacitance (pF)	DC breakdown voltage (V)	Transit Time (pS)	Junction area ($10^{-9}m^2$)	Si (substrate) layer width (μ)
Si/SiC DD hetero-structure	2.2	1.105	0.11	207.6	5.525	1.0	4.0

Table 2. Values of series resistance ($R_{S, total}$) of SiCDD hetero-structure IMPATT [bias current density = 8.0 x10^8 Am^{-2} and frequency =94GHz].

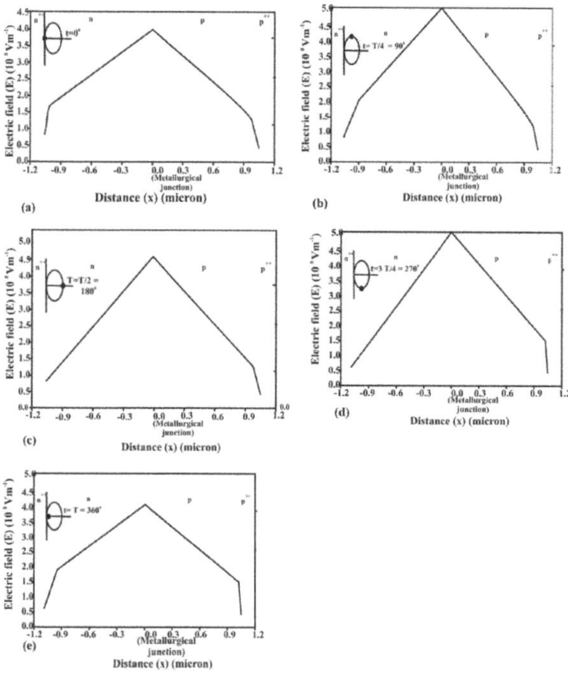

Figure 4. a-b): Plots of electric field profiles (snap-shots) at various points in time (phase angle 0^0 and 90^0) for the diode operating at a fundamental frequency 94 GHz, current density $8x10^8$ Am^{-2} and efficiency = 15%. (c-d): Plots of electric field profiles (snap-shots) at various points in time (phase angle 180^0 and 270^0) for the diode operating at a fundamental frequency 94 GHz, current density $8x10^8$ Am^{-2} and efficiency = 15%. Plots of electric field profile (snap-shot) at various points in time (phase angle 360^0) for the diode operating at a fundamental frequency 94 GHz, current density $8x10^8$ Am^{-2} and efficiency = 15%.

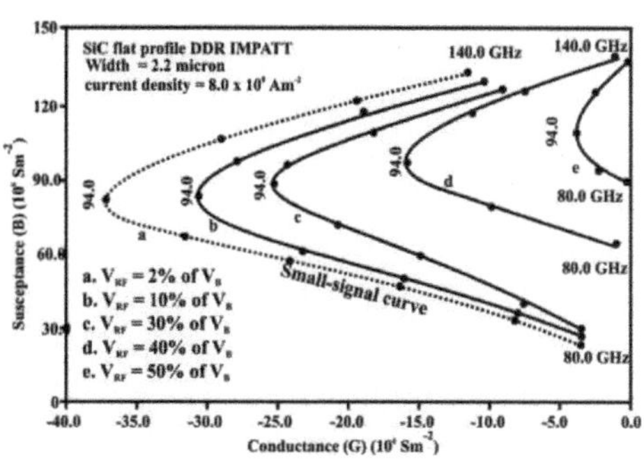

Figure 5. Diode admittance plots as a function of fundamental frequency and ac-voltage amplitude.

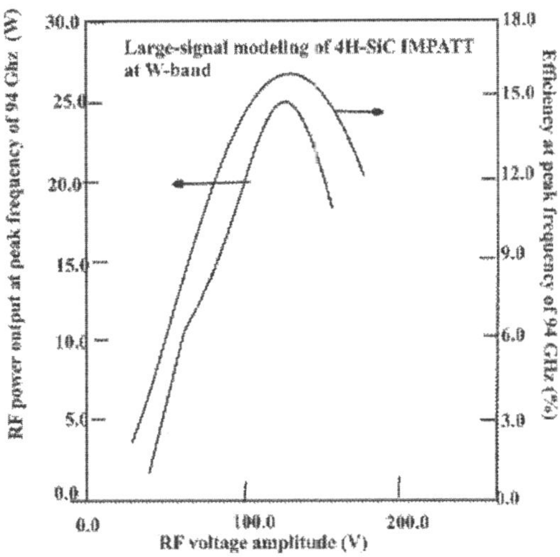

Figure 6. Plots of power generating efficiency and RF power with ac voltage amplitude of hetero-structure DDSiC IMPATTs for fundamental operating frequency 94 GHz (current density 8x10^8 Am^{-2})

4. Conclusion

The author has developed a generalized technique for large- signal simulation of DDR SiC IMPATT diode. This simulator is applicable for other WBG semiconductor based IMPATTs and also for different structures and doping profiles of the device. The validity of the proposed technique is verified from the simulated small-signal admittance plot. The results show for DC breakdown voltage of 207.0 V the large-signal (for ~ 50% voltage modulation) power output and efficiency are 25.0 W and 15.0%, respectively. To the best of author's knowledge, this is the first report on non-linear analysis of 4H-SiC IMPATTs at W-band.

Acknowledgements

Moumita Mukherjee wishes to acknowledge Defence Research and Development Organization (DRDO), Ministry of Defence, Govt. of India, and University Grants Commission (UGC), Govt. of India, for providing her financial assistance to carry out this study. The author is thankful to Director CMSDS, Dr. P. Datta, for his keen interest in this work. The author also gratefully acknowledges Prof. S. K. Roy, founder Director - CMSDS and Professor, IRPE, University of Calcutta, Prof. N. Mazumder, West Bengal University of Technology, India, Prof. J. P. Banerjee, former Director, CMSDS and Prof. D.N.Bose, Emeritus Professor,

Calcutta University, for their valuable suggestions and important comments during the development of the simulator.

Author details

Moumita Mukherjee*

Address all correspondence to: mou_mita_m@yahoo.com

CMSDS, a Centre of Drdo, Govt. of India and University of Calcutta, India

References

[1] Evans, W. J., & Haddad, G. I. (1968). A Large-Signal Analysis of IMPATT Diodes. *IEEE Trans. on Electron Devices*, ED-15(10), 708-717.

[2] Gupta, M. S., & Lomax, R. J. (1971). A self-consistent large-signal analysis of a Read-type IMPATT diode oscillator. *IEEE Trans. on Electron Devices*, ED-18, 544-550, August.

[3] Scharfetter, D. L., & Gummel, H. K. (1969). Large-Signal Analysis of a Silicon Read Diode Oscillator. *IEEE Trans. on Electron Devices*, ED-16(1), 64-77.

[4] Schroeder, W. E., & Haddad, G. I. (1970). Effect of harmonic and sub-harmonic signals on avalanche-diode oscillator performance. *Microwave Theory and Techniques*, MTT-18, 327-331.

[5] Sellberg, F. (1971). Large-signal theory for rectangular-voltage operation pf a uniform avalanche zone in IMPATT diodes. *Electron. Lett*, 7, 154-156.

[6] Electronic Archive: New Semiconductor Materials, Characteristics and Properties. http://www.ioffe.ru/SVA/NSM/Semicond.

Chapter 14

Applications of SiC-Based Thin Films in Electronic and MEMS Devices

Mariana Amorim Fraga, Rodrigo Sávio Pessoa,
Marcos Massi and Homero Santiago Maciel

Additional information is available at the end of the chapter

1. Introduction

The great development of thin film growth techniques has stimulated the industrial and academic researches about design, fabrication and test of thin film based devices. The replacement of the conventional bulk materials by thin films allows the fabrication of devices with smaller volume and weight, higher flexibility besides lower cost and good performance. It has been shown that the efficiency of thin film based devices is strongly dependent on their structural, electrical, mechanical and optical properties (Fraga, 2011a; Fraga 2012). At the same time that there is a trend in the miniaturization of electronic and electromechanical devices, there is also a considerable interest in the study of wide bandgap materials to replace the silicon as base material in these devices for harsh applications such as high temperatures and high levels of radiation (Fraga, 2012, Yeung, 2007).

Silicon carbide (SiC) has intrinsic properties that make it a material of great interest for microelectronic and MEMS (Micro-Electro-Mechanical Systems) applications. In the last years, there has been much debate in the literature about how the incorporation of dopant elements (such as nitrogen, oxygen, aluminum, boron, phosphorus, etc.) during the growth of SiC thin films by chemical vapor deposition (CVD) or physical vapor deposition (PVD) processes affects their properties. It has been noticed that the dopant incorporation allows controlling thin film properties such as optical bandgap and electrical conductivity, which are quite attractive because make possible to obtain semiconductor or insulator SiC-based films (Alizadeh, 2002; Medeiros, 2011).

In general, the use of amorphous SiC films has been preferred due to relatively their low growth temperature, which guarantees a larger compatibility with silicon-based technology (Hatalis, 1987).

Nowadays, SiC-based thin films, such as SiCN, SiCO, SiCNO, SiCB, SiCBN and SiCP, have been extensively used in electronic and MEMS devices either as a semiconductor or as an insulator, depending on the film composition. These films have been shown promising for applications in diodes, thin-film transistors (TFTs) and MEMS devices (Yih, 1994; Patil, 2003; Hwang, 1995; Fraga, 2011c).

The goal of this chapter is to discuss the role of in situ incorporation of nitrogen, oxygen, aluminum, boron, phosphorus and argon on the properties of SiC films. Special attention is given to the low temperature SiC growth processes. An overview on the applications of SiC-based thin films in electronic and MEMS devices is presented and discussed. Our recent researches on heterojunction diodes and MEMS sensors are emphasized.

2. Dopant incorporation during growth of SiC thin films

2.1. In situ doping

Most studies on SiC thin films, especially in their amorphous form, is not focused on intrinsic films. In general, the electrical properties of wide band gap semiconductor materials as the SiC are controlled by introducing dopants into the bulk material (Oliveira, 2002). Hence, determining the best material doping concentration is one important issue to be considered during a device development. SiC-based thin films with variable electrical conductivity, from semiconductor to insulator, have been produced by adjusting only the dopant concentration. This allows the use of these films in a variety of devices such as solar cells, different diode types, TFTs, MEMS sensors, among others (Vetter, 2006; Oliveira, 2004).

The main chemical elements used as SiC dopants are in group III (aluminum, Al, and boron, B) and group V (nitrogen, N, and phosphorus, P) of the periodic table. Therefore, n-type doping of SiC is commonly achieved by the use of nitrogen or phosphorus whereas p-type is by aluminium or boron. The in situ doping process is done by adding doping gas, such as nitrogen (N_2) or ammonia (NH_3) as nitrogen precursors, diborane (B_2H_6) as boron precursor, phosphine (PH_3) as phosphorus precursor and trimethylaluminum (TMA) as aluminium precursor, during SiC epitaxial growth (Miyajima, 2006). This process is a non-selective doping of SiC epitaxial layers grown on different substrate types (e.g. silicon, silicon-on-insulator (SOI) and quartz) and it has allowed the preparation of SiC films with high electrical conductivity and low defect density. In situ doping exhibits advantages when compared to post-deposition doping methods as ion implantation and thermal diffusion, such as the easy incorporation of dopants in CVD or PVD processes and the reduction of processing steps. It is noteworthy that the incorporation of dopants besides affecting the conductivity of the material also influences other properties namely Young's modulus, hardness, optical gap, transmittance, morphology, etc. (Murooka, 1996; Sundaram, 2004).

In situ doping of amorphous and crystalline SiC films is well-documented in the literature. Historically, the first papers on the electrical properties of a-SiC films were published during the 70's. In 1977, Anderson and Spear investigated the electrical properties of plasma enhanced chemical vapor deposition (PECVD) hydrogenated amorphous SiC (a-SiC:H) films. In the 80's, several papers on doping of a-SiC and a-SiC:H films and their potential applications were published (Tawada, 1982; Beyer 1985; Pereira 1985). Since then, numerous studies have demonstrated the great potential of the SiC-based thin films for electronic device applications.

2.1.1. Nitrogen incorporation

Among the different SiC-based films, the silicon carbonitride (SiCN) has been the most investigated due to its easy synthesis. Generally, SiCN films are produced by introducing nitrogen gas (N_2) during SiC film growth by CVD and PVD processes. The use of N_2 as doping gas is advantageous due to non-toxicity, low cost and high efficiency (Fraga, 2008). The control of N_2 gas flow during deposition process has been shown as a convenient and effective way to change the electrical properties of SiC films in order to obtain films with desired electrical conductivities for each application type. According to Alizadeh and Sundaram, for N_2/Ar ratios from 0.2 to 0.4, the N_2 gas acts like a dopant in a-SiC films prepared onto glass substrate by radiofrequency (RF) magnetron sputtering of a SiC target in N_2/Ar atmosphere, reducing their electrical resistivities from the range of 10^9 Ω.cm to 10^4 Ω.cm. However, for N_2/Ar ratios between 0.6 and 0.8 the film resistivity reach values in the range of 10^{10} Ω.cm, indicating the formation of insulator SiCN films. Other important correlations between N_2/Ar ratio and the properties of a-SiCN films were shown in their previous work: the bandgap and the percentage of optical transmission of these films increase with the N_2/Ar ratio increases. The electrical conductivity of sputtered a-SiCN films was also studied by Wu et al. where, in this work, the a-SiCN films were deposited onto quartz, glass and Si substrates at room temperature by RF reactive sputtering of a SiC target in $Ar/N_2/H_2/CH_4$ atmosphere. They observed that the dark conductivity decreases with increases in N_2 flow rate. Besides sputtering, other processes are being used to grown a-SiCN films. Gomez et al. used the electron cyclotron resonance (ECR) PECVD to prepare a-SiCN films using nitrogen, methane, and argon diluted silane as precursor gases. Yamamoto et al. investigated the correlation between nitrogen ion energy and the formed chemical bonds in a-SiCN films deposited on (100) Si substrates by nitrogen ion-assisted pulsed-laser ablation of a SiC target.

In situ nitrogen doping of crystalline SiC films have also been commonly reported. Wijesundara et al. investigated the nitrogen doping of polycrystalline 3C-SiC films grown on (100) Si substrates by LPCVD at various growth temperatures 650–850ºC using 1,3-disilabutane and NH_3 as precursors. They concluded that the electrical resistivity of the polycrystalline films is further controlled by adjusting the NH_3 flow rate in the reactor and that nitrogen-doped 3C-SiC films exhibit lower resistivities (around 0.02 Ω.cm) than the undoped (around 10 Ω.cm) obtained at 800ºC. The effects of N_2 flow rate and growth temperature on the electrical properties of nitrogen-doped 3C-SiC thin films, grown on Si_3N_4/p-Si (111) substrates by LPCVD at temperature 1100-1250ºC using organosilane-precursor trimethylsilane ($(CH_3)_3SiH$), were discussed by Cheng et al. It was observed that independent of the temperature process the film resistivity decreases continuously with N_2 flow increases. Liu et al. reported the synthesis of nitrogen-doped polycrystalline 3C-SiC thin films by LPCVD at 800ºC with optimized properties for MEMS applications: resistivity of 0.026 Ω.cm, residual stress of 254 MPa and strain of 4.5×10^{-4}. The deposition conditions to produce films with these characteristics were established through studies on the effects of the precursor gases flow rate, NH_3 and diclorosilane, on the material properties such resistivity, residual stress, strain, crystallinity and surface morphology. It is necessary to underline that for crystalline SiC films prepared by CVD processes, the site-competition epitaxy model has been shown as an efficient method to control in situ doping. This model is based on the variation of the Si/C ratio within the CVD reactor in order to control the dopant incorporation. In the case of the nitrogen doping in SiC, its incorporation is directly related to Si/C ratio (Larkin, 1997).

2.1.2. Phosphorus incorporation

In situ phosphorus doping, although little used, is another way used to obtain n-type SiC. Ruddell et al. employed a thermal CVD reactor for the deposition of phosphorus-doped SiC layers on Si substrates using silane/propane/phosphine gas chemistry over the temperature range 720-970ºC. SiC films with a phosphorus concentration of 5×10^{20} cm^{-3} and resistivity of 0.6 Ω.cm were obtained, which indicates a good efficiency of phosphorus doping. The correlations between the microwave power (from 150 to 900 W) and the properties of phosphorus-doped SiC:H films, prepared by ECR-CVD from a mixture of methane/silane/ hydrogen/ phosphine, were investigated by Yoon et al. The properties of n-type nanocrystalline hydrogenated cubic silicon carbide (nc-3C-SiC:H) prepared by hot-wire chemical vapor deposition (HWCVD), using phosphine and hexamethyldisilazane (HMDS) as dopants, at low temperatures around 300ºC were studied by Miyajima et al. In situ phosphorus doping during sputtering process was reported by Pereira et al. that performed the substitutional doping of RF sputtered amorphous SiC by adding controlled amounts of phosphine (PH_3) to the argon atmosphere at a constant substrate temperature of 200 °C. They observed that the conductivity of the SiC film increases about one order of magnitude when doped with phosphorus and in the presence of 0.5 m Torr of hydrogen. In recent work, Loubet et al. reported an epitaxy process based on a cyclical deposition-etch (CDE) technique to obtain ultra-low resistivity in situ phosphorus-doped silicon carbon (SiCP) layers for raised source/drain applications. Despite studies have demonstrating the efficiency of in situ phosphorus doping during SiC film growth, this process is not much used due to high toxicity and flammability of PH_3 gas. When used, in general, PH_3 is highly diluted (< 1%) in hydrogen because the risks in diluted form are less critical.

2.1.3. Aluminum incorporation

Considerable efforts have also been made to prepare p-type SiC films. In situ aluminum doping of SiC films is one of the processes used for this purpose. Wang et al. presented a doping method for growth of Al-doped single-crystalline 3C-SiC epilayers onto (100) Si substrates by atomic-layer epitaxy at 1000°C using silane, acetylene and TMA gases in a conventional LPCVD reactor. The hot-probe and Hall effect measurements confirmed that the

Al-doped SiC are of p-type conductivity. The film resistivity, measured at room temperature by a standard four-point probe system, was of 0.31 Ω.cm. Sha et al. discussed the photoluminescence properties of Al-doped SiC films deposited on Si substrates by RF magnetron sputtering of a single crystalline SiC target containing several pieces of aluminum on its surface. The influence of substrate temperature (300–390°C) on the aluminum doping of μc-SiC:H prepared by hot-wire CVD using TMA as gas dopant was discussed by Chen et al.. They concluded that a process temperature higher than 350°C is needed to obtain effective aluminum doping of μc-SiC:H thin films. An important observation on in situ aluminum doping in SiC is that according to the site-competition epitaxy model, the Al dopant incorporation has been found to be inversely related to the Si/C ratio within the CVD reactor. This behavior is opposite to that observed in nitrogen incorporation in SiC (Larkin, 1997).

2.1.4. Boron incorporation

In situ boron doping is other process used to obtain SiC films with p-type conductivity. Boron-doped sputtered a-SiC:H films, prepared onto corning glass and c-Si substrates maintained at temperatures of 125–250°C by magnetron sputtering of silicon in $Ar+H_2+B(CH_3)_3$ atmospheres, were reported by Uthanna et al. The highest values of dark conductivity and doping efficiency were achieved at a carbon content x= 0.04. It was also found that the film with x = 0.14, deposited at 175°C, has electrical properties as required for solar cell applications. Yoon et al. studied the effects of the diborane (B_2H_6) levels on the deposition rate, optical band gap and conductivity of boron-doped SiC:H films prepared using ECR PECVD technique from a mixture of methane, silane, hydrogen and diborane gases. It was observed that at a low microwave power of 150 W the band gap of the SiC:H film decreases as the diborane flow increases, whereas the films deposited at a high microwave power of 800 W remains relatively unaffected throughout the entire range of diborane levels investigated. Highly conductive boron-doped nanocrystalline SiC films with a low concentration of hydrogen-dilution (p-nc-SiC:H) grown by a mercury-sensitized photo-CVD method using silane, hydrogen, diborane and ethylene as a carbon source, were reported by Myong et al. These films were tested as window material for amorphous silicon solar cells and a good conversion efficiency of 10.4%, without use any back reflectors, was obtained. It has also been noticed the improvement of the Young's modulus of polycrystalline SiC film grown by LPCVD through the introduction of B_2H_6 in the precursor gas mixture (Murooka, 1996). This study concluded that the Young's modulus of SiC films increases with the addition of B_2H_6, and a maximum value of 600 GPa, which was 25% higher than in the case without B_2H_6, was reached at a source gas ratio B/Si=0.02.

Although boron-doped SiC films exhibit suitable properties for different applications, these films are still little used because, like the PH_3, the B_2H_6 dopant gas is toxic. An alternative method for the formation of SiCB films has been shown in the literature: the deposition of SiCB films using sputtering target containing boron in its composition. This method has also allowed the growth of quaternary compound SiCBN by introducing N_2 gas into deposition process. Optical properties of amorphous SiCBN thin films obtained, by co-sputtering from SiC and BN targets using N_2/Ar gas mixtures, were studied by Vijayakumar et al. It was found that the transmittance of the SiCBN films increases with nitrogen incorporation increases. Petrman et al. prepared SiBCN films by reactive magnetron sputtering of a $Si_{75}(B_4C)_{25}$ target in N_2/Ar gas mixtures. They showed the dependence of electrical resistivity and optical gap of SiBCN films on the N_2 content used in the N_2/Ar gas mixture.

2.2. Unintentional doping

The incorporation of dopants during SiC growth can also be unintentional, i.e., when elements presented during the deposition process are incorporated into film due to unwanted chemical reactions. Oxygen and hydrogen have been shown as the most common unintentional dopants of SiC films.

The contamination sources by hydrogen are well clear. It is known that the presence of hydrogen in the plasma and its consequent incorporation into the SiC film, in CVD or PVD deposition processes, is due to use of hydrogenated precursor gases as source of carbon (e.g. CH_4, C_2H_2, C_2H_4, C_2H_6, C_3H_8 and C_4H_{10}) and as source of silicon (e.g. SiH_4 and Si_2H_6). The hydrogen incorporation is significant, when the SiC films are produced at low temperature processes as PECVD and sputtering. This has motivated several studies on how the hydrogen incorporation affects the properties of SiC films and on the applications of a-SiC:H films in devices. Studies on PECVD and sputtered a-SiC:H films have shown that the hydrogen incorporation induces the formation of voids (Beyer, 1985) and increases the compressive stress (Kim, 1995). On the other hand, increasing hydrogen content in a-SiC:H films increases the optical gap (Shimada, 1979), improves the photoconductivity and increases the electrical resistivity. The optical and electrical changes due to hydrogen incorporation have stimulated the research on growth of a-SiC:H films by using the hydrogen dilution method, i.e, adding hydrogen gas (H_2) to the deposition process. This method is attractive because the optical gap of the a-SiC:H films can be varied by changing the H_2 flow rate while the other deposition parameters are kept constant. In recent years, a-SiC:H films have been successfully employed as electronic surface passivation of c-Si in photovoltaic applications exhibiting performance comparable to thermal SiO_2 and a-SiN_x that are the most used materials. As the photoconductivity of a-SiC:H films is high, another attractive application of these films is replacing Si:H top layer of pin solar cells (Vetter, 2006).

Regarding the unintentional contamination by oxygen, the sources have been more discussed in the literature. Many studies have showed that the oxygen contamination sources during the deposition of SiC films are mainly the residual gas in reactant chambers, possible air leak in the deposition system and adsorbed gas molecules on the reactor inner walls (Medeiros, 2011). Moreover, post-deposition film surface contamination by oxygen, that can occurs when the SiC film is exposed to atmospheric air, has also been described by different authors (Fraga, 2008). The oxygen incorporation in SiC films and the consequent formation of silicon oxycarbide (SiCO) can be interesting for different applications such as thin film anodes for lithium ion batteries (Shen, 2011) and as doping contacts for solar cell applications (Martins, 1996). The excellent physical and chemical properties identified in SiCO films have motivated studies on the growth and characterization of these films intentionally deposited through oxygen gas addition during their growth process. Amorphous SiCO films have been produced mainly by sputtering of a SiC target in O_2/Ar atmosphere or by PECVD using

$C_xH_y/Si_xH_y/O_2$ gas mixture which result at hydrogenated films (a-SiCO:H). It has been reported that these films exhibit mechanical and electrical properties strongly dependent of the oxygen-to-carbon ratio in their chemical composition. A potential application of SiCO and SiCO:H films is to replace SiO_2 in microelectronic devices because it possesses advantages over other lower dielectric materials since it is formed as an interface phase between SiC and SiO_2. Apart from applications in surface passivation, dielectric insulation and copper diffusion barrier, SiCO films have also been used in photodetectors and for oxygen detection in high temperature and corrosive environments. The literature also shows the synthesis of silicon oxycarbonitride (SiCNO) films by unintentional oxygen incorporation during low temperature reactive magnetron co-sputtering of silicon and graphite targets in mixed Ar/N_2 atmosphere (Medeiros, 2011). As others SiC-based films, SiCNO films exhibit high thermal stability, tunable bandgap characteristics and high gauge factor values (Cross, 2010; Terauds, 2010). A comparison among the physical, electrical, and reliability characteristics of SiC, SiCN, SiCO, SiCNO and SiN thin films was performed by Chen et al. It was observed that SiCNO films are the most appropriate to be used as copper diffusion barrier because exhibit more reliable electro-migration and stress-migration besides to present physical and electrical performance comparable to those of SiN films.

Although less studied, the argon is another atom that can be incorporated during the growth of SiC films. For this, bias assisted CVD or PVD technique with high potentials applied to the substrate holder allows tuning the argon ion (Ar^+) flux on the substrate surface and, consequently, the concentration of incorporated argon into film. Some studies on chemical composition of SiC films using RBS and XPS analysis have indicated argon content up to 8 at.% in film composition. A.K. Costa and co-workers investigated the influence of the substrate negative bias increases (from 0 to -100 V) on the properties of sputtered SiC films. They observed that the substrate bias leads to substantial argon incorporation into the SiC film, which results in the increase of the hardness. A negative characteristic was also observed. The microscopic examination of the film surface showed a large number of defects and pinholes as the argon content increases together with a reduction in the deposition rate due to the sputtering of the film surface by argon ions. In a recent work, Medeiros et al. showed that increasing the argon incorporated in Si_xC_y films, deposited onto (100) Si by dc magnetron co-sputtering technique under different negative substrate bias, promotes an increase of the elastic modulus and a reduction of the electrical resistivity. In addition, it is also noticed that co-sputtered Si_xC_y films grown at high negative substrate bias (between -100 and -300 V) are free of oxygen contamination.

3. Challenges and trends of the growth processes of SiC thin films

The main motivation to use thin films is simple: easy growth on a wide variety of substrates. In the case of the SiC, the use of thin films includes other motivations: (a) the cost of SiC bulk substrates is still high; (b) the defect density is relatively high and (c) the area of the substrate available is still small (Fraga, 2011d).

Nowadays, the SiC devices are categorized into two groups: one uses SiC bulks and the other uses SiC thin films grown on Si or insulator/Si substrates. In general, comparative studies show that devices based on SiC bulk substrates exhibit better performance than those thin film based (Harris, 1995). Then, the first challenge is to grown SiC thin films with properties as good as the bulk substrate. In addition, it is necessary to achieve high film growth rate as well as thickness uniformity and homogeneity when deposited on large-area Si wafers, which are important factors to reduce costs. For this, the influence of SiC growth process parameters, such as gases flow rate, substrate temperature, pressure and doping, have been evaluated and optimized.

The literature has shown that devices based on thin films, grown on the same conditions, often did not exhibit similar performance (Fraga, 2012). Identify and overcome the causes of the non-reproducibility is another challenge. The synthesis of high-quality SiC films with reproducible properties is fundamental for the advancement of the SiC thin film device technology. It is known that the properties of thin films are strongly dependent on the deposition conditions, i.e., the repeatable and precise control of deposition conditions is important to ensure the reproducibility film. The electrical resistivity and thickness measurements of films, deposited under same conditions, are the most used parameters to evaluate their reproducibility.

Extensive research has been done on growth of SiC films at low or high temperature process aiming to produce high quality films. Three methods have been most frequently used: CVD, PECVD and sputtering. Table 1 compares these methods. The most critical issue facing CVD SiC films for device applications is the high temperature necessary to assure the surface reactions and good deposition rate (Ong, 2006). The low temperature deposition is very important from the point of view of device integration. The synthesis of low-temperature deposited highly-conductive SiC films has been a goal of many research groups focused on the development of TFTs, solar cells and heterojunction bipolar transistors (Cheng, 1997). This has encouraged more studies on the optimization of low-temperature methods as PECVD and the different sputtering processes. PECVD SiC films have been deposited at a relatively low temperature (less than 300°C). The films present a good adhesion, high deposition rate and good uniformity. Sputtering process presents poor sidewall coverage due to the significant distance between the target and the substrate. Besides, the deposition is frequently made at room temperature and usually has a low deposition rate (Ong, 2006).

	CVD	**PECVD**	**Sputtering**
Cost	Fair	Fair	Fair
Uniformity	Fair	Fair	Fair
Substrate versatility	Good	Very good	Very good
Stress control	Poor	Very poor	Good
Throughput	Varies	Very good	Fair

Table 1. Comparison among the main methods used to grown SiC films.

To illustrate the comparison among PECVD and sputtering, we will discuss the results obtained in our works on SiC films grown by PECVD and RF magnetron sputtering. Regarding the throughput of the method, we have found high PECVD SiC film growth rate, from 24 to 36 nm/min, depending on SiH_4/CH_4 flow rate used (Fraga, 2007) and from 4.0 to 7.0 nm/min for films deposited by RF magnetron sputtering of a SiC target under different Ar/N_2 mixtures (Fraga, 2008). In order to evaluate the uniformity, the following tests were performed: (a) before the SiC film deposition, three small steps were created by placing strips on different points of the 2 inch p-type Si wafers, (b) after the deposition, each wafer was cut in fourth path equals, (c) the thicknesses and resistivities of the paths were measured by profilometry and four-points probe, respectively. It was found: thicknesses between 432 and 470 nm and resistivities between 12 and 16 Ω.cm for the pieces of PECVD SiC film, whereas for the sputtered films between thicknesses 337 and 385 nm and resistivities between 0.2 and 0.25 MΩ.cm. These results indicate that the throughput of PECVD is considerably greater than of RF magnetron sputtering. However, both methods present problems in terms of film uniformity. It is noteworthy that our tests were performed with 2 inch Si wafers whereas the semiconductor industries have been using up to 12 inch wafers. It is likely that the problems of uniformity are more significant in substrates with larger dimensions.

The residual stress control of SiC thin films is another issue, which can also be added to the challenges. It has been demonstrated that SiC film stress can affect the sensitivity, precision and functionality of thin-film based devices, thus, in some applications, is important to have a low residual stress.

SiC film deposition process	Young's modulus (GPa)	Reference
APCVD	450	Zorman, 1995
LPCVD	396	Fu, 2004
PECVD	88 to 153	El Khakani, 1994
PECVD	56	Flannery, 1998
PECVD	196	Cros, 1997
RF triode sputtering	231	El Khakani, 1994
RF magnetron sputtering	363	Singh, 2012
Co-sputtering	245 to 377	Medeiros, 2012

Table 2. Young's modulus reported in the literature for SiC films grown on Si substrates.

Initially, the research efforts on semiconductor and dielectric thin films were focused on their electrical properties in order to satisfy the demand in microelectronic devices industry. In general, studies on mechanical properties were limited to internal stress measurements (Tsuchiya, 2008). With the advent of microelectromechanical system (MEMS) technology, the thin films started to be used as mechanical structures, which made fundamental the knowledge of their mechanical and electromechanical properties. The high Young's (or elastic) modulus is the key mechanical property to use them as structural layer in MEMS devices. As can be observed in Table 2, crystalline and polycrystalline SiC films grown by atmospheric pressure chemical vapor deposition (APCVD) and LPCVD processes exhibit higher Young's modulus than the amorphous produced at low temperatures by PECVD and sputtering. In addition, it has been observed that the most of the SiC films still exhibit lower Young's modulus than the reported for SiC wafers which is in the range between 330 and 700 GPa depending on the polytype (Zorman and Parro, 2010).

One issue that should be considered is that, although the researches on SiC film growth have been mainly focused on the deposition of SiC on Si substrates, MEMS applications frequently require the growth of SiC thin films on sacrificial and insulating layers, as for example SiO_2 or Si_3N_4, grown on Si substrates. Thus, to evaluate the influence of substrate type on the properties of SiC films is another important point. Some works have investigated the growth and properties of SiC films on SiO_2, Si_3N_4 and poly-Si (Fleischman, 1998; Wu, 1999; Chen, 2000). A known drawback of SiC on insulator layers is the high stress caused by the large lattice and thermal mismatch between SiC and insulator, which become post-deposition annealing necessary to minimize this problem and improve the quality SiC film on insulator. Other particularities of SiC growth on insulator/Si substrates include: the effect of insulator layer thickness on the properties of SiC film and the choice of the suitable insulator for each device application. Chen et al. reported the SiC growth by CVD on the following substrates: thermally oxidized Si substrates with SiO_2 thicknesses of 30, 50, 70, and 100 nm, Si substrates with native oxide of 2 nm and 3 μm thick phosphosilicate glass (PSG). They observed that in the thickness range between 30 and 70 nm, the SiO_2 serves as a compliant layer which reduces the strain between SiC film and the substrate besides allows the growth of a more oriented SiC film. In relation to choice of the insulator, Si_3N_4 has shown more suitable than SiO_2 due to: (i) its higher dielectric constant that can reduce the leakage currents, and (ii) its thermal expansion coefficient is much closer to that of SiC than one of SiO_2, thus the stress in SiC film grown on Si_3N_4 will be lower than that on SiO_2 (Cheng, 2003). Nevertheless, most of the SiC thin-film MEMS devices reported in the literature use the SiO_2 as sacrificial layer and/or substrate electrical isolation (Chang, 2008; Mishra, 2009). One reason for this is the easy formation of SiO_2 achieved by thermal oxidation of Si substrates.

In our researches, we have explored the properties of SiC thin films grown on SiO_2/Si substrates by PECVD and RF magnetron sputtering for the development of strain gauges (Fraga, 2010) and pressure sensors (Fraga, 2011a; Fraga 2011c). Young's modulus was found to be 65 GPa for PECVD a-SiC film and 57 GPa for nitrogen-doped PECVD a-SiC film. These values are near to 56 GPa that was reported for PECVD a-SiC film grown on Si substrates (Flannery, 1998). On the other hand, the Young's modulus values found by us for RF-sputtered SiC and SiCN films on SiO_2/Si were considerably lower than those reported in the literature for films on Si substrates. We found Young's modulus of 40 GPa and 88 GPa for sputtered a-SiC and SiCN films respectively, whereas other authors have found values above 200 GPa for sputtered SiC film grown on Si (see Table 2) and 117 GPa for SiCN film grown on Si (Sundaram, 2004). For CVD processes, the following Young's modulus was found: 426 ± 100 GPa for SiC grown on SiO_2/Si by APCVD (Fleischman, 1998) and 426 GPa for SiC grown on Si_3N_4/Si by LPCVD (Cheng, 2002).

In general, the literature has shown that the substrate type influences the structure, morphology, electrical and mechanical properties of SiC films. However, although there are differences related to the substrate type, the properties, such as high Young's modulus, high chemical resistance and high thermal stability among others, that make the SiC film attractive for a variety of harsh environment MEMS device applications are maintained.

4. SiC thin film based electronic devices: diodes and TFTs

4.1. Diodes

4.1.1. Heterojunction diodes

A SiC/Si heterojunction diode is formed by the growth of a SiC thin film with opposite doping impurities to the Si substrate used. Thus, it forms a p-n heterojunction with principle similar to a homojunction. However, the energy band for a heterojunction diode is much more complicated than that of a p-n homojunction because uses two semiconductors of different band gaps (Li, 2006).

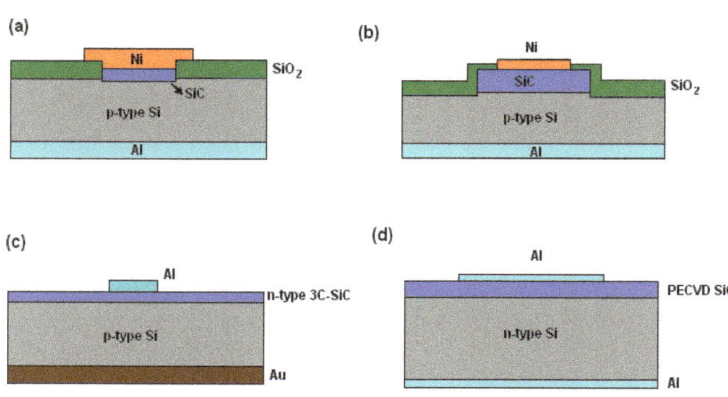

Figure 1. Schematic illustration of cross-section of heterojunction diode structures reported in literature: (a) and (b) Yih et al., (c) Chung and Ahn, (d) Oliveira et al.

The SiC/Si heterojunction diodes with high breakdown voltage and performance dependent on the quality of the SiC film used have been reported in the literature. Yih et al. developed SiC/Si heterojunction diodes using two different rapid thermal chemical vapor deposition (RTCVD) processes: one through the formation of crystalline β-SiC by propane carbonization of the Si substrate in regions unprotected by SiO_2 layer forming planar diodes, as shown in Figure 1 (a), and another by growing polycrystalline β-SiC through the decomposition of methylsilane (CH_3SiH_3) at 1300ºC forming mesa diodes, Figure 1 (b). Both diodes used Ni on the SiC film as ohmic contact and Al on Si as backside contact. These SiC/Si heterojunction diodes, for both device configurations, exhibit good rectifying properties. Reverse breakdown voltages of 50 V and 150 V were obtained for the planar and mesa heterojunction diodes, respectively. These results demonstrated the potential use of SiC/Si heterojunction for the fabrication of bipolar transistor.

A 3C-SiC/Si heterojunction diode fabricated by a more simple process was reported by Chung and Ahn. This diode was fabricated by the deposition of poly 3C-SiC thin films on p-type substrates using Ar, H_2 and HMDS gases in an APCVD system at 1100ºC. The ohmic contacts were prepared by the deposition of Al circle electrodes on poly 3C–SiC surface and an Au layer on the Si substrate side. Figure 1 (c) shows schematic diagram of the 3C-SiC/Si heterojunction diode formed. They concluded that p–n junction diode fabricated by poly 3C–SiC film has similar characteristics to single 3C–SiC p–n junction diode.

Although studies have shown that the deposition of SiC films on Si substrates at low temperatures processes results in heterojunctions with low breakdown voltage and high reverse leakage current, Oliveira et al. developed PECVD SiC/Si heterojunction diodes (Figure 1 (d)) with satisfactory electrical performance exhibiting good rectifying properties. An interesting conclusion of this work is that the post-deposition thermal annealing improves the electrical properties of the PECVD SiC films. It was observed that the heterojunction formed by annealed PECVD SiC film, at 550ºC for 120 min, has a leakage current approximately one order of magnitude smaller than that formed by as-deposited films.

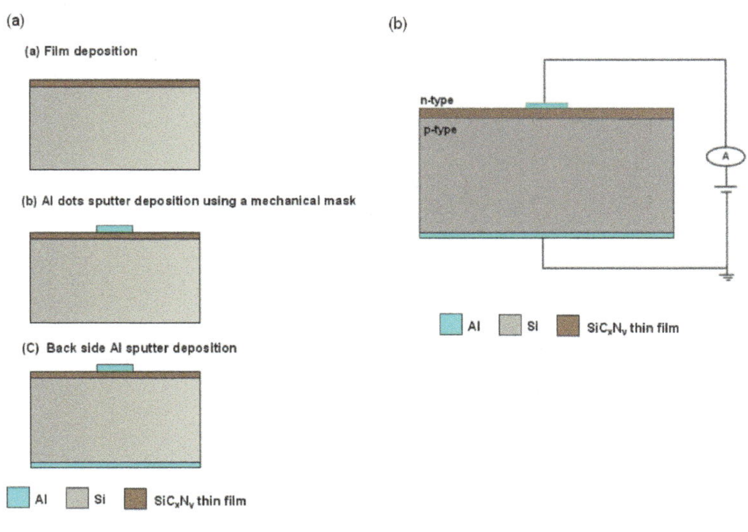

Figure 2. SiC_xN_y/Si heterojunction diode: (a) Sequence of fabrication and (b) schematic diagram of electrical characterization.

We have also studied heterojunctions formed at low temperatures using RF magnetron sputtering of a SiC target under different N_2/Ar gas flow ratio (from 0.1 to 0.3) to prepare a-SiC_xN_y film with different compositions on p-type (100) Si substrates (Fraga, 2011b). After deposition, the films were submitted to a thermal annealing at 1000ºC for 30 min. The n-type conductivity of the sputtered a-SiC_xN_y thin films was verified by hot probe technique. The electrical contacts were fabricated through deposition of Al dots on the a-SiC_xN_y surface and subsequently a layer of Al was sputtered on the back side of Si substrate (see Fig. 2 (a)). The a-SiC_xN_y film thicknesses are between 320 and 350 nm, whereas the Al layers have thicknesses around 225 nm. The motivation of this study was evaluate the influence of film composition on I-V characteristics of a-SiC_xN_y/Si heterojunction diodes. The I-V characteristics were measured using an Agilent B1500A semiconductor.

The analyses were performed in a voltage range from -10 to +10 V at room temperature. As can be observed in Figure 3, the N_2/Ar flow ratio increases from 0.1 to 0.2 reduced the electrical current in three orders of magnitude (from mA to μA). On the other hand, the current was not affected significantly increasing the N_2/Ar flow ratio from 0.2 to 0.3.

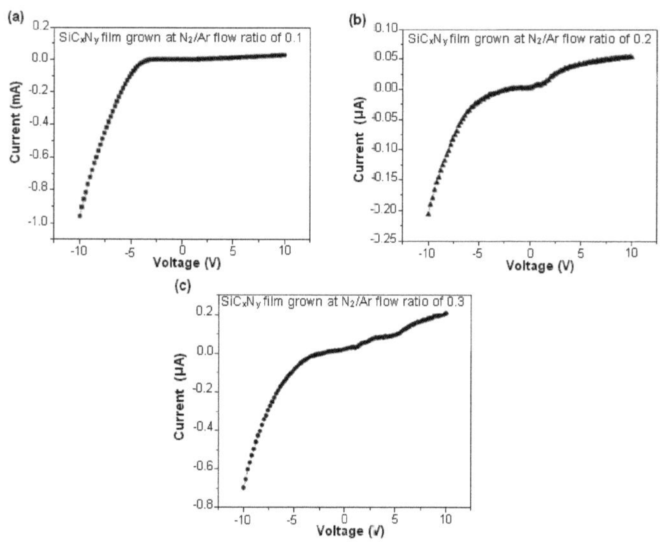

Figure 3. I-V characteristics of a-SiC_xN_y/Si heterojunction diodes at room temperature.

Regarding the temperature effects, the I-V characteristics at different temperatures of the a-SiC_xN_y/Si heterojunction diode, fabricated with the a-SiC_xN_y deposited at N_2/Ar of 0.1, is shown in Figure 4. It was observed that the electrical current increases when the temperature is increased from 35 to 135 ºC. However, at temperature of 160 ºC there is an almost overlap with the curve obtained at 135ºC.

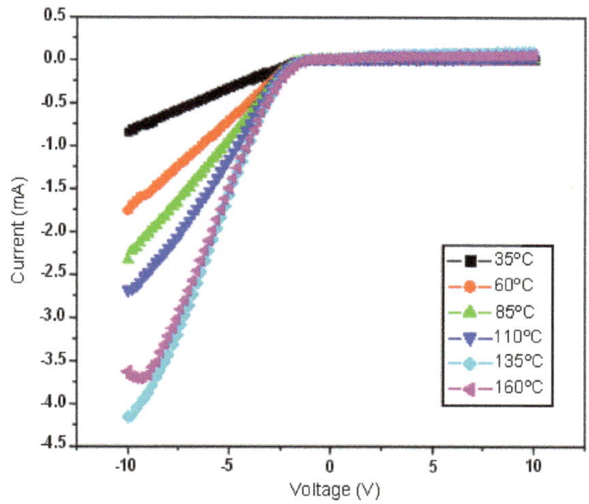

Figure 4. I-V characteristics of a-SiC_xN_y/Si heterojunction diode (N_2/Ar flow ratio= 0.1) at different temperatures.

4.1.2. Other diode types

SiC thin film electronic devices also include Schottky diode and light emitting diode (LED). Komiyama et al. fabricated Schottky diodes through the heteroepitaxial growth of 3C-SiC on a (001) Si substrate by introducing low-temperature growth (700–900ºC) of 3C-SiC using methylsilane single source, as an intermediate buffer layer, prior to the subsequent 3C-SiC active layer growth at a higher temperature (1150 ºC) using SiH_4 and C_3H_8 as precursors. The diode achieved a breakdown voltage of 190 V. A correlation between the film thickness and the leakage current of the Schottky diode was observed: on reverse bias the leakage current decreases when the 3C-SiC film thickness is increased. Figure 5 (a) shows the Schottky diode with Au,Al/poly 3C–SiC/SiO_2/Si substrate structure developed by Chung and Ahn. This diode exhibited a breakdown voltage of over 140 V together with a high leakage current. The authors suggested that the problem of the high leakage current is associated to random grooves, due to existence of anti-phase boundaries (APB) in the poly 3C-SiC film which demonstrates the dependence between the diode performance and the film characteristics.

Wahab et al. reported a Schottky diode formed by β-SiC thin films grown on (100) Si substrates, using reactive magnetron sputtering of a Si target in CH_4/Ar mixed plasma, with Au electrical contacts. Good electrical properties were observed such as ideality factor of 1.27 and leakage current density of 4 $\mu A/cm^2$.

A hot wire deposited a-SiC:H based thin film light emitting p–i–n diode was fabricated by Patil et al. The diode structure is formed by glass/TCO (SnO:F)/p-a-SiC:H/i-SiC:H/n-a-SiC:H/Al as illustrated in Figure 5 (b). The p-type a-SiC:H film was grown using SiH_4/C_2H_2/B_2H_6 gas mixture, the n-type using SiH_4/C_2H_2/PH_3 and the intrinsic using SiH_4/C_2H_2.

The deposition conditions of each film were optimized to obtain the p-, i- and the n-layers with desired electrical and optical properties. The layers exhibit the following bandgaps: 2.0 eV for p-type a-SiC:H, 2.06 for n-type a-SiC:H and 3.4 eV for intrinsic a-SiC:H.

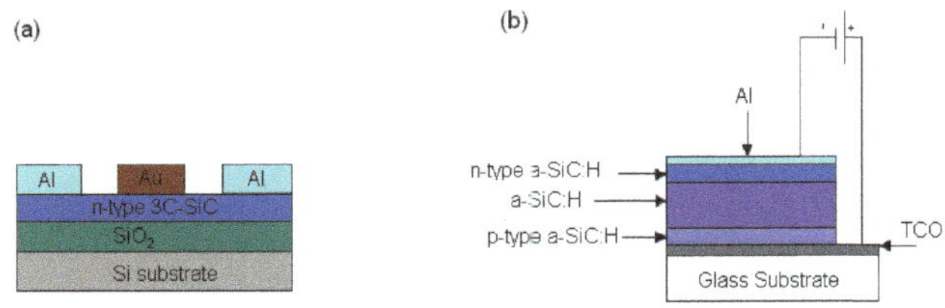

Figure 5. Schematic illustration of cross-section of device structures: (a) Schottky diode and (b) light emitting diode.

The diode was characterized and emits light in the visible region with low intensity. The authors attributed the low emission efficiency to the fact of the device be made in a single chamber and the same filament has been used to deposit all the a-SiC:H layers, which can has caused the contamination across the p–i interface.

4.2. Thin-film transistors (TFTs)

For many years the SiC thin film transistors prepared at low temperature have attracted special attention. In 1994, Hwang et al. developed two models of SiC submicron MOSFETs with vertical channel. The first model, as illustrated in Figure 6 (a), uses a sputtered SiC thin film, grown on a Si substrate at 600°C and annealed at 1300°C for 5 h under Ar atmosphere, as channel layer. It was observed that this structure can attain higher current, but the I-V characteristics can not be saturated because the channel depth is too large to be depleted. The second model fabricated (Figure 6 (b)) is formed by SiC thin film deposited by RF sputtering at 600°C on the sidewall of SiO_2 insulator. With this model, a complete saturation was achieved at drain voltage of 8 V for a 400 nm channel length. Furthermore, a drain breakdown voltage more than 16 V was achieved due to the wide bandgap of the SiC film used (2.2 eV). Both models were characterized under 600 K and the I-V curves do not show turn-off indicating that the TFTs fabricated can operate in this temperature range.

In 2006, Garcia et al. reported the first PECVD amorphous silicon carbide TFTs. The a-$Si_{1-x}C_x$:H films were deposited on glass substrates by PECVD at 300ºC using $SiH_4/CH_4/H_2$ gas mixture. Subsequently, n-type a-Si:H layer was deposited using $SiH_4/H_2/PH_3$ gas mixture and a photolithography was performed. Then, a PECVD SiO_2 layer, for the gate dielectric, was deposited using SiH_4/N_2O and a second photolithography was performed. The metal contacts were formed through the deposition of an Al layer followed by photolithography. Finally, an annealing in H_2 atmosphere at 350ºC was performed for 30 min. The TFT structure obtained is illustrated in Figure 6 (c). The TFT was tested at different temperatures. The drain current increased two orders of magnitude as temperature increased from 30ºC to

200ºC. This work compared this a-$Si_{1-x}C_x$:H TFT with a polycrystalline TFT fabricated by a KrF excimer laser annealing of the a-$Si_{1-x}C_x$:H films. The polycrystalline TFT exhibited output current at least an order of magnitude higher, when operated at room temperature, with respect to its amorphous being V_{GS} = 10 V for both.

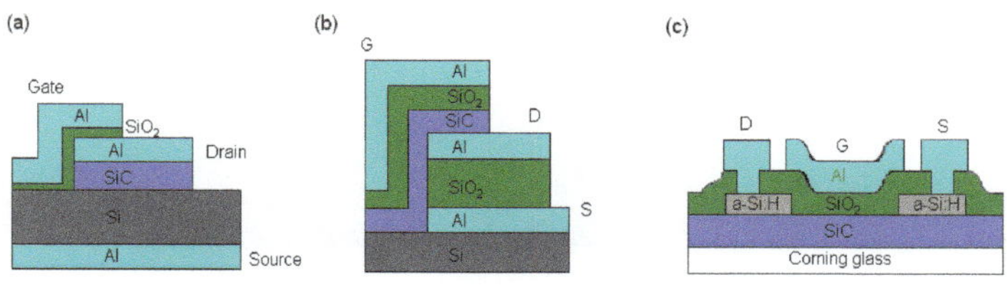

Figure 6. Schematic illustration of cross-section of TFT structures reported in literature: (a) and (b) Hwang et al., (c) Garcia et al.

5. SiC thin film based MEMS devices: sensors, RF MEMS and BioMEMS

5.1. Piezoresistive and capacitive sensors

It is well known that silicon piezoresistive sensors can not be used for high-temperature applications because of the p-n insulation of the piezoresistors. Several studies have demonstrated that the use SiC thin film piezoresistors is a good alternative for these applications due to their high gauge factor together with thermal stability. Ziermann et al. developed a piezoresistive β-SiC-on-silicon on insulator (SOI) pressure sensor with an on chip polycrystalline SiC thermistor for high operating temperatures. The test results from room temperature to 573 K demonstrated the capability of this sensor to monitor the cylinder pressure of combustion engines. We have studied the piezoresistive properties of amorphous SiC (a-SiC) films produced at low temperatures by PECVD and magnetron sputtering (Fraga, 2011d). Figure 7 (a) illustrates a piezoresistive pressure sensor with a-SiC piezoresistors developed by us.

In 2004, Young et al. proposed single-crystal 3C-SiC capacitive pressure sensors (see schematic illustration shown in Figure 7(b)) for sensing capabilities up to 400ºC. The fabrication of polycrystalline 3C-SiC capacitive pressure sensors was reported by Du et al. More recently, Chen and Mehregany reported the first all-SiC capacitive pressure sensor, incorporating a SiC diaphragm on a SiC substrate. Measurements of pressures up to 700 psi and temperatures up to 574°C were demonstrated. This shows that thin film-based technology has a lot to be developed to achieve the performance of sensors based on bulk materials.

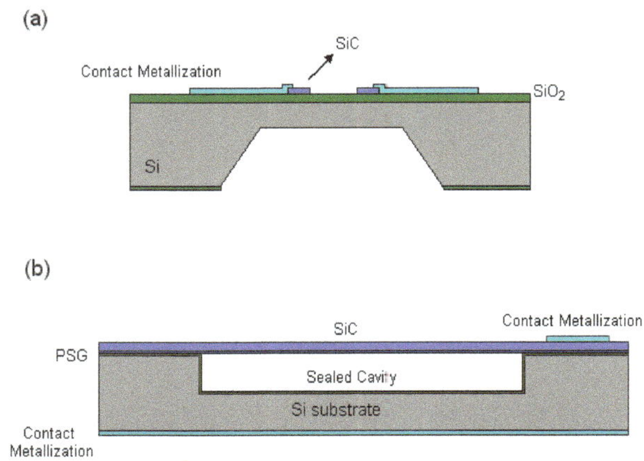

Figure 7. Examples of SiC film sensors shown in the literature: (a) piezoresistive and (b) capacitive.

5.2. RF MEMS

SiC films have been shown as a good alternative to the metal films in radiofrequency microelectromechanical systems (RF MEMS) applications, especially microbridge-based RF MEMS switches (Parro, 2008; Mishra, 2009) and MEMS resonators (Chang and Zorman, 2008).

Mishra et al. proposed a MEMS switch with low actuation voltage as illustrated in Figure 8 (a). This model uses a beam that is made of two materials: the SiC film to give mechanical stability and the Au to provide the conducting path to the ground. A process using four masks was employed to fabricate it: a high-resistivity p-type (100) Si wafer with 1.0 μm thermal SiO_2 was used as substrate, 800 nm of Au was deposited and patterned using lift-off to define the coplanar waveguide (CPW), to form the switch, a 1.5 μm of polyimide sacrificial layer was spun, soft-baked, and patterned to define the anchors. The anchors and the beam are made by depositing a 0.3 μm and 0.9 μm layers of SiC and Au, subsequent etching of the polyimide sacrificial layer. The switch exhibited an isolation of -40 db at 10 GHz and pull down voltage of 3 V.

The single-crystal and polycrystalline 3C-SiC lateral resonators were developed by Chang and Zorman. An illustration schematic of the cross-section of this resonator is shown in Figure 8 (b). The single crystalline (100) SiC film was produced by APCVD whereas the polycrystalline (111) SiC film by LPCVD. Both films were deposited on SiO_2/Si substrates. The experimental results showed that the 3C-SiC lateral resonators exhibit a resonant frequency similar to polysilicon devices and temperature coefficient of 22 ppm/ºC comparable to quartz oscillators (from 14 to 100 ppm/ºC), which confirm the potential of SiC films for RF MEMS applications.

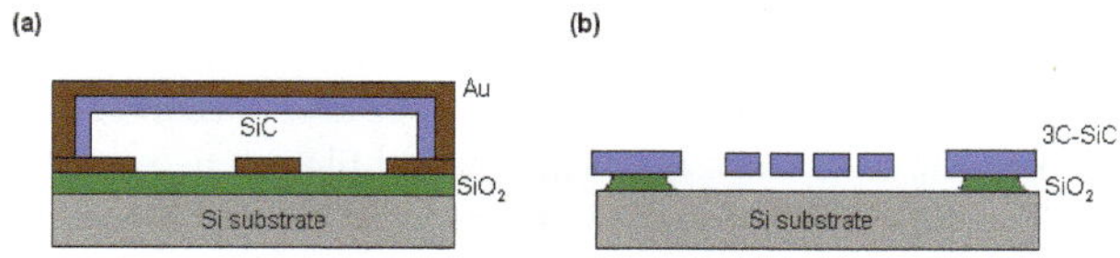

Figure 8. Examples of SiC thin film RF MEMS shown in the literature:

(a) switch and (b) lateral resonators.

5.3. BioMEMS

Biomedical or Biological Micro-Electro-Mechanical Systems (BioMEMS) are defined as systems or devices, which are constructed using micro/nanofabrication technology, for the analysis, delivery, processing, or for the development and construction of chemical and biological entities (Bashir, 2004). The first efforts in this field were directed to study the biocompatibility of common MEMS materials such as Si, SiO_2, Si_3N_4, polysilicon, SiC and SU-8. Kotzar et al. performed comparative studies among these materials and interesting conclusions were reached: (a) all materials were classified as non-irritants based on 1- and 12-week rabbit muscle implantations; (b) none of the materials were found to be cytotoxic in vitro using mouse fibroblasts; (c) only silicon nitride and SU-8 leached detectable non-volatile residues in aqueous physiochemical tests and (e) only SU-8 leached detectable non-volatile residues in isopropyl alcohol.

The biocompatibility of c-SiC and a-SiC films have been widely studied and promising results were reported (Santavirta, 1998; Kalnins, 2002; Coletti, 2007).

3C-SiC films grown on silicon substrates have been shown as a potential material for BioMEMS applications, especially for biosensing. Due to the mechanical strength, surface area-to-volume ratio, and extreme low mass, 3C-SiC BioMEMS structures have the potential to be mass sensors and resonators that are able to detect individual protein adsorption events (Zorman, 2012).

On the other hand, a-SiC based BioMEMS has been extensively developed and tested. Among the various bio-applications of a-SiC films can be mentioned: (i) coating material for implantable microsystems requiring hermetic sealing, owing to the fact that a-SiC is a excellent diffusion barrier material; (ii) membranes for microfluidics and Lab-on-a-Chip applications due to good a-SiC film chemical inertness property (Zorman, 2012).

6. Summary

In this chapter, the incorporation of dopant elements (such as N, P, B, Al and O) during the growth of SiC thin films has been described. The influence of each dopant type on the properties of SiC films was discussed. Furthermore, the challenges and trends related to the SiC thin film technology for device applications were discussed. The literature has shown that a-SiC and c-SiC films have maintained the most relevant properties observed in different poly-types of SiC substrates. However, to obtain high quality films is necessary to optimize the growth conditions. This review has indicated that using low temperature growth processes and in situ doping are the trend to produce high quality SiC thin films compatible with microelectronic and MEMS technologies. Particular attention should be directed to grow large area uniform SiC films on Si and insulator/Si substrates, which is essential to make low-cost devices.

Regarding the SiC-based thin film applications, it was showed that these films have been widely used in electronic and MEMS devices such as diodes, TFTs, sensors, RF MEMS and BioMEMS. It is important to underline that the use of SiC films in the amorphous or crystal-line form, doped or not, should be evaluated in terms of their properties. Much has been studied about the use of a-SiC films due to the fact of its ease of integration with the rising technologies, especially those that use temperature-sensitive substrates. Today, the area of bio-applications is a good example of this kind of requirement.

Acknowledgements

We would like to thank the Microfabrication Laboratory of the Brazilian Synchrotron Light Laboratory (LMF-LNLS) for the support in the development of the devices and sensors, Center of Semiconductor Components (CCS-UNICAMP) for the support in the characteriza-tion of the SiC films and the Electronics Laboratory of the Department of Precision Mechani-cal of the Faculty of Technology of Sao Paulo (FATEC-SP) for the support in the testing of the devices and sensors. This work was supported by FAPESP-MCT/CNPq-PRONEX (proc-ess nº 2011/50773-0).

Author details

Mariana Amorim Fraga[1,2*], Rodrigo Sávio Pessoa[2,3], Marcos Massi[2] and Homero Santiago Maciel[2,3]

*Address all correspondence to: mafraga@ita.br

1 Department of Sciences, Engineering and Mathematical Modeling, Regional University of Northwest Rio Grande do Sul State,, Brazil

2 Plasma and Processes Laboratory, Technological Institute of Aeronautics,, Brazil

3 IP&D, University of Vale do Paraiba,, Brazil

References

[1] Alizadeh, Z., & Sundaram, K. B. (2002). *Journal of Materials Science Letters*, 21.

[2] Anderson, D. A., & Spear, W. E. (1977). *Phil. Mag. B*, 35.

[3] Atwell, A. R., Okojie, R. S., Kornegay, K. T., Roberson, S. L., & Beliveau, A. (2003). *Sensors and Actuators A: Physical*, 104, 11-18.

[4] Bashir, R. (2004). *Advanced Drug Delivery Reviews,*, 56, 1565-1586.

[5] Beyer, W., Wagner, H., & Finger, F. (1985). *Journal of Non-Crystalline Solids*, 77-78, 856-860.

[6] Chang, W., & Zorman, C. (2008). *Sensors,*, 8, 5759-5774.

[7] Chen, J., Scofield, J., & Steckl, A. J. (2000). *Journal of The Electrochemical Society*, 147, 3845-3849.

[8] Chen, L., & Mehregany, M. (2007). A Silicon Carbide Capacitive Pressure Sensor for High Temperature and Harsh Environment Applications Solid-State Sensors, Actuators and Microsystems Conference Proceedings pages., 4.

[9] Chen, T., Yang, D., Carius, R., & Finger, F. (2011). *Thin Solid Films*, 519, 4516-4518.

[10] Cheng, K., Cheng, H., Lee, W., Lee, C., & Yew, T. (1997). Low Temperature Deposited Highly-Conductive N-type SiC Thin Film. MRS Proceedings, pages., 6.

[11] Cheng, L., Pan, M., Scofield, J., & Steckl, A. J. (2002). *Journal of Electronic Materials*, 31, 361-365.

[12] Cheng, L. (2003). SiC Thin-Films on Insulating substrates for Robust MEMS Applications. dissertation submitted to the Division of Research and Advanced Studies of the University of Cincinnati pages., 168.

[13] Chung, G. S., & Ahn, J. H. (2008). *Microelectronic Engineering,*, 85, 1772-1775.

[14] Coletti, C., Jaroszeski, M. J., Pallaoro, A., Hoff, A. M., Iannotta, S., & Saddow, S. E. (2007). Paper presented at Proceedings of the 29th Annual International Conference of the IEEE Engineering in Medicine and Biology Society. *Biocompatibility and wettability of crystalline SiC and Si surfaces*, 5849-5852.

[15] Cros, B., Gat, E., & Saurel, J. M. (1997). *Journal of Non-Crystalline Solids*, 209, 273-282.

[16] Cross, T. J., & Raj, R. (2010). Synthesis and Tribological Behavior of Silicon Oxycarbonitride Thin Films Derived From Poly (Urea) Methyl Vinyl Silazane. Progress in Nanotechnology Processing, John Wiley & Sons.

[17] Du, J., Ko, W. H., Mehregany, M., & Zorman, C. A. (2005). Proceedings of IEEE Sensor pages., 4.

[18] El Khakani, M. A., Chaker, M., Jean, A., Boily, S., Kieffer, J. C., O'Hern, M. E., Ravet, M. F., & Rousseaux, F. (1994). *Journal of Materials Research*, 9, 96-103.

[19] Flannery, A. F., Mourlas, N. J., Storment, C. W., Tsai, S., Tan, S. H., Heck, J., Monk, D., Kim, T., Gogoi, B., & Kovacs, G. T. A. (1998). *Sensors and Actuators A: Physical*, 70, 48-55.

[20] Fleischman, A. J. , Wei, X., Zorman, C. A. ., & Mehregany, M. (1998). *Materials Science Forum*, 264-268, 885-888.

[21] Fraga, M. A., Pessoa, R. S., Massi, M., Maciel, H. S., & Santos Filho., S. G. (2007). *Brazilian Journal of Vacuum Applications*, 26, 193-197.

[22] Fraga, M. A., Massi, M., Oliveira, I. C., Maciel, H. S. ., dos Santos Filho, S. G., & Mansano, R. D. (2008). *Journal of Materials Science: Materials in Electronics*, 19, 835-840.

[23] Fraga, M. A., Furlan, H., Massi, M., & Oliveira, I. C. (2010). *Microsystem Technologies*, 16, 925-930.

[24] Fraga, M. A. (2011a). *Materials Science Forum*, 679-680, 217-220.

[25] Fraga, M. A., Massi, M., & Furlan, H. (2011b). Influence of film composition on I-V characteristics of a-SiC_xN_y/Si heterojunction diodesBrazilian Physics Meeting.

[26] Fraga, M. A., Furlan, H., Massi, M., Oliveira, I. C., Mateus, C. F. R., & Rasia, L. A. . (2011c). *Microsystem Technologies*, 17, 477-480.

[27] Fraga, M. A., Pessoa, R. S., Maciel, H. S., & Massi, M. (2011d). Recent Developments on Silicon Carbide Thin Films for Piezoresistive Sensors Applications. Silicon Carbide- Materials, Processing and Applications in Electronic Devices (ed. M. Mukherjee), Intech 978-9-53307-968-4

[28] Fraga, M. A., Furlan, H., Pessoa, R. S., Rasia, L. A., & Mateus, C. F. R. (2012). Studies on SiC, DLC and TiO_2 thin films as piezoresistive sensor materials for high temperature application. Microsystem Technologies pages, in press., 7.

[29] Fu, X. A., Dunning, J., Zorman, C. A., & Mehregany, M. (2004). *Materials Science Forum*, 457-460, 1519-1522.

[30] Garcia, B., Estrada, M., Albertin, K. F., Carreno, M. N. P., Pereyra, I., & Resendiz, L. (2006). *Solid-State Electronics*, 50, 241-247.

[31] Harris, G. L. (1995). *Properties of Silicon Carbide*, Inspec, London, 0-85296-870-1.

[32] Hatalis, M. K., & Greve, D. W. (1987). *IEEE Electron Device Letters*, 8, 361-364.

[33] Hwang, J. D., Fang, Y. K., & Tsa, T. Y. (1995). *Solid-State Electronics*, 38, 275-278.

[34] Gomez, F. J., Prieto, P., Elizalde, E., & Piqueras, J. (1996). *Applied Physics Letters*, 69, 773-776.

[35] Kalnins, U., Erglis, A., Dinne, I., Kumsars, I., & Jegere, S. (2002). *Medical Science Monitor*, 8, 16-20.

[36] Kim, D. S., & Lee, Y. H. (1995). *Thin Solid Films*, 261, 192-201.

[37] Komiyama, J., Abe, Y., Suzuki, S., Kita, T., & Nakanishi, H. (2005). *Journal of Crystal Growth*, 275, 1001-1006.

[38] Kotzar, G., Freas, M., Abel, P., Fleischman, A., Roy, S., Zorman, C., Moran, J. M., & Melzak, J. (2002). *Biomaterials*, 23, 2737-2750.

[39] Larkin, D. J. (1997). *Physics State Solid (b)*, 202, 305-320.

[40] Li, S. S. (2006). p-n junction diodes, in Semiconductor Physical Electronics Springer 978-0-38728-893-2

[41] Liu, F., Carraro, C., Pisano, A. P., & Maboudian, R. (2010). *J. Micromech. Microeng*, 20.

[42] Loubet, N., Adam, T., Raymond, M., Liu, Q., Cheng, K. , Sreenivasan, R. ., Reznicek, A., Khare, P., Kleemeier, W., Paruchuri, V., Doris, B., & Sampson, R. (2012). *Thin Solid Films*, 520, 3149-3154.

[43] Martins, R., Vieira, M., Ferreira, I., Fortunato, E., & Guimarães, L. (1996). *Solar Energy Materials and Solar Cells*, 41-42, 493-517.

[44] Medeiros, H. S., Pessoa, R. S., Sagas, J. C., Fraga, M. A., Santos, L. V., Maciel, H. S. ., Massi, M., Sobrinho, A. S. S., & Costa, M. E. H. M. (2011). *Surface & Coatings Technology*, 206, 1787-1795.

[45] Medeiros, H. S., Pessoa, R. S., Fraga, M. A., Santos, L. V., Maciel, H. S. ., Massi, M., & Sobrinho, A. S. S. (2012). Argon incorporation on silicon carbide thin films deposited by bias co-sputtering technique. MRS Proceedings pages., 6.

[46] Mishra, B., Panigrahi, R., & Alex, Z. C. (2009). *International Journal of Recent Trends in Engineering*, 2, 120-122.

[47] Miyajima, S., Yamada, A., & Konagai, M. (2006). Properties of Nanocrystalline 3C-SiC:H and SiC:Ge:H Films Deposited at Low Substrate Temperatures. MRS Proc. pages., 910.

[48] Murooka, K., Higashikawa, I., & Gomei, Y. (1996). Applied Physics Letters pages., 69.

[49] Myong, S. Y., Lee, H. K. ., Yoon, E., & Lim, K. S. (2002). *Journal of Non-Crystalline Solids*, 298, 131-136.

[50] Okojie, R. S., Ned, A. A., Kurtz, A. D., & Carr, W. N. (1996). *a(6H)-SiC pressure sensors for high temperature applications, Micro Electro Mechanical Systems (MEMS'96) Proceedings,*, 146-149.

[51] Okojie, R. S., Ned, A. A., Kurtz, A. D., & Carr, W. N. (1998). *IEEE Trans. Elec. Dev.*, 45, 785-790.

[52] Oliveira, A. R. (2002). Dopagem elétrica de filmes finos de carbeto de silício amorfo hidrogenado (a-SiC:H) obtidos por PECVD. *Dissertação de Mestrado apresentada a Escola Politécnica da USP.*

[53] Oliveira, A. R., Pereyra, I., & Carreno, M. N. P. (2004). *Materials Science and Engineering B*, 112, 144-146.

[54] Ong, Y. Y., Chen, B. T., Tay, F. E. H., & Iliescu, C. (2006). *Journal of Physics: Conference Series 34*, 812-817.

[55] Parro, R. J., Scardelletti, M. C., Varaljay, N. C., Zimmerman, S., & Zorman, C. A. (2008). *Solid-State Electronics*, 52, 1647-1651.

[56] Patil, S. B., Kumbhar, A. A., Saraswat, S., & Dusane, R. O. (2003). *Thin Solid Films*, 430, 257-360.

[57] Pereira, J. M. T., Banerjee, P. K., & Mitra, S. S. (1985). *Thin Solid Films*, 127, 337-350.

[58] Petrman, V., Houska, J., Kos, S., Calta, P., & Vlcek, J. (2011). *Acta Materialia*, 59, 2341-2349.

[59] Ruddell, F. H., Mc Neill, D., Armstrong, B. M., & Gamble, H. S. (1991). Silicon carbide layers produced by rapid thermal vapor deposition. *Proc. SPIE*, 1361, 159.

[60] Santavirta, S., Takagi, M., Nordsletten, L., Anttila, A., Lappalainen, R., & Konttinen, Y. T. (1998). *Archives Orthopaedic Trauma Surgery*, 118, 89-91.

[61] Sha, Z. D., Wu, X. M., & Zhuge, L. J. (2005). *Physics Letters A*, 346, 186-192.

[62] Shen, J., & Raj, R. (2011). *Journal of Power Sources*, 196, 5945-5950.

[63] Shimada, T., Katayama, Y., & Komatsubara, K. F. (1979). *Journal of Applied Physics*, 80, 5530-5532.

[64] Shor, J. S., Goldstein, D., & Kurtz, A. D. (1993). *IEEE Trans. Elec. Dev.*, 40, 1093-1099.

[65] Singh, A. V., Chandra, S., Kumar, S., & Bose, G. (2012). *Journal of Micromechanical and Microengineering*, 22.

[66] Sundaram, K. B., & Alizadeh, J. (2000). Thin Solid Films, , 370, 151-154.

[67] Sundaram, K. B., Alizadeh, Z., Todi, R. M., & Desai, V. H. (2004). *Materials Science and Engineering: A*, 368, 103-108.

[68] Tawada, Y., Tsuge, K., Kondo, M., Okamoto, H., & Hamakawa, Y. (1982). *Journal of Applied Physics*, 53.

[69] Terauds, K., Sanchez-Jimenez, P.E., Raj, R., Vakifahmetoglu, C., & Colombo, P. (2010). *Journal of the European Ceramic Society*. 30, 2203-2207.

[70] Tsuchiya, T. (2008). Evaluation of Mechanical Properties of MEMS Materials and Their Standardization. in Reliability of MEMS: Testing of Materials and Devices (eds O. Tabata and T. Tsuchiya), Wiley-VCH Verlag GmbH & Co. KGaA 978-3-52731-494-2

[71] Uthanna, S., Schröder, B., & Oechsner, H. (1991). *Vacuum*, 42, 287-290.

[72] Vetter, M., Voz, C., Ferre, R., Martin, I., Orpella, A., Puigdollers, J., Andreu, J., & Alcubilla, R. (2006). *Thin Solid Films*, 511-512, 290-294.

[73] Vijayakumar, A., Todi, R. M., Warren, A. P., & Sundaram, K. B. (2008). *Diamond Related. Materials*, 17, 944-948.

[74] Wahab, Q., Karlsteen, M., Willander, M., & Sundgren, J. E. (1991). *Journal of Electronic Materials*, 20, 899-901.

[75] Wang, L., Dimitrijev, S., Han, J., Tanner, P., Lacopi, A., & Hold, L. (2011). *Journal of Crystal Growth*, 329, 67-70.

[76] Wijesundara, M. B. J., Gao, D., Carraro, C., Howe, R. T., & Maboudian, R. (2003). *Journal of Crystal Growth*, 259, 18-25.

[77] Wu, C. H., Zorman, C. A., & Mehregany, M. (1999). *Thin Solid Films*, 355-356, 179-183.

[78] Wu, X. C., Cai, R. Q. , Yan, P. X., Liu, W. M., & Tian, J. (2002). *Appl Surf. Sci*, 185, 262.

[79] Yamamoto, K., Koga, Y., & Fujiwara, S. (2001). *Diamond Rel. Mater.*, 10.

[80] Yeung, K. W., & Ong, C. W. (2007). *Sensors and Actuators A: Physical*, 137.

[81] Yih, P. H., Li, J. P., & Steckl, A. J. (1994). *IEEE Transactions on Electron Devices*, 41, 281-287.

[82] Young, D. J., Du, J., Zorman, C. A., & Ko, W. H. (2004). *IEEE Sensors Journal*, 4, 464-470.

[83] Yoon, S. F., Ji, R., Ahn, J., & Milne, W. I. (1997a). *Diamond Rel. Mater.*, 6, 48-54.

[84] Yoon, S. F., Ji, R., & Ahn, J. (1997b). *Journal of Non-Crystalline Solids*, 211, 173-179.

[85] Ziermann, R., Berg, J. V., Obermeier, E., Wischmeyer, F., Niemann, E., Moller, H., Eickhoff, M., & Krotz, G. (1999). *Materials Science and Engineering B*, 61-62, 576-578.

[86] Zorman, C. A., & Barnes, A. C. (2012). Silicon Carbide BioMEMS. Silicon Carbide Biotechnology First edition (ed. S. E. Saddow), Elsevier Inc. 978-0-12385-906-8

[87] Zorman, C. A., Fleischman, A. J., Dewa, A. S., Mehregany, M., Jacob, C., Nishino, S., & Pirouz, P. (1995). *Journal of Applied Physics*, 78, 5136-5138.

[88] Zorman, C. A., & Parro, R. J. (2010). Micro- and Nanomechanical Structures for Silicon Carbide MEMS and NEMS. in Silicon Carbide: Growth, Defects, and Novel Applications, 1eds P. Friedrichs, T. Kimoto, L. Ley and G. Pensl), Wiley-VCH Verlag GmbH & Co. KGaA 978-3-52741-002-6

Chapter 15

Silicon Carbide: A Biocompatible Semiconductor Used in Advanced Biosensors and BioMEMS/NEMS

Mahboobeh Mahmoodi and Lida Ghazanfari

Additional information is available at the end of the chapter

1. Introduction

In the last decade, there has been a tremendous development in the field of miniaturization of chemical and biochemical sensor devices. Microelectromechanical systems (MEMS) refer to microscopic devices that have a characteristic length of less than 1 mm but more than 100 nm and combine electrical and mechanical components. Nanoelectromechanical systems (NEMS) refer to nanoscopic devices that have a characteristic length of less than 100 nm and combine electrical and mechanical components. In mesoscale devices, if the functional components are on micro- or nanoscale, they may be referred to as MEMS or NEMS, respectively. These are referred to as an intelligent miniaturized system comprising sensing, processing, and/or actuating functions and combine electrical and mechanical components. The acronym MEMS originated in the USA. The term commonly used in Europe is micro system technology (MST) and in Japan, the term is micromachines. Another term generally used is micro/nanodevices.

MEMS/NEMS terms are also now used in a broad sense and include electrical, mechanical, fluidic, optical, and/or biological functions. MEMS/NEMS for optical applications are referred to as micro/nano optoelectromechanical systems (MOEMS/NOEMS). MEMS/NEMS for electronic applications are referred to as radio-frequency-MEMS/NEMS or RF-MEMS/RF-NEMS. MEMS/NEMS for biological applications are referred to as BioMEMS/BioNEMS. MEMS and emerging NEMS are expected to have a major impact on our lives, comparable to that of semiconductor technology, information technology, or cellular and molecular biology. MEMS/NEMS are used in electromechanical, electronics, information/communication, chemical, and biological applications [1]. To put the dimensions of MEMS and NEMS in perspective refer to Fig.1.

MEMS/NEMS need to be designed to perform expected functions in short durations, typically in the millisecond to picosecond range. Through MEMS it is possible to incorporate micro-scale types of devices such as motors, pumps, fluidic channels, sample preparation (including mixing or vaporization) chambers, and various types of sensors (including optical sensors) that will perform assorted tasks, such as monitoring. Again, one can do both monitoring in the physical sense and the chemical sense. SiC has been known for its outstanding mechanical and chemical properties making it equally attractive for MEMS and NEMS.

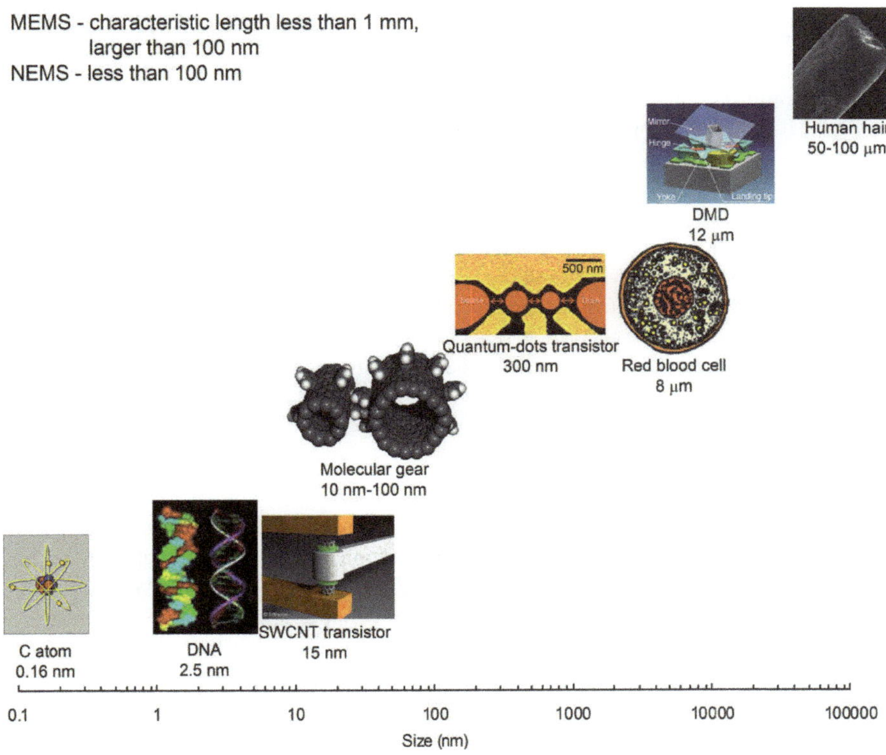

Figure 1. Dimensions of MEMS and NEMS in perspective. MEMS/NEMS examples shown are of a vertical single walled carbon nanotube (SWCNT) transistor (5 nm wide and 15 nm high), of molecular dynamic simulations of a carbon-nanotube based gear, quantum-dot transistor obtained from van der Wiel et al., and DMD obtained from www.dlp.com.

An ever-increasing demand for biomedical devices provides motivation for the development of advanced semiconducting materials for challenging applications ranging from disease detection to organ function restoration. The superior bioelectrical properties of silicon carbide (SiC) make it an ideal substrate for bioelectrodes thus allowing for an all-biocompatible, non-metalic biomedical system. Most of the studies conducted in the past on single-crystal SiC provide evidence of the attractive bio-potentialities of this material and hence suggest similar properties for crystalline SiC.

Recent interest has risen in employing these materials, tools and technologies for the fabrication of miniature sensors and actuators and their integration with electronic circuits to produce smart devices and systems. This effort offers the promise of: (1) increasing the performance and manufacturability of both sensors and actuators by exploiting new batch fabrication processes developed including micro stereo lithographic and micro molding techniques; (2) developing novel classes of materials and mechanical structures not possible previously, such as diamond-like carbon, silicon carbide and carbon nanotubes, micro-turbines and micro-engines; (3) development of technologies for the system level and wafer level integration of micro components at the nanometer precision, such as self-assembly techniques and robotic manipulation; (4) development of control and communication systems for MEMS, such as optical and wireless radio frequency, and power delivery systems, etc. The integration of MEMS, NEMS, interdigital transducers and required microelectronics and conformal antenna in the multifunctional smart materials and composites results in a smart system suitable for sending and control of a variety of functions in automobile, aerospace, marine and civil strutures and food and medical industries. The principle aim of this chapter is to present an overview of bioMEMS and bioNEMS technologies that utilize SiC as a key component in the structure. Therefore, this chapter focuses on reviewing examples where SiC is used in the mechanical components of micro /nanofabricated devices and biosensors for biomedical applications that can be used to further develop the technology.

2. SiC MEMS and NEMS

SiC is an excellent structural material for MEMS and NEMS applications due to its outstanding mechanical, chemical, and electrical properties combined with its compatibility with Si micromachining techniques. Nowhere is the limitation to silicon-based MEMS more apparent than in applications characterized by high temperature environments. The electronic properties of silicon place an upper limit on the operating temperature of electronic devices at roughly 250 °C for devices fabricated on conventional bulk silicon substrates and around 300 °C for devices built on silicon-on-insulator substrates. As for the mechanical properties, the upper limit for silicon-based micromechanical structures is around 500 °C, as seen in the plastic deformation of Si membranes subjected to deflecting loads at that temperature. The silicon surface is chemically active and will appreciably oxidize at temperatures above 800 °C. These material limitations require that silicon-based MEMS structures be enclosed in protective packaging to make them suitable for use in these conditions. In many cases, the packaging is so extensive that the benefits of using a silicon based MEMS device (i.e., low cost and small device size) are completely negated by the package. The desire to capitalize on MEMS technology for applications where the use of silicon is impractical has motivated the development of alternative semiconductors whose material properties are better suited for such applications [2].

Mostly, efforts to develop SiC for MEMS have focused on developing device technologies for harsh environment applications where Si, the dominant material in MEMS, is not well suited. Such environments include high temperature (>600 °C), high mechanical wear, high radiation, high oxidation, and harsh chemicals. Properties that make SiC particularly well suited for harsh environments include a wide electronic bandgap (ranging from 2.9 eV for 3C-SiC to 3.2 eV for 4H-SiC), high hardness (2,480 kg/mm^2), high resistance to chemical etching in acids and bases, slow oxidation rates, and very strong covalent Si–C bonds. SiC is of particular interest for use in MEMS-based microactuators, where its inert surface resists the deleterious effects of stiction and its high Young's modulus (~400 GPa) enables fabrication of mechanical resonators that can operate over a very wide frequency range, including the GHz range. Its chemical inertness, favorable mechanical properties, and biocompatibility make SiC particularly attractive for bioMEMS applications. Several comprehensive reviews of SiC MEMS technology, including material properties, processing techniques, and device examples have recently been published [2, 3].

The fabrication of bulk micromachined SiC MEMS structures using conventional Si bulk micromachining techniques is enabled by the ability of 3C-SiC films to be epitaxially grown on bulk Si substrates. These processes benefit tremendously from the chemical inertness of SiC with respect to Si etchants [2]. In applications where SiC surfaces are desired but bulk SiC substrates could be technically or economically constraining, 3C-SiC films grown on Si substrates provide a convenient and low-cost alternative to 6H- and 4H-SiC wafers. One of the earliest 3C-SiC MEMS structures to be fabricated was the freestanding membrane [4].

In general, the material properties of poly-SiC that are of importance to MEMS closely resemble its single crystalline counterpart. The poly-SiC films are not generally viewed as desirable for electronic devices for largely the same reasons as polysilicon. In terms of chemical inertness, fully crystallized poly-SiC is essentially as inert as 3C-SiC. Poly-SiC oxidizes slightly faster than 3C-SiC due to grain boundary diffusion of oxidants. The Young's modulus is typically reported to be around 400 GPa.

SiC, owing to its chemical inertness, is of interest as a structural material for biomedical microsystems. A recent study that investigated a wide range of MEMS materials (including Ti, Si, SiO2 and Si3N4) has shown that single crystal SiC is both biocompatible (to the extent of the tests performed) and sterilizable [2].

3. Biocompatibility

Biocompatibility is related to the behavior of biomaterials in various contexts. The term may refer to specific properties of a material without specifying where or how the material is used, or to more empirical clinical success of a whole device in which the material or materials are featured. The ambiguity of the term reflects the ongoing development of insights into how biomaterials interact with the human body and eventually how those interactions determine the clinical success of a medical device (such as pacemaker, hip replacement or stent). Modern medical devices and prostheses are often composed of more than one material so it might not always be sufficient to talk about the biocompatibility of a specific material. Cell-semiconductor hybrid systems represent an emerging topic of research in the biotechnological area with intriguing possible applications. A comprehensive understanding of the interactions governing such systems is the basis of present and future development of biologically interfaced device performance. To date, very little is known about the main processes that govern the communication between cells and the surfaces they adhere to. When cells adhere to an external surface an eterophilic binding is generated between the cell adhesion proteins and the surface molecules. After they adhere, the interface between them and the substrate becomes a dynamic environment where surface chemistry, topology, and electronic properties have been shown to play important roles. The biocompatibility single-crystal SiC was determined by culturing mammalian cells directly on SiC substrates and by evaluating the resulting cell adhesion quality and proliferation. The crystalline SiC is indeed a very promising material for bio-applications, with better bio-performance than crystalline Si. 3C-SiC, which can be directly grown on Si substrates, appears to be an especially promising bio-material: the Si substrate used for the epi-growth would in fact allow for cost-effective and straightforward electronic integration, while the SiC surface would constitute a more biocompatible and versatile interface between the electronic and biological world. The main factors that have been shown to define SiC biocompatibility are its hydrophilicity and surface chemistry. SiC surface morphology is shown to influence cell adhesion only when macropatterned, while SiC polytypism and doping concentration seem to have no influence on cell proliferation. The identification of the organic chemical groups that bind to the SiC surface, together with the calculation of SiC zeta potential in media, could be used to better understand the electronic interaction between cell and SiC surfaces. Using an appropriate cleaning procedure for the SiC samples before their use as substrates for cell cultures is also important. The cleaning chemistry may affect cell proliferation and emphasize the importance of the selection of an appropriate cleaning procedure for biosubstrates. SiC has been shown to be significantly better than Si as a substrate for cell culture, with a noticeably reduced toxic effect and enhanced cell proliferation. One of the possible drawbacks that may be associated with the use of SiC in vivo is related to the unclear and highly debated cytotoxic level of SiC particles. Nonetheless, the potential cytotoxicity of SiC particles does not represent a dramatic issue as much as it does for Si, since the great tribological properties of SiC make it less likely to generate debris.

Several papers have discussed testing silicon SiC in vitro. In one study the researchers tested SiC deposited from radiofrequency sputtering using alveolar bone osteoblasts and gingival fibroblasts for 27 days. The investigators reported that "Silicon carbide looks cytocompatible both on basal and specific cytocompatibility levels. However, fibroblast and osteoblast attachment is not highly satisfactory, and during the second phase of osteoblast growth, osteoblast proliferation is very significantly reduced by 30%". According to another paper, in a 48 h study using human monocytes, SiC had a stimulatory effect comparable to polymethacrylate. Cytotoxicity and mutagenicity has been performed on SiC-coated tantalum stents. Amorphous SiC did not show any cytotoxic reaction using mice fibroblasts L929 cell cultures when incubated for 24 h or mutagenic potential when investigated using Salmonella typhimurium mutants TA98, TA100, TA1535, and TA1537. An earlier study by the same authors of a SiC-coated tantalum stent reported similar results [5].

Kotzar et al. [6] evaluated materials used in microelectromechanical devices for biocompatibility. These included single crystal silicon, polysilicon (coating, chemical vapor deposition, CVD), single crystal cubic SiC (3C SiC or β-SiC, CVD), and titanium (physical vapor deposition). They concluded that the tested Si, SiC and titanium were biocompatible. Even though crystalline SiC biocompatibility has not been investigated in the past, information exists concerning the biocompatibility of the amorphous phase of this material (a-SiC). Materials commonly used in the fabrication and packaging of standard MEMS devices were recently evaluated for cytotoxicity using the ISO 10993 biocompatibility testing standards [7]. The material set comprised of: silicon (Si, 500 μm-thick), silicon dioxide (SiO_2, 0.5 μm-thick), silicon nitride (Si_3N_4, 0.2 μm-thick), polycrystalline silicon (polysilicon, 0.5 μm-thick), silicon carbide (SiC, 0.5 μm-thick), titanium (Ti, 0.5 μm-thick), and SU-8 (50 μm-thick) (Table 1).

The biocompatibility of the materials used in silicon-based devices, such as single crystalline silicon, polysilicon, silicon dioxide, silicone nitride and silicon carbide, were evaluated according to ISO 10993 standards by Kotzar et al. [6]. Using mouse fibroblasts in the tests, none of the materials were found to be cytotoxic. An in vivo tests based on implantation in rabbit muscle showed no sign of irritation. Only silicone nitride and SU-8 showed detectable nonvolatile residues. Furthermore, in vivo studies using Stainless Steel cages and Teflon cages reveal that silicon, silicon nitride, silicon dioxide, gold, and SU-8 are biocompatible. However, silicon and SU-8 have shown increased biofouling.

Material	Reactivity (0-4)
Positive Control (tin stabilized polyvinylchloride)	4
Negative Control (high density polyethylene)	0
Si (monocrystalline silicon)	0
Sio_2 (silicon dioxide)	0
Polysilicon (polycrystalline silicon)	0
Si_3N_4 (silicon nitride)	0
SiC (monocrystalline 3C silicon carbide)	0
Ti (sputtered titanium)	0
SU-8 (epoxy photoresist)	0

Table 1. In vitro cytotoxicity of MEMS materials [7]

4. Hemocompatibility

The interaction between blood proteins and the material is regarded as an important source of thrombogenesis. The adsorption of proteins is explained, from the thermodynamic point of view, in terms of the systems free energy or surface energy. However, adsorption itself does not induce thrombosis. Theories regarding correlations between thrombogenicity of a material and its surface charge or its binding properties proved not to be useful.

Thrombus formation on implant materials is one of the first reactions after deployment and may lead to acute failure due to occlusion and serve as a trigger for neointimal formation. Next to the direct activation by the intrinsic or extrinsic coagulation cascade, thrombus formation can also be initiated directly by an electron transfer process while fibrinogen is close to the surface. The electronic nature of a molecule can be defined as semiconductor or insulator. Contact activation is possible in the case of a metal since electrons in the fibrinogen molecule are able to occupy empty electronic states with the same energy. Therefore, the obvious way to avoid this transfer is to use a material with a significantly reduced density of empty electronic states within the range of the valence band of the fibrinogen. This is the case for the used silicon carbide coating.

Hemocompatibility leads to the following physical requirements: (1) to prevent the electron transfer, the solid must have no empty electronic states at the transfer level, i.e., deeper than 0.9 eV below Fermi's level. This requirements met by a semiconductor with a sufficiently large band gap (its valence band edge must be deeper than 1.4 eV below Fermi´s level) and a low density of states inside the band gap. (2) To prevent electrostatic charging of the interface (which may interfere with requirement 1) the electric conductivity must be higher than 10-3 S/cm. A material that meets these electronic requirements is silicon carbide in an amorphous, heavily n-doped, hydrogen-rich modification (a-SiC:H). The amorphous structure is required in order to avoid any point of increased density of electronic states, especially at grain boundaries.

At present, a-SiC:H is known for its high thromboresistance induced by the optimal barrier that this material presents for protein adhesion. These properties may translate into less protein biofouling and better compatibility for intravascular applications rather than Si. SiC has a relatively low level of fibrinogen and fibrin deposition when contacting blood. These proteins promote local clot formation; thus, the tendency not to adsorb them will resist blood clotting. It is now well established that SiC coatings are resistant to platelet adhesion and clotting both in vitro and in vivo [5]. In the Bolz et al. [8] study, a-SiC:H films were deposited using the glow discharge technique or plasma-enhanced chemical vapour deposition (PECVD). The technique provides the most suitable coating process due to its high inherent hydrogen concentration which satisfies the electronically active defects in the amorphous layers. They used fibrinogen as an example model for thrombogenesis in implants, although most haemoproteins are organic semiconductors. a-SiC:H coatings showed no time-dependent increase in the remaining protein concentration, confirming that no fibrinogen activation and polymerisation had taken place. These results support the electrochemical model for thrombogenesis at artificial surfaces and prove that a proper tailoring of the electronic properties leads to a material with superior hemocompatibility. The in vitro test showed that the morphology of the cells was regular. The a-SiC:H samples showed the same behaviour as the control samples. Blood and membrane proteins have similar band-gaps, because the electronic properties depend mainly on the periodicity of the amino acids, and the proteins differ only in the acid sequence, not in their structural periodicity. Apparently, similar reactions inducing a modification of proteins are responsible for the cell culture results.

A-SiC: H has superior hemocompatibility; its clotting time is 200 percent longer than to that of titanium and pyrolytic carbon. Furthermore, it has been shown that small variations in the preparation conditions cause a significant change in hemocompatibility. Therefore, it is of paramount importance to know the exact physical properties of the material in use. Amorphous silicon carbide can be deposited on any substrate material which is resistant to temperatures of approximately 250 °C. This property makes amorphous silicon carbide a suitable coating material for all hybrid designs of biomedical devices. The substrate material can be fitted to the mechanical needs, disregarding its hemocompatibility, whereas the coating ensures the hemocompatibility of the device. Possible applications are catheters or sensors in blood contact and implants, especially artificial heart valves.

Bolz and Schaldach [8] evaluated PECVD amorphous SiC for use on prosthetic heart valves. They showed decreased thrombogenicity of an amorphous layer of SiC compared to titanium. Several other studies showed that a hydrogen-rich amorphous SiC coating on coronary artery stents is anti-thrombogenic. Three studies showed a benefit that was attributed to the SiC-coated stent. In a direct comparison of the blood compatibility silicon wafers and SiC-coated (PECVD) silicon wafers, both appeared to provoke clot formation to a greater extent than diamond-like coated silicon wafers; silicon was worse than SiC-coated silicon. In conclusion, the hemocompatibility of SiC was demonstrated [5].

5. Microfabrication techniques

5.1. Material selection

MEMS or microelectromechanical systems, is the integration of mechanical elements, sensors, actuators, and electronics on a substrate, in which micro-fabrication technologies are used. The first step when designing a process flow for fabrication of a micromachined device is to choose the structural and other materials which are to be used in the process flow. Some of the properties of materials that are commonly used in MEMS fabrication are listed in Table 2. Traditionally, silicon and polysilicon have been used most often as the structural materials for MEMS. This was initially due to the wealth of existing knowledge on processing of silicon samples from microelectronic fabrication. Luckily for the micromachining engineers, silicon has several favourable mechanical properties in addition to its superb electrical specifications that have made it the material of choice for microelectronics. Nevertheless, there are numerous cases where other materials offer significant advantages over silicon. Examples include applications where a piezoelectric material is needed or when the devices are designed to work in harsh environments.

Materials	T_m (°C)	E (GPa)	σ_y (GPa)	v	ρ (Kg/m ^{3})	H (GPa)
Bulk silicon	1,415	160-200	-	0.22	2,330	5-13
Polysilicon	1,415	181-203	-	-	-	10-13
Silicon dioxide	1,700	70-75	8.4	0.17	2,200	15-18
Pyrex glass	-	64	-	0.2	2,230	8
Silicon nitride	1,800	210-380	14	0.25	3,100	24-27
Silicon carbide	-	300-400	21	0.19	3,210	-
CVD-diamond	661	800-1,100	0.2	0.07	3,530	-
Aluminum	1,772	70	0.137-0.170	0.33	2,700	-
Platinum	1,065	170	0.120	0.38	21,440	-
Gold	-	80	2.1	0.38-0.42	19,280	6.5
Stainless steal	-	200	0.045-0.345	0.3	7,900	-
Polyimide	410	7.5-15	0.042	0.35-0.45	1,420	-
Parylene-N	290	2.4	0.055	-	1,100	-
Parylene-C	380	2.7	0.062	-	1,290	-
Parylene-D	-	2.6	0.034	-	1,418	-
Hard backed SU-8		4-5		0.22	-	

Table 2. Mechanical Properties of Typical Functional Materials (T_m: Melting Temperature, E: Young's Modulus, σ_y: Yield Strength, v: Poisson's Ratio, ρ: Density, H: Knoop Hardness) [9]

SiC has also been used for the coating of other MEMS devices for increased wear resistance. On the other hand, the same advantages of SiC over silicon bring up challenges in deposition and etching of SiC films. The fabrication technologies differ depending on what needs to be accomplished. The aim is to incorporate electronics; they are normally fabricated using standard integrated circuit processes or sequences, such as Complementary Metal-Oxide Semiconductor (CMOS) technology or bipolar technology. On the mechanical side, the micromechanical components are fabricated using compatible micromachining processes that in essence selectively etch away parts of the silicon wafer or add new structural layers to form a mechanical device, or an electromechanical device if it has additional integrated electronics.

There are several types of methods used for micromechanical fabrications. Two of the common techniques are MUMPS, which applies to a poly-silicon and stands for Multi-User MEMS Processing System. A similar process called MUSIC is used for silicon carbide, another common substrate material used in MEMS. Through MEMS, it is possible to incorporate micro-scale types of devices such as motors, pumps, fluidic channels, sample preparation (including mixing or vaporization) chambers, and various types of sensors (including optical sensors) that will perform assorted tasks, such as monitoring. Monitoring can be performed in both the physical and chemical sense.

Therefore, MEMS can be thought of as an enabling technology that allows for the development of what we commonly call smart or intelligent systems, which operate without the need for external computing resources. This integrated microelectronics can process the information derived from the sensors and through some decision-making process direct actuators to respond by moving, positioning, regulating, pumping, and/or filtering, thereby controlling the environment for some desired outcome or purpose. Because MEMS devices are manufactured using batch fabrication techniques similar to those used over multiple decades in the integrated circuit industry, we see unprecedented levels of functionality, reliability, and sophistication being placed on small silicon chips at a relatively low cost. Therefore, there is significant potential for MEMS technologies.

Although crystalline SiC is a polymorphic material that exists in well over 100 distinct polytypes, only the cubic 3C-SiC, and the hexagonal 4H-SiC and 6H-SiC polytypes are technologically relevant for MEMS applications since they are the only configurations that can be produced as high-quality substrates and/or thin epitaxial films. At present, 6H-SiC and 4H-SiC are the only polytypes that are commercially available in large-area, integrated circuit (IC)-grade wafer form suitable for epitaxial growth of single crystalline films. In contrast, 3C-SiC is not widely available as bulk substrates, but single crystalline films can be epitaxially grown directly on Si wafers despite a significant mismatch in both lattice constant and thermal coefficient of expansion.

Single crystalline 3C-SiC piezoresistive pressure sensors have been fabricated using bulk micromachining for high temperature gas turbine applications. Bare silicon exhibits inadequate tribological performance. It needs to be coated with a solid and/or liquid overcoat or be surface treated (e.g., oxidation and ion implantation, commonly used in semiconductor manufacturing), which exhibits lower friction and wear. SiC films exhibit good tribological performance. Studies have been conducted on undoped polysilicon film, heavily doped (n+-type) polysilicon film, heavily doped (p+-type) single-crystal Si (100) and 3C-SiC (cubic or b-SiC) film [10].

5.2. Surface micromachining

Surface micromachining involves the monolithic fabrication of suspended microscale structures by selective removal of underlying thin film sacrificial layers. Surface micromachining is inherently an additive process utilizing thin film deposition techniques to produce both structural and sacrificial layers. As such, the primary function of the substrate is to provide mechanical support for the resulting device. Patterning of the thin film layers involves wet and dry etching techniques that are sensitive only to the chemical properties of the materials, and not their microstructure or crystallinity, thereby enabling a high degree of flexibility with respect to planar designs. In concept, there is no restriction on the structural and sacrificial materials to be used in the fabrication of a particular device as long as the materials are compatible with each other during the fabrication process. As such, surface micromachining is not constrained by the properties of the substrate and, thus, can accommodate an extremely wide range of materials, including SiC.

Like silicon, SiC thin films can be deposited by chemical vapor deposition (CVD), making it particularly well adapted as a "plug-and-play" substitute for polysilicon in surface micromachining. SiC films can be deposited by low pressure chemical vapor deposition (LPCVD), atmospheric pressure chemical vapor deposition (APCVD), and plasma-enhanced chemical vapor deposition (PECVD). Even for non- MEMS applications, CVD is the by far most common method to deposit SiC due at least in part to the availability of precursor gases as well as numerous process and equipment similarities to silicon CVD. The silicon carbide analog to polysilicon is polycrystalline 3C-SiC, hereafter referred to simply as poly-SiC. Poly-SiC is actually more versatile than polysilicon in that it can be deposited directly on SiO_2 and polysilicon. In essence, the process of fabricating MEMS structures in poly-SiC by surface micromachining mirrors that of polysilicon. The main differences are process used to deposit the SiC films, the selection of sacrificial layer material, and the etch recipes used to pattern the structural films. A significant breakthrough in the advancement of SiC surface micromachining was the development of reactive ion etching techniques that are highly selective to SiC, which when combined with MEMS-friendly SiC deposition techniques, allow SiC surface micromachining to follow directly from polysilicon micromachining.

A wide range of micromachined structures, such as lateral resonators, flow sensors, capacitive pressure sensors, micromotors, and microbridge resonators can be fabricated using the deposition, patterning, etching, and sacrificial release techniques commonly used in polysilicon surface micromachining [11]. Several groups have demonstrated surface micromachining using a-SiC films as structural layers. Examples include RF switches and accelerometers [12]. Amorphous-SiC films deposited by PECVD generally exhibit a very wide range of residual stress (typically compressive in as-deposited films) that exhibit a strong dependence on deposition conditions [13].

5.3. Bulk micromachining

Bulk micromachining can generally be defined as a process to fabricate suspended structures by selective bulk removal of the supporting substrate. Bulk micromachined structures can be comprised of the substrate material itself or thin films that are deposited directly onto the substrate. Unlike surface micromachining, the substrate in bulk micromachined devices is not merely a solid mechanical support, but rather forms a key component of the device structure.

In Si MEMS, direct wafer bonding has proven to be a key enabling process in the production of silicon-on-insulator (SOI) wafers for the realization of single crystalline Si MEMS devices. To create a SiC-on-insulator substrate, the Si wafer that was originally used as the substrate for SiC growth is removed by etching. The principal factor affecting yield is wafer bowing due to high tensile residual stress in the 3C-SiC films. The techniques developed for Si bulk micromachining, including photoelectrochemical etching, DRIE, and laser micromachining, have been successfully adapted for SiC albeit typically with much lower etch rates [13]. Among the first SiC MEMS structures to be routinely fabricated were diaphragms, cantilever beams, and related structures fabricated out of single crystalline 3C-SiC films by silicon anisotropic etching. Although conventional wet chemical techniques are not effective in etching structures into SiC substrates, several electrochemical etch processes have been demonstrated and used in the fabrication of bulk micromachined SiC MEMS devices from 6H- and 4H-SiC substrates. Examples of such structures include pressure sensors [14], accelerometers [15], and more recently, biosensors [16]. It is worth mentioning that, the good biocompatibility of devices made with common micromachining technologies allows the exploration of these technologies.

5.4. Encapsulation and protection

In addition to using capsules or shells for providing a hermetic or vacuum package for MEMS, it is increasingly attractive to use thin films to provide the necessary protection or encapsulation [17-19] (Fig.2). Thin films are attractive because they occupy a very small area, can be formed using a variety of techniques, and are compatible with wafer-level processing. In addition, they can take any shape or form. However, most thin film materials are either not hermetic, or are so thin that they can be compromised easily when exposed to the environmental conditions MEMS typically experience. Two categories of thin film materials can be identified: organic and inorganic materials. Organic materials include such films as epoxies, silicones, a variety of polymers including polyimides, polyurethanes, Parylene-C, etc. The majority of these films can be deposited at low temperatures, are quite conformal and their characteristics can be modified for different applications. However, most of these films are not hermetic and most are prone to moisture penetration, or can be attacked in harsh environments. In spite of this, these materials have found widespread use because they can be selectively used in applications which may not require very long-term operation, or where the conditions are controlled, or where the performance specifications are not highly restrictive. In fact, polymers are perhaps the most widely used material for packaging, albeit not hermetic or vacuum packaging. The second category of materials used for packaging and protection is in-organic materials. These materials include films such as silicon nitride, silicon carbide, polycrystalline diamond, metal thin films, tantalum oxide, or thin films of other materials that are resistant to environmental parameters. Semiconductor materials such as silicon or silicon carbide are quite attractive because they can be deposited readily and are resistant to many corrosive environments. The main challenge in using these materials is that they typically require a high temperature to achieve a reasonable deposition rate, and in some instances the films are not quite as conformal as required by some applications. Therefore, they have not been widely used for hermetic packaging, especially where hybrid components are involved. The discussion is limited to inorganic thin films since it is not possible to discuss the broad category of organic materials that are used nowadays in the encapsulation and packaging of microdevices [20].

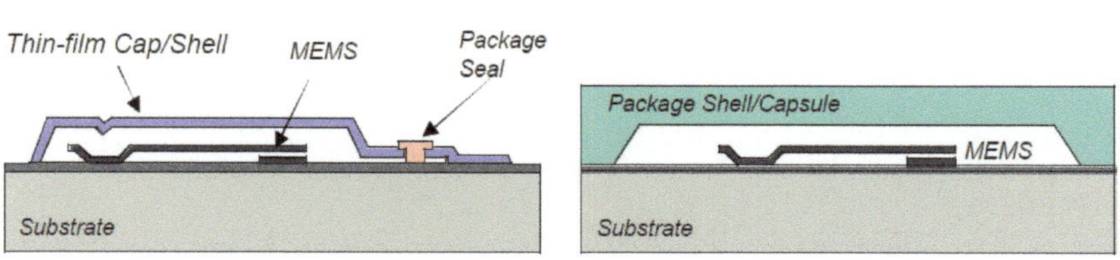

Figure 2. Structure of a MEMS package formed using a thin film capsule or shell, compared with a package formed using a bonded capsule or shell [20].

6. Implantable BioMEMS

To date, the majority of the development efforts in the MEMS field have focused on sophisticated devices to meet the requirements of industrial applications. However, MEMS devices for medical applications represent a potential multi-billion dollar market, primarily consisting of micro miniature devices with high functionality that are suitable for implantation. These implanted systems could revolutionize medical diagnostics and treatment modalities. Implantable muscle microstimulators for disabled individuals have already been developed. Precision sensors combined with integrated processing and telemetry circuitry can remotely monitor any number of physical or chemical parameters within the human body and thereby allow evaluation of an individual's medical condition.

MEMS processing technology is also being used to fashion functionally simple passive microdevices like retinal implants [21, 22]. In the future, in order to improve functionality and reduce size, ever increasing numbers of MEMS devices will have direct patient contact thus requiring that biocompatibility testing be performed on MEMS materials of construction. Kotzar et al. selected the following materials as MEMS materials of construction for implantable medical devices: (1) single crystal silicon (Si), (2) polycrystalline silicon (polysilicon), (3) silicon oxide (SiO_2), (4) silicon nitride (Si_3N_4), (5) single crystal cubic silicon carbide (3C-SiC or b-SiC), (6) titanium (Ti), and (7) SU-8 epoxy photoresist. Of these, polysilicon, Si_3N_4, and 3C-SiC were deposited by chemical vapor deposition (CVD), SiO_2 by thermal oxidation of Si, Ti by physical vapor deposition (PVD), and SU-8 by spin coating [23]. Many of the earlier studies examining the biocompatibility of SiC employed materials generated by fabrication methods suited to other implantable applications, such as RF sputtering. The results of the biocompatibility tests for these fabrication methods may not apply to materials fabricated using MEMS technology.

The Kotzar et al. study [6] results for SiC show that when the material is generated using MEMS fabrication techniques, it elicited no significant non-biocompatible responses to the test battery employed in this series. The biocompatibility testing discussed by Kotzar et al. did not uncover any MEMS material that was not biocompatible when subject to the processing, packaging, and sterilization methods. Amorphous-SiC (a-SiC) has the longest and most diverse track record of any SiC microstructure used in biomedical microdevices. Early work using a-SiC for medical applications focused on developing a-SiC films as corrosion-resistant coatings for macroscale structures, such as Ti alloy-based orthopedic implants and metallic coronary stents [24], thus setting the stage for its use in biomedical microsystems.

7. Biosensing

7.1. Biosensors

Most of the studies conducted in the past on Single-crystal SiC provide evidence of the attractive bio potentialities of this material and hence suggest similar properties for crystalline

SiC. The properties that make this material particularly promising for biosensing applications are: 1) the wide bandgap that, as mentioned before, increases the sensing capabilities of a semiconductor; 2) the chemical inertness that suggests the material's resistance to corrosion in harsh environments such as body fluids (e.g. SiC does not react with any known material at room temperature, the only efficient etch being molten KOH at 400-600 ºC); 3) the high hardness (5.8 GPa), high elastic modulus (424 GPa), and low friction coefficient (0.17) that make it an ideal material for smart-implants and in-vivo biosensors. Studies report the significant finding that SiC surfaces are better substrates for mammalian cell culture than Si in terms of both cell adhesion and proliferation. In the past, the fact that cells could be directly cultured on Si crystalline substrates led to a widespread use of these materials for biosensing applications [25-27].

Singh and Buchanan [28] studied silicon carbide carbon (SiC-C) composite fiber as an electrode material for neuronal activity sensing and for biochemical detection of electroactive neurotransmitters. The SiC-C electrode surface has nanosized pores which significantly increase the real surface area for higher charge densities for a given geometrical area. Neurotransmitters including dopamine and vitamin C were successfully detected using SiC-C composite electrodes. Researchers fabricated impedance and temperature sensors on bulk SiC for a biomedical needle that can be used for open heart surgery monitoring or graft monitoring of organs during transportation and transplantation. According to Godignon [29], other applications can be foreseen, such as DNA polymerase chain reaction (PCR), electrophoresis chips and cell culture micro-arrays. In DNA electrophoresis devices, the high critical electric field and high resistivity of semi-insulating SiC would be beneficial. In DNA PCR, it is the high thermal conductivity which could improve the device's behaviour. In addition, in most of these cases, the transparency of semi-insulating SiC can be used for optical monitoring of biological processes, such as the DNA reaction or the cell culture activity. Ghavami et al. used Field emission scanning electron microscopy (FE-SEM) and transmission electron microscopy (TEM) techniques to examine the structure of the SiCNP/GC modified electrode. The modified electrode shows excellent electrocatalytic activity toward guanine, adenine, thymine and cytosine. Differential pulse voltammetry (DPV) was proposed for simultaneous determination of four DNA bases. The effects of different parameters such as the thickness of the SiC layer, pulse amplitude, scan rate, supporting electrolyte composition and pH were optimized to obtain the best peak potential separation and higher sensitivity. The modified electrode can be used for simultaneous detection of purine and pyrimidine bases without any separation or pretreatment processes and may be used as a DNA biosensor in real samples [30].Caputo et al. [31] reported on biomolecule detection based on a two-color amorphous silicon photosensor. The device design has been optimized in order to maximize the spectral match between the sensor responses and the emission spectra of the fluorochromes. This optimization process has been carried out by means of a numerical device simulator, taking into account the optical and the electrical properties of the amorphous silicon materials. The development of minimally invasive and short-term implantable devices for on-line tissue monitoring is a field of increasing clinical and industrial interest, with applications in areas such as open-heart surgery or insulin control. Multisensory micro-needles have been developed to measure physiologically relevant intratissular parameters during cold transportation of transplantation organs. Si-based minimally invasive probes have already gone beyond the proof-of-concept stage and are currently undergoing Phase-I clinical trials [32]. Gabriel et al. [33] examine the feasibility of using SiC as a substrate for the development of minimally invasive multi-sensor micro-probes in the context of organ monitoring during transplantation. In particular, they make a thorough comparison of Si and SiC material mechanical and electrical properties. As illustrated in Figure 3, Si and semi-insulating SiC micro-needles for impedance and temperature measurement were fabricated using remarkably similar methods.

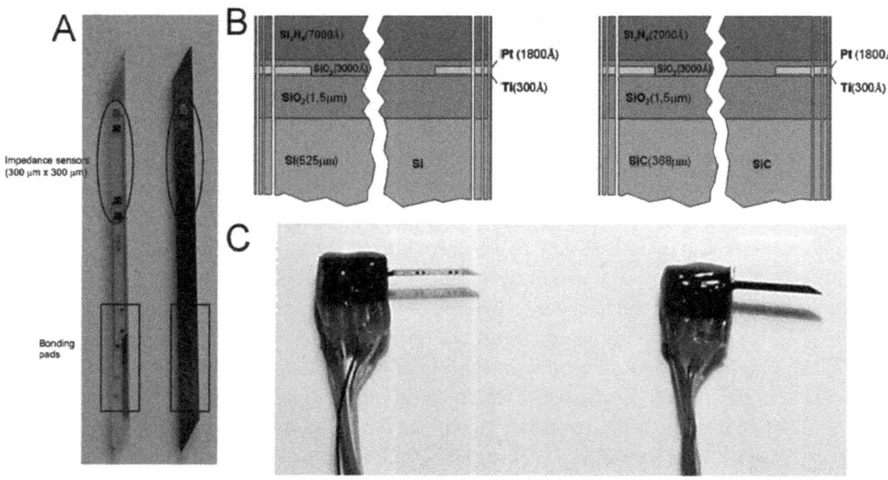

Figure 3. A) Needle-shaped Si (left) and SiC (right) probes for impedance and temperature monitoring. (B) Schematic drawing of the technological process for Si (left) and SiC (right) probe production. (C) Encapsulated SiC (left) and Si (right) devices [33].

Their results show that SiC outperforms Si in all respects, with a four times higher modulus of rupture for SiC devices and a 10-fold increase in the frequency range for electrical measurements in SiC-based probes. These results suggest that SiC should be preferably used over Si in all biomedical applications in which device breakage must be avoided or very precise electrical measurements are required [33]. 3C-SiC has a distinct advantage over the other SiC polytypes in that simple micromachining technique can be used to fabricate nanoscale 3C-SiC structures. In fact, the first SiC NEMS were demonstrated in 3C-SiC due to the ability to grow ultrathin 3C-SiC films on Si substrates combined with selective reactive ion etching processes that enable patterning and release of the nanostructures with essentially the same plasma [34]. Other groups have used similar techniques to create 3C-SiC and AlN NEMS structures [35]. Figure 4 shows SEM images of 3C-SiC NEMS microbridges. Naik et al. extended this work toward biosensing applications by demonstrating that 3C-SiC NEMS resonators were able to detect individual protein adsorption events by observing frequency shifts for each exposure event when the resonators were selectively exposed to bovine serum albumin (BSA) and β-amylase (200 kDa) [36].

Figure 4. Scanning electron micrographs of 3C-SiC NEMS microbridges [35].

Silicon carbide for chemical sensing devices has been demonstrated to be the best candidate for high temperature chemical gas sensors. The wide bandgap, combined with chemical inertness, result in SiC being the best material for gas sensing in harsh environments or at high temperatures. Thin film silicon carbide exhibits thermal conductivity on the same order of single crystalline silicon and has a fast thermal response [13]. MEMS-based microprobes for neural interfacing and biosensing applications are currently the subject of intense research due to the promise of achieving high functionality in a minimally invasive form-factor. A typical planar neural probe, for example, consists of a thin shank that supports multiple, thin film metallic electrodes. A common material for the shank is Si; however, concerns over the electrical performance, mechanical robustness and biological interactions of these structures currently limit their applicability in long-term deployment situations.

For neurostimulation applications, the electrode material must provide charge transfer under low interface impedance to avoid tissue trauma. Surface morphology plays an important role in the electrode's charge carrying capabilities. The real surface area of an electrode can be significantly different from its geometrical area. Higher real surface areas can deliver higher charge densities for a smaller electrode. For a microelectrode, size is a limiting factor, so there should be ways to increase real surface area. Real surface areas can be modified by treating the electrodes electrolytically. The SiC-C electrode surface as seen in the SEM images (Fig. 5) has nanosized pores which significantly increase the real surface area for higher charge densities for a given geometrical area. Thus high real surface area electrodes are highly desirable and were observed in electrolytically modified SiC-C electrode surfaces. Neurotransmitters including dopamine and vitamin C were successfully detected using SiC-C composite electrodes [37].

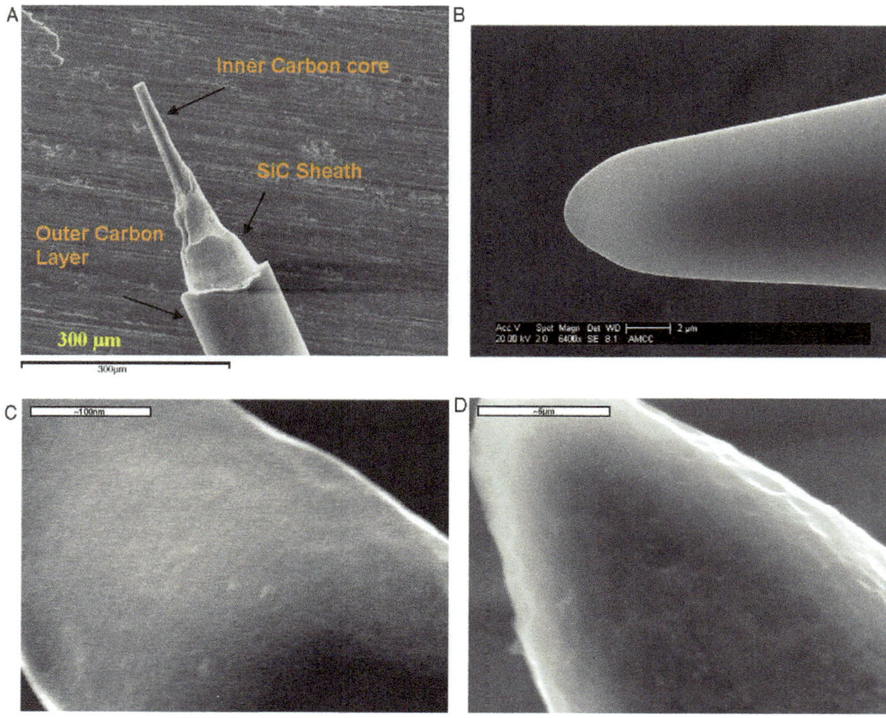

Figure 5. SEM pictures of (A) silicon carbide carbon composite fiber's etched section, (B) carbon recording tip, (C) low voltage etch (smooth surface), and (D) high voltage etch (high surface area surface) [37]

In many clinical settings, a decrease of the blood supply to body organs or tissues can have fatal consequences if it is not properly addressed promptly (e.g. mesenteric or myocardial ischemia). Sustained ischemia leads to hypoxia, a stressful condition for cells that can induce cell lysis (necrosis) and trigger programmed cell death (apoptosis) and, consequently, lead to organ failure. Aside from ischemic diseases, ischemia underlies other natural and clinically induced conditions, such as tumor growth, cold-preservation of grafts for transplantation or induced heart-arrest during open heart surgery. Therefore, the ability to monitor ischemia in clinical and experimental settings is becoming increasingly necessary in order to predict its irreversibility (e.g. in the transplantation setting), to develop drugs to prevent and revert its effects, and to develop vascular-targeting drugs for the treatment of growing tumors. To address these issues and to extend the utility of MEMS-based probe technology, a minimally invasive system for the continuous and simultaneous monitoring of tissue impedance has been developed, and experimental results have shown its reliability for early ischemia detection and accurate measurement of ischemic effects. This minimally invasive system consists of a small micro-machined silicon needle with deposited platinum electrodes for impedance measurement that can be inserted in biological tissues with minimal damage. High frequency impedance monitoring, based on both the phase and modulus components of impedance, has been correlated to the combined dielectric properties of the extracellular and intracellular compartments and insulating cell membranes, and can give complementary information on other effects of sustained ischemia. Moreover, multi-frequency monitoring of impedance has the advantage of yielding to more comprehensive empirical mathematical characterizations (i.e. the Cole model) that can provide additional information through the analysis of derived parameters and improve the reproducibility of results [5]. Gomez et al. [38] examined the feasibility of producing SiC-based needle-shaped impedance probes for continuous monitoring of impedance and temperature in living tissues. SiC needle-shaped impedance probes (Fig.6) were produced in standard clean room conditions.

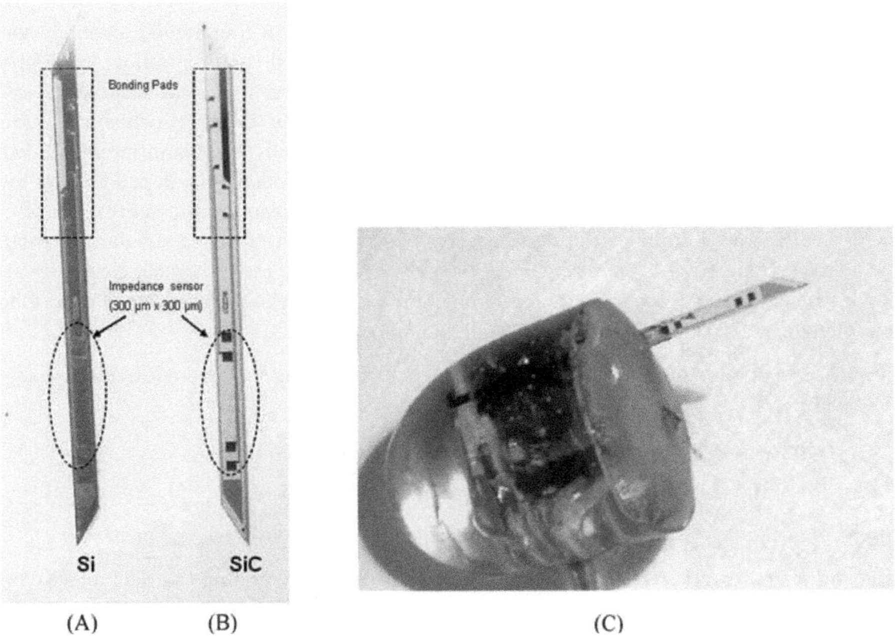

Figure 6. A) Needle-shaped Si probe for impedance; (B) Needle-shaped SiC probe for impedance; (C) Needle-shaped with packaging [38]

In vitro results obtained with SiC based impedance probes were compared with those obtained with Si-based probes, and they demonstrated that the use of SiC substrates was mandatory to extend the effective operation range of impedance probes beyond the 1 kHz range. In-vivo evaluation of SiC-based impedance probes was conducted on rat kidneys undergoing warm ischemia by dissecting and clamping the renal pedicles. A substantial rise in impedance modulus was shown throughout the ischemic period (5 to 50 min); this increase can be attributed to the occurrence of hypoxic edema as the result of cell swelling, which leads to a reduction of extracellular space, an increase in extracellular resistance, and cell-to-cell uncoupling. Upon unclamping of the renal artery (50 min), impedance modulus can be seen to return to its basal value, a fact that can be attributed in this experimental setting to a reversion from a short period of ischemia without substantial structural damage to the tissue. A fall in impedance modulus at low frequencies, however, has also been reported as a consequence of membrane breakdown and cell lysis due to sustained ischemia. It is in this respect that the multifrequency analysis of the phase component of impedance made possible by the use of SiC-based probes conveys useful complementary information [5]. Therefore, according to this set of materials, one can conclude that SiC would be considered a good candidate for biosensing applications.

7.2. Microelectrode arrays

The emerging field of monitoring biological signals generated during nerve excitation, synaptic transmission, quantal release of molecules and cell-to-cell communication, stimulates the development of new methodologies and materials for novel applications of bio-devices in basic science, laboratory analysis and therapeutic treatments. The electrochemical gradient results in a membrane potential that can be measured directly with an intracellular electrode. Extracellular signals are smaller than transmembrane potentials, depending on the distance of the signal source to the electrode. Over the last 30 years, non-invasive extracellular recording from multiple electrodes has developed into a widely-used standard method. A microelectrode array is an arrangement of several (typically more than 60) electrodes allowing the targeting of several sites for stimulation and extracellular recording at once. One can plan the realisation of four activities with the following tasks:

Task 1. Development of new biocompatible substrates favouring neuronal growth along specific pathways.

Task 2. Monitoring of electrical activity from neuronal networks.

Task 3. Resolution of cellular excitability over membrane micro areas.

Task 4. Detection of quantal released molecules by means of newly designed biosensors.

Task number 1 can be realized by means of SiC substrates, by plating the cells directly on the substrate or eventually with an additional proteic layer. For this aim, 3C-SiC films with controlled stoichiometry, different thickness and crystalline quality can be grown directly on silicon substrates or on silicon substrates previously 'carbonised'. The main objective of task number 2 is the realization of SiC microelectrode arrays whose dimensions will be compatible with the cellular soma (10-20 μm). In this structure, every element of the array is constituted by a doped 3C-SiC region, with metallic interconnections coated with amorphous silicon carbide, so that silicon carbide represents the only material interfaced to the biological environment. For the realization of Task number 3, the SiC array will be improved by constructing microelectrodes in the submicrometric range, in order to reveal electrical signals from different areas of the same cell. The objective of task number 4 is the construction of a prototype of SiC-electrodes array as a chemical detector for oxidizable molecules released during cell activities triggered by chemical substances (KCl or acetylcholine) on chromaffin cells of the adrenal gland. With respect to classical electrochemical methods requiring polarized carbon fibers with rough dimensions of 10 micrometers in diameter, the SiC multielectrode array should greatly improve the monitoring of secretory vesicle fusion to the plasma-membrane, allowing the spatial localization and temporal resolution of the event [5].

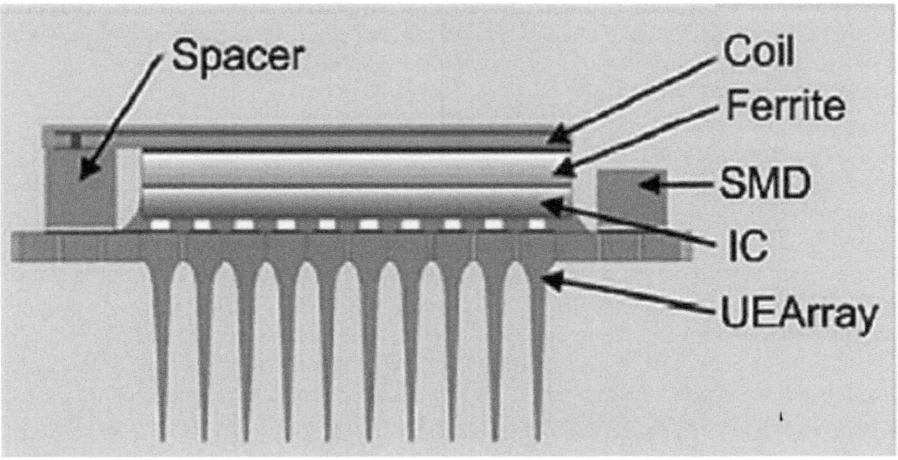

Figure 7. Schematic diagram of the Utah Electrode Array [41]

To demonstrate the feasibility of developing SiC-based devices on large-area SiC substrates, Godignon et al. recently reported the development of 4H-SiC-based microelectrode arrays on semi-insulating 4-in. diameter hexagonal substrates [39]. The microelectrode arrays consisted of a SiO_2/Si_3N_4 passivation layer sandwiched between microfabricated Pt electrodes and the SiC substrate. Carbon nanotubes (CNTs) were easily grown by rapid thermal chemical vapor deposition on the Pt electrodes using CH4 and H2 at 800 °C, conditions for which the SiC substrate is well suited. The detection of biological species using microarrays and lab on-a-chip systems is a powerful diagnostic tool that enables the acquisition of genetic, proteomic, and cellular information. Such approaches allow rapid analysis of disease diagnostics, drug discovery, or food and environmental analysis. In microarray applications, each pixel in the array is functionalised with well-defined probe molecules and a molecular recognition reaction occurs between the probe and the target molecules to be detected [40]. Hsu et al. [41] has recently developed a-SiC as a protective coating for MEMS-based Si penetrating microelectrodes. The technology driver for this research is an integrated, Si-based microneedle electrode array known as the Utah Electrode Array (UEA). Shown schematically in Figure 7, the UEA is a three-dimensional structure consisting of a 10 × 10 array of tapered silicon shanks that are bulk micromachined into a Si substrate. Each shank supports Ti/Pt/Ir electrodes that provide the electrical interface to nerve tissue. The array is integrated with Si-based IC's and packaged using a variety of conventional approaches that are adapted for this particular device. Hsu et al. have developed PECVD-based a-SiC:H coatings to encapsulate components of the UEA.

8. Biofiltration

Absorption filters based on hydrogenated amorphous silicon–carbon alloys are developed for application as fluorescence filters in microarray and lab-on-a-chip systems. Fluorescence is one of the most commonly used methods for the detection of proteins, cells and DNA in microarrays [42, 43]. The method uses an external light source, which excites the fluorophores that label the entities of interest. The integration of a fluorophore sensor at each pixel of a microarray would allow for a rapid and real-time detection of the biological recognition event, while potentially increasing sensitivity and portability, and reducing costs. Fluorescence detection, however, requires the development of efficient light management to prevent the excitation light from reaching the detector and at the same time allow the emission light to be transmitted to the detector. Lipovesk et al. [40] proposed a layer of hydrogenated amorphous silicon carbide (a-SiC:H) as a suitable optical filter which can be easily integrated with a-Si:H photosensors for on-chip detection of biomolecules labelled with Alexa Fluor 430 or PyMPO. Simple fabrication of absorbing a-SiC:H filters (single-layer, low cost, with low dependence on the incidence angle) presents an important advantage compared to other filtering solutions, such as interference filters [44] where a large number of layers need to be tuned accurately during deposition. The challenge is to optimise the filtering characteristics of the a-SiC:H filter in order to match the excitation/emission wavelength fingerprint of any selected fluorophore. Therefore, a-SiC:H filters of different carbon content resulting in appropriate optical properties have to be designed. The relation between the carbon content in the a-SiC:H film and its optical filtering characteristics must be determined to enable the fabrication of the optimal filter for detection of a selected fluorophore.

The filtering characteristics of all filters are compared to the excitation/emission properties of numerous biologically relevant natural fluorophores. For each fluorophore, the most suitable a- SiC:H filter is necessary in order to ensure the highest rejection ratio between the transmitted emission/excitation light, thus achieving the optimal sensitivity of the fluorescence measuring system. As a proof-of-concept, one of the filters is tested to demonstrate the detection of the reduced form of nicotinamide adenine dinucleotide (NADH), an enzyme cofactor and an important marker for cell metabolic activity [45]. MEMS devices fabricated from bulk 6H-SiC wafers leverage heavily on the outstanding mechanical durability of the substrate, the chemical inertness of the 6H-SiC surface, the commercial availability of high-quality wafer substrates and the versatile micromachining techniques available to render devices from them. As such, bulk SiC substrates offer nearly all the advantages of bulk Si substrates but with more robust properties. Like Si, porous SiC structures can be formed from 6H-SiC wafers by electrochemical etching. Under the proper conditions, the pore size and porosity of the resulting structures offer the potential to use porous SiC in biofiltration applications. For biofilteration applications, the porous material was formed by bulk electrochemical etching of p-type and n-type, 6H-SiCsubstrates [13]. Rosenbloom et al. reported the development of porous SiC membranes for use as protein filters [46]. The performance of the porous SiC membranes was evaluated using protein-containing solutions with proteins ranging in molecular weight from 17,000 to 80,000 Daltons (Da). It was found that the membranes were able to pass proteins with molecular weights of up to 29 kDa and were able to exclude proteins in excess of 45 kDa. Moreover, the porous SiC membranes exhibited lower protein absorption as compared to commercially available polymer-based membranes specifically designed for protein absorption, indicating the potential for SiC membranes in biofilteration applications [13].

Micromachined 3C-SiC membranes also provide excellent specimens to study the mechanical durability of 3C-SiC films since the adhesion of the film to the micromachined substrate is extremely high, due in part to the carbonization based growth process. Suspended 3C-SiC membranes have proven to be an attractive mechanical structure for micromachined pressure sensors owing to their chemical inertness, mechanical durability and high temperature stability. In comparison with Si membranes, SiC membranes are much easier to fabricate [2].

9. Biomedical imaging

In addition to its outstanding mechanical properties, 3C-SiC thin films have unique electro-optic properties that can be exploited for biomedical imaging. In conventional optical microscopy, diffraction effects limit the spatial resolution to one-half of the illuminating wavelength. A technique known as near-field scanning optical microscopy (NSOM) has recently emerged as a means to resolve images below the diffraction limit. NSOM utilizes a detector that is placed in very close proximity to a sample in order to detect evanescent fields associated with the surface of the specimen. Such characteristics make NSOM a potentially powerful tool for biological imaging. However, a drawback of NSOM is that the AFM-based probe significantly limits the ability to image biological specimens in fluidic environments [13, 47].

10. Microfluidics/Lab-on-a-Chip

An example of a wristwatch type biosensor based on microfluidics referred to as lab-on-a-chip system is shown in Fig. 8. The advantages of these systems are incorporating sample handling, separation, detection, and data analysis onto one platform. The chip relies on microfluidics and involves manipulation of tiny amounts of fluids in microchannels using microvalves. The test fluid is injected into the chip generally using an external pump or syringe. Some chips have been designed with an integrated electrostatically-actuated diaphragm type micropump. The sample, which can have volume measured in nanoliters, flows through microfluidic channels via an electric potential and capillary action using microvalves (having various designs including membrane type) for various analyses. The fluid is preprocessed and then analyzed using a biosensor.

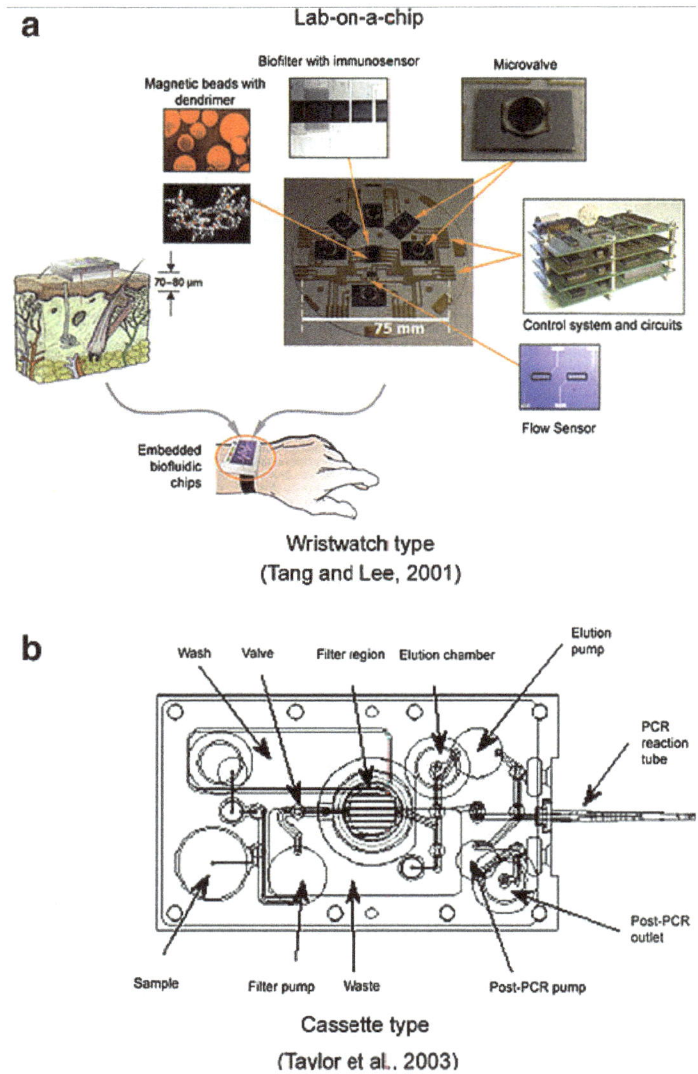

Figure 8. MEMS based biofluidic chip, commonly known as a lab-on-a chip that can be worn like a wristwatch [1].

The implementation of micropumps and microvalves allows for fluid manipulation and multiple sample processing steps in a single cassette. The three basic components of a mechanic valve are the actuator, the valve spring and the valve seat. The spring force keeps the valve shut in normally closed valves. In the case of normally open valves, the spring keeps the valve open and works against the actuator. A small spring constant can be realized with a soft material such as rubber. The solution with soft materials offers a further advantage of excellent sealing characteristics. The leakage ratio can be improved from three to four orders of magnitude compared to those made of hard material such as silicon, glass, or silicon nitride. If the valve is designed for bistable operation, there is no need for a valve spring because the two valve states are controlled actively. Since the non-powered state is undefined, a valve spring can still be considered for the initial, nonpowered state to assure safe operation. A bistable valve spring allows the valve seat to snap into its working position. In this case, the actuator needs to be powered in a short period to have enough force to trigger the position change. The force generated by the spring is then high enough to seal the valve inlet [48].

Valve seats represent a large challenge to microvalve design and fabrication. The valve seat should satisfy two requirements: low leakage and high resistance against particulate contamination. For a minimum leakage rate, the valve should be designed with a large sealing area, which must be extremely flat. Softer materials such as rubber or other elastomers are recommended for the valve seat. Resistance against particles can be realized in many ways. First, a hard valve seat can simply crush the particles. For this purpose, the valve requires actuators with large force such as piezostacks and hard coating layer for the valve seat. Second, the particles can be surrounded and sealed by a soft coating on the valve seat. Third, small particle traps such as holes or trenches can be fabricated on the valve seat or on the opposite valve base. A combination of the third measure with the first and second measures is recommended, so that tiny particles can be trapped and buried after being crushed by the large actuation force. For the first and third measures, the valve seat needs to be coated with hard, wear-resistant material such as silicon nitride, silicon carbide, or diamond [9].

Blood or other aqueous solutions can be pumped into the system where various processes are performed. If the adhesion between the microchannel surface and the biofluid is high, the biomolecules will stick to the microchannel surface and restrict flow. In order to facilitate flow, the microchannel surfaces with low bioadhesion are required. Fluid flow in polymer channels can produce triboelectric surface potential which may affect the flow. Polymers are known to generate surface potential and the magnitude of the potential varies from one polymer to another [1]. Conductive surface layers such as SiC can be deposited on the polymer channels to reduce triboelectric effects. Compared with its crystalline counterparts, amorphous-SiC is particularly attractive for microfluidics and related lab-on-a-chip applications because it can be deposited on a much wider range of substrate materials while retaining a high level of chemical inertness.

These features are the way for on-line monitoring of several processes in many application fields. A large new area is that of micrototal analysis systems (μTAS) or lab-on-a chip, where attempts are made to completely integrate biochemical systems on one silicon or glass chip. A currently emerging development is that of miniaturized integrated physical chemosensors and biosensors. Rather than applying a chemical interface to determine the biochemical properties of fluids, physical properties or phenomena in the fluid are used. Persisting problems with such a chemically sensitive interface layer, such as poor reproducibility, drift, ageing and contamination, are circumvented in this approach. All these devices require new technological approaches for the fabrication of small channels, novel integrated microdetectors, and other components. Most compounds used for biochemical analysis do not possess a fluorescent functionality, and thus labeling with a fluorescent marker is required. Recently, the on-chip integration of electrochemical and conductivity detection has been reported [49]. A schematical drawing of the chip is shown in Fig. 9.

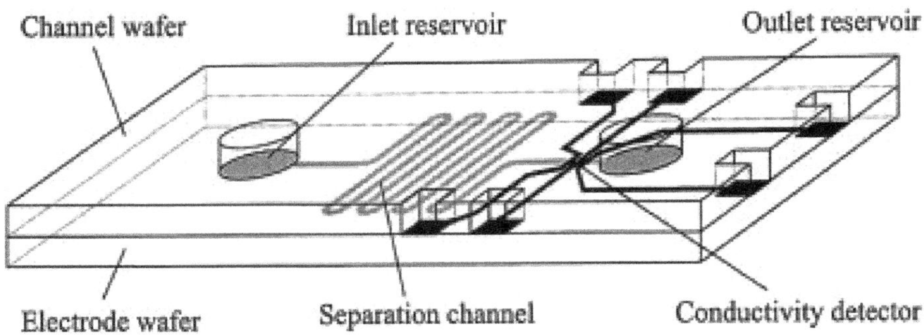

Figure 9. Drawing of a microchemical detector containing electrophoresis separation and conductivity detection [49]

Conductivity detection can be used for on-chip measurements. To avoid electrolysis and electrode fouling when the solution was in contact with the measurement electrodes, contactless conductivity detection was proposed. A four-electrode capacitively coupled (contactless) detector has been integrated on a Pyrex glass chip for detection of peptides (1 mM) and cations (5 mM K+, Na+, Li+). The Al electrode (500 nm Al/100 nm Ti) was deposited in a 600- nm-deep trench and was covered with a thin dielectric layer (30-nm SiC). The other parts of the channel were covered and insulated with Si_3N_4 (160 nm). This four-electrode configuration allows for sensitive detection at different background conductivities without the need for adjustment of measurement frequency. In contactless mode, the dielectric thickness should be small [50]. Iliescu et al. [51] recently explored the use of a-SiC membranes as structures for chip-based cell culturing. The optical properties of the SiC film supported the use of classical fluorescence microscopy and thus were ideal for cell culturing studies. NH_4F was used to reduce the native oxide on the SiC surfaces. The NH_4F surface treatment resulted in greater cell density on the a-SiC samples as compared with untreated surfaces. Collectively, these achievements show the potential for SiC in highly functional lab-on-a-chip devices.

11. Conclusions

Silicon Carbide is a wide-band-gap semiconductor biocompatible material that has the potential to improve biomedical applications. SiC devices offer higher power densities and lower energy losses, enabling lighter, more compact and more efficient products for biocompatible and long-term in vivo applications such as sensors. The main problem facing the medical community today is the lack of biocompatible materials that are also capable of electronic operation. Silicon carbide has gained favor in the biomedical microdevice community for its potential as a structural and packaging material. Currently, bioMEMS devices that utilize SiC are relatively simple in design. Continued development of processing techniques that make SiC compatible with polymeric and other temperature-sensitive substrates is critical if SiC is to play a significant role in next-generation implantable biomedical microdevices. Therefore, SiC will have matured to the point where it can take its place with Si and its derivatives (SiO_2 and Si_3N_4) in the toolbox of commonly used MEMS and NEMS materials. In the near future it is expected that the semiconductor structures will have a profound effect on the capabilities of BioMEMS. Not only will the quantum dots and quantum planar structures be a major player in this area, but it is also expected that nanoporous and especially nanorods and nanocolumn arrays will provide new directions for the development of chemical sensors and biosensors capable of tackling the modern challenges of direct chemical analysis.

Author details

Mahboobeh Mahmoodi[1*] and Lida Ghazanfari[2]

*Address all correspondence to: m.mahmoodi@aut.ac.ir

1 Department of Materials science and Engineering, Yazd Branch, Islamic Azad university, Yazd, Iran

2 Biomateials Group, Faculty of Biomedical Engineering, Amirkabir University of Technology, Tehran, Iran

References

[1] Bhushan B. Nanotribology and Nanomechanics of MEMS/NEMS and BioMEMS/ BioNEMS Materials and Devices, Microelectronic Engineering 2007; 84, 387–412.

[2] Zorman C. A., Parro R. J. Micro- and Nanomechanical Structures for Silicon Carbide MEMS and NEMS, Phys. Stat. Sol. B 2008; 245(7) 1404–1424.

[3] Cimalla V., Pezoldt J., Ambacher O. Group III Nitride and SiC Based MEMS and NEMS: Materials Properties, Technology and Applications J. Phy. D 2007; 40, 6386–434.

[4] Coletti C., Jaroszeski M. J., Pallaoro A., Hoff A. M., Iannotta S., Saddow S. E. Biocompatibility and Wettability of Crystalline SiC and Si Surfaces. In: Proceedings of 29th Annual International Conference of the IEEE EBMS, Lyon, France, August 23–26, 2007; p. 5849–52.

[5] Mahmoodi M., Ghazanfari L. Fundamentals of Biomedical Applications of Biomorphic SiC. In: Gerhardt R. (ed.) Properties and App lications of Silicon Carbide. In-Tech;2011. p 297-345. Avaliable from http://www.intechopen.com/books.

[6] Kotzara G., Freasa M., Abelb P., Fleischmarc A., Royc S., Zormand C., Morane J. M., Melzak J. Evaluation of MEMS Materials of Construction for Implantable Medical Devices, Biomat.

[7] Ferrari M. BioMEMS and Biomedical Nanctechnology, Volume III, Therapeutic Micro /Nano Technology, Springer 2006.

[8] Bolz A., Schaldach M. Biomaterials Haemocompatibility Optimization of Implants by Hybrid Structuring. Med. Biol. Eng. Comput. 1993; 31, 123-130.

[9] Nguyen N. T., Wereley S. T. Fundamentals and Applications of Microfluidics Artech House, 2002; 471.

[10] Bhushan B., S. Sundararajan, Li X., Zorman C.A., Mehregany M. Tribology Issues and Opportunities in MEMS, Kluwer Academic, Dordrecht, Netherlands 1998; 407-430.

[11] Mehregany M., Zorman C. A. SiC MEMS: Opportunities and Challenges for Applications in Harsh environments. Thin Solid Films 1999; 355–356:518–24.

[12] Parro R. J., Scardelletti M. C., Varaljay N. C., Zimmerman S., Zorman C. A. Amorphous SiC as a Structural Layer in Microbridge Based RF MEMS Switches for Use in Software Defined Radio. Solid-State Electron 2008;52, 1647–51.

[13] Zorman C. A., Barnes A. C. Silicon Carbide BioMEMS, Elsevier Inc. Chapter 10, 2012.

[14] Okojie R. S., Ned A. A., Kurtz A. D. Operation of a 6H-SiC Pressure Sensor at 500 °C. Sens Actuators A 1998; 66, 200–4.

[15] Atwell A. R., Okojie R. S., Kornegay K. T., Roberson S. L., Beliveau A. Simulation, Fabrication and Testing of Bulk Micromachined 6H-SiC High-g Piezoresistive Accelerometers. Sens Actuators A 2003;104, 11–18.

[16] Zhao F., Islam M. M. Fabrication of Single Crystal Silicon Carbide MEMS/NEMS for Bio Sensing and Harsh Environments. In: Proceedings of MEMS 2011. Cancun, Mexico, January 23, 2011; p. 261–3.

[17] Kageyama, et al. Resonating Microstructures in Microshells with HF Permeable Polycrystalline Silicon and Vacuum Sealing Thin Films, Digest, The 10th IEEE Int. Conf. On Solid-State Sensors and Actuators (Transducers '01), Sendai, Japan, June 1999; 340-343.

[18] Stark B.H., Najafi K. An Ultra-Thin Hermetic Package Utilizing Electroplated Gold, Technical Digest, 11th IEEE Int. Conf. Solid-State Sensors and Actuators (Transducers '01), Munich, Germany June 2001; 194-197.

[19] Stark B. H., Najafi K. A Wafer-Level Vacuum Packaging Technology Utilizing Electroplate Nickel, ASME Congress, MEMS And Nanotechnology Symposium, New Orleans, USA, Nove. 17-22, 2002.

[20] Najafi K. Micropackaging Technologies for Integrated Microsystems: Applications to MEMS and MOEMS, Micromachining and Microfabrication Process Technology VIII, Proceedings of SPIE 2003; 4979.

[21] Miller S., Rizzo J. F., Wyatt J. L. Development of a Silicon Retinal Implant: Long-Term Biocompatibility of Materials, Invest Ophthalmol Visual Sci. 1995; 36(4) S746.

[22] Zrenner E., Kohler K., Haemmerle H., Nisch W., Stelzle M. Biocompatibility and Long Term Stability of Silicon Photodiode Arrays Implanted in the Subretinal Space, Soc. Neurosci. Abstr. 1999; 25(1–2) 219.

[23] Peachey N. S., Chow A.Y. Subretinal Implantation of Semiconductor-Based Photodiodes: Progress and Challenges, J. Rehab. Res. Dev. 1999; 36(4) 371–6.

[24] Amon M., Bolz A., Schaldach M. Improvement of Stenting Therapy with a Silicon Carbide Coated Tantalum Stent. J. Mater. Sci. Mater. Med. 1996; 7, 273–8.

[25] Thewes R., Paulus C., et al. Integrated Circuits for the Biology-to-Silicon Interface, ESSDERC Proceedings 2004; 19-28.

[26] Rizzo J. F., Wyatt J., Kelly S., Shire D. Perpetual Efficacy of Electrical Stimulation of Human Retina with a Microelectrode Array During Short-Term Surgical Trials, Invest.Ophtalm. Vis. Sci. 2003; 44, 5362-9.

[27] Humayun M. S., et al. Visual Perception in a Blind Subject with a Chronic Microelectr. Retinal Prosthesis, Vis. Res. 2003; 43, 2537-81.

[28] Singh S. h., Buchanan R. C. SiC-C Fiber Electrode for Biological Sensing. Mater. Sci. and Eng. C 2007; 27, 551–557.

[29] Godignon P. New Generation of SiC Based Biodevices Implemented on 4 Wafers. Mater. Sci. Forum 2005; 483–485. 1097-1100.

[30] Ghavami R., Salimi A., Navaee A. SiC Nanoparticles-Modified Glassy Carbon Electrodes for Simultaneous Determination of Purine and Pyrimidine DNA Bases, Biosensors and Bioelectro. 2011; 26, 3864–3869.

[31] Caputo D., De Cesare G., Nascetti A. Scipinotti R. Two-Color Sensor for Biomolecule Detection. Sensor Letters, 2008; 6, 4, 542-547.

[32] Mai Y., Atkins A.G. Further Comments on J-Shaped Stress-Strain Curves and the Crack Resistance of Biological Materials, J. Phys. D: Appl. Phys. 1989; 22, 48–54.

[33] Gabriela G., Erilla I., Carob J., Gomeza R., Rierab D., Villaa R., Godignon P. Manufacturing and Full Characterization of Silicon Carbide-Based Multi-Sensor Micro-Probes for Biomedical Applications, Microelectronics Journal 2007; 38, 406–415.

[34] Yang Y. T., Ekinci K. L., Huang X. M. H., Schiavone L. M., Zorman C. A., Mehregany M., et al. Monocrystalline Silicon Carbide NEMS. Appl Phys Lett 2001; 78, 162–4.

[35] Cimalla V., Neibelschutz F., Tonish K., Foerster C. h., Brueckner K., Cimalla I., et al. Nanoelectromechanical Devices for Sensing Applications. Sens Actuators B 2007;126, 24–34.

[36] Naik A. K., Hanay M. S., Hiebert W. K., Feng X. L., Roukes M. L.. Towards Single-Molecule Nanomechanical Mass Spectroscopy. Nat Nanotechnol 2009; 4, 445–50.

[37] Singh S., Buchanan R. C. SiC-C Fiber Electrode for Biological Sensing, Materials Science and Engineering C 2007; 27, 551–557.

[38] Gomez R , Ivorra A., Villa R., Godignon P., Mill'an J., Erill I.,Sola A., Hotter G., Palacios L. A SiC Microdevice for the Minimally Invasive Monitoring of Ischemia in Living Tissues, Biomed Microdevices 2006; 8, 43–49.

[39] Godigon P., Martin I., Gabriel G., Gomez R., Placidi M., Villa R. New Generation of SiC Based Biodevices Implemented on 4 wafers. Mater. Sci. Forum. 2010;645–648:1097–100.

[40] Lipovsek B., Joskowiak A., Krc J., Topic M., Prazeres D. M. F., Chu V., Conde J. P. Characterisation of Hydrogenated Silicon–Carbon Alloy Filters with Different Carbon Composition for On-Chip Fluorescence Detection of Biomolecules, Sensors and Actuators A 2010; 163, 96–100.

[41] Hsu J. M., Tathireddy P., Rieth L., Norman A. R., Soltzbacher F. Characterization of a-SiC:H Thin Films as an Encapsulation material for Integrated Silicon Based Neural Interface Devices. Thin Solid Films 2007;516, 34–41.

[42] Salitermann S. S. Fundamentals of BioMEMS and Medical Microdevices, SPIE, Bellingham, WA, USA, 2006.

[43] Pennathur S., Fygenson D. K. Improving Fluorescence Detection in Lab on Chip Devices, Lab on a Chip 2008; 8, 649–652.

[44] Born M., Wolf E. Principles of Optics, Cambridge University Press, Cambridge 1999.

[45] Fixe F., Prazeres D. M. F., Chu V., Conde J. P. An on-Chip Thin Film Photodetector for the Quantification of DNA Probes and Targets in Microarrays, Nucleic Acids Research 2004; 32 e70.

[46] Rosenbloom A. J., Sipe D. M., Shishkin Y., Ke Y., Devaty R. P., Choyke W. J. Nanoporous SiC: a Candidate Semi-Permeable Material for Biomedical Applications. Biomed. Microdevices 2004; 6, 261–7.

[47] Taubner T., Korobkin D., Urzhumov Y., Shvets G., Hillenbrand R. Near-Field Microscopy Through a SiC Superlens. Science 2006; 313, 1595.

[48] Zahn J. D. Methods in Bioengineering, Biomicrofabrication and Biomicrofluidics, Artech House 2010.

[49] Berthold A., Laugere F., Schellevis H., de Boer C. R., Laros M., Guijt R. M., Sarro P. M., Vellekoop M. J. Fabrication of a Glass-Implemented Microcapillary Electrophoresis Device with Integrated Contactless Conductivity Detection, Electrophoresis 2002; 23, 3511–3519. 2002; 23, 2737–2750.

[50] Li P. C. H. Microfluidic Lab-on-a-Chip for Chemical and Biological Analysis and Discovery, Taylor & Francis 2005; 94.

[51] Iliescu C., Chen B., Poenar D. P., Lee Y. Y. PECVD Amorphous Silicon Carbide Membranes for Cell Culturing. Sens. Actuators B 2008; 129, 404–11.

Chapter 16

Radiation Response of Silicon Carbide Diodes and Transistors

Takeshi Ohshima, Shinobu Onoda, Naoya Iwamoto,
Takahiro Makino, Manabu Arai and
Yasunori Tanaka

Additional information is available at the end of the chapter

1. Introduction

Silicon Carbide (SiC) is regarded as a promising candidate for electronic devices used in harsh radiation environments (Rad-hard devices) such as in space, accelerator facilities and nuclear power plants [1-5]. In order to apply SiC to such rad-hard devices, we have to know the radiation response of the characteristics of SiC devices, because semiconductor devices show destructive and non-destructive malfunctions and/or degradation their characteristics due to irradiation. For radiation effects on semiconductor devices, three major effects, Single Event Effects (SEEs), Total Ionizing Dose (TID) effect, and Displacement Damage Dose (DDD) effects are known.

When charged particles such as heavy ions are irradiated into semiconductors, dense charge (electron-hole pairs) is generated in semiconductors along to the ion track. The malfunctions of electronic devices such as LSIs and power devices caused by charge generated by charged particles are called SEEs. The SEEs occur even by only one particle incidence, and there are both nondestructive (soft errors) and destructive (hard errors) SEE failures [6-8]. The soft errors arise if the amount of charge collected by devices is large enough to reverse or flip the data state of a memory cell, register, latch, or flip-flop. Since the soft errors are not destructive, the function of semiconductor devices can be recovered by writing new data to the bit and/or resetting of devices. For example, the Single Event Upset (SEU) and the Multiple Bit Upset (MBU) in a Static Random Access Memory (SRAM) and a Dynamic Random Access Memory (DRAM), the Single Event Functional Interrupt (SEFI) in Field Programmable Gate Array (FPGA) or DRAM control circuitry are known as the soft errors. Recently, the Single

Event Transient (SET) arises as a serious issue for analog electronics and digital logic cells. In general, the SETs in analog electronics are referred to as ASETs, and those in digital combinatorial logic are referred to as DSETs. In contrast, the Single Event Latch-up (SEL), the Single Event Burnout (SEB), and the Single Event Gate Rupture (SEGR) in power electronic devices are known as the hard errors.

Electron-hole pairs are induced in insulator layers of Metal-Insulator-Semiconductor (MIS) structure devices, such as Metal-Oxide-Semiconductor (MOS) devices by irradiation, and as a result, charge trapped in insulator (oxide) and/or traps near the interface between oxide and semiconductor (interface traps) are generated. Since such charge trapped in insulator and interface traps act give harmful influence to transport properties of semiconductors, the electrical characteristics of MIS devices are degraded by their generation [9, 10]. For example, the shift of threshold voltage (V_T) and the decrease in the channel mobility (μ_{ch}) are observed in MOS field effect transistors (FETs). This radiation effect is called the TID effect, and in general, the value of the TID effects gradually increases with increasing dose of radiations because the amount of radiation-induced charge in insulator and interface traps increases with increasing dose.

When energetic particles are irradiated into semiconductor crystals, atoms at lattice sites are scattered into non-lattice sites (knock-on effects). As a result, vacancies and interstitials are created in semiconductor crystals. This is the origin of the DDD effect. However, in reality, the structure of residual defects is not so simple and a wide variety of defects such as divacancies, vacancy clusters, and vacancy-impurity complexes exists in crystals because generated vacancies and interstitials thermally diffuse and finally they become stable defects. In general, such defects act as scattering/recombination centers to free carriers, and as a result, the electrical characteristics of semiconductors devices are degraded. In the case of the DDD effect, similar to the TID effect, the degradation of the characteristics of semiconductor devices becomes larger with increasing fluence of radiation. The degradation of the electrical performance of solar cells installed in space satellites is known as one of the examples of the DDD effect [11-14].

In this chapter, the effects of radiation on the electrical characteristics of SiC devices are described from the point of view of the TID effect and the SEEs.

2. Gamma-ray irradiation effects on SiC MOSFETs

Figure 1 shows the change in the subthreshold region of drain current (I_D) – gate voltage (V_G) curves (subthreshold curves) for n-channel 6H-SiC MOSFETs by gamma-ray irradiation. The bias of 12 V was applied to drain (V_D) during measurements. The gate oxide of the MOSFETs was formed using pyrogenic oxidation (H_2:O_2 = 1:1) at 1100°C. The mark „+" on the each line indicates the value of V_T. As shown in the figure, the value of V_T shifts to the negative voltage side, and also the I_D-V_G curves stretches after irradiation. This suggests that charge in oxide and interface traps are generated by gamma-ray irradiation.

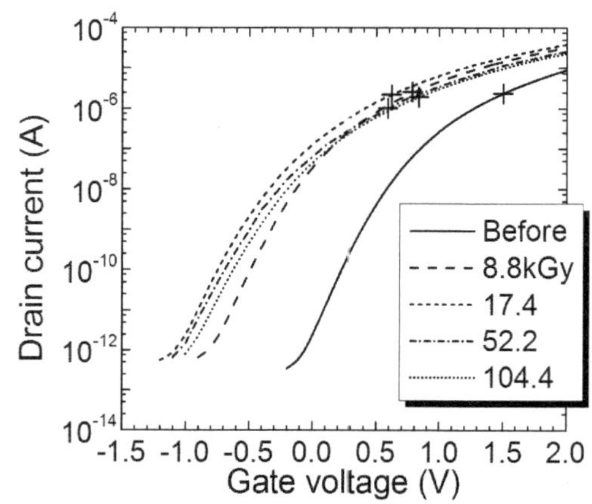

Figure 1. Change in the subthreshold region of $I_D - V_G$ curves (subthreshold curves) for n-channel 6H-SiC MOSFETs by gamma-ray irradiation. The bias of 12 V was applied to drain during measurements. The „+" mark on the each line indicates the value of V_T.

According to Mcwhorter and Winokur [9], the density of charge trapped in gate oxide (ΔN_{OX}) and interface traps (ΔN_{IT}) generated by irradiation can be estimated from the shift of subthreshold curves using a following analysis. Since charge trapped in gate oxide does not respond to bias applied to gate, the entire subthreshold curve is simply shifted by the generation of charge trapped in gate oxide. On the other hand, since the charge state of interface traps depends on Fermi level (thus, the value of the bias applied to gate oxide), the subthereshold curve is stretched by the generation of interface traps. This behavior can be expressed as

$$\Delta V_T = \Delta V_{OX} + \Delta V_{IT} \tag{1}$$

where ΔV_T, ΔV_{OX} and ΔV_{IT} are the shift of the threshold voltage by irradiation, the voltage shifts due to the generation of oxide-trapped charge and interface traps, respectively. Also, since the charge state of interface traps is assumed to be neutral at midgap state, at which Fermi level corresponds to the intrinsic Fermi level, the shift of the midgap voltage (ΔV_{MID}) due to irradiation is caused by oxide-trapped charge. Thus,

$$\Delta V_{MID} = \Delta V_{OX} \tag{2}$$

Since the subthreshold curve between V_{MID} and V_T is stretched by the generation of interface traps, ΔV_{IT} is determined as

$$\Delta V_{IT} = \left(V_T - V_{MID}\right)_{post} - \left(V_T - V_{MID}\right)_{pre} \tag{3}$$

where „post“ and „pre“ mean after and before irradiation, respectively.

In order to obtain the value of V_{MID}, firstly, the drain current corresponding to the midgap condition (I_{MID}) is estimated. In the subthreshold region, I_D is expressed as the formula [15]

$$I_D = 2^{1/2}\mu(W/L)(qN_A L_B/\beta)(n_i/N_A)^2 \exp(\beta\phi_s)(\beta\phi_s)^{-1/2} \tag{4}$$

where N_A, n_i, ϕ_s and L_B are the acceptor (or donor) concentration in the channel region of a MOSFET, the intrinsic carrier concentration, band bending at the surface and the Debye length given by $L_B = (\varepsilon_s/(\beta q N_A))^{1/2}$, respectively. Here, β is equal to $q/k_B T$, where q and k_B are the electron charge and the Boltzmann constant, respectively. At the midgap condition, ϕ_s is equal to $(k_B T/q)\ln(N_A/n_i)$. Thus, I_{MID} can be estimated from eq. (4) using ϕ_s for $(k_B T/q)\ln(N_A/n_i)$. Then, the value of V_{MID} can be obtained from the value of V_G at I_{MID} on subthreshold curves. It should be mentioned that for the determination of V_{MID} for SiC, it is necessary to linearly extrapolate the lower position of the subthreshold curve down to the lower part of the curve, since the value of I_{MID} is of the order of 10^{-30} A.

The value of ΔN_{OX} and ΔN_{IT} is estimated from

$$\Delta N_{OX} = \Delta V_{OX}\ C_{OX}/q \tag{5}$$

$$\Delta N_{IT} = \Delta V_{IT}\ C_{OX}/q \tag{6}$$

where C_{OX} is equal to ε_{OX}/t_{OX}, and ε_{OX} and t_{OX} are the relative dielectric constant of SiO_2 and the thickness of gate oxide, respectively.

Figure 2 (a) shows the ΔV_T as a function of absorbed dose for n-channel 6H-SiC MOSFETs. The triangles, circles and squares represent results obtained from MOSFETs of which gate oxide was fabricated by dry (Dry) and pyrogenic (Pyro) oxidations at 1100°C and pryogenic oxidation followed by hydrogen annealing at 700°C for 30 min at a pressure of 20 Torr (H_2), respectively. For the details of the fabrication process of those MOSFETs, please see Ref. [16, 17]. For the Dry SiC MOSFETs, the ΔV_T slightly shifts to the positive voltage side above 50 kGy although the value does not change below 30 kGy. For the Pyro SiC MOSFETs, the value of ΔV_T shifts to the negative voltage side, and the negative shift become smaller above 30 kGy. For the H_2 SiC MOSFETs, the ΔV_T shows the negative shift around 20 kGy, however, the voltage shift terns to the positive above 30 kGy. Since the value of ΔV_T is affected by the generation of charge trapped in gate oxide and interface traps, for understanding the behavior of ΔV_T, it is necessary to know the information on ΔV_{OX} and ΔV_{IT}. Therefore, the value of ΔV_{OX} and ΔV_{IT} is derived from the subthreshold curves using Eqs. (1) – (4). The absorbed dose dependence of ΔV_{OX} and ΔV_{IT} is shown in Figs. 3 (a) and (b), respectively. The values of ΔV_{OX} for the Dry and the Pyro SiC MOSFETs show the negative voltage shift and the shift becomes large with absorbed dose. These results indicate that for the Dry and the Pyro SiC

MOSFETs, the positive charge is trapped in gate oxide by gamma-ray irradiation and the trapped charge increases with increasing absorbed dose. Since the shift of ΔV_{OX} for the Pyro SiC MOSFETs is larger than that for the Dry SiC MOSFETs, the value of trapped charge for the Pyro SiC MOSFETs is larger than that for the Dry SiC MOSFETs. On the other hand, ΔV_{OX} for the H_2 SiC MOSFETs shows complicated behaviors although the shift is very slight even after 530 kGy irradiation. Thus, firstly the ΔV_{OX} shifts to the negative voltage side at doses below 40 kGy. However, the value shows a positive voltage shift around 60 kGy although a negative shift appears at 180 kGy. Then, finally, the shift becomes positive again after irradiation at 530 kGy.

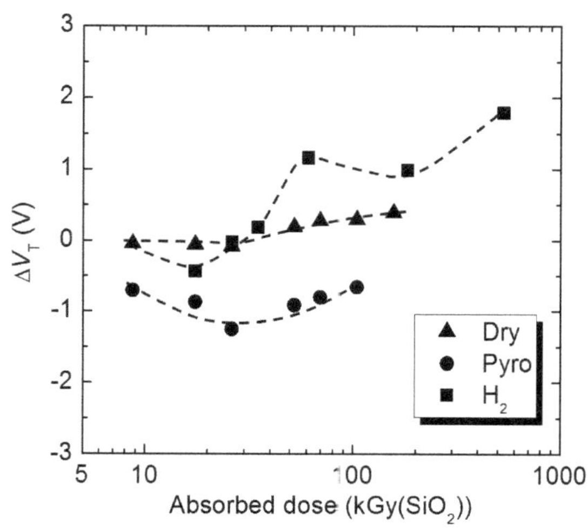

Figure 2. ΔV_T as a function of absorbed dose for n-channel 6H-SiC MOSFETs. The triangles, circles and squares represent results obtained from MOSFETs of which gate oxide was fabricated by dry (Dry) and pyrogenic (Pyro) oxidations at 1100°C and pryogenic oxidation followed by hydrogen annealing at 700°C for 30 min at a pressure of 20 Torr (H_2), respectively.

These behaviors indicate that both positively and negatively charges are generated in gate oxide for the H_2 SiC MOSFETs by gamma-ray irradiation. It was reported from the change in capacitance – voltage characteristics of 6H-SiC MOS capacitors due to gamma-ray irradiation that negative and positive trapped charges were generated near SiO_2/SiC interface and in oxide at 40 nm from the interface, respectively [18]. Although the mechanism of H_2-anneling effect on the gate oxide and the interface between oxide and SiC has not yet been clarified, since the initial value of V_T decreased by H_2 annealing [19], the large shift of ΔV_T to the positive voltage side and the unique behavior of ΔV_{OX} might occurs due to the reduction of H_2-annealing effects by gamma-ray irradiation. Also, it should be noticed that a part of interface traps might be detected as oxide-trapped-charge in this analysis since interface traps in the middle region of the band gap of 6H–SiC have extremely long charge release times at RT, and they act just as charge trapped in oxide [20]. In contrast to ΔV_{OX}, the values of ΔV_{IT}

for all SiC MOSFETs show the positive voltage side and their shifts become larger with increasing absorbed dose although the absolute values depend on the fabrication process of gate oxide, as shown in Fig. 3 (c).

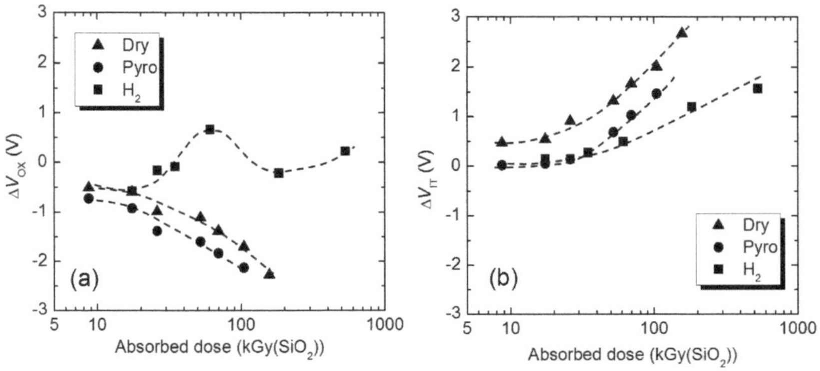

Figure 3. (a) ΔV_{OX} and (b) ΔV_{IT} as a function of absorbed dose for n-channel 6H-SiC MOSFETs. The triangles, circles and squares represent results obtained from MOSFETs of which gate oxide was fabricated by dry (Dry) and pyrogenic (Pyro) oxidations at 1100°C and pryogenic oxidation followed by hydrogen annealing at 700°C for 30 min at a pressure of 20 Torr (H_2), respectively.

The values of ΔN_{OX} and ΔN_{IT} are estimated from Figs. 3 (b) and (c), respectively, using Eq. (5)/(6). Figures 4 (a) and (b) show ΔN_{OX} and ΔN_{IT}, respectively, for the Dry (triangles), the Pyro (circles) and the H_2 (squares) SiC MOSFETs as a function of absorbed dose. For comparison, the reported results of Si MOSFETs are also plotted in the figures (upside-down triangles) [9]. The value of ΔN_{OX} for the Dry SiC MOSFETs is slightly smaller than that of the Pyro SiC MOSFETs and both values increase with increasing absorbed dose with an exponent of 2/3. It is also found that Si MOSFETs show the 2/3 power-low dependence, although the value of ΔN_{OX} for the Si MOSFETs is larger than that for the SiC MOSFETs [9]. On the other hand, the change in ΔN_{OX} for the H_2 MOSFETs due to irradiation show a different behavior from others, and the value is in order of 10^{11} /cm^2 even after irradiation at 530 kGy. These results indicate that the characteristics of gate oxide fabricated by H_2-annealing differ from those by non-annealing. However, it should be noticed that ΔN_{OX} estimated in this analysis is a value subtracting a positive component from a negative component. Thus, if both positive and negative components are almost the same value, the net number of ΔN_{OX} is small. Therefore, from this result, we cannot simply conclude that the quality of gate oxide fabricated by H_2-annealing is higher than that of gate oxide fabricated by non-annealing or not.

For ΔN_{IT}, the H_2 SiC MOSFETs have lower values than the other MOSFETs at absorbed doses above 30 kGy. The characteristics of SiC MOS devices were reported to be degraded by carbon related compounds remaining around the interface between SiO_2 and SiC [21]. Since such compounds might also act as precursors of radiation-induced interface traps, it is assumed that H_2 annealing to gate oxide of SiC MOSFETs reduces residual compounds near the interface. For the absorbed dose dependence cf ΔN_{IT}, the H_2 and the Dry SiC MOSFETs have the 2/3 power-low dependence although ΔN_{IT} for the Pyro SiC MOSFETs increases with increasing the absorbed dose with an exponent of approximately 3/2. The 2/3 power-law dependence is also reported in Si of which gate oxide was formed using dry oxidation [9]. The power-law dependence comes from the generation mechanism of interface traps, and the structural and/or electrical properties of the interface between SiO_2 and SiC for the H_2 and the Dry SiC MOSFETs are different from those for the Pyro SiC MOSFETs. Therefore, it is suggests that the characteristics of the interface between SiO_2 and SiC formed by pyrogenic oxidation followed by H_2-annealing are similar to those formed by dry oxidation.

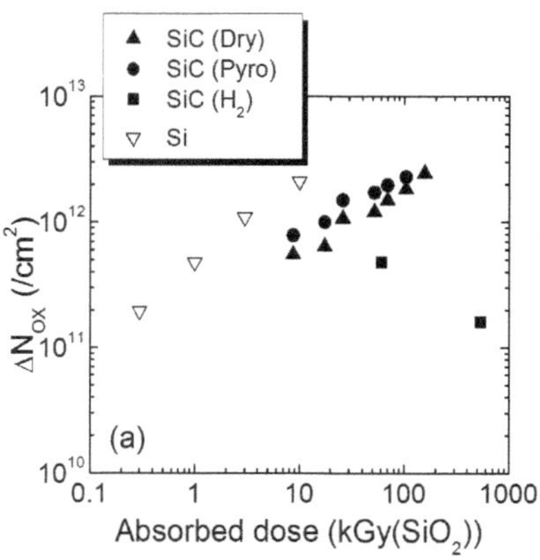

Figure 4. a) ΔN_{OX} and (b) ΔN_{IT} for Dry (triangles), Pyro (circles) and H_2 (squares) SiC MOSFETs as a function of absorbed dose. For comparison, the reported results of Si MOSFETs are also plotted in the figures (upside-down triangles) [9].

The μ_{ch} for Si MOSFETs is known to decrease with increasing absorbed dose [10]. In order to confirm this for SiC MOSFETs, μ_{ch} for the H_2 SiC MOSFETs were plotted as a function of absorbed dose (Fig. 5). For comparison, the result reported for Si MOSFETs are also plotted in the figure [9]. The μ_{ch} for the H_2 SiC MOSFETs does not change up to 20 kGy and the value decreases with increasing absorbed dose above 60 kGy. Then, the value of μ_{ch} reduces to be 50 % of the initial value at 530 kGy. On the other hand, μ_{ch} for the Si MOSFETs decreases with increasing absorbed dose and becomes 50 % of the initial value by irradiation at 10 kGy. Although the initial value of μ_{ch} for Si MOSFETs (600 cm^2/Vs) is much higher than the initial value of μ_{ch} for the H_2 SiC MOSFET (~ 50 cm^2/Vs), the value for Si MOSFETs is assumed to be almost zero after irradiation at 100 kGy whereas the H_2 SiC MOSFETs still keep 25 cm^2/Vs of μ_{ch} even after irradiation at 530 kGy. In addition, it is mentioned that the stability of their electrical performance against irradiation is also important for Rad-hard devices. Therefore, it can be concluded that SiC MOSFETs are quite tolerant against radiation in comparison with Si MOSFETs. For the degradation mechanism of μ_{ch}, Ohshima et al. reported [17] that the relationship between the decrease of μ_{ch} and ΔN_{IT} for SiC MOSFETs was described by the same relationship reported for Si MOSFETs ($\mu_{ch} = \mu_0 /(1 + \alpha \Delta N_{IT})$) [10], where μ_0

and α are the initial value of the channel mobility and a constant (= $7.0\pm1.3\times10^{-13}$), respectively. This suggests that μ_{ch} for SiC MOSFETs as well as Si MOSFETs can be explained in term of carrier scattering in the channel region by interface traps generated by gamma-ray irradiation. Since interface traps located in the middle of the band gap behave just like charge trapped oxide for SiC, ΔN_{IT} obtained in this analysis means the net density of interface traps which act as carrier scattering centers. It was reported that the channel mobility of 6H–SiC MOSFETs is affected by acceptor-like traps existing near the conduction band edge [22]. Although the relationship between interface traps induced by irradiation and intrinsic interface traps has not yet been clarified in the case of SiC MOS devices, it is assumed that the radiation resistance of SiC MOSFETs might be improved by the reduction of initial interface traps generated near the conduction band edge.

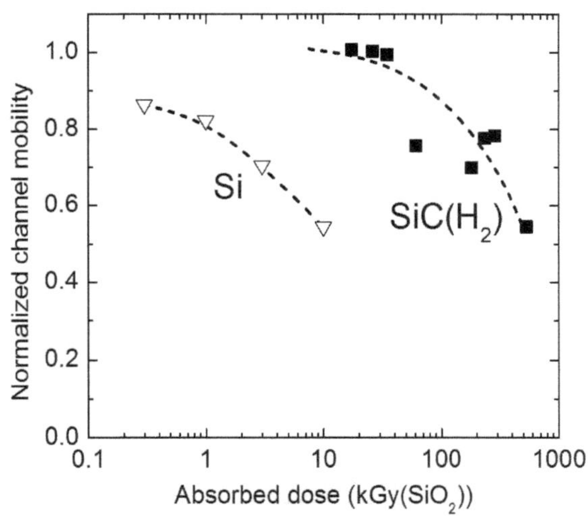

Figure 5. μ_{ch} for H_2 SiC MOSFETs as a function of absorbed dose. For comparison, the result reported for Si MOSFETs are also plotted in the figure [9]. The value of the channel mobility is normalized by the initial value.

Next, the effects of the surface morphology on μ_{ch} of SiC MOSFETs irradiated with gamma-rays will be discussed. In this study, MOSFETs were fabricated on n-type 6H-SiC epitaxial layers using the same fabrication process except the procedures of high temperature annealing after implantation [23]. Thus, although all samples were annealed at 1650°C for 3 min in an Ar atmosphere, the surface of one series of samples was covered with carbon films (C-coating) during the annealing to avoid the degradation of the surface morphology [24], and the other series of samples were annealed without the carbon coating (non-coating). After the annealing, the carbon films were removed by the oxidation at 800°C for 30 min in O_2 gas. Gate oxide of both series of the MOSFETs were formed by pyrogenic oxidation (H_2:O_2 = 1:1) at 1100°C for 30 min. For the details of the fabrication process, please see Ref. [23]. The initial values of μ_{ch} for C-coating and non-coating SiC MOSFETs are 41 and 44 cm^2/Vs, respectively. For the surface morphology, the values of root mean square (RMS) for the C-coating and non-coating SiC are obtained to be 0.67 and 1.36 nm, respectively, from AFM measurements, whereas the RMS was 0.25 nm before annealing.

Figures 6 (a) and (b) show μ_{ch} and ΔN_{IT}, respectively, for C-coating (squares) and non-coating (circles) SiC MOSFETs as a function of absorbed dose. As shown in the figure, no significant decrease or slight increase in μ_{ch} is observed for the C-coating SiC MOSFETs. The value of ΔN_{IT} for the C-coating SiC MOSFETs is estimated to be less than 4x10^{11} /cm^2, and no significant increase in ΔN_{IT} is observed up to 3 MGy. In contrast, μ_{ch} for the non-coating SiC MOSFETs decreases with increasing absorbed dose above 2 MGy. In the absorbed region that μ_{ch} decreases, ΔN_{IT} increases with increasing absorbed dose, and the value becomes of the order of 10^{12} /cm^2 by irradiation above 2 MGy. As above-mentioned, μ_{ch} is degraded by the generation of interface traps. Therefore, the decrease in μ_{ch} for the non-coating SiC MOSFETs can be interpreted in terms of the generation of interface traps. Also, it was reported by Kimoto [24] the channel mobility can be affected by the surface roughness. So, the higher radiation resistance obtained for the C-coating SiC MOSFETs compared to non-coating ones is caused by the less surface roughness.

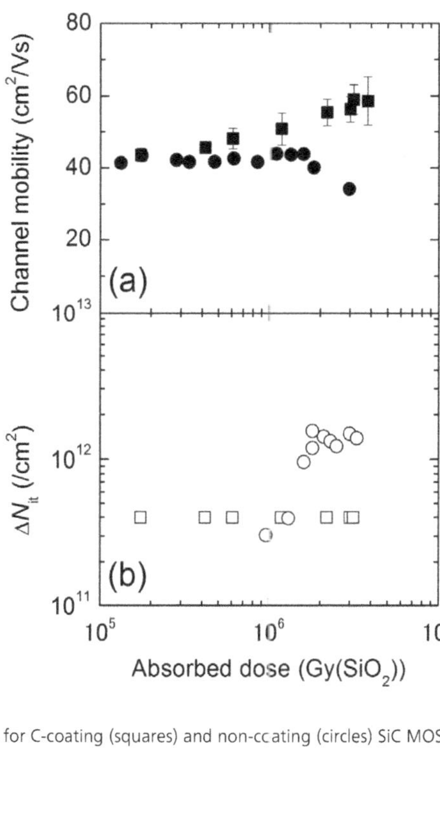

Figure 6. a) μ_{ch} and (b) ΔN_{IT} for C-coating (squares) and non-coating (circles) SiC MOSFETs as a function of absorbed dose.

3. Radiation hardness of SiC devices

In this section, the change in the electrical characteristics of SiC transistors such as Static Induction Transistors (SITs), Metal-Semiconductor (MES) FETs and MOSFETs due to gamma-ray irradiation will be compared to Si MOS FETs from the point of view of the radiation hardness. Figure 7 shows ΔV_T for SiC SITs [25], SiC MESFETs [26], C-coating+Dry SiC MOSFETs and C-coating+Pyro SiC MOSFETs as a function of absorbed dose. All transistors were irradiated with gamma-rays at RT. During gamma-ray irradiation, no bias was applied to any electrodes of the transistors. For comparison, the results reported for Si MOSFETs are also plotted in the figure [9]. No significant change in ΔV_T for All SiC transistors is observed up to 10^5 Gy whereas the Si MOSFETs show obvious degradation in ΔV_T. This indicates that those SiC transistors have extremely high radiation resistance compared to the Si MOSFETs. The value of ΔV_T for both the SiC MOSFETs shifts to the negative voltage side in high dose regions, and the shift for the C-coating+Dry ones is larger than that for the C-coating+Pyro ones. Thus, the C-coating+Pyro MOSFETs have higher radiation resistance than the C-coating+Dry MOSFETs. For the SiC MESFETs, the shift of ΔV_T to the negative voltage side increases with increasing in absorbed doses regions between 4×10^5 and 2×10^6 Gy, and the maximum shift of -0.75 V is observed at 2×10^6 Gy. However, the negative shift becomes smaller with increasing absorbed dose above 3×10^6 Gy and the value of ΔV_T becomes -0.27 V after irradiation at 10^7 Gy. For the SiC SITs, although the positive shift is observed for ΔV_T above 10^6 Gy, the value is relatively small (0.45V at 7×10^6 Gy) compared to other SiC transistors. Thus, it can be concluded that the radiation hardness of the SiC SITs and the MESFETs is higher than that of the SiC MOSFETs. Since SITs and MESFETs do not have gate oxide, such high radiation resistance to gamma-rays can be observed. However, it should be noticed that the characteristics of SiC SITs and MESFETs are also affected by TID effects since the SiC SITs and the MESFETs is covered with a insulator (oxide) for the surface termination, and charge is trapped in such insulator. In addition, in such a high absorbed dose region, the displacement damage effect by Compton electrons also occurs and the characteristics of devices are degraded.

Next, the change in the electrical characteristics of the SiC SITs by gamma-ray irradiation is expressed. The SiC SITs have an on-resistance of 0.15 Ω and a blocking voltage of 900 V at V_G of -10 V before irradiation [27, 28]. Since the SiC SITs were developed as power devices, two Si power devices with similar current and voltage ratings, Si MOSFET (17N80C3) and Si IGBT (5J301), were also irradiated with gamma-rays for comparison. The SiC SITs mounted in TO220 packages were irradiated with gamma-rays at absorbed dose rate of 8.8 kGy/h at RT. During irradiation, no bias was applied to electrodes. The shift of the breakdown voltage for the SiC SITs (squares), the Si MOSFETs (triangles) and the Si IGBT (upside-down triangles) as a function of absorbed dose is shown in Fig. 8. The blocking characteristics for the SiC SITs and the Si ones (IGBTs and MOSFETs) were measured under V_G at 10 V and 0V, respectively. No significant change in the breakdown voltage for the SiC SITs and the Si IGBT is observed up to 10^7 Gy. For the Si MOSFETs, the shift of the breakdown voltage increases with absorbed dose above 4×10^5 Gy, and the large shift of -500 V is observed at 10^7 Gy. It was also reported [25] that no significant increase in the leakage current for the SiC SITs (of the order of 10^{-6} A) was observed where the leakage current for the Si MOSFETs increased to 10^{-4} A level after irradiation 10^7 Gy.

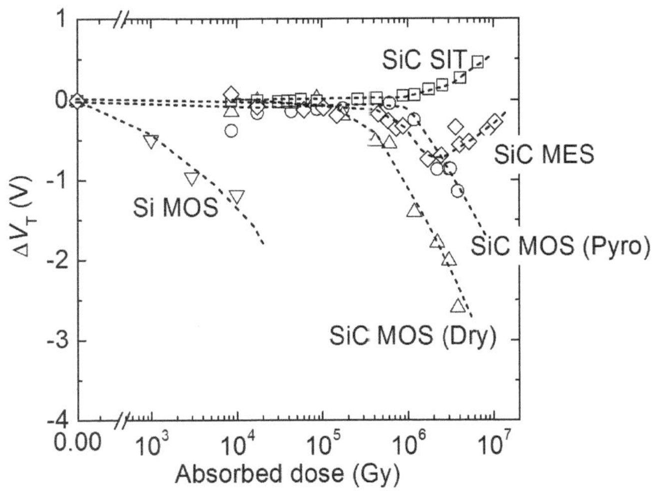

Figure 7. Change in ΔV_T for SiC SITs (squares), SiC MESFETs (diamonds), C-coating+Dry SiC MOSFETs (triangles) and C-coating+Pyro SiC MOSFETs (circles) as a function of absorbed dose. All transistors were irradiated with gamma-rays at RT. During gamma-ray irradiation, no bias was applied to any electrodes of the transistors. For comparison, the results reported for Si MOSFETs (upside-down triangles) are also plotted in the figure [9].

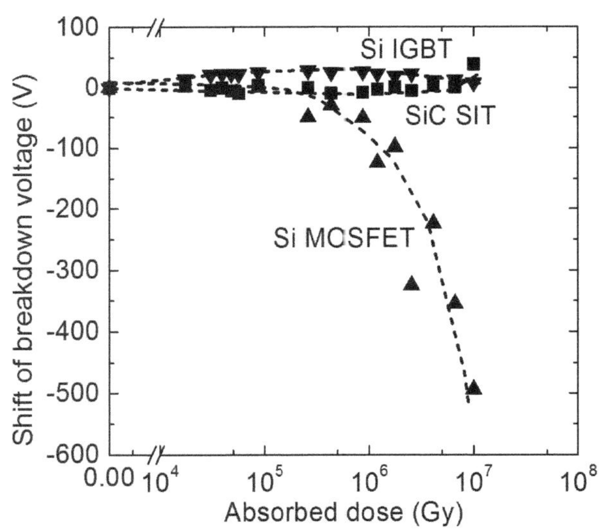

Figure 8. Shift of the breakdown voltage from the initial value for SiC SITs (squares), Si MOSFETs (triangles) and Si IGBT (upside-down triangles) as a function of absorbed dose. The blocking characteristics for SiC SITs and Si ones (IGBTs and MOSFETs) were measured under V_G at 10 V and 0V, respectively.

The on-state characteristics were measured under V_G at +2.5 V for the SiC SITs and at +15 V for the Si transistors (IGBTs and MOSFETs). Then, the on-voltage was defined as the value of V_D at I_D of 10 A. Figure 9 shows the shift of the on-voltage for the SiC SITs (squares), the Si MOSFETs (triangles) and the Si IGBT (upside-down triangles) as a function of absorbed dose. The shift of on-voltage for the SiC SITs and the Si MOSFETs due to gamma-ray irradiation shows a very stable behavior up to 10^7 Gy, whereas the on-voltage for the Si IGBTs remarkably increases after irradiation at 8×10^5 Gy (from 2.3 to more than 20 V). It was reported [29] that the displacement damage effect induced by Compton electrons degrades the gain for Si bipolar transistors. So, the result obtained from the Si IGBT is interpreted in terms of the majority carrier removal in the drift region (low doping region) due to the displacement damage effect. For the SiC SITs and the Si MOSFETs, since the doping concentration in the drift region is not low, the displacement damage effect might not be observed and as a result, on-voltage shows almost constant values up to 10^7 Gy. Although the stable on-voltage behavior is obtained for the SiC MOSFETs, the large fluctuation of V_T was reported due to the TID effect. Considering gamma-ray irradiation effects on the breakdown voltage, the on-voltage, and V_T, the characteristics of only the SiC SITs show the stable behaviors up to 10^7 MGy. Thus, we can conclude that the SiC SITs have extremely high radiation resistance, they have an enough potential for electronic devices used in harsh radiation environments such as nuclear power plants, space, and so on.

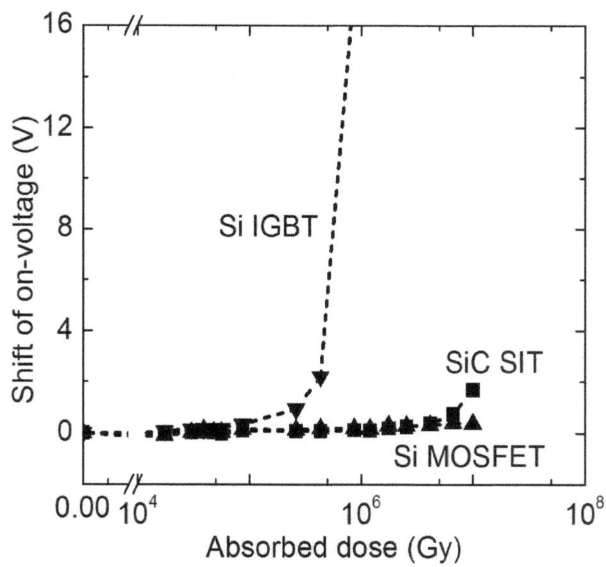

Figure 9. Shift of the on-voltage from the initial value for SiC SITs (squares), Si MOSFETs (triangles) and Si IGBT (upside-down triangles) as a function of absorbed dose.

4. Charge induced in SiC diodes by Ion irradiation

Since destructive or/and non-destructive malfunctions called SEEs occurs in electronic devices by charge (electron-hole pairs) generated by charged particle incidence, especially heavy ions. The SEEs on semiconductor devices are one of the most major issues for space applications. On the other hand, for high energy physics using accelerators with high luminosity, such as J-PARC and Super-LHC, Rad-hard particle detectors are expected to be developed. For the development of Rad-hard particle detectors as well as Rad-hard devices for space applications, it is important to clarify the behavior of charge generated in devices by charged particle incidence. In a previous study [30], Nava et al. reported that the Charge Collection Efficiency (CCE) obtained from 4H-SiC Schottky diodes by alpha particle incidence was estimated to be 100 %. It was also reported that 4H-SiC Schottky diodes could detect X-rays from radio isotopes [31,32]. Besides, the neutron detection by SiC diodes was investigated previously [33, 34]. As for light ions and X-rays irradiation into SiC, relatively large number of studies has been already reported. On the other hand, from the point of view of SEEs, study of ion irradiation on electronic devices using heavy ions is important. In this section, charge induced in SiC diodes by heavy ion incidence is reviewed on the basis of our previous studies [35-40].

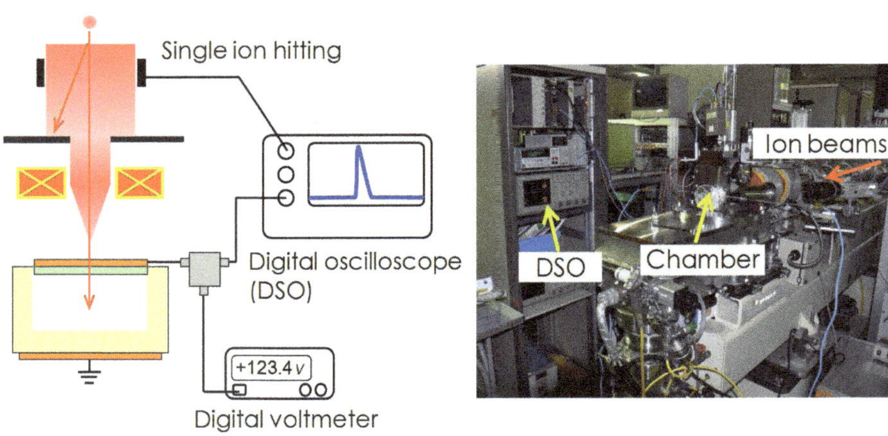

Figure 10. Schematic set-up of the TIBIC system installed at JAEA Takasaki and photo of the TIBIC system.

In order to obtain the information on charge induced in electronic devices, Ion Beam Induced Charge (IBIC) measurements is thought to be one of the useful methods. However, the decrease in collected charge during IBIC measurements should be considered for the accurate evaluation of charge induced by ion beams, since the device characteristics are degraded by radiation damage created in samples by ion incidence [41]. Therefore, single-ion hit Transient Ion Beam Induced Current (TIBIC) was developed at JAERI Takasaki in order to realize the evaluation of ion-induced current with minimizing the influence of damage

[42]. Figure 10 shows the schematic set-up of the TIBIC system installed at JAEA Takasaki and the photo of the TIBIC system. The TIBIC collection system connects with a heavy ion microbeam line from the 3MV Tandem accelerator, and consists of a single event triggering system and a fast switch beam shutter system. The transient current signals induced by ions can be detected using a digital sampling oscilloscope (Tektronix 3 GHz TDS694C or 15 GHz TDS6154C). The details of the single ion hit TIBIC collection system are described in Ref. [43]. Since the TIBIC system connects with a beam scanning system, spatial images of transient current signals can be obtained.

Figure 11 shows TIBIC signals obtained from 6H-SiC n^+p diodes with applied bias of 30, 90 or 150 V. Si ions with 12 MeV were used as probe beams. In this study, the 6H-SiC n^+p diodes with 100 - 300 μm diameters were fabricated on p-type substrates with p-type epitaxial layers (Al doping concentration between $8x10^{14}$ and $3.5x10^{15}$ /cm^3). The n^+ region was formed by three-fold implantation (60, 90, 140 keV) of phosphorus (P) ions at 800°C and subsequent annealing at 1650°C for 3 min in argon (Ar) atmosphere. The thickness and a mean P concentration of the implanted layer are ~100 nm and $5x10^{19}$ /cm^3, respectively. During the annealing, the sample surface was covered with a carbon film to avoid the degradation of the surface morphology [24]. The details of the diode fabrication process are described elsewhere [40]. The peak height of the TIBIC signals increases with increasing applied bias, and the value becomes to 0.50 from 0.19 mA when applied bias increases to 150 from 30 V. The fall-time, which is defined as the time from 90 % to 10 % of the current transient, shorten with increasing applied reverse bias, and the value decreases to 0.48 from 0.98 ns when applied bias increases to 150 from 30 V. These results can be interpreted in terms of an increase of the electric field in the depletion layer due to increasing applied bias. It is also mentioned the leakage currents of the diodes were in order of 10^{-11} A at an applied reverse bias of 150 V, and no significant differences in I -V characteristics between before and after TIBIC measurements were observed.

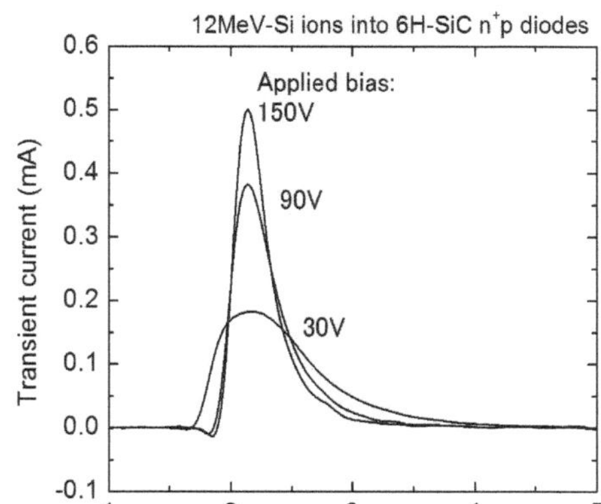

Figure 11. TIBIC signals obtained from 6H-SiC n^+p diodes with applied bias of 30, 90 or 150 V. Si ions with 12 MeV were used as probe beams.

By the integration of a TIBIC signal, charge collected by a diode can be estimated. Charge collected by the 6H-SiC n^+p diodes as a function of applied bias is shown in Fig. 12. In this study, Si ions with different energies were applied as probe beams, and the value of energy of Si ions are described in the figure. Charge collected by the diodes increases with increasing applied bias, and the value of collected charge saturates in a higher bias region. For example, the saturation is observed above 40 and 60 V for 15 and 18 MeV, respectively. Charge generated in the depletion region of a diode can be collected by its electric field (Drift component). On the other hand, charge generated in deeper than the depletion region diffuses, and only charge reaching the repletion region can be collected by a diode (Diffusion component) whereas some generated carriers recombine during diffusion. Thus, if the depletion region is shorter than the projection range of ions, the decrease in collected charge is observed due to the recombination of generated carriers during diffusion. Since ions with higher energy have a longer projection range, the results obtained in Fig. 12 can be qualitatively interpreted in terms of the drift and the diffusion components. However, in reality, since an extended drift region is temporarily created in a deeper region than the depletion region, the saturation of collected charge occurs even in the case that the depletion region is shorter than the ion projection range [44].

At a bias of 150V, the depletion region is estimated to be 7 μm, and this is longer than the ion projection range of Si ions at 18 MeV which is estimated to be 4.8 μm by a Monte Carlo simulation code, SRIM [45]. Thus, at a bias of 150 V, all charge generated in the 6H-SiC diodes by Si ion incidence can be collected by the electric field in the depletion layer. The CCE for the 6H-SiC diodes is estimated from the value of charge collected at a bias of 150 V. Here, the value of CCE is defined as

$$\left(Q_{exp} / Q_{ideal}\right) \text{ x } 100 \qquad (7)$$

where Q_{exp} and Q_{ideal} are the value of charge experimentally obtained at 150 V and the ideal value of charge generated in SiC, respectively. The value of Q_{ideal} is obtained by the equation

$$Q_{ideal} = \left(E_{ion} / E_{e-h}\right) \text{x} e \qquad (8)$$

where E_{ion}, $E_{e\text{-}h}$ and e are the energy of incident ions, the generation energy of an electron-hole (*e-h*) pair and electron charge, respectively. In this study, the value of $E_{e\text{-}h}$ in 6H-SiC is assumed to be 7.8 eV (= 2.8Eg) on the analogy of $E_{e\text{-}h}$ in Si because the value of the energy for 6H-SiC has not been determined yet. It should be mentioned that the energy loss in the top Al electrode, the n^+ region and by non-ionizing collisions and also the decay of signal in the measurement system are not considered in this estimation, and the reduction of the CCE due to those effects is estimated to be between 8 and 14 %. The value of the CCE for the SiC n^+p diodes probed by Si ions at energies of 6, 12, 15 and 18 MeV is estimated to be 74, 83, 86 and 88 %, respectively. Since the effect of the energy loss in the Al electrode and the n^+ region on the reduction of the CCE value decreases with increasing ion energy, the experimental result that higher CCE value is observed by higher energy ion incidence is reasonable. However, even after considering energy loss in those regions, the value of the CCE for 6MeV is not comparable to that for 12, 15 and 18 MeV. This suggests that the CCE is degraded by another effect in the case of 6 MeV-Si.

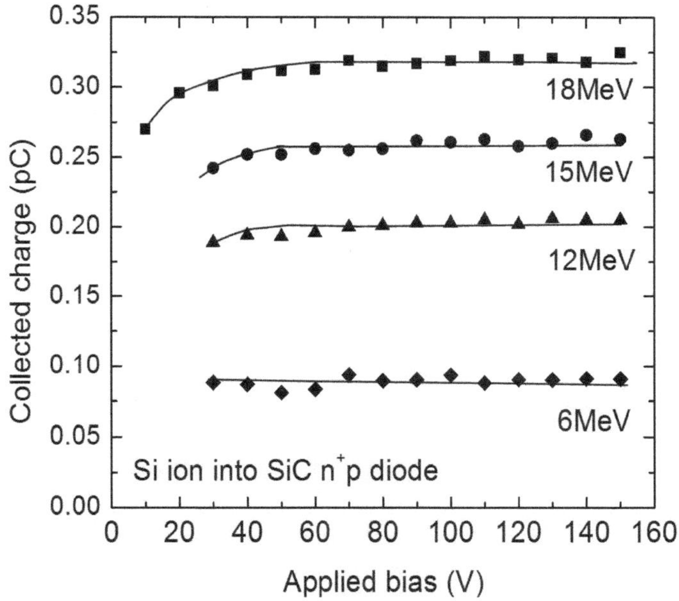

Figure 12. Charge collected by 6H-SiC n^+p diodes as a function of applied bias. Si ions with different energies were applied as probe beams, and the value of energy of Si ions are described in the figure.

In order to understand the degradation of the CCE due to not energy loss near the surface regions, the effect of ion species on the CCE was investigated. Figure 13 shows the relationship between ions species with the same energy (12 MeV) and the value of the CCE. The value of the CCE is obtained from the integration of TIBIC signals for the 6H-SiC n^+p diodes at a bias of 150 V. The CCE for the diodes probed by O ions is estimated to be 90 %, and this value is the highest of all ion species in Fig. 13. With increasing atomic number, the value of the CCE decreases. The CCE of 42 % is observed by Au ion incidence. The degradation of the CCE for SiC diodes by Au ion incidence was also reported [36]. Zajic et al. suggested that high density of e-h pairs is generated by heavy ions, and generated e-h pairs are easy to recombine in such dense plasma [46].

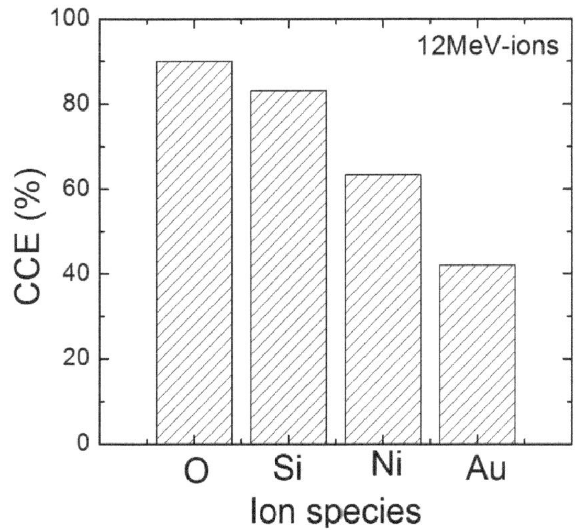

Figure 13. Relationship between ions species with the same energy (12 MeV) and the value of CCE. The value of CCE is obtained from the integration of TIBIC signals for 6H-SiC n^+p diodes at a bias of 150 V.

The carrier density generated in SiC, and the distributions of e-h pairs are calculated on the basis of Kobetich and Katz (KK) theorem [47]. In this calculation, the KK model improved using empirical equations reported by Fageeha et al. [48] was applied since the KK model overestimates the density of e-h pairs at the core of the ion track. The calculated results of the density of e-h pairs generated in SiC by (Left) 12 MeV-O and (Right) -Au ion irradiation are shown in Fig. 14. In the case of 12 MeV-O ion incidence, the radius of the ion track at the sample surface and projection range of ions are estimated to be ≈ 40 nm and 5.2 μm, respectively. On the other hand, the ion track radius at the surface and the ion range for 12 MeV-Au ions are estimated to be ≈ 2 nm and 1.9 μm, respectively. Since the energy (12 MeV) is the same for both O and Au ions, the total number of e-h pairs generated in the ion track region is the same between O and Au ions. Thus, the density of e-h pairs in SiC irradiated with Au ions is much higher than that irradiated with O ions, and the estimated density of e-h pairs in SiC irradiated with 12 MeV-Au ions is a several orders of magnitude higher than that in SiC irradiated with 12 MeV-O ions. In such a high density of e-h pairs, the ambipolar effect occurs easily and the electric field temporarily weakens. As a result, the amount of the recombination between electrons and holes increases. For the dynamics of carriers generated in SiC by heavy ion incidence, please see Ref. [44]. The result obtained in this study indicates that it is important to consider the decrease in the CCE for SiC particle detectors when heavy ions are detected. From the point of view of SEEs in SiC, the decrease in collected charge is thought to be one of the advantages for the development of Rad-Hard devices. The similar charge collection behaviours have been also obtained for SiC p^+n diodes, although only results obtained from SiC n^+p diodes were introduced in this article [39].

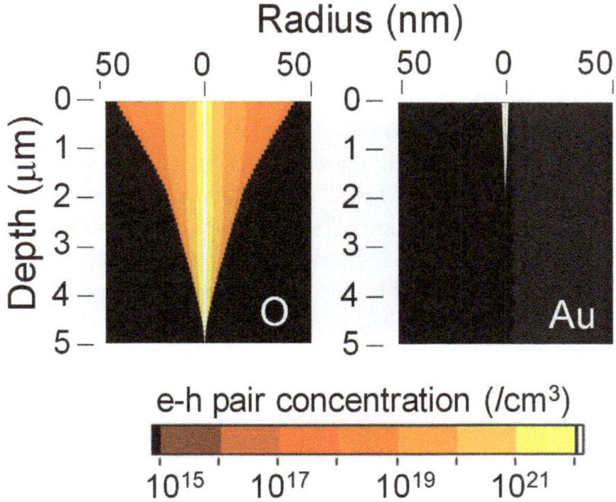

Figure 14. Calculated results of the e-h density generated in SiC by (Left) 12 MeV-O and (Right) -Au ion irradiation.

For the effects of ion incidence on MOS capacitors fabricated on SiC, it was reported that the peak amplitude of TIBIC signals decreased and the fall time increased with increasing number of incident ions [49-51]. Furthermore, the peak of TIBIC signals can be refreshed to its original value by applying a forward bias of + 1V to the gate electrode. From the measurement of the capacitance of SiC MOS capacitors during O ion irradiation, the value of capacitance was found to increase with increasing number of incident ions. This indicates that the depletion length of the MOS capacitors becomes shorten with increasing number of incident ions. Since large amounts of charge are induced by heavy ion incidence and some of them might flow to the interface between SiO_2 and SiC, the degradation of TIBIC signals can be explained by a change in the net bias applied to the gate oxide due to the creation of the inversion region or/and charging up deep traps. The refreshment of TIBIC signals by applying a forward bias can be also interpreted in terms of releasing charge from the interface or/and deep traps. For the effects of heavy ion irradiation on 6H-SiC MOSFETs, Onoda et al. Reported from experimental results and their simulation using the Technology Computer Aided Design (TCAD) [52] that the charge collection behaviours were affected by drift, funnelling, diffusion, and recombination, and especially, the enhancement of transient currents was observed due to the parasitic bipolar action. It was also reported that the enhanced charge collection was observed for 4H-SiC MESFETs by heavy ion incidence [26]. According to device simulations using the TCAD, it was concluded that the enhanced charge collection effect can be interpreted in terms of not only the bipolar action but also the channel modulation effects. For the DDD effect in SiC devices, it was reported that the value of the CCE for SiC n^+p diodes and the majority carrier concentration in them decreased with increasing gamma-rays, electrons or protons and the damage factor of the CCE and the carrier removal rate can be scaled by Non Ionizing Energy Loss (NIEL) [53-55].

5. Summary

In order to develop Rad-hard devices based on SiC, the radiation response of SiC devices have to be understood. In this chapter, effects of gamma-rays and swift heavy ions on SiC devices were reviewed. Firstly, the gamma-ray irradiation effects on SiC MOSFETs were introduced, and the degradation of their characteristics was discussed on the basis of charge generated in gate oxide and interface traps by irradiation. Then, the radiation resistance of SiC transistors, MOSFETs, MESFETs and SITs was compared to Si transistors. SiC transistors showed higher radiation resistance than Si transistors, and SiC SITs could be operated up to 10 MGy. This indicates that SiC SITs have extremely high radiation tolerance from the point of view of TID effects. Charge generated in 6H-SiC $n^{+}p$ diodes by heavy ion incidence was evaluated using TIBIC. The signal peak of the transient current increased, and the fall-time decreased with increasing applied reverse bias. The high CCE values were observed when ions with relatively light mass such as O and Si ions were applied as probe ions. However, the CCE decreased with increasing atomic number, and the value reduced to approximately 40 % when 12 MeV-Au ions were applied as probe ions. From the calculation based on the modified KK model, it was found that the density of e-h pairs in SiC irradiated with heavy ions, such as Ni and Au, is much higher than that in SiC irradiated with O and Si ions. Therefore, the decrease in the CCE by the irradiation of ions with heavy mass was interpreted in terms of the recombination of e-h pairs in plasma.

Acknowledgements

This study of gamma-ray irradiation effects on SiC SIT was supported by the Strategic Promotion Program for Basic Nuclear Research by the Ministry of Education, Culture, Sports, Science and Technology of Japan. Also, the study of charge induced in SiC pn diodes and MOS capacitors by heavy ion incidence was partially supported by the Ministry and Education, Science, Sports and Culture, Grant-in-Aid for Scientific Research (B), 2006, 18360458 and (B), 2009, 21360471, respectively.

Author details

Takeshi Ohshima[1*], Shinobu Onoda[1], Naoya Iwamoto[1], Takahiro Makino[1],
Manabu Arai[2] and Yasunori Tanaka[3]

*Address all correspondence to: ohshima.takeshi20@jaea.go.jp

1 Japan Atomic Energy Agency (JAEA), Japan

2 New Japan Radio Co., Ltd. (NJRC), Japan

3 National Institute of Advanced Industrial Science and Technology(AIST), Japan

References

[1] Ohshima, T., Yoshikawa, M., Itoh, H., Takahashi, T., Okumura, H., Yoshida, S., & Nashiyama, N. (1996). Effects of Gamma-ray Irradiation and Thermal Annealing on Electrical Characteristics of SiC MOSFETs. Inst. *Phys. Conf. Ser.*, 142, 801-804.

[2] Ohshima, T., Yoshikawa, M., Itoh, H., Aoki, Y., & Nashiyama, I. (1998). Generation of Interface Traps and Oxide-Trapped Charge in 6H-SiC Metal-Oxide-Semiconductor Transistors by Gamma-ray Irradiation. *Jpn. J. Appl. Phys.*, 37, L 1002-L1004.

[3] Lee, K. K., Ohshima, T., & Itoh, H. (2002). Radiation Response of p-Channel 6H-SiC MOSFETs Fabricated Using Pyrogenic Conditions. *Materials Science Forum*, 389-393, 1097-1100.

[4] Nishijima, T., Ohshima, T., & Lee, K. K. (2002). Investigation of the Radiation Hardness on Semiconductor Devices using the Ion Micro-Beam. *Nucl. Instrum. Meth. B*, 190, 329-334.

[5] Lee, K. K., Ohshima, T., & Itoh, H. (2003). Performance of Gamma Irradiated p-Channel 6H-SiC MOSFETs: High Total Dose. *IEEE Trans. Nucl. Sci.*, 50, 194-200.

[6] Binder, D., Smith, E. C., & Holman, A. B. (1984). Satellite Anomalous from Galactic Cosmic Rays. *IEEE Trans. Nucl. Sci.*, NS-22, 2675-2680.

[7] Dodd, P. E., Musseau, O., Shaneyfelt, M. R., Sexton, F. W., D'hose, C., Hash, G. L., Martinez, M., Loemker, R. A., Leray, J-L., & Winokur, P. S. (1988). Impact of Ion Energy on Single-Event Upset. *IEEE Trans. Nucl. Sci.*, 45, 2483-2491.

[8] Sexton, F. W. (2003). Destructive Single-Event Effects in Semiconductor Devices and ICs. *IEEE Trans. Nucl. Sci.*, 50, 603-621.

[9] McWhorter, P. J., & Winokur, P. S. (1986). Simple Technique for Separating the Effects of Interface Traps and Trapped-Oxide Charge in Metal-Oxide-Semiconductor Transistors. *Appl. Phys. Lett.*, 48, 133-135.

[10] Sexton, F. W., & Schwank, J. R. (1985). Correlation of Radiation Effects in Transistors and Integrated Circuits. *IEEE Trans. Nucl. Sci.*, NS-32, 3975-3981.

[11] Ohshima, T., Morita, Y., Nashiyama, I., Kawasaki, O., Hisamatsu, T., Nakao, T., Wakow, Y., & Matsuda, S. (1996). Mechanism of Anomalous Degradation of Silicon Solar Cells Subjected to High-Fluence Irradiation. *IEEE Trans. Nucl. Sci.*, 43, 2990-2997.

[12] Ymamaguchi, M. (2001). Radiation-Resistant Solar Cells for Space Use. *Sol. Energy Mater. Sol. Cells.*, 68, 31-53.

[13] Imaizumi, M., Sumita, T., Kawakita, S., Aoyama, K., Anzawa, O., Aburaya, T., Hisamatsu, T., & Matsuda, S. (2005). Results of Flight Demonstration of Terrestrial Solar Cells in Space. *Prog. Photovolt: Res. Appl.*, 13, 93-102.

[14] Sato, S., Ohshima, T., & Imaizumi, M. (2009). Modeling of Degradation Behavior of InGaP/GaAs/Ge Triple-Junction Space Solar Cell Exposed to Charged Particles. *J. Appl. Phys.*, 105, 044504-1-6.

[15] Brew, J. R. (1981). *Applied Solid State Science.*, ed. Dawon Kahhny. Academic: New York.

[16] Ohshima, T., Yoshikawa, M., Itoh, H., Aoki, Y., & Nashiyama, I. (1999). Gamma-ray Irradiation Effects on 6H-SiC MOSFET. *Mater. Sci. & Engineer. B*, 480, 61-62.

[17] Ohshima, T., Itoh, H., & Yoshikawa, M. (2001). Effect of Gamma-ray Irradiation on the Characteristics of 6H Silicon Carbide Metal-Oxide-Semiconductor Field Effect Transistor with Hydrogen-Annealed Gate Oxide. *J. Appl. Phys.*, 90, 3038-3041.

[18] Yoshikawa, M., Saitoh, K., Ohshima, T., Itoh, H., Nashiyama, I., Takahashi, Y., Ohnishi, K., Okumura, H., & Yoshida, Si. (1998). Generation Mechanisms of Trapped Charges in Oxide Layers of 6H-SiC MOS Structures Irradiated with Gamma-Rays. *Mater. Sci. Forum*, 264-268, 1017-1020.

[19] Ohshima, T., Yoshikawa, M., Itoh, H., Kojima, K., Okada, S., & Nashiyama, I. (2000). Influence of Post-Oxidation Annealing on Electrical Characteristics in 6H-SiC MOSFETs. *Mater. Sci. Forum*, 338-342, 1299-1302.

[20] Yoshikawa, M., Saitoh, K., Ohshima, T., Itoh, H., Nashiyama, I., Yoshida, S., Okumura, H., Takahashi, Y., & Ohnishi, K. (1996). Depth Profile of Trapped Charges in Oxide Layer of 6H-SiC Metal-Oxide-Semiconductor Structures. *J. Appl. Phys.*, 80, 282-287.

[21] Hornetz, B., Michel, H. J., & Halbritter, J. (1994). ARXPS Studies of SiO2SiC Interfaces and Oxidation of 6H SiC Single-Crystal Si-(001) and C-(001) Over-Bar Surfaces. *J. Mater. Res.*, 9, 3088-3094.

[22] Schorner, R., Friedrichs, P., & Peters, D. (1999). Detailed Investigation of n-Channel Enhancement 6H-SiC MOSFET's. *IEEE Trans. Electron Devices*, 46, 533-541.

[23] Hishiki, S., Iwamoto, N., Ohshima, T., Itoh, H., Kojima, K., & Kawano, K. (2009). Effects of Fabrication Process on the Electrical Characteristics of n-channel MOSFETs Irradiated with Gamma-Rays. *Mater. Sci. Forum*, 600-603, 707-710.

[24] Negoro, Y., Katsumonot, K., Kimoto, T., & Matsunami, H. (2004). Electronic Behaviors of High-Dose Phosphorus-Ion Implanted 4H-SiC(0001). *J. Appl. Phys.*, 96, 224-228.

[25] Tanaka, Y., Onoda, S., Takatsuka, A., Ohshima, T., & Yatsuo, T. (2010). Radiation Hardness Evaluation of SiC-BGSIT. *Mater. Sci. Forum*, 645-648, 941-944.

[26] Onoda, S., Iwamoto, N., Ono, S., Katakami, S., Arai, M., Kawano, K., & Ohshima, T.
(2009). Transient Response of Charge Collection by Single Ion Strike in 4H-SiC MES- 9
FETs. *IEEE Trans. Nucl. Sci.*, 56, 3218-3222.

[27] Tanaka, Y., Yano, K., Okamoto, M., Takatsuka, A., Fukuda, K., Kasuga, M., Arai, K., & Yatsuo, T. (2006). Fabrication of 700 V SiC-SIT with Ultra-Low On-Resistance of 1.01 mΩcm^2. *Mater. Sci. Forum*, 527-529, 1219-1222.

[28] Tanaka, Y., Okamoto, M., Takatsuka, A., Arai, K., Yatsuo, T., Yano, K., & Kasuga, M. (2006). 700-V 1.0-m Ωcm^2 buried gate SiC-SIT (SiC-BGSIT). *IEEE Electron Device Lett.*, 27, 908-910.

[29] Gover, J. E., & Srour, J. R. (1986). Basic Radiation Effects in Nuclear Power Electronics Technology. *Sandia Natinal Labs Report SAND85-0776.*

[30] Nava, F., Vittone, E., Vanni, P., Verzellesi, G., Fuochi, P. G., Lanzieri, C., & Glaser, M. (2003). Radiation Tolerance of Epitaxial Silicon Carbide Detectors for Electrons, Protons and Gamma-rays. *Nucl. Instr. and Meth. A*, 505, 645-655.

[31] Bertuccio, G. (2005). Prospect for Energy Resolving X-ray Imaging with Compound Semiconductor Pixel Detectors. *Nucl. Instr. and Meth. A*, 546, 232-241.

[32] Lees, J. E., Bassford, D. J., Fraser, G. W., Horsfall, A. B., Vassilevski, K. V., Wright, N. G., & Owens, A. (2007). Semi-Transparent SiC Schottky Diodes for X-ray Spectroscopy. *Nucl. Instr. Meth. A*, 578, 226-234.

[33] Giudice, A. Lo., Fasolo, F., Durisi, E., Manfredotti, C., Vittone, E., Fizzotti, F., Zanini, A., & Rosi, G. (2007). Performances of 4H-SiC Schottky Diodes as Neutron Detectors. *Nucl. Instr. Meth. A*, 583, 177-180.

[34] Flammang, R. W., Seidel, J. G., & Ruddy, F. H. (2007). Fast Neutron Detection with Silicon Carbide Semiconductor Radiation Detectors. *Nucl. Instr. Meth. A*, 578, 177-179.

[35] Ohshima, T., Satoh, T., Oikawa, M., Yamakawa, T., Onoda, S., Wakasa, T., Laird, J. S., Hirao, T., Kamiya, T., Itoh, H., Kinoshita, A., Tanaka, R., Nakano, I., Iwami, M., & Fukushima, Y. (2005). Characterization of Charge Generated in Silicon Carbide n^+p Diodes using Transient Ion Beam-Induced Current. *Nucle. Instrum. Meth. A*, 541, 236-240.

[36] Ohshima, T., Satoh, T., Oikawa, M., Onoda, S., Hishiki, S., Hirao, T., Kamiya, Yokoyama T., Sakamoto, A., Tanaka, R., Nakano, I., Wagner, G., & Itoh, H. (2007). Degradation of Charge Collection Efficiency Obtained for 6H-SiC n^+p Diodes Irradiated with Gold Ions. *Mater. Sci. Forum*, 556-557, 913-916.

[37] Onoda, S., Ohshima, T., Hirao, T., Hishiki, S., Iwamoto, N., Kojima, K., & Kawano, K. (2009). Transient Response to High Energy Heavy Ions in 6H-SiC n^+p Diodes. *Mater. Sci. Forum*, 600-603, 1039-1042.

[38] Onoda, S., Iwamoto, N., Murakami, M., Ohshima, T., Hirao, T., Kojima, T., Kawano, K., & Nakano, I. (2009). Charge Collection Properties of 6H-SiC Diodes by Wide Variety of Charged Particles up to Several Hundreds MeV. *Mater. Sci. Forum*, 615-617, 861-864.

[39] Ohshima, T., Iwamoto, N., Onoda, S., Kamiya, T., & Kawano, K. (2009). Comparative Study of Transient Current induced in SiC p+n and n+p Diodes by Heavy Ion Micro Beams. *Nucl. Instrum. Meth. B*, 267, 2189-2192.

[40] Ohshima, T., Iwamoto, N., Onoda, S., Wagner, G., Itoh, H., & Kawano, K. (2011). Charge Generated in 6H-SiC n^+p Diodes by MeV Range Heavy Ions. *Surface & Coatings Tech.*, 206, 864-868.

[41] Hirao, T., Nashiyama, I., Kamiya, T., & Nishijima, T. (1995). Effects of Micro-Beam Induced Damage on Single-Event Current Measurements. *Nucl. Instr. Meth. B*, 104, 508-514.

[42] Japan Atomic Energy Agency. (2012). *Takasaki Advanced Radiation Research Institute*, http://www.taka.jaea.go.jp/index_e.html, accessed 1 July.

[43] Laird, J. S., Hirao, T., Mori, H., Onoda, S., Kamiya, T., & Itoh, H. (2000). Development of a New Data Collection System and Chamber for Microbeam and Laser Investigations of Single Event Phenomena. *Nucl. Instr. Meth. B*, 181, 87-94.

[44] Iwamoto, N., Onoda, S., Makino, T., Ohshima, T., Kojima, K., Koizumi, A., Uchida, K., & Nozaki, S. (2011). Transient Analysis of an Extended Drift Region in a 6H-SiC Diode Formed by a Single Alpha Particle Strike and Its Contribution to the Increased Charge Collection. *IEEE Trans Nucl. Sci.*, 58, 305-313.

[45] Ziegler, J. F., Biersack, J. P., & Ziegler, M. D. (2008). *SRIM, The Stopping and Range of Ions in Matter*, SRIM Co., Chester, Maryland, USA.

[46] Zajic, V., & Thieberger, P. (1999). Heavy Ion Linear Energy Transfer Measurements during Single Event Upset Testing of Electronic Devices. *IEEE Trans. Nucl. Sci.*, 46, 59-69.

[47] Kobetich, E. J., & Katz, R. (1968). Energy Deposition by Electron Beams and δ Rays. *Phy. Rev.*, 170, 391-396.

[48] Fageeha, O., Howard, J., & Block, R. C. (1994). Distribution of Radial Energy Deposition around the Track of Energetic Charged-Particles in Silicon. *J. Appl. Phys.*, 75, 2317-2321.

[49] Ohshima, T., Iwamoto, N., Onoda, S., Makino, T., Deki, M., & Nozaki, S. (2011). Refreshable Decrease In Peak Height Of Ion Beam Induced Transient Current From Silicon Carbide Metal-Oxide-Semiconductor Capacitors. *AIP conference proceedings 1336: Application of Accelerators in Research and Industry: 21th International Conference.*, August, Fort Worth, Texas, USA, 660-664.

[50] Ohshima, T., Iwamoto, N., Onoda, S., Makino, T., Nozaki, S., & Kojima, K. (2011). Oxygen Ion Induced Charge in SiC MOS Capacitors Irradiated with Gamma-rays. 1
Mater. Sci. Forum, 679-680370-373.

[51] Makino, T., Iwamoto, N., Onoda, S., Ohshima, T., Kojima, K., & Nozaki, S. (2012). Peak Degradation of Heavy-Ion Induced Transient Currents in 6H-SiC MOS Capacitors. *Mater. Sci. Forum*, 717-720469-472.

[52] Onoda, S., Makino, T., Iwamoto, N., Vizkelethy, G., Kojima, K., Nozaki, S., & Ohshima, T. (2010). Charge Enhancement Effects in 6H-SiC MOSFETs Induced by Heavy Ion Strike. *IEEE Trans. Nucl. Sci.*, 57, 3373-3379.

[53] Onoda, S., Ohshima, T., Hirao, T., Mishima, K., Hishiki, S., Iwamoto, N., Kojima, K., & Kawano, K. (2007). Decrease of Charge Collection due to Displacement Damage by Gamma Rays in a 6H-SiC diode. *IEEE Trans. Nucl. Sci.*, 54, 1953-1960.

[54] Onoda, S., Iwamoto, N., Hirao, T., Kawano, K., Kojima, K., & Ohshima, T. (2009). Reduction of Effective Carrier Density and Charge Collection Efficiency in SiC Devices due to Radiations. *AIP conference proceedings 1099: Application of Accelerators in Research and Industry: 20th International Conference*, 10-15 August 2008, Fort Worth, Texas, USA, 1010-1013.

[55] Iwamoto, N., Onoda, S., Ohshima, T., Kojima, K., Koizumi, A., Uchida, K., & Nozaki, S. (2010). Charge Collection Efficiency of 6H-SiC $P^{+}N$ Diodes Degraded by Low-Energy Electron Irradiation. *Mater. Sci. Forum*, 645-648921-924.

Permissions

The contributors of this book come from diverse backgrounds, making this book a truly international effort. This book will bring forth new frontiers with its revolutionizing research information and detailed analysis of the nascent developments around the world.

We would like to thank Yasuto Hijikata, for lending his expertise to make the book truly unique. He has played a crucial role in the development of this book. Without his invaluable contribution this book wouldn't have been possible. He has made vital efforts to compile up to date information on the varied aspects of this subject to make this book a valuable addition to the collection of many professionals and students.

This book was conceptualized with the vision of imparting up-to-date information and advanced data in this field. To ensure the same, a matchless editorial board was set up. Every individual on the board went through rigorous rounds of assessment to prove their worth. After which they invested a large part of their time researching and compiling the most relevant data for our readers. Conferences and sessions were held from time to time between the editorial board and the contributing authors to present the data in the most comprehensible form. The editorial team has worked tirelessly to provide valuable and valid information to help people across the globe.

Every chapter published in this book has been scrutinized by our experts. Their significance has been extensively debated. The topics covered herein carry significant findings which will fuel the growth of the discipline. They may even be implemented as practical applications or may be referred to as a beginning point for another development. Chapters in this book were first published by InTech; hereby published with permission under the Creative Commons Attribution License or equivalent.

The editorial board has been involved in producing this book since its inception. They have spent rigorous hours researching and exploring the diverse topics which have resulted in the successful publishing of this book. They have passed on their knowledge of decades through this book. To expedite this challenging task, the publisher supported the team at every step. A small team of assistant editors was also appointed to further simplify the editing procedure and attain best results for the readers.

Our editorial team has been hand-picked from every corner of the world. Their multi-ethnicity adds dynamic inputs to the discussions which result in innovative outcomes. These outcomes are then further discussed with the researchers and contributors who give their valuable feedback and opinion regarding the same. The feedback is then collaborated with the researches and they are edited in a comprehensive manner to aid the understanding of the subject.

Apart from the editorial board, the designing team has also invested a significant amount of their time in understanding the subject and creating the most relevant covers. They scrutinized every image to scout for the most suitable representation of the subject and create an appropriate cover for the book.

The publishing team has been involved in this book since its early stages. They were actively engaged in every process, be it collecting the data, connecting with the contributors or procuring relevant information. The team has been an ardent support to the editorial, designing and production team. Their endless efforts to recruit the best for this project, has resulted in the accomplishment of this book. They are a veteran in the field of academics and their pool of knowledge is as vast as their experience in printing. Their expertise and guidance has proved useful at every step. Their uncompromising quality standards have made this book an exceptional effort. Their encouragement from time to time has been an inspiration for everyone.

The publisher and the editorial board hope that this book will prove to be a valuable piece of knowledge for researchers, students, practitioners and scholars across the globe.

List of Contributors

T. S. Argunova
Loffe Physical-Technical Institute, RAS, St. Petersburg, Russia
X-ray Imaging Center, Department of Materials Science and Engineering, Pohang University of Science and Technology, Pohang, Republic of Korea

M. Yu. Gutkin
Institute of Problems in Mechanical Engineering, RAS, St. Petersburg, Russia
Department of Physics of Materials Strength and Plasticity, St. Petersburg State Polytechnical University, St. Petersburg, Russia
Department of Theory of Elasticity, St. Petersburg State University, St. Petersburg, Russia

J. H. Je
X-ray Imaging Center, Department of Materials Science and Engineering, Pohang University of Science and Technology, Pohang, Republic of Korea

V. G. Kohn
Russian Research Center 'Kurchatov Institute', Moscow, Russia

E. N. Mokhov
Ioffe Physical-Technical Institute, RAS, St. Petersburg, Russia

Sadafumi Yoshida
Advanced Power Electronics Research Center, National Institute of Advanced Industrial Science and Technology (AIST), Japan

Yasuto Hijikata and Hiroyuki Yaguchi
Graduate School of Science and Engineering, Saitama University, Japan
Division of Mathematics Electronics and Informatics, Graduate School of Science and Engineering, Saitama University, Japan

Kair Kh. Nussupov and Nurzhan B. Beisenkhanov
Kzakh-British Technical University, Kazakhstan

Hitoshi Habuka
Department of Chemical and Energy Engineering, Yokohama National University, Japan

Zhongchang Wang
WPI Research Center, Advanced Institute for Materials Research, Tohoku University, Japan

Ciprian Iliescu
Institute of Bioengineering and Nanotechnology, Singapore

Daniel P. Poenar
Microelectronics Centre, School of Electrical and Electronics Engineering, Nanyang Technogical University, Singapore

Heiji Watanabe and Takuji Hosoi
Department of Material and Life Science, Graduate School of Engineering, Osaka University, Suita, Osaka 565-0871, Japan

Shuhei Yagi
Division of Mathematics Electronics and Informatics, Graduate School of Science and Engineering, Saitama University, Japan

Sanjeev Kumar Gupta, Jitendra Singh and Jamil Akhtar
CSIR-Central Electronics Engineering Research Institute (CEERI), Pilani, India

John Rozen
IBM T.J. Watson Research Center, Yorktown Heights, NY, USA

Yintang Yang and Baoxing Duan
School of Microelectronics, Xidian University, China

Xianjun Zhang
College of Automation & Electronic Engineering, Qingdao University of Science & Technonology, China

Malek Gassoumi and Hassen Maaref
Laboratoire de Micro-Optoelectroniques et Nanostructures, Universite de Monastir, Monastir 5000, Tunisia

Moumita Mukherjee
CMSDS, A Centre of Drdo, Govt. of India and University of Calcutta, India

Mariana Amorim Fraga
Department of Sciences, Engineering and Mathematical Modeling, Regional University of Northwest Rio Grande do Sul State, Brazil
Plasma and Processes Laboratory, Technological Institute of Aeronautics, Brazil

Rodrigo Sávio Pessoa and Homero Santiago Maciel
Plasma and Processes Laboratory, Technological Institute of Aeronautics, Brazil
IP&D, University of Vale do Paraiba, Brazil

Marcos Massi
Plasma and Processes Laboratory, Technological Institute of Aeronautics, Brazil

Mahboobeh Mahmoodi
Department of Materials science and Engineering, Yazd Branch, Islamic Azad University, Yazd, Iran

Lida Ghazanfari
Biomaterials Group, Facult of Biomedical Engineering, Amirkabir University of Technology, Tehran, Iran

Takeshi Ohshima, Shinobu Onoda, Naoya Iwamoto and Takahiro Makino
Japan Atomic Energy Agency (JAEA), Japan

Manabu Arai
New Japan Radio Co., Ltd. (NJRC), Japan

Yasunori Tanaka
National Institute of Advanced Industrial Science and Technology (AIST), Japan